VOLUME ONE HUNDRED AND TWENTY THREE

INTERNATIONAL REVIEW OF NEUROBIOLOGY

Structure to Function of G Protein-Gated Inwardly Rectifying (GIRK) Channels

INTERNATIONAL REVIEW OF NEUROBIOLOGY

VOLUME 123

SERIES EDITORS

R. ADRON HARRIS

Waggoner Center for Alcohol and Drug Addiction Research
The University of Texas at Austin
Austin, Texas, USA

PETER JENNER

Division of Pharmacology and Therapeutics
GKT School of Biomedical Sciences
King's College, London, UK

VOLUME ONE HUNDRED AND TWENTY THREE

INTERNATIONAL REVIEW OF NEUROBIOLOGY

Structure to Function of G Protein-Gated Inwardly Rectifying (GIRK) Channels

Edited by

PAUL A. SLESINGER

Department of Neuroscience,
Friedman Brain Institute,
Icahn School of Medicine at Mount Sinai,
New York, New York, USA

KEVIN WICKMAN

Department of Pharmacology,
University of Minnesota,
Minneapolis, Minnesota, USA

ELSEVIER

AMSTERDAM • BOSTON • HEIDELBERG • LONDON
NEW YORK • OXFORD • PARIS • SAN DIEGO
SAN FRANCISCO • SINGAPORE • SYDNEY • TOKYO
Academic Press is an imprint of Elsevier

Academic Press is an imprint of Elsevier
225 Wyman Street, Waltham, MA 02451, USA
525 B Street, Suite 1800, San Diego, CA 92101-4495, USA
125 London Wall, London, EC2Y 5AS, UK
The Boulevard, Langford Lane, Kidlington, Oxford OX5 1GB, UK

First edition 2015

ISBN: 978-0-12-802458-4
ISSN: 0074-7742

For information on all Academic Press publications
visit our website at http://store.elsevier.com/

CONTENTS

Contributors *ix*
Preface *xi*

1. Unifying Mechanism of Controlling Kir3 Channel Activity by G Proteins and Phosphoinositides **1**

Diomedes E. Logothetis, Rahul Mahajan, Scott K. Adney, Junghoon Ha, Takeharu Kawano, Xuan-Yu Meng, and Meng Cui

1. Introduction 2
2. Regulation of K_{ACh} by G Proteins 3
3. Molecular Constituents of Kir3 Channels and Physiological Roles 4
4. Dependence of K_{ACh} Channel Activity on PIPs 5
5. Physiological Significance of Distinct Membrane-Delimited Signaling Mechanisms 9
6. Molecular Mechanism of PIP_2 Control of Kir Channel Gating 12
7. Posttranslational Modification Can Control Activity Through Changes in Channel–PIP_2 Interactions 17
8. The Kir3 Channel Site of Gβγ Action 18

References 21

2. The Roles of Gβγ and Gα in Gating and Regulation of GIRK Channels **27**

Nathan Dascal and Uri Kahanovitch

1. Introduction 28
2. Interaction of Gα and Gβγ With GIRK 32
3. Roles of Gβγ and Gα in GIRK's Evoked and Basal Activity 41
4. Basal Activity of GIRKs 51
5. Multimolecular Signaling Complex and the Mechanism of GIRK Activation 60

Acknowledgments 68
References 68

3. RGS Redundancy and Implications in GPCR–GIRK Signaling **87**

Craig A. Doupnik

1. RGS Proteins—Background and Significance 88
2. Direct "Membrane-Delimited" GPCR-to-Ion Channel Signaling 89

3. RGS Actions on GPCR–GIRK Channel Gating Kinetics 90
4. RGS Functional Redundancy—Emerging Evidence from Retinal Neurons 94
5. RGS Functional Redundancy in Neuronal GPCR–GIRK Channel Signaling 96
6. RGS Functional Redundancy in Cardiac GPCR-GIRK Channel Signaling 100
7. RGS Actions on G_q-Mediated Changes in GIRK Channel Activity 101
8. Conclusions 102
References 105

4. Structural Insights into GIRK Channel Function 117

Ian W. Glaaser and Paul A. Slesinger

1. Introduction 118
2. Selectivity and Permeation 121
3. Rectification 128
4. Gating 130
5. Conclusions 144
Acknowledgments 145
References 145

5. Localization and Targeting of GIRK Channels in Mammalian Central Neurons 161

Rafael Luján and Carolina Aguado

1. Introduction 162
2. Molecular Organization and Heterogeneity of GIRK Channels 163
3. Neuroanatomical Approaches to the Study of GIRK Channel Distribution 164
4. Regional Distribution of GIRK Channel Subunits in the CNS 169
5. Cellular Distribution of GIRK Channel Subunits 175
6. Subcellular Distribution of GIRK Channel Subunits 179
7. Developmental Aspects of GIRK Channel Distribution 190
8. Conclusions and Future Perspectives 193
Acknowledgments 194
References 194

6. GIRK Channel Plasticity and Implications for Drug Addiction 201

Ezequiel Marron Fernandez de Velasco, Nora McCall, and Kevin Wickman

1. Introduction 202
2. GIRK Channels in the Mesocorticolimbic System 207

3. Drug-Dependent Plasticity of GIRK Channel-Dependent Signaling 211
4. Mechanisms Underlying the Plasticity of GIRK Channel-Dependent Signaling 216
5. Concluding Thoughts and Future Directions 223
Acknowledgments 223
References 223

7. GIRK Channels: A Potential Link Between Learning and Addiction 239

Megan E. Tipps and Kari J. Buck

1. Introduction 240
2. GIRK Channels 241
3. GIRK Channel Modulation by Drugs of Abuse 245
4. Neuronal Plasticity 251
5. Summary and Future Directions 262
References 263

8. Behavioral and Genetic Evidence for GIRK Channels in the CNS: Role in Physiology, Pathophysiology, and Drug Addiction 279

Jody Mayfield, Yuri A. Blednov, and R. Adron Harris

1. Introduction 280
2. GIRK Channels in CNS Disorders 281
3. GIRK Channels in Addiction 291
4. Therapeutic Potential of GIRK Channel Modulators 300
5. Concluding Remarks and Future Directions 302
Acknowledgments 303
References 303

Index *315*
Contents of Recent Volumes *321*

CONTRIBUTORS

Scott K. Adney
Department of Physiology and Biophysics, School of Medicine, Virginia Commonwealth University, Richmond, Virginia, USA

Carolina Aguado
Instituto de Investigación en Discapacidades Neurológicas (IDINE), Departamento de Ciencias Médicas, Facultad de Medicina, Universidad Castilla-La Mancha, Campus Biosanitario, Albacete, Spain

Yuri A. Blednov
Waggoner Center for Alcohol and Addiction Research, The University of Texas at Austin, Austin, Texas, USA

Kari J. Buck
Department of Behavioral Neuroscience, Oregon Health & Science University, and Portland Alcohol Research Center, Portland VA Medical Center, Portland, Oregon, USA

Meng Cui
Department of Physiology and Biophysics, School of Medicine, Virginia Commonwealth University, Richmond, Virginia, USA

Nathan Dascal
Department of Physiology and Pharmacology and Sagol School of Neuroscience, Tel Aviv University, Tel Aviv, Israel

Craig A. Doupnik
Department of Molecular Pharmacology & Physiology, University of South Florida College of Medicine, Tampa, Florida, USA

Ian W. Glaaser
Department of Neuroscience, Friedman Brain Institute, Icahn School of Medicine at Mount Sinai, New York, New York, USA

Junghoon Ha
Department of Physiology and Biophysics, School of Medicine, Virginia Commonwealth University, Richmond, Virginia, USA

R. Adron Harris
Waggoner Center for Alcohol and Addiction Research, The University of Texas at Austin, Austin, Texas, USA

Uri Kahanovitch
Department of Physiology and Pharmacology and Sagol School of Neuroscience, Tel Aviv University, Tel Aviv, Israel

Takeharu Kawano
Department of Physiology and Biophysics, School of Medicine, Virginia Commonwealth University, Richmond, Virginia, USA

Diomedes E. Logothetis
Department of Physiology and Biophysics, School of Medicine, Virginia Commonwealth University, Richmond, Virginia, USA

Rafael Luján
Instituto de Investigación en Discapacidades Neurológicas (IDINE), Departamento de Ciencias Médicas, Facultad de Medicina, Universidad Castilla-La Mancha, Campus Biosanitario, Albacete, Spain

Rahul Mahajan
Department of Physiology and Biophysics, School of Medicine, Virginia Commonwealth University, Richmond, Virginia, USA

Ezequiel Marron Fernandez de Velasco
Department of Pharmacology, University of Minnesota, Minneapolis, Minnesota, USA

Jody Mayfield
Waggoner Center for Alcohol and Addiction Research, The University of Texas at Austin, Austin, Texas, USA

Nora McCall
Department of Neuroscience, University of Minnesota, Minneapolis, Minnesota, USA

Xuan-Yu Meng
Department of Physiology and Biophysics, School of Medicine, Virginia Commonwealth University, Richmond, Virginia, USA

Paul A. Slesinger
Department of Neuroscience, Friedman Brain Institute, Icahn School of Medicine at Mount Sinai, New York, New York, USA

Megan E. Tipps
Department of Behavioral Neuroscience, Oregon Health & Science University, Portland, Oregon, and Department of Pharmacology, University of Minnesota, Minneapolis, Minnesota, USA

Kevin Wickman
Department of Pharmacology, University of Minnesota, Minneapolis, Minnesota, USA

PREFACE

In 1921, Dr. Otto Loewi published a landmark study examining the idea that neurons signal to cardiac cells by releasing chemical transmitters. In this experiment, Dr. Loewi isolated two frog hearts, one with the vagus nerve attached and the other one free of the nerve. He electrically stimulated the vagus nerve and observed the heart beat slowing down. He then removed the saline from the heart with the stimulated vagus nerve and applied it to the other isolated heart. Like the electrical stimulation, the heart rate for the isolated heart also slowed, demonstrating that a chemical, not electrical, signal altered the heart rate. Dr. Loewi referred to this chemical as *vagusstoff* or "vagus stuff." Drs. Loewi and Dale later received the Nobel Prize for Medicine in 1936 for their work on chemical transmission. In the ensuing years, it was determined that vagusstoff is the neurotransmitter acetylcholine and that stimulation of G protein-coupled muscarinic receptors slows the heart rate by opening a potassium channel, called the cardiac muscarinic potassium channel (K_{ACh}).

In the late 1980s, there was intense focus on which subunit of the G protein, Gα or Gβγ, was responsible for activating K_{ACh} channels. In the 1990s, the genes encoding the K_{ACh} channel were identified, and since then, the channels have been referred to as G protein-gated inwardly rectifying potassium (GIRK) channels. In addition to the heart, GIRK channels are abundantly expressed in the brain, where they control neuronal excitability. Over the last 20 years, there has been considerable interest in determining how GIRK channels operate (i.e., gating, structure), what regulates their activity, how their activity affects the heart and brain, and elucidating their role in human diseases. In this volume of *International Review of Neurobiology*, the reader is guided from the early days investigating the G protein regulation of GIRK channels to our current knowledge on the role of GIRK channels in human diseases.

Chapter 1 reviews the molecular studies on G protein and phosphoinositide regulation of GIRK channels, providing mechanistic insights into how channel–phosphoinositide interactions are used to fine-tune the gating of GIRK channels. The intricacies of G protein regulation of GIRK channels, involving both the Gα subunit and the Gβγ dimer, are discussed in Chapter 2. While Gβγ subunits are known to directly activate GIRK channels, Gα subunits play an important role in regulating GIRK channel activity. As details on the complexity of G protein signaling were revealed, a model for GIRK channel signaling has emerged in which GIRK channels are embedded in a dynamic signaling complex along with G proteins, G protein-coupled receptors, and other proteins, including regulator of G protein signaling (RGS) proteins. A large and diverse set of RGS proteins has been described, and Chapter 3 reviews the different types of RGS proteins and how they affect GIRK channel signaling.

During the last ~17 years, there has been considerable advance in determining the structure of ion channels, including GIRK channels, with atomic resolution. In Chapter 4, the results of structure–function experiments involving GIRK channels are discussed in the context of recent high-resolution atomic structures of inwardly rectifying K^+ channels. These structures have provided valuable insights into the structural mechanisms underlying permeation, rectification, phosphatidylinositol 4,5-bisphosphate regulation, alcohol activation, and Gβγ activation. Chapter 5 reviews the cellular and subcellular localization of the four distinct GIRK channel subunits in mammalian brain neurons and discusses implications of channel subtype diversity for neuronal physiology.

While the contribution of GIRK channels to cardiac physiology is relatively well understood, the relevance of GIRK channels to neurophysiology and behavior is less clear. Over the last 20 years, however, there have been significant advances in our understanding of the role neuronal GIRK channels play in normal and disease states. In Chapter 6, the impact of drugs of abuse, such as cocaine, methamphetamine, and morphine, on inhibitory neurotransmission mediated by GIRK channels is reviewed, with a focus on the subcellular mechanisms linking changes in GIRK channel activity with addiction. Chapter 7 reviews the evidence for GIRK channel signaling contributing to learning and memory mechanisms underlying cognitive dysfunctions and drug addiction. Lastly, the behavioral and genetic evidence from animal models and genetic studies in humans linking GIRK channels with central nervous system disorders are discussed in Chapter 8.

Since the days of the discovery that *vagusstoff* slows the heart rate, we have learned significantly more about GIRK channels. This volume in *International Review of Neurobiology* provides a comprehensive compendium on the physiology and function of GIRK channels.

Paul A. Slesinger
Kevin Wickman

CHAPTER ONE

Unifying Mechanism of Controlling Kir3 Channel Activity by G Proteins and Phosphoinositides

Diomedes E. Logothetis[1], Rahul Mahajan, Scott K. Adney, Junghoon Ha, Takeharu Kawano, Xuan-Yu Meng[2], Meng Cui
Department of Physiology and Biophysics, School of Medicine, Virginia Commonwealth University, Richmond, Virginia, USA
[1]Corresponding author: e-mail address: delogothetis@vcu.edu

Contents

1. Introduction 2
2. Regulation of K_{ACh} by G Proteins 3
3. Molecular Constituents of Kir3 Channels and Physiological Roles 4
4. Dependence of K_{ACh} Channel Activity on PIPs 5
5. Physiological Significance of Distinct Membrane-Delimited Signaling Mechanisms 9
6. Molecular Mechanism of PIP_2 Control of Kir Channel Gating 12
7. Posttranslational Modification Can Control Activity Through Changes in Channel–PIP_2 Interactions 17
8. The Kir3 Channel Site of Gβγ Action 18
References 21

Abstract

The question that started with the pioneering work of Otto Loewi in the 1920s, to identify how stimulation of the vagus nerve decreased heart rate, is approaching its 100th year anniversary. In the meantime, we have learned that the neurotransmitter acetylcholine acting through muscarinic M2 receptors activates cardiac potassium (Kir3) channels via the βγ subunits of G proteins, an important effect that contributes to slowing atrial pacemaker activity. Concurrent stimulation of M1 or M3 receptors hydrolyzes PIP_2, a signaling phospholipid essential to maintaining Kir3 channel activity, thus causing desensitization of channel activity and protecting the heart from overinhibition of

[2] Present Addresses: School for Radiological and Interdisciplinary Sciences (RAD-X), Collaborative Innovation Center of Radiation Medicine of Jiangsu Higher Education Institutions, and Jiangsu Provincial Key Laboratory of Radiation Medicine and Protection, Soochow University, Suzhou 215123, China.

International Review of Neurobiology, Volume 123
ISSN 0074-7742
http://dx.doi.org/10.1016/bs.irn.2015.05.013

pacemaker activity. Four mammalian members of the Kir3 subfamily, expressed in heart, brain, endocrine organs, etc., are modulated by a plethora of stimuli to regulate cellular excitability. With the recent great advances in ion channel structural biology, three-dimensional structures of Kir3 channels with PIP_2 and the Gβγ subunits are now available. Mechanistic insights have emerged that explain how modulatory control of activity feeds into a core mechanism of channel–PIP_2 interactions to regulate the conformation of channel gates. This complex but beautiful system continues to surprise us for almost 100 years with an apparent wisdom in its intricate design.

1. INTRODUCTION

In this review, we will present our current understanding of the control of Kir3 channel activity by phosphoinositides (PIPs) aided by other molecules, such as G proteins, with a historical perspective of its development.

When the vagus nerve is stimulated, chemical transmission of neural signals occurs via release of acetylcholine (ACh) to slow heart rate (Loewi, 1921; Loewi & Navratil, 1926). An increase in K^+ conductance, K_{ACh}, found in atrial myocytes and cardiac pacemaking cells, underlies the hyperpolarization effect of ACh that precedes the decrease in heart rate (Burgen & Terroux, 1953; Trautwein & Dudel, 1958). With the advent of the patch-clamp technique (Hamill, Marty, Neher, Sakmann, & Sigworth, 1981), K_{ACh} single-channel characteristics were determined, showing short open times (1–2 ms) and a conductance of ~40 pS (Sakmann, Noma, & Trautwein, 1983). A year later it was shown that the effect of ACh on K_{ACh} channels is mediated locally, in a membrane-delimited fashion, and not by diffusible second messengers (Soejima & Noma, 1984). In their cell-attached experiments, Soejima and Noma perfused the inside of the pipette with solutions containing ACh to demonstrate that K_{ACh} activation would only occur when ACh was perfused in the patch pipette and not in the bath (Fig. 1). The ability to unequivocally identify K_{ACh} has led to hundreds of publications revealing multiple ways of controlling the activity of these important channels directly or allosterically through interactions of the channel protein with other regulatory proteins, metal ions, ethanol, and membrane lipids. In this review, we will focus on how fundamental protein–lipid interactions lie at the core of controlling the ion channel gates and how other modulatory signals feed onto this central control mechanism to adjust channel activity.

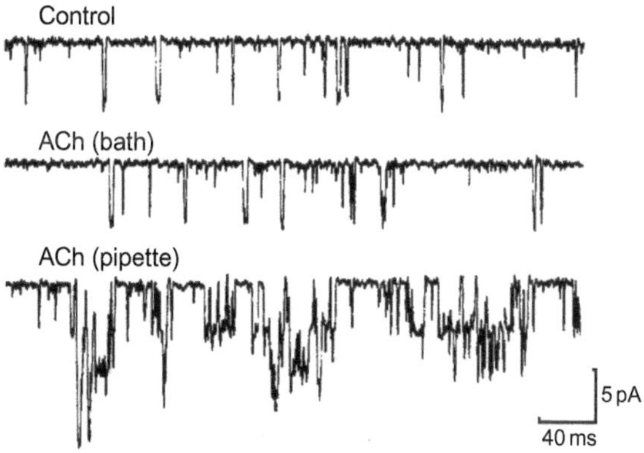

Figure 1 The membrane-delimited activation of K_{ACh} is occluded by the patch pipette. The top trace (control) was obtained before application of ACh (basal activity), the middle trace during perfusion of the bath with ACh. The bottom trace is a record obtained during application of ACh through the patch pipette only. Channel activity increased markedly only when ACh was perfused directly on the outer surface of the patch being recorded and up to three overlapping unitary amplitudes can be seen. *Adapted from Soejima and Noma (1984).*

2. REGULATION OF K_{ACh} BY G PROTEINS

In 1985, several reports established the involvement of GTP-binding (G) proteins in transducing the ACh signal to K_{ACh} (Breitwieser & Szabo, 1985; Endoh, Maruyama, & Iijima, 1985; Pfaffinger, Martin, Hunter, Nathanson, & Hille, 1985; Sorota, Tsuji, Tajima, & Pappano, 1985). The involvement of G proteins to stimulate K_{ACh} in response to ACh was also demonstrated using excised membrane patches from atrial cells with ACh or adenosine on the external side of the patch and GTP on the cytoplasmic side, an effect prevented by prior pertussis toxin treatment of the cells (Kurachi, Nakajima, & Sugimoto, 1986). Purified G proteins applied to inside-out patches of atrial cell membranes could induce K_{ACh} activity (Yatani, Codina, Brown, & Birnbaumer, 1987). In fact, the βγ subunits of G proteins (Gβγ) were found to be responsible for K_{ACh} activation, identifying the K_{ACh} channel as the first effector protein to be activated by nanomolar concentrations of the Gβγ subunits (Logothetis, Kurachi, Galper, Neer, & Clapham, 1987). Later in the same year, it was reported that picomolar concentrations of active Gα subunits, activated with a nonhydrolyzable GTP analog (GTPγS), could induce K_{ACh} activation

(Codina, Yatani, Grenet, Brown, & Birnbaumer, 1987), igniting a controversy that took several years to settle in favor of Gβγ, which is now accepted as the stimulatory component of G proteins (Ito et al., 1992; Logothetis, Kim, Northup, Neer, & Clapham, 1988, Reuveny et al., 1994; Wickman et al., 1994; reviewed by Stanfield, Nakajima, & Nakajima, 2002). Yet, a role for Gα subunit in the regulation of K_{ACh} activity is not fully understood and has not been definitively excluded, as activated Gα has been reported by other labs to exhibit weak stimulatory (Ito et al., 1992; Logothetis et al., 1988) or inhibitory effects (Schreibmayer et al., 1996; Vivaudou et al., 1997). Chapter "The Roles of Gβγ and Gα in Gating and Regulation of GIRK Channels" by Dascal and Kahanovitch explores in detail the role of Gα in regulating G protein-sensitive K^+ channels.

3. MOLECULAR CONSTITUENTS OF KIR3 CHANNELS AND PHYSIOLOGICAL ROLES

In 1993, the first molecular component of K_{ACh}, Kir3.1 (also referred to as GIRK1), was cloned from heart cDNA libraries, either through PCR, using degenerate primers based on the related Kir2.1 (Kubo, Baldwin, Jan, & Jan, 1993) or by expression cloning in *Xenopus* oocytes (Dascal et al., 1993). Two additional cDNAs from a brain library were obtained by low stringency hybridization using Kir3.1 as the probe (Lesage et al., 1994). Kir3.2 (or GIRK2) and Kir3.3 (or GIRK3) are expressed mainly in the CNS (Jelacic, Sims, & Clapham, 1999; Lesage et al., 1994, 1995). In 1995, Kir3.4 (also referred to as GIRK4 or CIR) was identified as the second molecular component of K_{ACh} (Krapivinsky et al., 1995). Although Kir3.4 are expressed as homomers in heart (e.g., Bender et al., 2001; Corey & Clapham, 1998), it is their coexpression into heteromers with Kir3.1 that gives rise to a conductance that behaves like K_{ACh} both biophysically and in response to G protein subunit modulation. In the brain, Kir3 channels can exist as homomers (e.g., Kir3.2a and Kir3.2c: Inanobe et al., 1999). However, several neurons coexpress several Kir3 subunits (Kir3.1–Kir3.4) (e.g., Chen, Ehrhard, Goldowitz, & Smeyne, 1997; Cruz et al., 2004; Karschin, Dissmann, Stuhmer, & Karschin, 1996) and form multiple heteromeric combinations (e.g., Kir3.1/3.2: Inanobe et al., 1999; Liao, Jan, & Jan, 1996; Kir3.2/3.3: Cruz et al., 2004; Inanobe et al., 1999; Jelacic, Kennedy, Wickman, & Clapham, 2000; Kir3.2/3.4: Lesage et al., 1995; Spauschus et al., 1996; reviewed by Hibino et al., 2010). Although Kir3.1 expression alone yields nonfunctional channels mostly trapped in the endoplasmic reticulum, heteromeric combinations of Kir3.2 and Kir3.4 with Kir3.1 produce greatly enhanced currents compared to homomeric assemblies (Chan, Sui, Vivaudou, & Logothetis, 1996). Single point mutations on the pore helices of Kir3 channels have yielded high activity homomers [Kir3.1(F137S): Chan et al., 1996; Vivaudou et al., 1997; Kir3.2(E152D): Yi, Lin, Jan, & Jan, 2001; Kir3.4(S143T): Vivaudou et al., 1997] that have been used extensively in the literature to compare characteristics and responses of homomers versus heteromers. Knockdown or knockout strategies in mice have revealed the contribution of Kir3 subunits to physiological processes. Thus, Kir3.2 knockdown has shown that it is critically involved in the formation of IPSPs in hippocampal and cerebellar neurons (Lüscher, Jan, Stoffel, Malenka, & Nicoll, 1997; Signorini, Liao, Duncan, Jan, & Stoffel, 1997). Kir3.4 knockout did not alter resting heart rates of mice but showed decrease in heart rate response to parasympathetic stimulation. Moreover, knockout mice lost the beat-to-beat fluctuations known as heart rate variability that reflects the balance between sympathetic and parasympathetic controls (Wickman, Nemec, Gendler, & Clapham, 1998). In the brain, Kir3.4 knockouts exhibited impaired performance in spatial learning and memory tasks (Wickman, Karschin, Karschin, Picciotto, & Clapham, 2000).

4. DEPENDENCE OF K_{ACh} CHANNEL ACTIVITY ON PIPs

In 1998, two reports from different laboratories showed that stimulation of G protein-sensitive channels depended on the phosphoinositide PIP_2 (Huang, Feng, & Hilgemann, 1998; Sui, Petit-Jacques, & Logothetis, 1998). Maintenance of Kir3.1/3.4 activity induced by G proteins or intracellular Na^+ depended on hydrolyzable ATP in a way that could be mimicked by PIP_2. A PIP_2-specific antibody or stimulation of phospholipase C (PLC) inhibited channel activity, and $G\beta\gamma$ or Na^+ could restore activity only in the presence of PIP_2 (Sui et al., 1998). In fact, $G\beta\gamma$ and Na^+ served to stabilize channel PIP_2 interactions (Huang et al., 1998; Zhang, He, Yan, Mirshahi, & Logothetis, 1999). These results were quite surprising and begged the question of how PIPs controlled channel activity.

PIPs are phosphorylated forms of phosphatidyl inositol (PI). PI is comprised of two fatty acid chains (arachidonyl is 20-carbon long and unsaturated, while stearyl is 18-carbon long and saturated) linked by a glycerol (G) moiety to a water-soluble inositol headgroup (Fig. 2). Specialized lipid kinases (19 in number) and phosphatases (28 in number), respectively, add and remove phosphates at specific positions of the inositol ring to give rise to seven phosphorylated species: three different isomers of

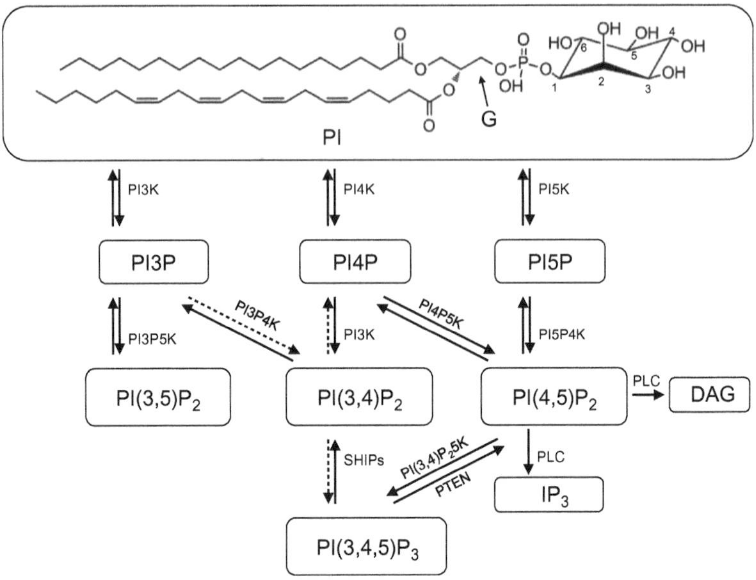

Figure 2 Metabolism of phosphoinositides. Phosphatidyl inositol (PI) is phosphorylated by specific kinases at the 3′, 4′, or 5′ positions to yield monophosphorylated PIPs. PIP kinases recognize the monophosphorylated species and phosphorylate them to yield the diphosphorylated PIPs. PIP_3 (phosphatidylinositol 3,4,5-trisphosphate) can be formed by phosphorylation of $PI(4,5)P_2$ (phosphatidylinositol 4,5-bisphosphate) by PI3K. The reverse reactions are catalyzed by phosphatases acting at the 3′, 4′, or 5′ positions (5′ phosphatase, e.g., SHIPs; 3′ phosphatase, e.g., PTEN). Phospholipase C (PLC) hydrolyzes $PI(4,5)P_2$ into inositol trisphosphate (IP_3) and diacylglycerol (DAG). *Modified from Logothetis et al. (2015).*

monophosphorylated PIPs, three different isomers of diphosphorylated PIPs, and one species of a triphosphorylated PIP. The most abundant PIPs in the plasma membrane are the predominant $PI(4,5)P_2$ and its precursor PI(4)P that comprise about 0.5–1% of the total phospholipid pool. It has been well appreciated since the 1970s that PLC activation results in hydrolysis of PIP_2 into inositol trisphosphate (IP_3) and diacylglycerol (DAG) (Fig. 2). These two ubiquitous hydrolysis products of PIP_2 serve as important intracellular signals: IP_3 induces release of calcium from endoplasmic reticulum stores, while DAG activates membrane or membrane-associated proteins, such as protein kinase C (PKC). On the other hand, since the 1990s, it has been appreciated that the added phosphates to PI confer a negative charge to PIPs that allows them to engage in direct electrostatic interactions with basic regions of proteins. These interactions range from nonspecific to highly specific. A number of membrane-associated proteins utilize interactions with PIPs to localize themselves to the plasma membrane and control events such as vesicular fusion to the plasma membrane, endocytosis, actin cytoskeleton dynamics, actions of AKAPs (A-kinase anchoring proteins), and recruitment to the membrane of GEFs and GAPs that modulate intracellular Gα (small G) protein activity (Logothetis et al., 2015). Ion channels and transporters have been studied extensively with regard to their interactions with PIPs and their functional implications, since the pioneering work of Hilgemann (Hilgemann & Ball, 1996), which motivated the studies on PIP_2 dependence of Kir3 channels mentioned earlier (Huang et al., 1998; Sui et al., 1998; Zhang et al., 1999). It appears that electrostatic interactions are critical for the proper function of ion channels and transporters by mechanisms that seem to be well conserved across the ion channel superfamily (e.g., Logothetis et al., 2015; Logothetis, Petrou, Adney, & Mahajan, 2010). It appears that control of transmembrane protein function by PIPs is not limited to ion channels and transporters but extends to other transmembrane proteins, such as single-pass growth factor receptors, like the epidermal growth factor receptor (Logothetis et al., 2015; Michailidis et al., 2011). Although the effects of PIPs on membrane-associated and transmembrane proteins have been studied extensively in the past couple decades, many questions remain unanswered regarding the details of how the lipid–protein interactions regulate protein activity.

Kir channels demonstrate a range of affinities for PIP_2 but, as is the case with many other transmembrane proteins, PIP_2 interaction appears to be necessary for activation of all members of this class of channels (Rohács et al., 2003). With regard to Kir3 channels, not only G protein stimulation but also other activators such as intracellular Na^+ or Mg^{2+} ions shown to be G protein-independent ways of gating Kir3 channels, all have a requirement of intact PIP_2 levels to exert their stimulatory effects (Petit-Jacques, Sui, & Logothetis, 1999). Interestingly, Kir3 channels demonstrate a relatively weak affinity for PIP_2 and are unique among the inward rectifiers in that PIP_2 alone is not sufficient for channel activity but requires the copresence of intracellular activators such as Gβγ, ions (i.e., Na^+ or Mg^{2+}), and ethanol. Zhang et al. (1999) used a chimeric strategy between Kir3.4 and Kir2.1 (a channel that can be activated by PIP_2 alone) to show that specific amino acid differences between the two channels mostly accounted for their differences in affinity for PIP_2. A point mutation in Kir3.4 to the corresponding residue found in Kir2.1 (Kir3.4-I229L) was sufficient to strengthen the affinity of the Kir3.4 channel so that it could now be activated by PIP_2 directly, without requiring the copresence of G protein subunits or ions (Fig. 3A). In fact,

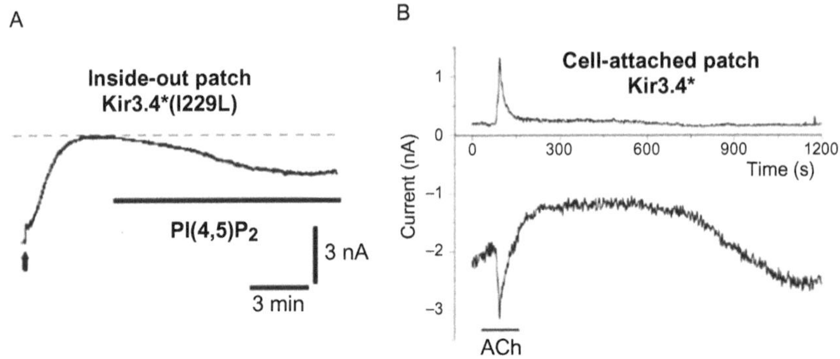

Figure 3 The membrane-delimited regulation of Kir3 activity by PIP_2 hydrolysis is not occluded by the patch pipette. (A) Transition from a cell-attached to an inside-out macropatch (at arrow) recording from a *Xenopus* oocyte in a Mg^{2+}-containing bath solution shows rundown of Kir3.4*(I229L) activity. The I229L mutation serves to strengthen channel–PIP_2 interactions. The kinetics of this rundown activity is slower than control (not shown). Perfusion of $PI(4,5)P_2$ could reactivate Kir3.4* currents of this mutant demonstrating the membrane-delimited signaling of PIP_2 activation. Again PIP_2 fails to activate the control channel (not shown) in the absence of other gating molecules (e.g., Gβγ, Na^+, ethanol) that serve to strengthen interactions of the channel with PIP_2. (B) Bath application of ACh outside the patch pipette activates M1 receptor and inhibits the active homomeric Kir3.4* channel currents (Vivaudou et al., 1997) recorded in a cell-attached patch from a *Xenopus* oocyte. ACh was applied to the cell via the bathing solution. *Panel (A): adapted from Zhang et al. (1999) and panel (B): adapted from Zhang et al. (2003).*

both the Gβγ subunits, and the Na^+ and Mg^{2+} ions were shown to increase the channel affinity for PIP_2, suggesting that they gated the channel, at least in part, by strengthening channel–PIP_2 interactions (Huang et al., 1998; Zhang et al., 1999). A crystal structure for the analogous point mutation in Kir2.2 (I223L) suggests that this simple point mutation changes the cytosolic gate of the channel into a conformation resembling a PIP_2-bound state (Hansen, Tao, & MacKinnon, 2011). However, this hypothesis has not been tested for the Kir3.2 channel that has also been crystallized (Whorton & MacKinnon, 2011, see below), as no crystal structures exist of the Kir3.2 (I234L) mutant.

Under physiological conditions that do not cause PIP_2 hydrolysis, it is likely that there is enough PIP_2 in the intact membrane available to interact with ion channels and support their activity. How might channel activity be inhibited when conditions change and PIP_2 is hydrolyzed through stimulation of PLC activity (i.e., via Gq or tyrosine kinase (TK) receptor signaling)? Since application of PIP_2 to the internal surface could modulate Kir currents in inside-out patches and PIP_2 is natively found exclusively in the inner leaflet of the plasma membrane, a likely mechanism of membrane-delimited current regulation by PIP_2 is clearly suggested. The experiments in Fig. 3 tested whether the Soejima and Noma experiment (Fig. 1) that established membrane-delimited signaling for G protein signaling could also be applied for PIP_2 signaling (Zhang et al., 2003). The Kir3.4* mutant I229L strengthens channel interactions with PIP_2, such that the channel is constitutively activated by PIP_2 alone, like other Kir channels. Excision into an inside-out patch yields current rundown, while perfusion of PIP_2 in the bath activates the current (Fig. 3A), establishing the membrane-delimited nature of the signaling. Unlike Gi protein signaling, we can see that in a cell-attached recording, PIP_2 hydrolysis triggered by ACh application and Gq stimulation in the bath solution decreases basal Kir3.4 activity in a reversible manner (Fig. 3B) (the outward current spike upon ACh application indicates that ACh is indeed hydrolyzing PIP_2, as its hydrolysis product IP_3 is causing Ca^{2+} release from internal stores and activation of a transient calcium-activated Cl^- current endogenous to *Xenopus* oocytes, where this experiment was performed). Thus, PIP_2 hydrolysis outside the patch allows the higher concentration of PIP_2 within the patch to diffuse across the seal of the pipette with the membrane patch or activated PLC to diffuse into the patch, resulting in inhibition of the Kir3.4 current. The inability of G protein signaling to diffuse across the seal in the same manner as PIP_2 may mean that the G protein signaling system operates as a macromolecular complex of the GPCR/G proteins/channel, and it is not possible for such a complex to diffuse through the seal between the pipette and the cell membrane.

5. PHYSIOLOGICAL SIGNIFICANCE OF DISTINCT MEMBRANE-DELIMITED SIGNALING MECHANISMS

How does PIP_2 hydrolysis affect K_{ACh}? Kobrinsky, Mirshahi, Zhang, Jin, and Logothetis (2000) showed that ACh in rat atrial myocytes activates Gq signaling through M3-type muscarinic receptors, resulting in PIP_2 hydrolysis. Inhibition of Gq signaling or manipulations to strengthen channel–PIP_2 interactions attenuated current desensitization that occurred shortly following the activation of the current by the PTX-sensitive G proteins. Keselman, Fribourg, Felsenfeld, and Logothetis (2007) compared the kinetics of PIP_2 hydrolysis via a fluorescence reporter with those of G protein-activated Kir3 currents. Isolated pleckstrin homology (PH) domains from PLCδ1 were labeled with variants of the green fluorescence protein (cyan: CFP; yellow: YFP) and expressed in HEK-293 cells along with muscarinic receptors and Kir3.4* channels. Under resting conditions, the labeled PH domains bind PIP_2 in the plasma membrane and are sufficiently close (<100 nM), by virtue of being concentrated in the two-dimensional plane of the membrane, to undergo fluorescence resonance energy transfer (FRET). With FRET, the energy from excitation of CFP at ~430 nM is transferred to YFP, emitting a fluorescence signal at ~530 nM. CFP emission at ~475 nM, on the other hand, is reduced (Fig. 4A). Upon perfusion of ACh, PIP_2 hydrolysis (to IP_3 and DAG via PLCβ1 activation) causes translocation of the labeled PH domains to the cytosol, causing them to move farther away from each other in the three-dimensional space of the cytosol and therefore to decrease the FRET signal. In fact, the CFP emission increases as energy is no longer transferred to YFP, while YFP emission decreases (Fig. 4B). Keselman and colleagues used this FRET approach to monitor PIP_2 hydrolysis and relate it to its effects on Kir3 currents in recordings where both signals were measured simultaneously (Fig. 4C). When Kir3.4 channels were transfected in HEK-293 cells with M2 (Gi-coupled) receptor and the labeled CFP- and YFP-PH domains (black traces), there was no FRET change and the current showed minimal decline during the ACh perfusion. In contrast, when M1 (Gq-coupled) and M2 receptors were cotransfected together with the channel and labeled PH domains (red traces, gray in the print version), there was a clear FRET decrease that correlated closely with a decrease in the current. These results supported the conclusions from the previous study (Kobrinsky et al., 2000), in which activation of the Gq pathway induced Kir3 current desensitization by hydrolysis of PIP_2.

Thus, two membrane-delimited signaling mechanisms converge at the level of the channel to modulate Kir3 activity (Fig. 5). Activation of Gi/o-coupled muscarinic receptors rapidly opens Kir3 channels via a Gβγ-mediated mechanism, while the Gq-coupled muscarinic receptors more slowly inhibit Kir3 currents via PIP_2 depletion through its hydrolysis. In heart, the same extracellular message, ACh, serves to activate both types of receptors that trigger intracellular events resulting in first the activation of the channel (via Gβγ) and therefore inhibition of pacemaker activity but second to partially inhibit this activation (via PIP_2 hydrolysis) limiting the magnitude and duration of the stimulatory effect. Since Gβγ serves to strengthen channel–PIP_2 interactions (Huang et al., 1998; Zhang et al., 1999), the question of how this occurs molecularly had to await a structural understanding of how these modulators interacted with the channel. Neuronal calcium channels demonstrate some parallels to K_{ACh} and have been shown to be

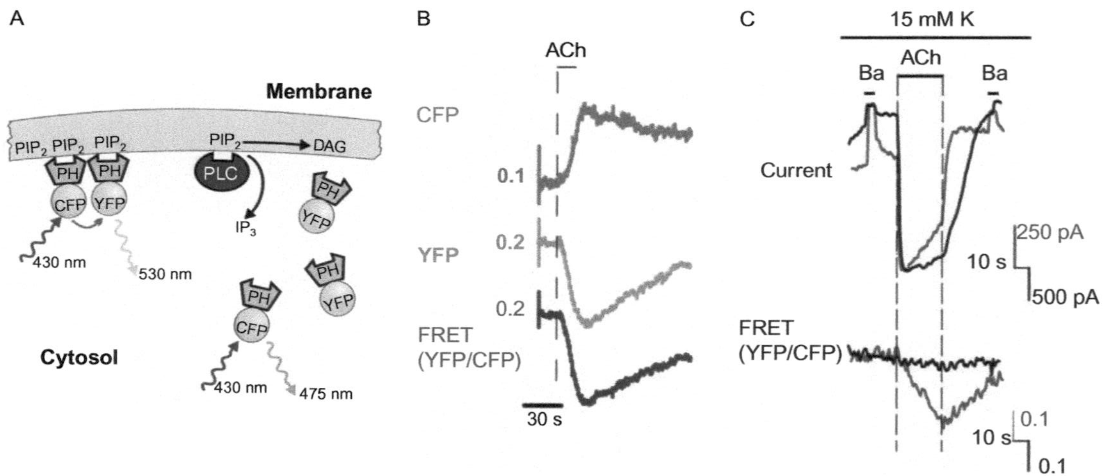

Figure 4 Simultaneous measurement of Kir3 currents and PIP_2 hydrolysis indicates that current inhibition is due to PIP_2 hydrolysis. (A) The pleckstrin homology (PH) domains from PLCδ can be tagged with eCFP or eYFP and cotransfected in a HEK-293 cell. These domains interact with PIP_2 in the plasma membrane where they get concentrated, and if excited at ~430 nm they emit at ~475 nm (eCFP) or at ~530 nm if eCFP and eYFP are less than 100 Å away, as fluorescence resonance energy transfer (FRET) occurs. When PIP_2 is hydrolyzed by stimulation of PLC, the tagged PH domains fall into the cytosol and FRET decreases. (B) Representative traces showing changes in fluorescence intensity in response to ACh. Top panel, changes in CFP emission as monitored at 480 nm (blue, dark gray in the print version, trace); middle panel, changes in YFP emission as monitored at 535 nm (yellow, gray in the print version, trace); and bottom panel, changes in the ratio of YFP over CFP and calculated 535 nm/480 nm (red, black in the print version, trace). Fluorescence was monitored on a single cell. HEK-293 cells were transfected with M1R, the eCFP-PH domain, and the eYFP-PH domain of PLCδ. (C) Representative traces showing effects of ACh on Kir3 current (upper panel) and PIP_2 hydrolysis as monitored by changes in FRET (lower panel) in cells expressing M2R alone (black traces) or both M1R and M2R (red, gray in the print version, traces). Current and FRET measurements were performed simultaneously on a single cell. HEK-293 cells were transfected with Kir3, MRs, the eCFP-PH and eYFP-PH domain of PLCδ. *Adapted from Keselman et al. (2007).*

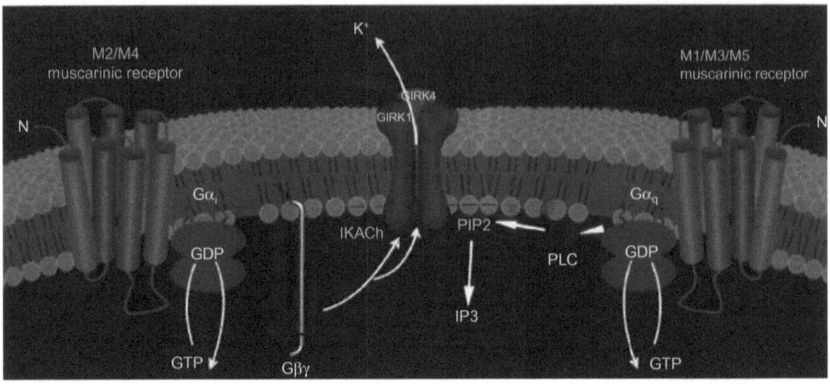

Figure 5 A cartoon of ACh signaling to cardiac Kir3 (GIRK1/GIRK4) channels through two distinct pathways. ACh released by the vagus nerve can act in atrial and pacemaker cells of the heart through either Gi-coupled muscarinic receptors (M2R or M4R) or through Gq-coupled muscarinic receptors (M1R, M3R, or M5R). M2R stimulation activates Kir3 channels via the Gβγ subunits, while M3R stimulation activated PLC to hydrolyze PIP_2 and limit the ability of Gβγ to continue to activate the channel. Thus, Kir3 channels integrate two distinct signals to activate and desensitize a tightly timed response to slow heart rate (see Kobrinsky et al., 2000).

both Gβγ- and PIP_2-sensitive (Hibino et al., 2010). Determining how integration of these two membrane-delimited signaling mechanisms serve physiologic roles in other systems, such as brain and endocrine system, and with other channels and transporters (e.g., voltage-gated K^+, Ca^{2+}, Kir6.2, or transmembrane proteins) remains an active area of research.

6. MOLECULAR MECHANISM OF PIP_2 CONTROL OF KIR CHANNEL GATING

Since PIP_2 seems to be absolutely required for the activity of all Kir channels (and most ion channels in general), we will discuss PIP_2-dependent channel gating first, as this seems to highlight a conserved gating machinery utilized in this family of channels that other modulators (e.g., G proteins and others—see below) could affect as well.

A number of crystallographic complexes of membrane-associated proteins with PIPs have revealed two major generalizations (Rosenhouse-Dantsker & Logothetis, 2007).

(1) All PIP-binding sites contain basic residues with at least one lysine and in most of the structures at least one arginine that engage in electrostatic interactions with the negatively charged PIPs; at least one residue with an aromatic ring, mostly tyrosine or histidine but tryptophan or phenylalanine residues are also seen; small polar residues such as serine or asparagine are also encountered.

(2) The affinity of PIP binding seems to be directly related to the number of specific contacts made between the protein and the PIP, such that specificity may be determined by the relative binding affinity of the various PIPs. Thus, low specificity appears to correlate with a low binding affinity as well.

Kir3 channels show low specificity (i.e., preference for binding to $PI(4,5)P_2$ vs. other PIPs with phosphates at positions other than 4 and 5 of the inositol ring) and also show weak affinity compared to most other Kir channels (Rohács et al., 2003). The only Kir channel with lower specificity and affinity than Kir3 is the Kir6 (or K_{ATP}) channel (Rohács et al., 2003).

Nishida, Cadene, Chait, and MacKinnon (2007) obtained a high-resolution 3D structure of a Kir3.1 chimera between most of the cytosolic domains of the mammalian Kir3.1 channel and most of the transmembrane domains of a bacterial Kir channel (Fig. 6A). They captured in two crystal forms the cytosolic or G-loop gate (see Pegan et al., 2005) in "open" (or dilated) and "closed" (or constricted) conformations (Nishida et al., 2007). Although they were not able to record activity from this chimeric protein, Leal-Pinto et al. (2010) managed to functionally reconstitute the chimeric protein in planar lipid bilayers by including PIP_2 in the internal side of the lipid bilayer. The EC_{50} for diC8-PIP_2 activation of the Kir3.1 chimera was 17 μM.

Molecular dynamics simulation studies by Meng, Zhang, Logothetis, and Cui (2012) suggest that PIP_2 engages in salt bridge interactions with a dozen basic Kir3.1 residues (K49, R52, R66, K79, R81, K183, K188, K189, R190, R219, R229, R313) (Fig. 6B) (also see Logothetis et al., 2015). Interactions of PIP_2 with W80 have been reported by another modeling study on the Kir3.1 chimera (Stansfeld, Hopkinson, Ashcroft, & Sansom, 2009). The types of residues identified are consistent with those seen in membrane-associated cocrystal structures with PIPs (see above, generalization #1) (Rosenhouse-Dantsker & Logothetis, 2007). Mutation of these residues and their effects on PIP_2 sensitivity has not been probed yet in Kir3 channels. Mutation of the equivalent positioned basic residues to neutral ones in Kir2.1 showed that all but two residues decreased PIP_2 sensitivity (Lopes et al., 2002). The distribution of these residues is insightful: in the N terminus (K49, R52), surrounding the slide helix (R66, K79, R81), the C linker (K183, K188, K189, R190—connecting the C terminus and

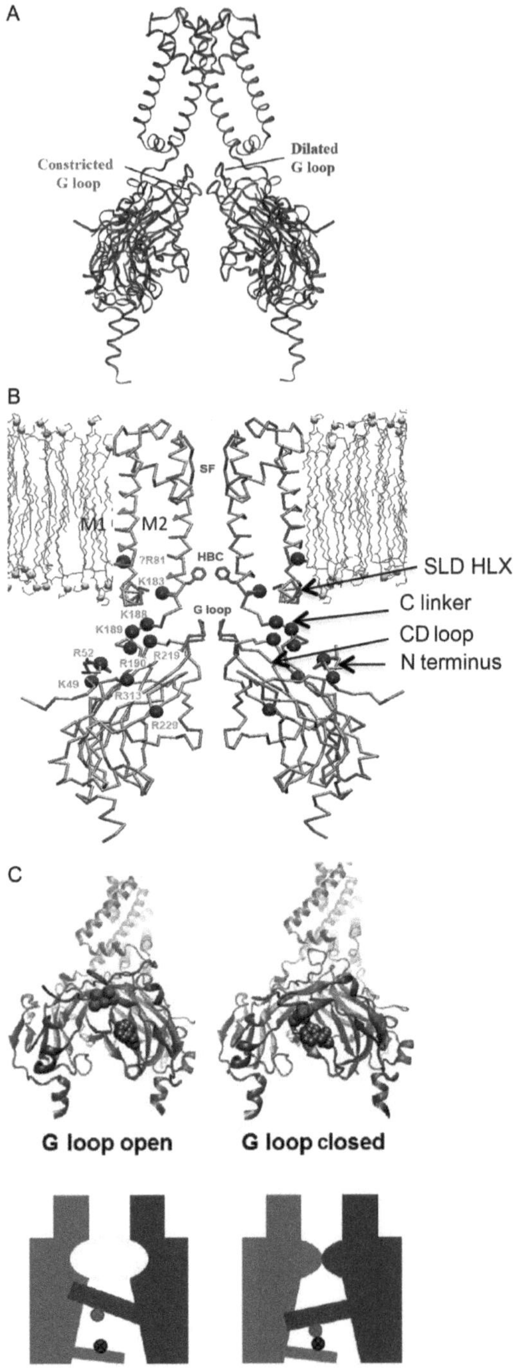

Figure 6 See legend on the opposite page.

the inner transmembrane helix), the CD loop (R219, R229), and the end of the G-loop gate (R313). The regulatory role of the CD loop in channel gating is best illustrated by the actions of intracellular Na^+ ions. Kir3.2 and Kir3.4 channels contain the motif DLR(K/N)SH in their CD loops that binds intracellular Na^+, which as discussed earlier can stimulate Kir3 channel activity independently of G proteins (Ho & Murrell-Lagnado, 1999; Inanobe, Nakagawa, Matsuura, & Kurachi, 2010; Rosenhouse-Dantsker et al., 2008; Whorton & MacKinnon, 2011; Zhang et al., 1999). Rosenhouse-Dantsker et al. (2008) showed that Na^+ is in part coordinated by the side chains of the Asp (D228 in Kir3.2 and D223 in Kir3.4) and His residues. Kir3.1 contains N217 in the corresponding position of the critical Asp residue and thus it lacks Na^+ sensitivity. R219 is predicted to interact with PIP_2 transiently (Meng et al., 2012). These studies showed that in the absence of Na^+, the Asp and Arg residues form a salt bridge that is broken in the presence of Na^+, as it engages the Asp residue to coordinate it. This presumably frees the Arg residue in the CD loop to interact with PIP_2 and activate the channel.

The simulations by Meng et al. (2012) of the Kir3.1 chimera in the presence of $PI(4,5)P_2$ identified an intermediate state between the closed and open conformations of the G-loop gate. This intermediate conformational state reveals a large upward movement of the LM loop to interact with the N terminus of the adjacent subunit through formation of the βA–βM interface. The movement of the N terminus to form the βA–βM interactions liberates the CD loop that was previously interacting with the N terminus to interact with the G loop and stabilize it in the open conformation. These

Figure 6 The Kir3.1 chimera captured with its cytosolic gate in closed and open conformations. (A) Conformations of the cytosolic G-loop gate of two opposing subunits in the "constricted" or closed (red, gray in the print version) and "dilated" or open (blue, black in the print version) are shown. (B) A pose of two opposing Kir subunits that shows the permeation pathway and three gates (two transmembrane and one cytosolic). Ten experimentally identified PIP_2-sensitive residues are shown as blue (black in the print version) spheres, while the regions that these residues are found in have been labeled. (C) (Top) Two residues marking the LM loop (L333, red, dark gray in the print version) and DE loop (F243, gray) are depicted as spheres in both panels of two adjacent channel subunits with the G loop in open and closed conformations, respectively. (Bottom) Schematized versions of channel conformations are shown in the top panels. Adjacent channel subunits are illustrated in gray and dark red (dark gray in the print version). G-loop state is illustrated as open or closed by state of constriction in the central pore. Specific residues (filled circles) of GIRK1 L333 (red, gray in the print version) and F243 (black) are drawn along with conformation of LM loop and the DE–LM loop cleft.

structural changes induced by the presence of PIP_2 in the simulations are consistent with the transition of closed to open G-loop conformations captured by the crystal structures (Nishida et al., 2007) (Fig. 6C). How opening of the G-loop gate leads to subsequent opening of the inner helix gate (F192 seen in Fig. 7) and eventually the selectivity helix gate is not understood presently but is an active area of research.

In 2011, two atomic resolution structures of Kir2.2 and Kir3.2 channels bound to PIP_2 were reported (Hansen et al., 2011; Whorton & MacKinnon, 2011). Six basic channel residues were found to interact directly with PIP_2, five of which were common to both channels. For Kir3.2, these residues were distributed in the N terminus K64 (corresponding to the Kir3.1-R52), the post-slide helix K90 and R92 (corresponding to the Kir3.1 K79 and R81 residues), the cytosolic end of M2 K194 (corresponding to

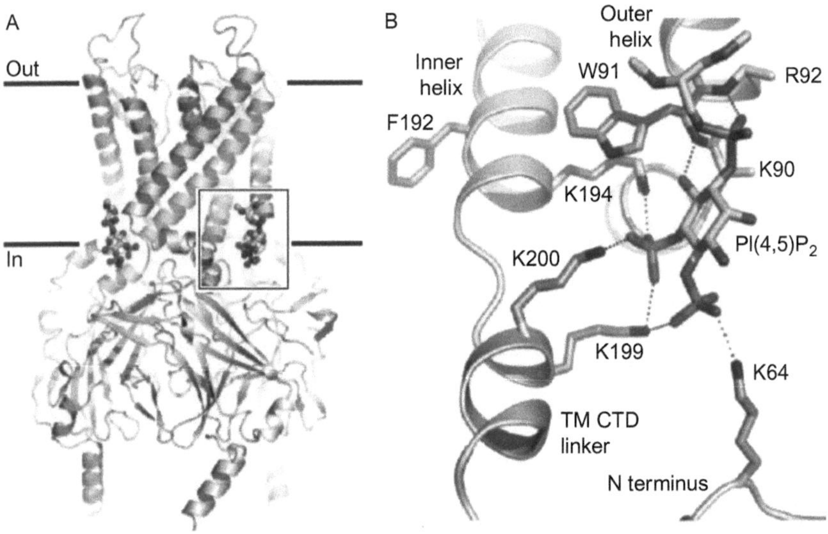

Figure 7 PIP_2 binds at the interface between the transmembrane and cytoplasmic domains and is coordinated by several positively charged residues. (A) The PIP_2-binding site is shown in the context of the whole channel (gray). The thick black lines indicate the approximate boundary of the plasma membrane, and the black box highlights the region of the close-up view in B. (B) A close-up view shows the PIP_2 molecules colored yellow (light gray in the print version), orange (gray in the print version), and red (dark gray in the print version) (for carbon, phosphorus, and oxygen atoms, respectively) interacting with Kir3.2. The main coordinating residues are shown as sticks. Residues Lys90 and Arg92 were modeled as alanines due to a lack of electron density, but probably still contribute to the positive electrostatics of the binding site. The important gating residue Phe192 is also shown for reference. *Adapted from Whorton and MacKinnon (2011).*

the Kir3.1 K183), and the C linker K199 and K200 (corresponding to the Kir3.1 K188 and K189) (Fig. 7). The Kir3.1 residues identified by the simulations but not by the crystal structure of Kir3.2 (i.e., N terminus K49, preslide helix R66, the C linker R190, the CD loop R219 and R229, and the G-loop gate R313) possibly represent transitional states of interactions leading to the stable state of channel–PIP_2 interactions captured by the crystal structure. Alternatively, PIP_2 interaction differences between the two channel subunits may be due to underlying structural differences between them. The Kir3.2 residue R201 (corresponding to Kir3.1-R190) appears to have an intimate role in channel gating (Whorton & MacKinnon, 2011).

7. POSTTRANSLATIONAL MODIFICATION CAN CONTROL ACTIVITY THROUGH CHANGES IN CHANNEL–PIP_2 INTERACTIONS

Figure 8 depicts the cytosolic surface of a Kir3 channel as it faces the inner leaflet of the membrane. The transmembrane domains of the channel are not shown for clarity. Channel sites that have been implicated to affect the sensitivity of these Kir3 channels to phosphorylation by protein kinase A (PKA) and PKC or TKs are shown (Logothetis et al., 2007). PIP_2-interacting residues are shown in cyan (light gray in the print version). The proximity of modulatory sites to those of PIP_2 is tantalizing, strengthening the notion that they all converge onto a common gating mechanism to regulate channel activity. Functional evidence also supports this notion. Du et al. (2004) provided evidence in Kir channels that distinct modulators, such as Gq and PKC activation, Gβγ subunits, Mg^{2+}, and pH, all depend on PIP_2 sensitivity to exert their effects. Mutations causing either stronger or weaker interactions with PIP_2 changed significantly the modulatory action of these signals. Additional studies on the PKA and PKC modulation of Kir3 channels corroborated these conclusions and suggested that strategic placement of negatively charged phosphate groups via phosphorylation can exert either stimulatory or inhibitory effects by altering the same gating mechanism utilized by PIP_2 (Keselman et al., 2007; Lopes et al., 2007). The precise mechanism by which a particular phosphorylated site may exert its effects on Kir3 channel gating is not currently understood but is an active area of research.

The principle has been demonstrated in another PIP_2-sensitive channel, the SK2–CaM (calmodulin) complex, where phosphorylation of CaM(T79) has been shown to directly compete with PIP_2 for two basic residues that normally coordinate it, one on the channel and one on CaM (Logothetis

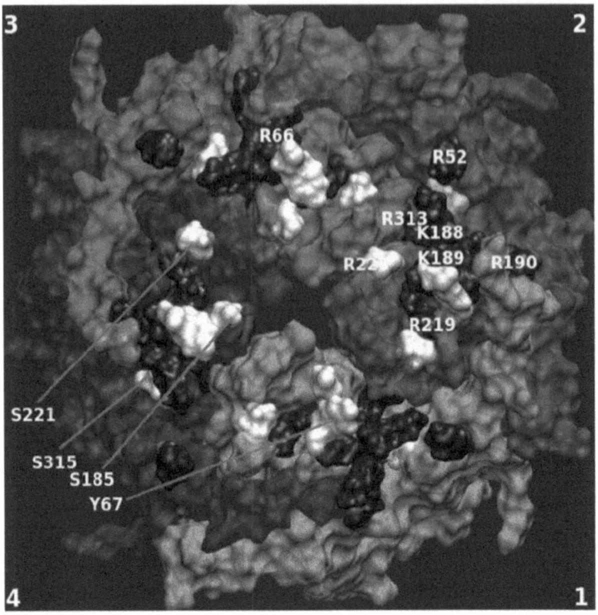

Figure 8 Localization of phosphorylation sites of the Kir3.1 channel relative to its PIP_2-sensitive residues. Putative phosphorylation sites as suggested by functional evidence are shown in the Kir3.1 structure highlighted in orange (gray in the print version), in addition to the cyan (light gray in the print version)-highlighted residues that affect channel–PIP_2 interactions. The surface of the Kir3.1 shown is the cytosolic surface that faces the membrane. The transmembrane domains have been removed in order to visualize the residues that interact with PIP_2 and the adjacent phosphorylation sites. *Adapted from Logothetis, Lupyan, and Rosenhouse-Dantsker (2007).*

et al., 2015; Zhang et al., 2014). These results have suggested that diverse modulators of channel activity share a common gating mechanism through adjustments in interactions of the channel with PIP_2 (Logothetis et al., 2015, 2007).

8. THE KIR3 CHANNEL SITE OF Gβγ ACTION

In 2013, two independent studies reported the long-awaited complexes of Gβγ with a cocrystal structure with Kir3.2 (Whorton & MacKinnon, 2013) and a computational model with Kir3.1 (Mahajan et al., 2013). The two studies identified Gβγ to interact between two subunits making contacts within a cleft generated by the LM loop of one subunit and the DE loop of the adjacent subunit. Other interaction contacts between Gβγ and the two channel subunits involved the βN strand and the βK-L loop. There was good agreement between the two complex structures with interesting differences.

Whorton and MacKinnon (2013) combined individually purified mouse Kir3.2 and Gβγ with an eight-carbon long acyl chain PIP_2 and incubated them at room temperature before crystal trials that were conducted in a high salt solution containing Na^+. Crystals of the complex diffracted at 3.5 Å resolution capturing the channel in a conformation that may represent an intermediate between the closed conformation and a partially open conformation previously determined by the same group (Whorton & MacKinnon, 2011). The complex included a PIP_2 molecule and a Na^+ ion, revealing a smaller contact surface (~700 Å) than that seen in other Gβγ effectors for which complex structures have been determined (Reuveny, 2013).

Mahajan et al. (2013) used a multistage docking strategy to generate models of the complex between Gβγ and the Kir3.1 channel (also see Mahajan & Logothetis, 2015). The predictions of the models were tested experimentally using electrophysiological and biochemical techniques providing a compelling picture of how Gβγ interacts with human Kir3.1 channels to stimulate their activity. A largely hydrophobic contact surface (~1800 Å) was seen between Kir3.1 and Gβ. The footprint of Gβγ onto the Kir3.1 cytosolic domain predicted by these models was in excellent agreement with previously published NMR data (Yokogawa, Osawa, Takeuchi, Mase, & Shimada, 2011). Gβ blades 7, 1, and 2 of Gβγ interacted with two adjacent subunits of the channel tetramer.

The footprint of Gβγ on either Kir3.1 or Kir3.2 predicted by the differing techniques involved the same channel regions. The mouse Kir3.2 shows only a 55.4% identity with the human Kir3.1, and the predicted specific interactions of the two channels with Gβγ show not only many similarities but also several interesting differences. The pattern of interactions between Gβγ and either Kir3.1 or Kir3.2 suggests that common interactions between Gβγ with these Kir3 channel subunits may serve a fundamental role by which Gβγ activates Kir3 channels, while differences could be representative of differing interactions in the two channels perhaps underlying some of their functional differences or be due to differing degrees of channel activation captured by the two techniques.

Interestingly, several residues previously identified by different studies to be important were identified in these structural studies (Finley, Arrabit, Fowler, Suen, & Slesinger, 2004; Ford et al., 1998; He et al., 2002; He, Zhang, Mirshahi, & Logothetis, 1999; Mirshahi, Robillard, Zhang, Hébert, & Logothetis, 2002; Zhang et al., 2003, 1999). In particular, the work by Mahajan et al. (2013) proposes residue interactions key to Gβγ stimulation of Kir3 activity [Kir3.1-L333 (He et al., 1999) and the Gα-interacting Gβ-L55, Gβ-K89 (Ford et al., 1998)] interact at the DE–LM cleft between two adjacent channel subunits and stabilize the channel in an overall conformation where the cleft is opened and the LM loop is in a "raised" conformation (the LM loop apart from the DE loop) (Fig. 9A).

We propose that the model of a cascade of switching interactions involving changing interactions among the G loop, CD loop, N terminus of an adjacent subunit, and LM loop predicted by simulations in the presence of PIP_2 and Kir3.1 chimera by Meng et al. (2012) extends further down the channel to the DE loop. In the closed state, the downward moved LM loop closely interacts with the adjacent DE loop but this interaction

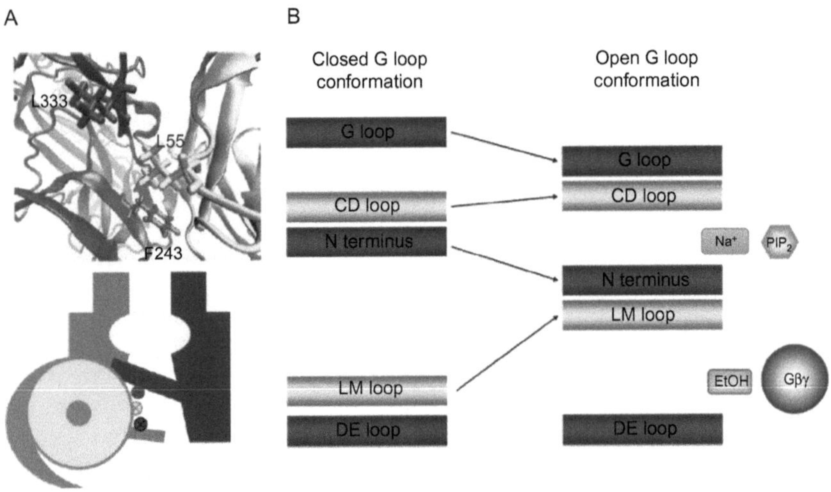

Figure 9 Sites of interaction of Gβγ-, PIP_2-, and other Kir3-gating molecules. (A) Close-up view of the cleft between LM and DE loops. Cartoon depictions of two adjacent channel subunits are in dark red (black in the print version) and gray, while Gβ1 is in yellow (light gray in the print version) and Gγ2 in orange (gray in the print version). Specific residues are highlighted in stick representation: Kir3.1-L333 (red, dark gray in the print version), Kir3.1-F243 (gray), and Gβ-L55 (yellow, light gray in the print version). (B) Summary of the major results of Meng et al. (2012) and their extension to include the DE loop and LM–DE loop cleft (Mahajan et al., 2013). Transitioning from the closed to open, the secondary structure elements switch their close interactions from the adjacent element on one side to the element on the other side. Na ions help PIP_2 to stabilize the conformation on the right by direct interactions with the CD loop and N terminus. We propose that Gβγ works through a similar mechanism by stabilizing the same overall conformation but by direct interactions with a different part of the channel. Its proposed site of action at the DE–LM loop cleft is shared with the site of ethanol (EtOH) action.

switches in the open state so the LM loop instead moves upward to interact closely with the adjacent N terminus. Thus, this model proposes that PIP_2 and Gβγ modulate the same cascade of switching interactions but their sites of action are distinct. PIP_2 acts close to the pore and gates by directly interacting with residues of the CD loop and N terminus. Intracellular Na^+ is chelated by and therefore effects changes in residues adjacent to the PIP_2-binding site involving the CD loop. Gβγ acts at the level of the LM and DE loops. Ethanol, which activates the channel, also interacts with the channel at the LM–DE loop cleft (Aryal, Dvir, Choe, & Slesinger, 2009), supporting the importance of this cleft in channel activation (Fig. 9B). Thus, multiple modulatory inputs (e.g., Gβγ, ethanol, and Na^+) feed into the central PIP_2-controlled gating mechanism to adjust channel activity.

REFERENCES

Aryal, P., Dvir, H., Choe, S., & Slesinger, P. A. (2009). A discrete alcohol pocket involved in GIRK channel activation. *Nature Neuroscience, 12*(8), 988–995.

Bender, K., Wellner-Kienitz, M. C., Inanobe, A., Meyer, T., Kurachi, Y., & Pott, L. (2001). Overexpression of monomeric and multimeric GIRK4 subunits in rat atrial myocytes removes fast desensitization and reduces inward rectification of muscarinic K^+ current (I_{KACh}). Evidence for functional homomeric GIRK4 channels. *The Journal of Biological Chemistry, 276*(31), 28873–28880.

Breitwieser, G. E., & Szabo, G. (1985). Uncoupling of cardiac muscarinic and β-adrenergic receptors from ion channels by a guanine nucleotide analogue. *Nature, 317*(6037), 538–540.

Burgen, A. S. V., & Terroux, K. G. (1953). On the negative ionotropic effect in the cat's auricle. *The Journal of Physiology, 120*, 449–464.

Chan, K. W., Sui, J. L., Vivaudou, M., & Logothetis, D. E. (1996). Control of channel activity through a unique amino acid residue of a G protein-gated inwardly rectifying K^+ channel subunit. *Proceedings of the National Academy of Sciences of the United States of America, 93*(24), 14193–14198.

Chen, S. C., Ehrhard, P., Goldowitz, D., & Smeyne, R. J. (1997). Developmental expression of the GIRK family of inward rectifying potassium channels: Implications for abnormalities in the weaver mutant mouse. *Brain Research, 778*, 251–264.

Codina, J., Yatani, A., Grenet, D., Brown, A. M., & Birnbaumer, L. (1987). The α subunit of the GTP binding protein Gk opens atrial potassium channels. *Science, 236*(4800), 442–445.

Corey, S., & Clapham, D. E. (1998). Identification of native atrial G-protein-regulated inwardly rectifying K^+ (GIRK4) channel homomultimers. *The Journal of Biological Chemistry, 273*(42), 27499–27504.

Cruz, H. G., Ivanova, T., Lunn, M. L., Stoffel, M., Slesinger, P. A., & Luscher, C. (2004). Bi-directional effects of $GABA_B$ receptor agonists on the mesolimbic dopamine system. *Nature Neuroscience, 7*, 153–159.

Dascal, N., Lim, N. F., Schreibmayer, W., Wang, W., Davidson, N., & Lester, H. A. (1993). Expression of an atrial G-protein-activated potassium channel in *Xenopus* oocytes. *Proceedings of the National Academy of Sciences of the United States of America, 90*(14), 6596–6600.

Du, X., Zhang, H., Lopes, C., Mirshahi, T., Rohacs, T., & Logothetis, D. E. (2004). Characteristic interactions with phosphatidylinositol 4,5-bisphosphate determine regulation of Kir channels by diverse modulators. *The Journal of Biological Chemistry, 279*(36), 37271–37281.

Endoh, M., Maruyama, M., & Iijima, T. (1985). Attenuation of muscarinic cholinergic inhibition by islet-activating protein in the heart. *The American Journal of Physiology, 249*, H309–H320.

Finley, M., Arrabit, C., Fowler, C., Suen, K. F., & Slesinger, P. A. (2004). βL-βM loop in the C-terminal domain of G protein-activated inwardly rectifying K^+ channels is important for Gβγ subunit activation. *The Journal of Physiology, 555*(Pt. 3), 643–657.

Ford, C. E., Skiba, N. P., Bae, H., Daaka, Y., Reuveny, E., Shekter, L. R., et al. (1998). Molecular basis for interactions of G protein βγ subunits with effectors. *Science, 280*(5367), 1271–1274.

Hamill, O. P., Marty, A., Neher, E., Sakmann, B., & Sigworth, F. J. (1981). Improved patch-clamp techniques for high-resolution current recording from cells and cell-free membrane patches. *Pflügers Archiv, 391*(2), 85–100.

Hansen, S. B., Tao, X., & MacKinnon, R. (2011). Structural basis of PIP_2 activation of the classical inward rectifier K^+ channel Kir2.2. *Nature, 477*(7365), 495–498.

He, C., Yan, X., Zhang, H., Mirshahi, T., Jin, T., Huang, A., et al. (2002). Identification of critical residues controlling G protein-gated inwardly rectifying K^+ channel activity through interactions with the βγ subunits of G proteins. *The Journal of Biological Chemistry, 277*(8), 6088–6096.

He, C., Zhang, H., Mirshahi, T., & Logothetis, D. E. (1999). Identification of a potassium channel site that interacts with G protein βγ subunits to mediate agonist-induced signaling. *The Journal of Biological Chemistry, 274*(18), 12517–12524.

Hibino, H., Inanobe, A., Furutani, K., Murakami, S., Findlay, I., & Kurachi, Y. (2010). Inwardly rectifying potassium channels: Their structure, function, and physiological roles. *Physiological Reviews, 90*(1), 291–366, Review.

Hilgemann, D. W., & Ball, R. (1996). Regulation of cardiac Na^+, Ca^{2+} exchange and KATP potassium channels by PIP2. *Science, 273*(5277), 956–959.

Ho, I. H., & Murrell-Lagnado, R. D. (1999). Molecular mechanism for sodium-dependent activation of G protein-gated K^+ channels. *The Journal of Physiology, 520*(Pt. 3), 645–651.

Huang, C. L., Feng, S., & Hilgemann, D. W. (1998). Direct activation of inward rectifier potassium channels by PIP_2 and its stabilization by Gβγ. *Nature, 391*(6669), 803–806.

Inanobe, A., Nakagawa, A., Matsuura, T., & Kurachi, Y. (2010). A structural determinant for the control of PIP_2 sensitivity in G protein-gated inward rectifier K^+ channels. *The Journal of Biological Chemistry, 285*(49), 38517–38523.

Inanobe, A., Yoshimoto, Y., Horio, Y., Morishige, K. I., Hibino, H., Matsumoto, S., et al. (1999). Characterization of G-protein-gated K^+ channels composed of Kir3.2 subunits in dopaminergic neurons of the substantia nigra. *The Journal of Neuroscience, 19*(3), 1006–1017.

Ito, H., Tung, R. T., Sugimoto, T., Kobayashi, I., Takahashi, K., Katada, T., et al. (1992). On the mechanism of G protein beta gamma subunit activation of the muscarinic K^+ channel in guinea pig atrial cell membrane. Comparison with the ATP-sensitive K^+ channel. *The Journal of General Physiology, 99*, 961–983.

Jelacic, T. M., Kennedy, M. E., Wickman, K., & Clapham, D. E. (2000). Functional and biochemical evidence for G-protein-gated inwardly rectifying K^+ (GIRK) channels composed of GIRK2 and GIRK3. *The Journal of Biological Chemistry, 275*(46), 36211–36216.

Jelacic, T. M., Sims, S. M., & Clapham, D. E. (1999). Functional expression and characterization of G-protein-gated inwardly rectifying K^+ channels containing GIRK3. *The Journal of Membrane Biology, 169*(2), 123–129.

Karschin, C., Dissmann, E., Stuhmer, W., & Karschin, A. (1996). IRK(1–3) and GIRK(1–4) inwardly rectifying K^+ channel mRNAs are differentially expressed in the adult rat brain. *The Journal of Neuroscience, 16*, 3559–3570.

Keselman, I., Fribourg, M., Felsenfeld, D. P., & Logothetis, D. E. (2007). Mechanism of PLC-mediated Kir3 current inhibition. *Channels (Austin, Tex.), 1*(2), 113–123.

Kobrinsky, E., Mirshahi, T., Zhang, H., Jin, T., & Logothetis, D. E. (2000). Receptor-mediated hydrolysis of plasma membrane messenger PIP_2 leads to K^+-current desensitization. *Nature Cell Biology, 2*(8), 507–514.

Krapivinsky, G., Gordon, E. A., Wickman, K., Velimirović, B., Krapivinsky, L., & Clapham, D. E. (1995). The G-protein-gated atrial K^+ channel I_{KACh} is a heteromultimer of two inwardly rectifying K^+-channel proteins. *Nature, 374*(6518), 135–141.

Kubo, Y., Baldwin, T. J., Jan, Y. N., & Jan, L. Y. (1993). Primary structure and functional expression of a mouse inward rectifier potassium channel. *Nature, 362*(6416), 127–133.

Kurachi, Y., Nakajima, T., & Sugimoto, T. (1986). Acetylcholine activation of K^+ channels in cell-free membrane of atrial cells. *The American Journal of Physiology, 251*(3 Pt. 2), H681–H684.

Leal-Pinto, E., Gómez-Llorente, Y., Sundaram, S., Tang, Q. Y., Ivanova-Nikolova, T., Mahajan, R., et al. (2010). Gating of a G protein-sensitive mammalian Kir3.1 prokaryotic Kir channel chimera in planar lipid bilayers. *The Journal of Biological Chemistry, 285*(51), 39790–39800.

Lesage, F., Duprat, F., Fink, M., Guillemare, E., Coppola, T., Lazdunski, M., et al. (1994). Cloning provides evidence for a family of inward rectifier and G-protein coupled K^+ channels in the brain. *FEBS Letters, 353*(1), 37–42.

Lesage, F., Guillemare, E., Fink, M., Duprat, F., Heurteaux, C., Fosset, M., et al. (1995). Molecular properties of neuronal G-protein-activated inwardly rectifying K^+ channels. *The Journal of Biological Chemistry, 270*(48), 28660–28667.

Liao, Y. J., Jan, Y. N., & Jan, L. Y. (1996). Heteromultimerization of G-protein-gated inwardly rectifying K^+ channel proteins GIRK1 and GIRK2 and their altered expression in weaver brain. *The Journal of Neuroscience, 16*(22), 7137–7150.

Loewi, O. (1921). Über humorale übertragbarkeit der Herznervenwirkung. *Pflügers Archiv, 189*, 239–242.

Loewi, O., & Navratil, E. (1926). Über humorale Übertragbarkeit der Herznervenwirkung. *Pflügers Archiv, 214*(1), 678–688.

Logothetis, D. E., Kim, D. H., Northup, J. K., Neer, E. J., & Clapham, D. E. (1988). Specificity of action of guanine nucleotide-binding regulatory protein subunits on the cardiac muscarinic K^+ channel. *Proceedings of the National Academy of Sciences of the United States of America, 85*(16), 5814–5818.

Logothetis, D. E., Kurachi, Y., Galper, J., Neer, E. J., & Clapham, D. E. (1987). The βγ subunits of GTP-binding proteins activate the muscarinic K^+ channel in heart. *Nature, 325*(6102), 321–326.

Logothetis, D. E., Lupyan, D., & Rosenhouse-Dantsker, A. (2007). Diverse Kir modulators act in close proximity to residues implicated in phosphoinositide binding. *The Journal of Physiology, 582*(Pt. 3), 953–965, Review.

Logothetis, D. E., Petrou, V. I., Adney, S. K., & Mahajan, R. (2010). Channelopathies linked to plasma membrane phosphoinositides. *Pflügers Archiv, 460*(2), 321–341, Review.

Logothetis, D. E., Petrou, V. I., Zhang, M., Mahajan, R., Meng, X. Y., Adney, S. K., et al. (2015). Phosphoinositide control of membrane protein function: A frontier led by studies on ion channels. *Annual Review of Physiology*, 77, 81–104, Review.

Lopes, C. M., Remon, J. I., Matavel, A., Sui, J. L., Keselman, I., Medei, E., et al. (2007). Protein kinase A modulates PLC-dependent regulation and PIP2-sensitivity of K^+ channels. *Channels (Austin, Tex), 1*(2), 124–134.

Lopes, C. M., Zhang, H., Rohacs, T., Jin, T., Yang, J., & Logothetis, D. E. (2002). Alterations in conserved Kir channel-PIP_2 interactions underlie channelopathies. *Neuron, 34*(6), 933–944.
Lüscher, C., Jan, L. Y., Stoffel, M., Malenka, R. C., & Nicoll, R. A. (1997). G protein-coupled inwardly rectifying K^+ channels (GIRKs) mediate postsynaptic but not presynaptic transmitter actions in hippocampal neurons. *Neuron, 19*(3), 687–695, Erratum in *Neuron, 19*(4), 945.
Mahajan, R., Ha, J., Zhang, M., Kawano, T., Kozasa, T., & Logothetis, D. E. (2013). A computational model predicts that Gβγ acts at a cleft between channel subunits to activate GIRK1 channels. *Science Signaling, 6*(288), ra69.
Mahajan, R., & Logothetis, D. E. (2015). Mechanism of G protein regulation of K^+ channels. In J. Zheng & M. C. Trudeau (Eds.), *Handbook of ion channels* (pp. 505–518). Taylor and Francis Books, LLC.
Meng, X. Y., Zhang, H. X., Logothetis, D. E., & Cui, M. (2012). The molecular mechanism by which PIP_2 opens the intracellular G-loop gate of a Kir3.1 channel. *Biophysical Journal, 102*(9), 2049–2059.
Michailidis, I. E., Rusinova, R., Georgakopoulos, A., Chen, Y., Iyengar, R., Robakis, N. K., et al. (2011). Phosphatidylinositol-4,5-bisphosphate regulates epidermal growth factor receptor activation. *Pflügers Archiv, 461*(3), 387–397.
Mirshahi, T., Robillard, L., Zhang, H., Hébert, T. E., & Logothetis, D. E. (2002). Gβ residues that do not interact with Gα underlie agonist-independent activity of K^+ channels. *The Journal of Biological Chemistry, 277*(9), 7348–7355.
Nishida, M., Cadene, M., Chait, B. T., & MacKinnon, R. (2007). Crystal structure of a Kir3.1-prokaryotic Kir channel chimera. *The EMBO Journal, 26*(17), 4005–4015.
Pegan, S., Arrabit, C., Zhou, W., Kwiatkowski, W., Collins, A., Slesinger, P. A., et al. (2005). Cytoplasmic domain structures of Kir2.1 and Kir3.1 show sites for modulating gating and rectification. *Nature Neuroscience, 8*(3), 279–287, Erratum in *Nature Neuroscience, 8*(6), 835.
Petit-Jacques, J., Sui, J. L., & Logothetis, D. E. (1999). Synergistic activation of G protein-gated inwardly rectifying potassium channels by the βγ subunits of G proteins and Na^+ and Mg^{2+} ions. *The Journal of General Physiology, 114*(5), 673–684.
Pfaffinger, P. J., Martin, J. M., Hunter, D. D., Nathanson, N. M., & Hille, B. (1985). GTP-binding proteins couple cardiac muscarinic receptors to a K channel. *Nature, 317*(6037), 536–538.
Reuveny, E. (2013). Structural biology: Ion channel twists to open. *Nature, 498*(7453), 182–183.
Reuveny, E., Slesinger, P. A., Inglese, J., Morales, J. M., Iñiguez-Lluhi, J. A., Lefkowitz, R. J., et al. (1994). Activation of the cloned muscarinic potassium channel by G protein βγ subunits. *Nature, 370*(6485), 143–146.
Rohács, T., Lopes, C. M., Jin, T., Ramdya, P. P., Molnár, Z., & Logothetis, D. E. (2003). Specificity of activation by phosphoinositides determines lipid regulation of Kir channels. *Proceedings of the National Academy of Sciences of the United States of America, 100*(2), 745–750.
Rosenhouse-Dantsker, A., & Logothetis, D. E. (2007). Molecular characteristics of phosphoinositide binding. *Pflügers Archiv, 455*(1), 45–53, Review.
Rosenhouse-Dantsker, A., Sui, J. L., Zhao, Q., Rusinova, R., Rodríguez-Menchaca, A. A., Zhang, Z., et al. (2008). A sodium-mediated structural switch that controls the sensitivity of Kir channels to PtdIns(4,5)P_2. *Nature Chemical Biology, 4*(10), 624–631.
Sakmann, B., Noma, A., & Trautwein, W. (1983). Acetylcholine activation of single muscarinic K^+ channels in isolated pacemaker cells of the mammalian heart. *Nature, 303*(5914), 250–253.
Schreibmayer, W., Dessauer, C. W., Vorobiov, D., Gilman, A. G., Lester, H. A., Davidson, N., et al. (1996). Inhibition of an inwardly rectifying K^+ channel by G-protein alpha-subunits. *Nature, 380*(6575), 624–627.

Signorini, S., Liao, Y. J., Duncan, S. A., Jan, L. Y., & Stoffel, M. (1997). Normal cerebellar development but susceptibility to seizures in mice lacking G protein-coupled, inwardly rectifying K^+ channel GIRK2. *Proceedings of the National Academy of Sciences of the United States of America, 94*, 923–927.

Soejima, M., & Noma, A. (1984). Mode of regulation of the ACh-sensitive K-channel by the muscarinic receptor in rabbit atrial cells. *Pflügers Archiv, 400*(4), 424–431.

Sorota, S., Tsuji, Y., Tajima, T., & Pappano, A. J. (1985). Pertussis toxin treatment blocks hyperpolarization by muscarinic agonists in chick atrium. *Circulation Research, 57*(5), 748–758.

Spauschus, A., Lentes, K. U., Wischmeyer, E., Dissmann, E., Karschin, C., & Karschin, A. (1996). A G-protein-activated inwardly rectifying K^+ channel (GIRK4) from human hippocampus associates with other GIRK channels. *The Journal of Neuroscience, 16*(3), 930–938.

Stanfield, P. R., Nakajima, S., & Nakajima, Y. (2002). Constitutively active and G-protein coupled inward rectifier K^+ channels: Kir2.0 and Kir3.0. *Reviews of Physiology, Biochemistry and Pharmacology, 145*, 47–179, Review.

Stansfeld, P. J., Hopkinson, R., Ashcroft, F. M., & Sansom, M. S. (2009). PIP_2-binding site in Kir channels: Definition by multiscale biomolecular simulations. *Biochemistry, 48*(46), 10926–10933.

Sui, J. L., Petit-Jacques, J., & Logothetis, D. E. (1998). Activation of the atrial K_{ACh} channel by the βγ subunits of G proteins or intracellular Na^+ ions depends on the presence of phosphatidylinositol phosphates. *Proceedings of the National Academy of Sciences of the United States of America, 95*(3), 1307–1312.

Trautwein, W., & Dudel, J. (1958). Mechanism of membrane effect of acetylcholine on myocardial fibers. *Pflügers Archiv, 266*(3), 324–334, German.

Vivaudou, M., Chan, K. W., Sui, J. L., Jan, L. Y., Reuveny, E., & Logothetis, D. E. (1997). Probing the G-protein regulation of GIRK1 and GIRK4, the two subunits of the K_{ACh} channel, using functional homomeric mutants. *The Journal of Biological Chemistry, 272*(50), 31553–31560.

Whorton, M. R., & MacKinnon, R. (2011). Crystal structure of the mammalian GIRK2 K^+ channel and gating regulation by G proteins, PIP_2, and sodium. *Cell, 147*(1), 199–208.

Whorton, M. R., & MacKinnon, R. (2013). X-ray structure of the mammalian GIRK2-βγ G-protein complex. *Nature, 498*(7453), 190–197.

Wickman, K. D., Iñiguez-Lluhl, J. A., Davenport, P. A., Taussig, R., Krapivinsky, G. B., Linder, M. E., et al. (1994). Recombinant G-protein βγ subunits activate the muscarinic-gated atrial potassium channel. *Nature, 368*(6468), 255–257.

Wickman, K., Karschin, C., Karschin, A., Picciotto, M. R., & Clapham, D. E. (2000). Brain localization and behavioral impact of the G-protein-gated K^+ channel subunit GIRK4. *The Journal of Neuroscience, 20*(15), 5608–5615.

Wickman, K., Nemec, J., Gendler, S. J., & Clapham, D. E. (1998). Abnormal heart rate regulation in GIRK4 knockout mice. *Neuron, 20*(1), 103–114.

Yatani, A., Codina, J., Brown, A. M., & Birnbaumer, L. (1987). Direct activation of mammalian atrial muscarinic potassium channels by GTP regulatory protein Gk. *Science, 235*(4785), 207–211.

Yi, B. A., Lin, Y. F., Jan, Y. N., & Jan, L. Y. (2001). Yeast screen for constitutively active mutant G protein-activated potassium channels. *Neuron, 29*(3), 657–667.

Yokogawa, M., Osawa, M., Takeuchi, K., Mase, Y., & Shimada, I. (2011). NMR analyses of the Gβγ binding and conformational rearrangements of the cytoplasmic pore of G protein-activated inwardly rectifying potassium channel 1 (GIRK1). *The Journal of Biological Chemistry, 286*(3), 2215–2223.

Zhang, H., Craciun, L. C., Mirshahi, T., Rohács, T., Lopes, C. M., Jin, T., et al. (2003). PIP_2 activates KCNQ channels, and its hydrolysis underlies receptor-mediated inhibition of M currents. *Neuron, 37*(6), 963–975.

Zhang, H., He, C., Yan, X., Mirshahi, T., & Logothetis, D. E. (1999). Activation of inwardly rectifying K^+ channels by distinct PtdIns(4,5)P_2 interactions. *Nature Cell Biology*, *1*(3), 183–188.

Zhang, M., Meng, X. Y., Cui, M., Pascal, J. M., Logothetis, D. E., & Zhang, J. F. (2014). Selective phosphorylation modulates the PIP_2 sensitivity of the CaM-SK channel complex. *Nature Chemical Biology*, *10*(9), 753–759.

CHAPTER TWO

The Roles of Gβγ and Gα in Gating and Regulation of GIRK Channels

Nathan Dascal[1], Uri Kahanovitch
Department of Physiology and Pharmacology and Sagol School of Neuroscience, Tel Aviv University, Tel Aviv, Israel
[1]Corresponding author: e-mail address: dascaln@tauex.tau.ac.il

Contents

1. Introduction 28
2. Interaction of Gα and Gβγ With GIRK 32
   2.1 Interaction of Gβγ and GIRK Revealed by Structural Studies 32
   2.2 The Roles of C- and N-Terminal Segments Lacking from the Crystal Structures 33
   2.3 GIRK–Gα Interactions 35
3. Roles of Gβγ and Gα in GIRK's Evoked and Basal Activity 41
   3.1 Agonist-Evoked Current, I_{evoked} 42
4. Basal Activity of GIRKs 51
   4.1 Physiology of I_{basal} 51
   4.2 Heterologous Expression Studies: GPCR-Independent I_{basal} and the Role of Gβγ 55
   4.3 I_{basal} and I_{evoked} are Interrelated: The Roles of $G\alpha^{GDP}$ and Gβγ 56
5. Multimolecular Signaling Complex and the Mechanism of GIRK Activation 60
   5.1 Preformed (Permanent) Signaling Complex in the GPCR–GIRK Cascade: Pro and Contra 60
   5.2 GIRK Channel and G Protein Subunits: Unique Association and Stoichiometry Rules 63
   5.3 Specificity of GPCR–GIRK Signaling: Relative Roles of Gα and Gβγ 65
Acknowledgments 68
References 68

Abstract

G protein-gated K^+ (GIRK, or Kir3) channels mediate inhibitory neurotransmission via G protein-coupled receptors (GPCRs) in heart and brain. The signaling cascade involves activation of GPCR by an agonist, activation of a G protein followed by rearrangement or dissociation of activated $G\alpha^{GTP}$ from Gβγ, and activation of GIRK by Gβγ. Gβγ is the main transducer of GPCR activating signal to the GIRK channel. It promotes channel opening by direct binding to GIRK's cytosolic domain formed by the N- and C-terminal segments of the GIRK's four subunits. Gβγ's interaction with, and activation of, the GIRK channels are well understood and reviewed elsewhere; however, several important details involving distal parts of the cytosolic domain remain incompletely understood. $G\alpha_{i/o}$ also binds to GIRKs and has been implicated in regulating channel's gating, in concert with

International Review of Neurobiology, Volume 123
ISSN 0074-7742
http://dx.doi.org/10.1016/bs.irn.2015.06.001

Gβγ. Among known functions of Gα, best-described (though not well understood) are selectivity of signaling (only $G_{i/o}$ proteins normally couple to GIRKs) and regulation of the basal activity of GIRKs (I_{basal}). A role for a direct effect of the activated $G\alpha_{i/o}^{GTP}$ in GIRK gating has also been proposed but remains elusive. This chapter discusses the mechanisms of signaling within the essential cascade, from GPCR to the heterotrimeric G protein and to the channel. The focus is on the role of Gα and on the relationships between Gα and Gβγ in channel regulation, their role in specific signaling from GPCRs to GIRKs, and the role of stoichiometry and cooperativity of G protein–GIRK interactions in channel's function.

1. INTRODUCTION

G protein-gated K^+ channels (GIRK, or Kir3), a subfamily of tetrameric inwardly rectifying K^+ channel family (Kir1–7; Fig. 1A), mediate inhibitory neurotransmission via G protein-coupled receptors (GPCRs) in heart and brain. GIRKs are the first discovered and prototypic direct effectors of Gβγ (chapter "Unifying Mechanism of Controlling Kir3 Channel Activity by G Proteins and Phosphoinositides" by Logothetis et al.), which operate within a now classical and prototypic signaling cascade: from GPCRs to G proteins and then directly to the GIRK channel (Dascal, 1997; Hibino et al., 2010; Inanobe & Kurachi, 2014; Wickman & Clapham, 1995; Yamada, Inanobe, & Kurachi, 1998). For brevity, we will term it the GPCR–GIRK cascade. This cascade is a major transducer of inhibitory actions of many neurotransmitters (e.g., acetylcholine (ACh), dopamine, GABA (via the $GABA_B$ receptors), opioids, serotonin, adenosine, etc.). GIRK channels and the GPCR–GIRK cascade that play important roles in the regulation of heartbeat and cardiac pacemaking (Dascal, 1997; Kovoor et al., 2001; Mesirca et al., 2014; Noma & Trautwein, 1978) are involved in the regulation of neuronal excitability and plasticity, pain and analgesia, and have been implicated in the etiology of disorders such as Down's syndrome, epilepsy, and drug and alcohol abuse (Kobayashi & Ikeda, 2006; Luján, Marron Fernandez de Velasco, Aguado, & Wickman, 2014; Lujan, Maylie, & Adelman, 2009; Luscher & Slesinger, 2010), arrhythmia and atrial fibrillation (AF) and remodeling (Voigt, Abu-Taha, Heijman, & Dobrev, 2014), and hypertension caused by primary aldosteronism (Al-Salameh, Cohen, & Desailloud, 2014; Williams, Monticone, & Mulatero, 2015). Many physiological and pathological aspects of GPCR–GIRK cascade are discussed in detail in other chapters and will be only briefly touched here.

The GPCR–GIRK signaling is membrane delimited (Breitwieser & Szabo, 1985; Kurachi, Nakajima, & Sugimoto, 1986; Logothetis,

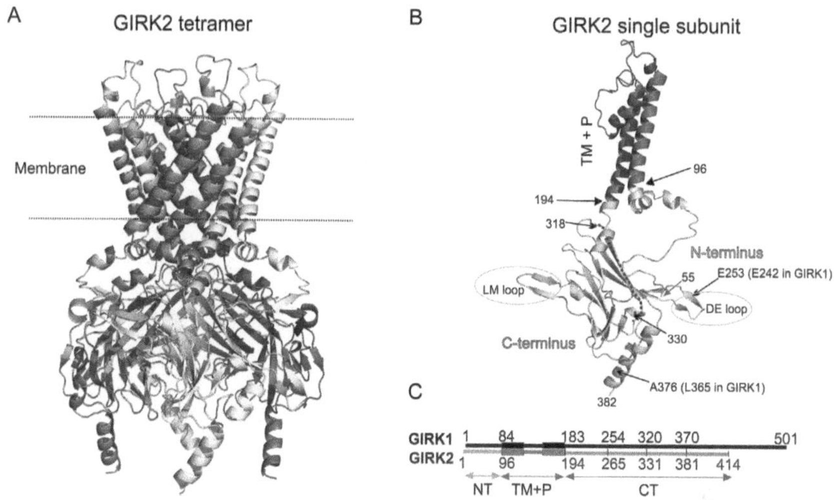

Figure 1 ***The structure of the GIRK channel.*** (A) The full tetrameric GIRK2 heterotetramer in the preopen state. The figure was drawn using PyMOL with the coordinates from Whorton and MacKinnon (2013) (accession code 4KFM). The four subunits are shown as ribbons of different colors. The membrane is outlined approximately. (B) The structure of one GIRK2 subunit. The domains are colored differently: NT is yellow (light gray in the print version), CT is green (light gray in the print version), and the transmembrane + pore domain (TM + P) is magenta (dark gray in the print version). The protein used for crystallization (Whorton & MacKinnon, 2013) was truncated at its ends. It begins with I55 and ends with N382. For orientation, also shown are the numbers of amino acid residues that lie at the transition from NT to TM (L96) and from TM to the CT (K194). The dashed blue (dark gray in the print version) line between a.a. 318 and 330 denotes the segment identified by Clancy et al. (2005) as crucial for $G\alpha_o$ interaction. Alanine 376 corresponds to L365 in GIRK1, where the major $G\alpha_{i3}{}^{GTP\gamma S}$-interacting segment starts (Mase, Yokogawa, Osawa, & Shimada, 2012). Another residue important for interaction with $G\alpha_{i3}{}^{GTP\gamma S}$ in GIRK1, E242 (Mase et al., 2012), is conserved in GIRK2 (E253). Yellow (light gray in the print version) ellipses denote the major Gβγ-interacting determinants, loops DE and LM (Whorton & MacKinnon, 2013). (C) A linear scheme of GIRK1 and GIRK2 subunits. The two subunits have been aligned so that the transmembrane domains coincide. The numbering of GIRK2 corresponds to the GIRK2A isoform. NT, N-terminal cytosolic part; CT, C-terminal cytosolic part; TM, transmembrane domains M1 and M2; P, the pore region. The coloring of the segments in GIRK2 corresponds to the color code in (B).

Kurachi, Galper, Neer, & Clapham, 1987; Pfaffinger, Martin, Hunter, Nathanson, & Hille, 1985; Soejima & Noma, 1984), which means that all participants are either intrinsic membrane or membrane-attached proteins, and there are no diffusible cytosolic intermediates. A general scheme of the cascade involves several steps: binding of agonist to the GPCR; activating interaction of the GPCR with the target $G_{i/o}$ protein which causes GDP–GTP exchange at the $G\alpha_{i/o}$ subunit and partial (Bunemann, Frank, & Lohse, 2003; Gales et al., 2006) or full separation (Bondar & Lazar, 2013; Gibson & Gilman, 2006; Lambert, 2008) of Gβγ from Gα; and binding of Gβγ to the cytosolic domain of GIRK, which promotes the closed-open transition (activation) of the channel (Dascal, 2001; Hibino et al., 2010; Wickman & Clapham, 1995). GIRK gating crucially depends on the membrane phospholipid PIP_2, an essential gating agent of all Kir channels (Hansen, 2015; Huang, Feng, & Hilgemann, 1998; Logothetis et al., 2015; Wang, Whorton, & MacKinnon, 2014) (chapters "Unifying Mechanism of Controlling Kir3 Channel Activity by G Proteins and Phosphoinositides" by Logothetis et al. and "Structural Insights into GIRK Channel Function" by Glaaser and Slesinger). In addition, the channels are also activated by direct binding of Na ions to a well-defined binding site present in GIRK2 and GIRK4 but not in GIRK1 (Ho & Murrell-Lagnado, 1999; Sui, Chan, & Logothetis, 1996; Sui, Petit-Jacques, & Logothetis, 1998). The crucial Na^+-binding site is also conserved in GIRK3 but the functional role of this site in GIRK3 has not yet been examined. Direct gating by Na^+ shares with Gβγ the requirement for PIP_2 and may involve similar structural rearrangements of channel's protein during the gating process (Logothetis et al., 2015; Wang et al., 2014; Whorton & MacKinnon, 2011). Depletion of PIP_2 closes GIRK channels and makes them unresponsive to Gβγ or alcohol (Huang et al., 1998; Kobrinsky, Mirshahi, Zhang, Jin, & Logothetis, 2000). In healthy cells and at normal physiological levels of PIP_2, however, transmitters and hormones regulate GIRKs through Gβγ.

Despite the apparent simplicity, the GPCR–GIRK pathway is versatile and features diversity and specificity that imply a high level of complexity in signaling. Coupling selectivity of G protein to GIRK is a striking example. In native cells, only GPCRs coupled to $G_{i/o}$G proteins activate GIRK. G_s-coupled GPCRs can activate GIRKs under nonphysiological (via overexpression) conditions (Lim, Dascal, Labarca, Davidson, & Lester, 1995; Wellner-Kienitz, Bender, & Pott, 2001), but those coupled to G_q never activate GIRKs (Lei et al., 2000; Rusinova, Mirshahi, & Logothetis, 2007; Sharon, Vorobiov, & Dascal, 1997). In many cases, GPCR–GIRK cascade specificity is thought to depend on the formation of multiprotein signaling complexes, comprising all or some of the essential partners of the cascade (GPCR, heterotrimeric G protein and the GIRK), and additional auxiliary and scaffolding proteins and even protein kinases, phosphatases, and other enzymes (Doupnik, 2008; Luján et al., 2014; Nagi & Pineyro, 2014; Nikolov & Ivanova-Nikolova, 2004b; Padgett & Slesinger, 2010; Raveh, Riven, & Reuveny, 2009; Zylbergold, Ramakrishnan, & Hebert, 2010).

The composition and mode of function of such complexes are the subject of extensive research. Additional levels of versatility and complexity, discussed elsewhere, are added by the regulation or modulation of the channel by its direct interaction with G protein receptor kinases (Raveh, Cooper, Guy-David, & Reuveny, 2010), alcohols (Bodhinathan & Slesinger, 2014; Kobayashi et al., 1999; Lewohl et al., 1999), and additional factors such as metabolites of arachidonic acid (Rogalski & Chavkin, 2001). Furthermore, the gating or interaction with Gβγ are regulated by direct phosphorylation by serine/threonine protein kinase A (PKA) (Medina et al., 2000; Mullner et al., 2000) and protein kinase C (PKC) (Hill & Peralta, 2001; Sharon et al., 1997; Stevens, Shah, Pinnock, & Lee, 1999), and by tyrosine phosphorylation (Ippolito, Temkin, Rogalski, & Chavkin, 2002; Rogalski, Appleyard, Pattillo, Terman, & Chavkin, 2000). In many cases, the amino acid residues crucial for the regulation have been mapped and mechanisms of regulation have been proposed. A more complete understanding of the basic mechanism of the GPCR–GIRK signaling complex will provide further insights into these complex regulations in the future.

The historical controversy whether it is Gα or Gβγ that activates the channel has been resolved in favor of Gβγ in the early 1990s (chapter "Unifying Mechanism of Controlling Kir3 Channel Activity by G Proteins and Phosphoinositides" by Logothetis et al.), but other roles for Gα became apparent. Multiple lines of evidence indicate that Gα is not merely a "passive" donor of Gβγ, but acts in concert with Gβγ, regulating or modulating various aspects of function of the channel. Both Gβγ (see chapter "Structural Insights into GIRK Channel Function" by Glaaser and Slesinger) and $G\alpha_{i/o}$ (Clancy et al., 2005; Huang, Slesinger, Casey, Jan, & Jan, 1995; Ivanina et al., 2004; Mase et al., 2012) interact with multiple binding sites on cytosolic N- and C-terminal domains of GIRK subunits. Several roles for Gα, some potentially related to the direct Gα–GIRK interaction, have been proposed and investigated over the last two decades. First, $G\alpha_{i/o}$ in its GDP-bound form ($G\alpha^{GDP}$) forms the Gαβγ heterotrimer, which is the donor of Gβγ for GIRK activation. In addition, $G\alpha_{i/o}{}^{GDP}$ is an essential regulator of the constitutive (basal) activity of GIRKs, I_{basal}, primarily through regulating the availability of Gβγ for the channel (Chan, Sui, Vivaudou, & Logothetis, 1997; Leaney & Tinker, 2000; Peleg, Varon, Ivanina, Dessauer, & Dascal, 2002; Rishal, Porozov, Yakubovich, Varon, & Dascal, 2005; Rubinstein, Peleg, Berlin, Brass, & Dascal, 2007). Second, several works have reported that either certain activated $G\alpha^{GTP}$ may enhance or inhibit the Gβγ-mediated activation of GIRKs (Leal-Pinto et al., 2010; Schreibmayer et al., 1996), implying a *bona fide* role for $G\alpha^{GTP}$ as a modulator of the neurotransmitter-evoked GIRK response, I_{evoked}

(Berlin et al., 2011). However, so far this kind of regulation remains controversial (Wang et al., 2014) and requires further enquiry. Third, a role for Gα as an anchor to dock Gβγ at the GIRK channel before activation by GPCR has been suggested (Clancy et al., 2005).

This chapter discusses the mechanisms of signaling within the essential composition of the GPCR–GIRK cascade, i.e., its three main elements: the GPCR, the heterotrimeric G protein, and the channel. The focus is on the role of Gα and on the relationships between Gα and Gβγ in channel regulation. The detailed mechanism of regulation by Gβγ and the involvement of additional regulatory elements such as RGS are addressed only briefly and are covered in detail in other chapters.

2. INTERACTION OF Gα AND Gβγ WITH GIRK

2.1 Interaction of Gβγ and GIRK Revealed by Structural Studies

The Gβγ, the inactive $G\alpha_{i/o}{}^{GDP}$ and the activated $G\alpha_{i/o}$ ($G\alpha_{i/o}{}^{GTP}$) all bind to the cytosolic domain of GIRK subunits. Cytosolic N-terminal (NT) and C-terminal (CT) domains of GIRK subunits (Fig. 1) encompass several Gβγ and $G\alpha_i$-binding segments. Crystal structures of partly truncated cytosolic domains of GIRK1 and GIRK2 (Inanobe, Matsuura, Nakagawa, & Kurachi, 2007; Nishida, Cadene, Chait, & Mackinnon, 2007; Nishida & MacKinnon, 2002; Pegan et al., 2005; Whorton & MacKinnon, 2011) and a GIRK2–Gβγ complex (Whorton & MacKinnon, 2013) have been determined, and the binding of Gβγ to GIRK subunits is well characterized (see chapter "Structural Insights into GIRK Channel Function" by Glaaser and Slesinger for details). In brief, functional, biochemical, fluorescence or bioluminescence resonance energy transfer (FRET or BRET), and crystallographic data unequivocally show that one GIRK tetramer binds up to four Gβγ molecules (Clancy et al., 2005; Corey & Clapham, 2001; Ito et al., 1992; Ivanova-Nikolova, Nikolov, Hansen, & Robishaw, 1998; Riven, Iwanir, & Reuveny, 2006; Sadja, Alagem, & Reuveny, 2002; Whorton & MacKinnon, 2013). The Gβγ interaction sites have been best studied in GIRK1 and GIRK2 subunits and mapped with high precision. NMR studies with a truncated GIRK1 cytosolic domain (Yokogawa, Osawa, Takeuchi, Mase, & Shimada, 2011), the crystal structure of a "preopen" GIRK2–Gβγ complex (Whorton & MacKinnon, 2013), and protein docking models of a GIRK1–Gβγ complex (Mahajan et al., 2013) suggest binding of Gβγ at GIRK intersubunit interfaces, with multiple point of contact between Gβγ and GIRK. Particularly important (Whorton & MacKinnon, 2013), as also suggested by previous mapping and structure–function studies (Finley, Arrabit, Fowler, Suen, & Slesinger, 2004), is the βL–βM loop, which contains a leucine residue (L333 in GIRK1, L344 in GIRK2, L339 in GIRK4; Fig. 1B) previously identified as highly important for GIRK gating and to some extent for Gβγ binding (Finley et al., 2004; He, Zhang, Mirshahi, & Logothetis, 1999; Ivanina et al., 2003). The GIRK-interacting surface of Gβ largely coincides with the interface at which Gα binds to Gβγ, supporting previous structure–function studies. These results strongly support the notion that, in order to activate GIRK, Gβγ needs to separate from Gα (partly or fully) (Albsoul-Younes et al., 2001; Ford et al., 1998); the Gαβγ heterotrimer cannot activate GIRK (Ford et al., 1998; Wickman & Clapham, 1995). Because crystal structure captures a certain static conformation (pose) of the channel–Gβγ complex, it is plausible that additional poses of GIRK and Gβγ, with partly distinct sets of contact points, exist. Indeed, several mutagenesis-based studies indicate that additional amino acid residues in Gβ may participate in regulating GIRK basal or evoked activities (Albsoul-Younes et al., 2001; Mirshahi, Robillard, Zhang, Hebert, & Logothetis, 2002; Zhao et al., 2003). For a detailed discussion on structural and functional aspects of Gβγ interaction with the channel, refer chapters "Unifying Mechanism of Controlling Kir3 Channel Activity by G Proteins and Phosphoinositides" by Logothetis et al. and "Structural Insights into GIRK Channel Function" by Glaaser and Slesinger.

2.2 The Roles of C- and N-Terminal Segments Lacking from the Crystal Structures

The purified GIRK1 and GIRK2 proteins used for NMR, crystallography, and modeling did not contain the initial segment (amino acid residues (a.a.) 1–40 in GIRK1) and the segment close to the plasma membrane (a.a. 64–84 in GIRK1), and the distal part of the CT (distal CT) of GIRK1 and GIRK2 (Fig. 1B) (These segments reduce the stability of the recombinant protein produced in bacterial or yeast cells, possibly because of their high mobility or for other reasons.). Whereas the distal CT truncation in GIRK2 was small, about 30 a.a. (Whorton & MacKinnon, 2013), GIRK1 distal CT lacks >120 a.a. (see Nishida et al., 2007; Nishida & MacKinnon, 2002). The structural elements "outside" the available crystal structures, especially the distal CT of GIRK1 but potentially also the NT, may carry out important functions and need to be taken into account when devising conceptual models of channel function.

2.2.1 The Distal C-Terminus of GIRK1

The unique distal CT of GIRK1, which is longer than in all other subunits (see Fig. 1C, for an example), takes part in several aspects of GIRK regulation. First, it regulates both basal and neurotransmitter-evoked activity of the channel (Chan et al., 1997; Kahanovitch et al., 2014; Wydeven, Young, Mirkovic, & Wickman, 2012). In particular, a single residue within the distal CT domain (Q404) selectively influences I_{evoked}. It is not clear whether it affects the interaction with Gβγ or (possibly allosterically) the coupling between Gβγ binding and gate opening (Wydeven et al., 2012). The mechanisms by which the distal CT of GIRK1 regulates I_{basal} will be discussed in detail later.

Second, when purified as a separate protein, the distal CT of GIRK1 does not bind Gβγ by itself, or shows a very weak binding (Huang, Jan, & Jan, 1997; Ivanina et al., 2003). Nevertheless, the distal CT contains a structural determinant which imparts high-affinity binding of Gβγ to the rest of GIRK1's C-terminus (Huang et al., 1997) and to the full cytosolic domain of GIRK1 (Kahanovitch et al., 2014). The distal CT is essential for the recruitment of Gβγ by GIRK1-containing channels to the plasma membrane, observed in a heterologous system (see below). The truncation of distal CT does not impair activation by Gβγ; therefore, it is safe to assume that it does not participate in the formation of the "activation" site through which the binding of Gβγ initiates channel opening (Kahanovitch et al., 2014). Therefore, the distal CT of GIRK1 has been proposed to participate in the formation of the high-affinity Gβγ binding site in GIRK1, which is hypothesized to be important for anchoring of Gβγ (Kahanovitch et al., 2014). These findings support earlier models that propose the existence of separate Gβγ-binding sites for basal and evoked activity (He et al., 1999), or for Gβγ anchoring and activation (Rubinstein et al., 2009).

Third, the last 20 a.a. segment of the distal CT of GIRK1 noncompetitively suppresses the activation of GIRK1/4 by Gβγ (Luchian et al., 1997), and removal of distal CT increased the open probability (P_o) in the presence of Gβγ (Kahanovitch et al., 2014). Therefore, it has been hypothesized that this is part of an intrinsic inhibitory gate (a "lock") which helps to maintain the closed conformation of the GIRK1-containing channels (Luchian et al., 1997; Rubinstein et al., 2009). However, detailed structural information on separate anchoring and activation sites and on the hypothetical "lock" is still lacking, and these ideas await further structural and protein modeling studies. Finally, the distal CT of GIRK1 is indispensable for a synergistic enhancement of GIRK1–Gα_i binding by Gβγ (see below).

2.2.2 *The Initial Segment of the NT*

The NTs of several GIRK subunits play important roles in channel biogenesis and trafficking, which are involved in GIRK regulation of neuronal plasticity (Chung, Qian, Ehlers, Jan, & Jan, 2009; Ma et al., 2002). There is less information regarding the role of initial part of the NT, which is missing in the crystal crystallography and NMR studies, in channel gating. Phosphorylation Tyr_{12} of GIRK1 has been reported to regulate channel's I_{basal} by unmasking a previously occluded GAP (GTPase-activating protein) domain that accelerates the GTPase activity of channel-associated Gα (Clayton, Xu, & Chavkin, 2009; Ippolito et al., 2002). Although direct biochemical support for a GAP activity of GIRK1 is lacking, this interesting hypothesis deserves further consideration and testing, with regard to the role and interactions of Gα with GIRK.

2.3 GIRK–Gα Interactions

The interaction between GIRKs and Gα subunits is less well characterized than for Gβγ; there are no crystal structures of GIRK–Gα or GIRK–Gαβγ complexes. GIRK–Gα interactions are best characterized for $G\alpha_{i/o}$, and rest mainly upon biochemical (protein–protein interaction) data and one work that utilized NMR spectroscopy. The interactions were usually studied with either "inactive" (GDP-bound) $G\alpha^{GDP}$ or the "active" $G\alpha^{GTP\gamma S}$ (GTPγS is a nonhydrolyzable analog of GTP, which remains tightly bound to Gα and is used in biochemical and functional studies to preserve structure and function of the activated Gα.).

2.3.1 *Mapping GIRK–Gα Interaction Sites*

Whereas it is certain that NT and CT of GIRK subunits bind $G\alpha_{i/o}$, the mapping is incomplete; there are substantial variations in mapping data obtained for different subunits and/or with alternative methods (Table 1). Both $G\alpha^{GDP}$ and $G\alpha^{GDP}\beta\gamma$ bind the entire cytosolic NT of GIRK1 and GIRK2; in the latter, the interaction site has been narrowed down to the second half of NT (Huang et al., 1995; Ivanina et al., 2004; Peleg et al., 2002). Later works also demonstrated interactions of *in vitro* translated or purified $G\alpha_{i3}$, $G\alpha_{i1}$, and $G\alpha_o$ with the C-terminal domains of GIRK1 and GIRK2 (Clancy et al., 2005; Ivanina et al., 2004). GTPγS and GDP forms of $G\alpha_{i3}$ and $G\alpha_{i1}$ interact with essentially the same segments in GIRK1. The best $G\alpha_i$-binding peptide in GIRK1-CT was a 320–370 a. a. segment (Ivanina et al., 2004), which is largely exposed to the cytosol

Table 1 Binding Sites for Gα on GIRK Subunits

Protein	Gα	Method	Region	Comments	Citation
H_6-fused recombinant parts of GIRK1	Bovine brain purified $G\alpha^{GDP}$, $G\alpha^{GDP}\beta\gamma$	Pull down	Full N-terminus	No binding to full CT	Huang et al. (1995)
GST-fused recombinant parts of GIRK1	*i.v.t.* $G\alpha_{i3}$ and $G\alpha_{i1}$	Pull down	320–370	Binding better in GDP than GTPγS; $G\alpha_{i3}$ binds better than $G\alpha_{i1}$	Ivanina et al. (2004)
GST-fused recombinant parts of GIRK2	*i.v.t.* $G\alpha_{i3}$ and $G\alpha_{i1}$		46–96 (2d half of NT), 310–380		
GST-fused recombinant parts of GIRK2	Purified recombinant Gα and Gβγ subunits	Coimmunoprecipitation and pull down	G318, C321, A323, I328, T329, S330	$G\alpha^{GDP}\beta\gamma$ and $G\alpha^{GTP\gamma S}$ bind here	Clancy et al. (2005)
i.v.t. cytosolic parts of GIRK1 and GIRK2	Purified GST-fused $G\alpha_{i3}$	Pull down	Whole cytosolic domains (NT + CT, no TM)	Gβγ enhances interaction in GIRK1, not in GIRK2	Rubinstein et al. (2009)
Purified cytosolic domain of GIRK1 (no distal CT)	Purified $G\alpha_{i3}{}^{GTP\gamma S}$	NMR	E242, V358, L365, L366, M367, S368, S369, L371, I372, A373		Mase et al. (2012)

H_6, 6-histidine tag; GST, glutathione-*S*-transferase; *i.v.t.*, *in vitro* translated (produced in rabbit reticulocyte lysate).

in the GIRK1 crystal (Inanobe et al., 2007; Nishida et al., 2007; Nishida & MacKinnon, 2002).

In the CT of GIRK2, interaction with $G\alpha_{i3}$ or $G\alpha_o$ was located to a.a. 310–380 (Ivanina et al., 2004), in line with the GIRK1 data (to compare a.a. numbers in GIRK1 and GIRK2, see Fig. 1C). Using mutagenesis followed by biochemical and functional tests, Slesinger and colleagues (Clancy et al., 2005) further identified a shorter segment (a.a. 318–330; dotted blue (dark gray in the print version) line in Fig. 1B) and, within it, a cluster of six a.a. (Table 1) which were at variance with a distinct member of the family of inward rectifier channels, IRK1 (Kir2.1), which is not G protein-dependent and does not bind Gα. Mutations of these six a.a. to their IRK1 counterparts abolished binding of $G\alpha_o$ to GIRK2-CT, whereas binding of Gβγ persisted. In the full channel, the six a.a. mutation diminished the surface expression of GIRK2 and did not allow the measurement of GIRK2 currents. Nevertheless, functional tests with a single a.a. mutation, C321V, indicated a large decrease in agonist-evoked currents, supporting the role of the 318–330 segment in GIRK2 regulation by $G\alpha_{i/o}$. Notably, in the $GIRK2_{C321V}$ mutant, both I_{basal}, Gβγ- and ethanol-dependent activation remained intact (Clancy et al., 2005). In GIRK2, both I_{basal} and ethanol activation are largely Gβγ-independent (Bodhinathan & Slesinger, 2014; Rubinstein et al., 2009). Taken together, these findings suggest that the $GIRK2_{C321V}$ channel is functional and that Gβγ gating is intact, and point to a specific role for Gα in GPCR-induced GIRK2 activation by the heterotrimeric Gαβγ (Clancy et al., 2005). However, this role is not entirely clear, as we will further discuss below.

2.3.2 GIRK1–Gα Interaction by NMR Spectroscopy

Shimada and colleagues (Mase et al., 2012) used NMR spectroscopy to map the $G\alpha_{i3}{}^{GTP}$/GIRK1 interaction surface as well as conformational changes resulting from the binding. The purified GIRK1 protein was a slightly extended version of the GIRK1's cytosolic domain reported in the first crystal structure (Nishida & MacKinnon, 2002), but still lacked much of NT and the last ~127 a.a of the CT. The affinity of GIRK1–$G\alpha_{i3}{}^{GTP}$ interaction was estimated to be ~fourfold lower than that of Gβγ, supporting an independent semiquantitative estimate from a pull-down study (Berlin et al., 2011). The ~1 mM K_D in solution was estimated to convert to nM to μM affinity in the submembrane space, which is indicative of a medium-affinity interaction. The NMR results suggest that the main site of interaction with

$G\alpha_{i3}{}^{GTP}$ is a stretch of nine a.a. from L365 to A373 (which was the last a.a. in the recombinant protein used in this study) (Mase et al., 2012). The first five out of these nine a.a. are part of the αA helix seen in the crystal structure of the truncated GIRK1 (Nishida et al., 2007; Nishida & MacKinnon, 2002). It is not known whether the αA helix extends to a.a. 370–373 because they are not present in the crystal structure. The region corresponding to a.a. 365–373 of GIRK1 is not fully available in the crystal of GIRK2; at least two a.a. are lacking. The available part in GIRK2 (a.a. 376–382, Fig. 1B) is also an α-helix, but sequence conservation in this region in all GIRK subunits is low. In addition, glutamate E242, located in the main part of the CT (Fig. 2B) appears to participate in $G\alpha_{i3}$ binding (Mase et al., 2012; see Table 1). This glutamate is highly conserved both in GIRKs and in G protein-insensitive Kir channels such as IRK1 (Kir2.1) and ROMK1 (Kir1.1) and may be of importance in gating. In all, the $G\alpha^{GTP}$-binding regions in GIRK1 *do not* overlap with the Gβγ-binding surface estimated either from crystallography of GIRK2 (Whorton & MacKinnon, 2013), NMR analysis of GIRK1 (Yokogawa et al., 2011), and GIRK1–Gβγ protein modeling and structure–function studies (Mahajan et al., 2013).

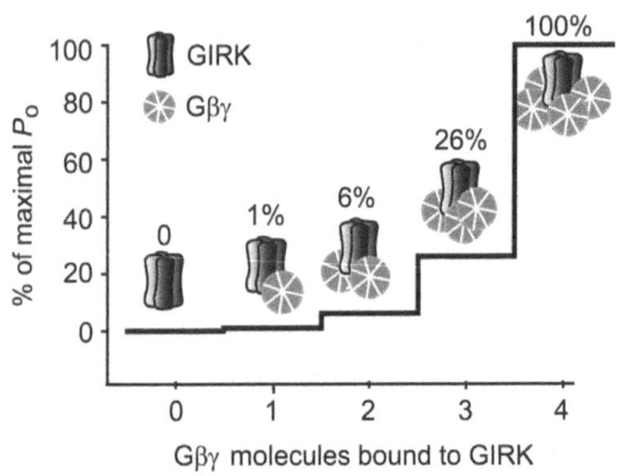

Figure 2 ***Cooperativity in GIRK1/2 gating by Gβγ.*** Bound Gβγ contributes to channel opening in a graded manner. The graded contribution is manifested in incremental changes in P_o when each additional Gβγ molecule binds. With one Gβγ bound, the P_o is only 1% of maximal attainable P_o. With two Gβγ bound to activation sites, the P_o is 6% of maximal; with three Gβγ, it is 26% of maximum; and four Gβγ molecules interacting with the channel simultaneously produce the maximal activation, $P_{o,max}$. P_o for each state was calculated by normalizing published P_o values of each of the four gating modes, corresponding to one to four Gβγ occupied state, to $P_{o,max}$ (corresponding to four Gβγ-occupied channel) (Ivanova-Nikolova et al., 1998). The experimental data have been obtained for the GIRK1/4 channel. Very close values have been were obtained by Sadja et al. (2002).

NMR showed that, in $G\alpha_{i3}^{GTP}$, mainly the GTPase domain (the $\alpha2/\alpha3$ domain largely corresponding to the switch 2 region) and also the helical domain participate in GIRK1 binding. Since switch 2 domain is masked by $G\beta\gamma$ in the $G\alpha^{GDP}\beta\gamma$ heterotrimer, $G\alpha^{GDP}$ must bind to GIRK1 in a different way from $G\alpha^{GTP}$. Although NMR assigned a less prominent role to the helical domain of $G\alpha$ (Mase et al., 2012), the importance of the latter is supported by an independent structure–function study (Rusinova et al., 2007). The proposed model suggests simultaneous binding of both $G\alpha^{GTP}$ and $G\beta\gamma$ to the GIRK1 tetramer (Mase et al., 2012), supporting a ternary complex between GIRK and both arms of the GPCR-activated heterotrimeric G protein (Berlin et al., 2010; Huang et al., 1995; Peleg et al., 2002; Rebois & Hebert, 2003; Rebois et al., 2006; Slesinger, Reuveny, Jan, & Jan, 1995). Furthermore, the RGS contact site of $G\alpha$ does not overlap that of GIRK, suggesting the possibility of a larger and specifically regulated signaling complex centered on GIRK (see chapter "RGS Redundancy and Implications in GPCR–GIRK Signaling" by Doupnik).

Although the issue of $G\alpha$–GIRK1 stoichiometry has not been explicitly discussed, a close examination of the proposed GIRK1–$G\alpha_{i3}^{GTP}$ interaction model (Mase et al., 2012) reveals that two $G\alpha$-binding αA helixes of two adjacent GIRK1 subunits are simultaneously engaged in contact with one $G\alpha$ molecule. One αA binds the $\alpha2/\alpha3$ domain and the other one with the helical domain of $G\alpha_{i3}$. This leaves only two more GIRK1 subunits' C-terminal αA sites, which can interact with one additional $G\alpha^{GTP}$. Therefore, according to this scenario, GIRK1 tetramer may interact with only two $G\alpha^{GTP}$. This contrasts with the well-established GIRK:$G\beta\gamma$ stoichiometry of 1:4. The functional consequences of such unequal stoichiometry will be discussed below.

2.3.3 The Role of NT and Distal CT in GIRK–$G\alpha$ Interactions

The interaction of separate NT and CT segments of GIRK1 and GIRK2 with $G\alpha_{i3}$ is stronger in GDP than in GTPγS (Ivanina et al., 2004), but *in vitro* translated full cytosolic domain of GIRK1 interact equally well with $G\alpha_{i3}^{GDP}$ and $G\alpha_{i3}^{GTP\gamma S}$ (Berlin et al., 2010, 2011). Furthermore, the interaction of $G\alpha_{i3}$ in either GDP or GTPγS with the full cytosolic domain is strongly enhanced by $G\beta\gamma$, and this effect is absent in a truncated GIRK1 version missing the distal CT (GIRK1Δ121) and in GIRK2 which has a shorter distal CT (Fig. 1C) (Kahanovitch et al., 2014; Rubinstein et al., 2007). Moreover, the presence of the NT is also essential for the $G\beta\gamma$-induced potentiation of $G\alpha_{i3}$ binding to GIRK1; $G\beta\gamma$ did not enhance

$G\alpha_{i3}$ binding to the full-length CT of GIRK1 (Berlin et al., 2010). Similar results were obtained with $G\alpha_{i1}$ (Tal Keren Raifman & N. D., unpublished results). And yet, the separate distal CT did not detectably bind $G\alpha_i$ in a pull-down assay (Ivanina et al., 2004). Interestingly, the enhancement did not appear to be mutual, since addition of $G\alpha_{i3}$ did not enhance the binding of GIRK1 cytosolic domain to Gβγ (Berlin et al., 2010). The physiological role of the Gβγ regulation of Gα–GIRK1 interaction is unclear, but it has been proposed to take part in Gα or Gαβγ docking to GIRK (Berlin et al., 2011; Rubinstein et al., 2009). Overall, the results of protein–protein interaction studies strongly indicate the importance of the three-dimensional fold formed by the full-length NT and CT domains of GIRK1 for a full spectrum of channel's interactions with Gα and Gβγ.

2.3.4 Comparing the Binding/Mutagenesis and NMR Data: Are They Discrepant?

One important point of agreement between the results of biochemical/mutagenesis and NMR mapping is that Gα and Gβγ interactions sites in the core of the C-terminus of GIRK1 and GIRK2 do not overlap. However, the crucial Gα binding sites mapped by NMR versus the structure–function approaches do not coincide. The Gα-binding segment in GIRK2, a.a. 318–330 (Clancy, Boyer, & Slesinger, 2007) is not fully conserved in GIRK1 (a.a. 307–319): 4 a.a. differences out of 13, including last 3 of the 6 crucial a.a. However, the first three important a.a. including the cysteine C_{321} (C_{310} in GIRK1) are conserved. Nevertheless, these residues have not been found as Gα-interacting in GIRK1 by the NMR spectroscopy (Mase et al., 2012).

These disparities may stem from the difference in methodologies. However, we think that a more plausible explanation is a divergent Gα interaction with GIRK1 and GIRK2 subunits. Indeed, the best Gα-binding peptide from GIRK1, a.a. 320–370, identified by the pull-down method (Ivanina et al., 2004) contains most of the Gα-binding sites found by NMR, but not the 13 a.a. site analogous to that of GIRK2 found by Clancy et al. (2005) (The GIRK2 Gα-binding peptide from the work of Ivanina et al. (2004), a.a. 310–380, is too large to provide any additional insight; it includes all putative sites.). Mase et al. (2012) had stressed that the main $G\alpha_{i3}$-binding site in a.a. 365–373 of GIRK1 is not conserved in GIRK2, suggesting different $G\alpha_{i/o}$ binding determinants. These differences may confer distinct functional properties, although it is not known how. Nonetheless, the possibility remains that, for GIRK1, the conserved part of the 13 a.a. site also interacts with Gα; coexistence of several binding determinants for Gα binding, each being best detected by a different method, is a valid option. We also stress that the interactions of Gα with the other two GIRK subunits, three and four, have not been studied at all. Whereas GIRK4 is highly homologous to GIRK2, GIRK3 has a unique long NT and a short CT; how this may affect interactions with Gβγ and Gα is unknown. In the absence of crystal structures of full-length GIRK1 and a GIRK–Gα complex, protein modeling and standard structure–function approaches will probably remain the main tools to address these issues, for now. The importance of such seemingly mundane studies becomes apparent when considering the involvement of Gα in regulating both basal and evoked activity of GIRK, which are highly important in health and disease (see below and also chapters "Localization and Targeting of GIRK Channels in Mammalian Central Neurons" by Luján and Aguado; "GIRK Channel Plasticity and Implications for Drug Addiction" by de Velasco et al.; "GIRK Channels: A Potential Link Between Learning and Addiction" by Tipps and Buck; and "Behavioral and Genetic Evidence for GIRK Channels in the CNS: Role in Physiology, Pathophysiology, and Drug Addiction" by Mayfield et al.).

3. ROLES OF Gβγ AND Gα IN GIRK'S EVOKED AND BASAL ACTIVITY

Activation by agonist is the main process by which GIRK channels are known to regulate the activity of excitable cells. GIRKs are the natural effectors of all classical small-molecule neurotransmitters and many neuropeptides. They are activated by cholinergic muscarinic, adrenergic, dopaminergic, serotonergic, metabotropic glutamatergic, $GABA_B$, cannabinoid and opioid receptors, and many others. The mechanisms and physiological roles of activation of GIRKs by plethora of GPCRs have been extensively reviewed (Guyon, 2014; Kobayashi & Ikeda, 2006; Luján et al., 2014; Luscher & Slesinger, 2010; Nagi & Pineyro, 2014; Padgett & Slesinger, 2010) and are treated here only generally.

Thus, traditionally, GIRKs have been considered as operated exclusively by inhibitory neurotransmitters which elicit the GIRK's evoked response (I_{evoked}). However, recent studies have revealed that GIRKs often possess substantial basal activity, I_{basal}, which is involved in regulation of neuronal and cardiac excitability (Luscher, Jan, Stoffel, Malenka, & Nicoll, 1997; Torrecilla et al., 2002; Wiser et al., 2006), and is also involved in certain cardiovascular pathologies (Al-Salameh et al., 2014; Voigt et al., 2014). Here, we reiterate the common scheme of GIRK activation by agonist/GPCR that produces the Gβγ-dependent I_{evoked}, and discuss in more detail the physiological role and mechanisms of I_{basal}, the interrelation between I_{basal} and I_{evoked}, mechanisms of specificity of signaling in the GPCR–GIRK cascade, and the roles of $G\alpha^{GDP}$ and $G\alpha^{GTP}$.

Most of these issues are under intensive investigation and some remain incompletely understood or controversial.

3.1 Agonist-Evoked Current, I_{evoked}

3.1.1 Activation

The classical scheme of the activation of GIRK is well established and has been extensively reviewed (see Section 1). When GPCR binds an agonist, it catalyzes GDP–GTP exchange at the cognate Gα, which leads to full or partial dissociation of Gα and Gβγ (Lambert, 2008). Gβγ then binds to GIRK, which leads to conformational changes, an increased interaction with PIP_2, opening of GIRK's gate(s), and K^+ flux through the pore (see chapter "Unifying Mechanism of Controlling Kir3 Channel Activity by G Proteins and Phosphoinositides" by Logothetis et al.). These processes determine the kinetics of activation and the peak amplitude of the response.

3.1.2 Deactivation

The cessation of the current after agonist removal (deactivation) is the result of agonist unbinding from the GPCR which is followed by GTP hydrolysis by the Gα subunit, reassociation of $G\alpha^{GDP}$ with Gβγ, and channel closure (Dascal, 1997; Gilman, 1987; Mark & Herlitze, 2000). Changes in deactivation may lead to profound alterations in the inhibitory control of the cell by the neurotransmitters (Luján et al., 2014; Ostrovskaya et al., 2014). The kinetics of deactivation are governed largely by the kinetics of agonist unbinding (Benians, Leaney, Milligan, & Tinker, 2003) and by the rate of GTP hydrolysis (Breitwieser & Szabo, 1988; Doupnik, Davidson, Lester, & Kofuji, 1997; Saitoh, Kubo, Miyatani, Asano, & Nakata, 1997), and are profoundly regulated by RGSs and related auxiliary proteins (see chapter "RGS Redundancy and Implications in GPCR–GIRK Signaling" by Doupnik).

3.1.3 Desensitization

I_{evoked} of GIRK is also characterized by desensitization (decay of the response observed in the constant presence of agonist). Desensitization of GIRK's I_{evoked} usually displays fast and slow components, with time courses of several seconds to tens of seconds (Dascal, 1997). The slow components largely reflect cellular processes related to GPCR phosphorylation by G protein-dependent kinases (GRK), recruitment of β-arrestins, and GPCR uncoupling and internalization (Ferguson, 2001; Shukla, Xiao, & Lefkowitz, 2011). The fast desensitization appears to mainly reflect the intrinsic properties of the G protein cycle (Chuang, Yu, Jan, & Jan, 1998; Leaney et al., 2004; Sickmann & Alzheimer, 2003) but may also depend on

GIRK subunit stoichiometry and composition, and involve multiple additional mechanisms in specific cases, such as PIP_2 depletion, depletion of submembrane K^+, direct interaction with GRK, and other factors (see Bender et al., 2001, 2004; Kobrinsky et al., 2000; Meyer et al., 2000; Padgett & Slesinger, 2010; Raveh et al., 2010).

We next focus on issues related to activation kinetics and the magnitude of the evoked response, where an interplay between Gα and Gβγ plays a major role.

3.1.4 Activation Kinetics

Activation kinetics determine the onset of the inhibitory effect of GIRK-activating neurotransmitters in cells. Moreover, in conjunction with other measurements, the kinetics are a sensitive indicator of the basic molecular mechanisms underlying the response. Generally, activation kinetics of neurotransmitter-induced GIRK currents are much slower than for neurotransmitter response mediated by ionotropic receptors-channels, such as nicotinic or glutamatergic ones (Hille, 2002). Ligand binding is diffusion-limited and also may change in the course of interaction with the GPCR–G protein complex (Ambrosio & Lohse, 2012). Therefore, ligand may become rate-limiting at low doses; activation kinetics reach maximal speed at saturating doses of agonist (Bunemann, Bucheler, Philipp, Lohse, & Hein, 2001; Marcaggi, Mutoh, Dimitrov, Beato, & Knöpfel, 2009; Zhang, Dickson, & Doupnik, 2004).

When an agonist is applied at a saturating dose using a fast perfusion system, GIRK current starts after a lag and then rises along a sigmoidal or close-to-exponential time course (Pfaffinger et al., 1985; Pott, 1979; Sodickson & Bean, 1996). At 35–37 °C, the lag usually lasts 50–100 ms (Pott, 1979; Sodickson & Bean, 1996). The exponential activation time constant (τ_{act}) or time-to-peak of I_{evoked} are usually in the order of 200–800 ms, but vary for different neurotransmitters and in different cells, even when measured in the same experiment. For instance, in hippocampal CA3 slice neurons, τ_{act} was ~220 ms for $GABA_B$ and ~500 ms for 5HT1A- and adenosine-mediated GIRK currents, and about four times slower for somatostatin-induced I_{evoked} (Sodickson & Bean, 1998). Similarly, in atrial myocytes, half-time of activation of the GIRK current was four times slower for adenosine than for acetylcholine ACh (1.7 vs. 0.4 s) (Wellner-Kienitz, Bender, Meyer, Bunemann, & Pott, 2000). Overexpression of the $G\alpha_{i/o}$-coupled adenosine receptors, A1R, greatly accelerated the rise time of adenosine I_{evoked}, suggesting that the amount of this GPCR was rate-limiting, and implying a collision-type coupling between the receptor and the G protein in this case (Wellner-Kienitz et al., 2000). It is important to realize that the surface densities of GPCRs in living cells are sometimes rather low, a few molecules/μm^2 in the plasma membrane (e.g., Falkenburger, Jensen, & Hille, 2010), which suggests that this factor may become rate-limiting for many G protein effectors, including GIRKs, if activated by low-abundance GPCRs.

The relatively slow activation time of GIRK (250–300 ms, at best, for the lag and rising phase taken together) is the consequence of the final speed of biochemical processes that precede channel activation in the GPCR–GIRK cascade. The kinetics of most stages of the GPCR–GIRK cascade that precede channel opening have been dissected in living cells mainly using intra and intermolecular FRET or BRET with genetically engineered, fluorescently labeled GPCRs, G proteins and effectors, and using NMR and molecular dynamics simulations (for review, see Lohse, Hein, et al., 2008; Lohse, Nikolaev, et al., 2008; Lohse, Maiellaro, & Calebiro, 2014). Conformational changes between inactive and active states in a GPCR occur within microseconds (Nygaard et al., 2013). The macroscopic rate constants of agonist-induced conformational changes are 1 ms for rhodopsin (Hofmann et al., 2009) and, for most other GPCRs studied, 10–50 ms (Chachisvilis, Zhang, & Frangos, 2006; Hoffmann et al., 2005; Marcaggi et al., 2009; Vilardaga, Bunemann, Krasel, Castro, & Lohse, 2003). These fast kinetics imply that activation of GPCR is not rate-limiting for GIRK's I_{evoked}. On the other hand, macroscopic rates of G protein activation by a ligand-bound GPCR (rearrangement or dissociation of Gα and Gβγ) are much slower, within several hundreds of milliseconds to 1–2 s, depending on GPCR and Gα species (Bunemann et al., 2003; Gales et al., 2005, 2006; Hein, Frank, Hoffmann, Lohse, & Bunemann, 2005; Jensen, Lyssand, Hague, & Hille, 2009). It is still unclear to what extent some of these time constant depend on the potentially diffusion-limited encounters between activated receptors and the G proteins (Lohse et al., 2014). Nevertheless, since these kinetics are reproducibly obtained with robustly overexpressed GPCRs (which should alleviate this concern; Hille, Dickson, Kruse, & Falkenburger, 2014), it appears that the rate of G protein activation is the rate-limiting step in effector activation, including GIRK (Hein & Bunemann, 2009; Jensen et al., 2009).

In all, in the case of activation of GIRK, the main rate-limiting factor is the kinetics of activation of G proteins rather than speed of activation of GIRK by the Gβγ (Lohse, Hein, et al., 2008; Lohse, Nikolaev, et al., 2008). In support, τ_{act} of heterologously expressed GIRK1/2 correlates with the GPCR-catalyzed GDP release rates from $G\alpha_{i/o}$ ($G\alpha_o$ being faster than

Gα$_i$), and τ_{act} changed in accord with changes in GDP release properties of Gα$_o$/Gα$_{i2}$ chimeras (Zhang et al., 2004). The pertussis toxin (PTX)-insensitive member of the G$_{i/o}$ family, G$_z$, characterized by very slow GDP dissociation and GTP hydrolysis rates (Fields & Casey, 1997), produces very slow GIRK activation when expressed in *Xenopus* oocytes (Vorobiov, Bera, Keren-Raifman, Barzilai, & Dascal, 2000). Further, GIRK activation can also be accelerated by RGS proteins (Chuang & Chuang, 2012; Jeong & Ikeda, 2001). The latter can be partly due to a kinetic scaffolding mechanism (Ross, 2008), because the activation time of an effector response elicited by multiple G protein cycles depends both on the on-rate of G protein activation (largely determined by the GDP dissociation) and by the off-rate of the process, mainly determined by the GTP hydrolysis. The latter is accelerated by the RGSs, making more Gα^{GDP} available for re-entry into the cycle (Ross, 2008; Zhong et al., 2003). However, RGS proteins also appear to act by a variety of additional, specific mechanisms, by augmenting GPCR–GIRK coupling (Chuang & Chuang, 2012) or by helping scaffolding the components of the macromolecular signaling complex (Hollinger & Hepler, 2002) (see chapter "RGS Redundancy and Implications in GPCR–GIRK Signaling" by Doupnik).

The above considerations suggest that binding of Gβγ to GIRK and channel opening are not rate-limiting in the activation of most GIRKs. However, this may not be always the case, at least for aberrant (mutated) channels. Alterations in the N-terminal region of GIRK1 (Slesinger et al., 1995) and at the junction between the TM and the cytoplasmic domains, which is believed to participate in transducing movements of the cytosolic domains to the gate, can slow down speed of activation of GIRK1/4 channels, suggesting that gating-associated movements within the channel molecule may become rate-limiting in some cases (Sadja & Reuveny, 2009). Subunit-specific kinetic aspects of GIRKs have not been studied in detail. However, the importance of GIRK1 subunit for fast activation has been demonstrated (Sadja et al., 2002). We consistently observe slower activation kinetics of I_{evoked} in GIRK2 homotetramers than in GIRK1/2 heterotetramers (unpublished results). Whether this is the result of differences in G protein coupling or intrinsic channel gating properties remains to be investigated.

3.1.5 The Magnitude of I_{evoked}: Cooperative Regulation by Gβγ

Assuming a constant level of phosphatidylinositol diphosphate (PIP_2) in an intact healthy cell (Logothetis, Jin, Lupyan, & Rosenhouse-Dantsker, 2007), Gβγ is the main gating factor for GIRKs, and the extent of activation is a function of the concentration of Gβγ available for GIRK activation. The gating parameter that changes upon channel activation by agonist or Gβγ is P_o, the channel's open probability. The single-channel current, i_{single}, and the number of functional channels in the plasma membrane, N, remain unchanged, at least on the timescale of a several second-long acute response to neurotransmitter (reviewed in Dascal, 1997; Stanfield, Nakajima, & Nakajima, 2003). On the macroscopic level, the relation between whole-cell I_{evoked} and the steady-state P_o measured on single-channel level is generally applicable, due to the relatively slow desensitization (compared to activation). This relation is described by a simple equation, $I = i_{single} \cdot P_o \cdot N$ (Hille, 2002).

On the molecular level, the Gβγ-dependent P_o is determined largely by the number of free Gβγ subunits interacting with each channel. Early works of Kurachi and collaborators demonstrated cooperative activation of cardiac GIRK (I_{KACh}) by G proteins, with a Hill coefficient of ~2.4–3. They proposed that GIRK is activated by several G protein molecules in a cooperative process (Ito et al., 1991; Kurachi, Ito, & Sugimoto, 1990; Yamada et al., 1993). The presence of four Gβγ binding sites in a GIRK channel has been corroborated by biochemical and structural studies (Corey, Krapivinsky, Krapivinsky, & Clapham, 1998; Whorton & MacKinnon, 2013). Lower Hill coefficient values are often observed for direct activation of GIRKs by Gβγ in excised patches (e.g., 1.3 in *Xenopus* oocytes for GIRK1/4; Luchian et al., 1997); 2.03 in locus coeruleus neurons (Albsoul-Younes et al., 2001). To note, in a dose–response experiment such as those done by Kurachi and colleagues (Yamada et al., 1993), the Hill coefficient provides a minimal estimate of stoichiometry and does not provide a good estimate of the extent of positive cooperativity in binding (Weiss, 1997). So far there are no direct biochemical data on positive cooperativity in Gβγ binding (when occupancy of one Gβγ binding site increases the affinity of Gβγ binding to another site). However, there is no doubt regarding the general cooperative nature of gating of GIRK by Gβγ. Works of Ivanova-Nikolova, Breitwieser, and colleagues (Ivanova-Nikolova & Breitwieser, 1997; Ivanova-Nikolova et al., 1998; Nikolov & Ivanova-Nikolova, 2004a, 2004b) demonstrated modal character of GIRK1/4 gating. Their single-channel analysis of GIRK1/4 channels activated by ACh or purified Gβγ in excised patches of atrial cells provided strong evidence for a graded contribution of each bound Gβγ (Fig. 2). With only one Gβγ the P_o is very low, 1% of maximal open probability, $P_{o,max}$ ($P_{o,max}$ is achieved when all four Gβγ binding sites are occupied). Occupation of each additional site increases P_o in a more than additive manner, corroborating cooperativity either in binding or in gating transitions. Very similar conclusions have been reached by Reuveny and colleagues (Sadja et al., 2002) who used an independent approach, by monitoring the activation of chimeric GIRK–IRK channels by agonist or purified Gβγ. The cooperative nature of GIRK gating is in good agreement with considerations based on structural studies: several molecules of Gβγ may be needed to produce sufficient mechanical strain to cause relative rotation of GIRK's cytoplasmic domains, which is then translated into gate opening (Whorton & MacKinnon, 2013). Based on the assumption of cooperative action of four Gβγ molecules, Kurachi and collaborators (Hosoya, Yamada, Ito, & Kurachi, 1996; Murakami, Inanobe, & Kurachi, 2013; Murakami, Suzuki, Ishii, Inanobe, & Kurachi, 2010) have developed an allosteric Monod–Wyman–Changeux-type model which adequately describes the magnitude and kinetics (but not I_{basal}), and the effect of RGS proteins, for the cardiac I_{KACh}.

3.1.6 Role of $G\alpha_{i/o}{}^{GTP}$ in Regulating Agonist- and Gβγ-Evoked Responses

The first and most obvious role of $G\alpha_{i/o}$ is as Gβγ donor for GIRKs (Fig. 3). $G\alpha^{GDP}$ "sequesters" Gβγ before activation by GPCR/agonist, thus regulating the GIRK's basal current. $G\alpha^{GDP}$ may also play a role in signaling complexes and specificity. These issues are discussed later.

After GPCR activation by agonist and the exchange of GDP for GTP at Gα, both Gβγ and $G\alpha^{GTP}$ are present in the channel's immediate vicinity. As discussed in chapter "Unifying Mechanism of Controlling Kir3 Channel Activity by G Proteins and Phosphoinositides" by Logothetis et al., direct activation of GIRK by $G\alpha_{i/o}{}^{GTP}$ has been ruled out in a series of studies by several groups (Ito et al., 1992; Logothetis et al., 1987; Reuveny et al., 1994; Schreibmayer et al., 1996; Wickman et al., 1994). Nevertheless, since $G\alpha_{i/o}{}^{GTP}$ can directly interact with GIRKs at binding sites distinct from those of Gβγ, modulation of GIRKs by a direct interaction with $G\alpha_{i/o}{}^{GTP}$ has been hypothesized. However, so far there is no consensus on such modulation.

In *Xenopus* oocyte studies, expression of Gβγ together with GIRKs usually produced near-maximal activation which was not further enhanced by activation of a $G_{i/o}$-coupled GPCR (Lim et al., 1995; Reuveny et al., 1994). This suggested that activated $G\alpha_{i/o}{}^{GTP}$ does not significantly affect GIRK's on top of maximal activation by Gβγ. However, it is not known whether this is the case with different GIRK subunit compositions. Interestingly, GIRK1/4 still showed sizeable activation by agonist after coexpression of Gβγ in *Xenopus* oocytes, hinting at a possible enhancing effect of

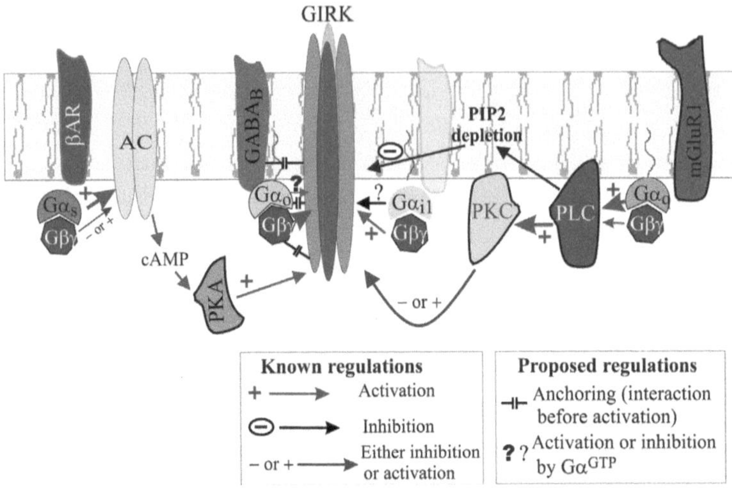

Figure 3 ***Regulation of GIRK by the different Gα subunits and Gα-initiated signaling pathways.*** Some $G\alpha_{i/o}$ such as $G\alpha_{i3}$ and $G\alpha_o$ are strongly coupled to GIRK and are fast, efficient donors of Gβγ. Both $G\alpha_{i/o}$ of this subclass, Gβγ, and some GPCRs such as $GABA_B$ receptor, directly interact with GIRKs and have been hypothesized to form stable signaling complexes with the channel. Other $G\alpha_i$, in particular $G\alpha_{i1}$, show weaker direct interaction with some GIRK subunits. It is not definitely known whether any of the $G\alpha_{i/o}$ subunits, in their activated, GTP-bound form, affect the channel directly. Activation of G_q (in the example shown, by the metabotropic glutamate receptor mGluR1) produces $G\alpha_q{}^{GTP}$ which (often synergistically with Gβγ) activates phospholipase C (PLC) and eventually inhibits GIRKs via a dual pathway by depleting PIP_2 and by activating protein kinases C (PKC) (some PKC types enhance GIRKs activity). $G\alpha_q$ was also proposed to inhibit GIRK by a direct interaction (Koike-Tani et al., 2005). Activation of G_s-coupled GPCRs, for example, by a β-adrenergic receptor (βAR), is followed by activation of adenylyl cyclases (AC) by $G\alpha_s{}^{GTP}$ (synergistically or antagonistically with Gβγ for some AC subtypes; Sadana & Dessauer, 2009). This is followed by elevation of the level of cAMP and PKA-induced phosphorylation of defined residues in GIRKs, which enhances channel's opening (Treiber et al., 2013). Upon overexpression of $G\alpha_s$ or βAR, G_s can also open GIRK through direct donation of Gβγ.

$G\alpha_{i/o}{}^{GTP}$ for this subunit composition. However, it could also be that saturating Gβγ effect has not been achieved in these experiments (Vivaudou et al., 1997).

In excised patches of atrial cells or in cells heterologously expressing GIRK1/4 or GIRK1/2, direct application of $G\alpha_{i/o}{}^{GTP}$ also did not consistently modulate Gβγ-induced channel activity. In comparison, addition of $G\alpha^{GDP}$ reduced Gβγ-induced activation, likely by sequestering the free Gβγ (Ito et al., 1992; Logothetis et al., 1987; Peleg et al., 2002;

Reuveny et al., 1994). In one study, $G\alpha_{i1}^{GTP\gamma S}$ was reported to attenuate Gβγ activation of a GIRK in excised patches of *Xenopus* oocytes (Schreibmayer et al., 1996) but this occurred only with very low expression levels of GIRK, suggesting potential interference with an unidentified endogenous factor of limited abundance (Peleg et al., 2002; Yakubovich, Pastushenko, Bitler, Dessauer, & Dascal, 2000).

Two recent studies took the advantage of highly purified proteins (same as used for crystallography), and exploited the most direct approach: incorporating purified GIRK channels into lipid bilayers and testing the effects of purified Gα and Gβγ added to the solution (Leal-Pinto et al., 2010; Wang et al., 2014). Either KirBac/GIRK1 chimera (homotetramer of monomers comprising the transmembrane domain of bacterial KirBac1 fused with the partial cytosolic domain of GIRK1; Nishida et al., 2007) or purified recombinant GIRK2 homotetramer (Whorton & MacKinnon, 2011) were used. Both studies (Leal-Pinto et al., 2010; Wang et al., 2014) confirmed the absolute necessity of PIP_2 for channel gating. In both cases, added Gβγ activated the channel. Unexpectedly, for the chimeric KirBac/GIRK1 chimera, either Gβγ or $G\alpha_{i1}^{GDP}$ added alone substantially inhibited the PIP_2-preactivated channel; activation by Gβγ required $G\alpha_{i/o}^{GTP\gamma S}$ as a necessary cofactor (Leal-Pinto et al., 2010). In contrast, with reconstituted GIRK2, both PIP_2 and Gβγ were essential for channel activation and neither could activate the channel alone. $G\alpha_{i1}^{GDP}$ inhibited the Gβγ-induced activity, whereas $G\alpha_{i1}^{GTP\gamma S}$ was without effect (Wang et al., 2014). The reasons for the discrepancies are unknown and may include differences in methodological details or channel constructs (a GIRK1/KirBac chimera vs. GIRK2), or the use of $G\alpha_{i1}$ which appears to be less well coupled to GIRKs than other $G\alpha_{i/o}$ species (see below). In summary, the role of $G\alpha^{GTP}$ in regulating the magnitude of Gβγ activation of GIRK remains unclear and requires further study.

Studies from our laboratory suggest that $G\alpha_{i/o}^{GTP}$ may regulate the kinetics of activation of GIRK1/2 channels. In whole-cell recordings in *Xenopus* oocytes, τ_{act} of I_{evoked} (elicited by acetylcholine (ACh) acting on muscarinic m2 receptors, m2R) was accelerated by coexpression of $G\alpha_{i3}$, whereas $G\alpha_{i1}$ was less effective (Ivanina et al., 2004). Since $G\alpha_{i1}$ and $G\alpha_{i3}$ display similar rates of GDP dissociation (Linder, Ewald, Miller, & Gilman, 1990; Zhang et al., 2004), we hypothesized that the accelerating effect of $G\alpha_{i3}$ is related to its GDP-bound form, due to an enhanced formation of GIRK–Gαβγ macromolecular signaling complex (Ivanina et al., 2004). However, later, using fluorescently labeled Gα constructs showing different kinetic properties in coupling to GIRK1/2, we were able to separate between the G$\beta\gamma$-sequestering function of Gα_{i3} (as Gα^{GDP}) and its function as accelerator of GIRK activation. We proposed that it is G$\alpha_{i3}{}^{GDP}$ that accelerates the activation of GIRK1/2, although it is not clear whether this is a consequence of the direct interaction between Gα and the channel (Berlin et al., 2011).

In conclusion, there is strong evidence that Gα^{GTP} is not a stand-alone activator of GIRK. It also appears that in living cells Gα^{GTP} does not appreciably regulate the amplitude of GIRK currents. However, it may play a role in the regulation of activation kinetics. There are hints to possible differences in regulation by Gα^{GTP} related to GIRK subunit composition, or to Gα species, which are worth pursuing. Such differences, if confirmed, may be related to selectivity and specificity of signaling in the GPCR–GIRK cascades which may eventually be of high relevance for understanding and treatment of GIRK-related disorders and rational drug design.

3.1.7 Regulation of GIRKs by Activation of PTX-Insensitive Gα

Whereas G_s and G_q normally do not activate GIRKs, as discussed in Section 5.3, GPCRs coupled to these G protein subclasses significantly regulate GIRK activity via activated G$\alpha_s{}^{GTP}$ and G$\alpha_q{}^{GTP}$ (Fig. 3). The archetypal G_s–adenylyl cyclase–PKA cascade enhances GIRK's P_o through phosphorylation of the channel subunits on defined C-terminal amino acids residues (Medina et al., 2000; Mullner, Steinecker, Gorischek, & Schreibmayer, 2009; Mullner et al., 2000; Mullner, Yakubovich, Dessauer, Platzer, & Schreibmayer, 2003; Rusinova et al., 2009). Phosphorylation by PKA has been proposed to serve as an "on-switch" for channel activation by G$\beta\gamma$ itself, with dephosphorylation by protein phosphatase 2A turning off the channel (Medina et al., 2000). Schreibmayer and colleagues (Treiber et al., 2013) studied PKA phosphorylation of homotetrameric GIRK4 and GIRK1* (a pore mutant $\text{GIRK1}_{\text{F137S}}$ which, unlike the wild-type GIRK1, is able to form functional channels that reach the plasma membrane; Chan, Sui, Vivaudou, & Logothetis, 1996). Both homotetrameric channels were phosphorylated on several serine and threonine residues, and their currents were potentiated by PKA. Mutations of phosphorylation sites showed that phosphorylation of both GIRK1 and GIRK4 subunits is necessary for a full effect of PKA on the GIRK1/4 heterotetramer (Treiber et al., 2013). The β-adrenergic–PKA cascade has been also reported to regulate the deactivation of cardiac GIRK, I_{KACh}, by an indirect mechanisms involving phosphorylation of RGS10 (Bender et al., 2008), but this pathway remains less well characterized.

Activated G_q is a strong inhibitor of GIRKs, acting through at least two interrelated pathways (Fig. 3): by depleting PIP_2 (Kobrinsky et al., 2000; Lei,

Jones, Talley, Garrison, & Bayliss, 2003; Lei, Talley, & Bayliss, 2001) and by activating classical, Ca^{2+}-dependent protein kinases C (cPKC) which phosphorylate defined amino acids in the C-termini of GIRK subunits (Hill & Peralta, 2001; Makary et al., 2011; Mao et al., 2004; Mark, Ruppersberg, & Herlitze, 2000; Nikolov & Ivanova-Nikolova, 2004a, 2004b; Sharon et al., 1997; Stevens et al., 1999). In contrast, a "novel" Ca^{2+}-independent nPKC, PKCε, enhances GIRK activity (Makary et al., 2011; Voigt et al., 2007).

4. BASAL ACTIVITY OF GIRKs

Besides the well-recognized role of GIRKs as mediators of inhibitory neurotransmission initiated by $G_{i/o}$-coupled GPCRs, studies of the last decade suggest an important role for basal (constitutive) activity of GIRK channels, I_{basal}. Basal activity may be either constitutive, i.e., independent on activation of GPCRs, or it can arise from the presence of an ambient activator such as neurotransmitter of local hormone. In the latter case, it can actually be classified as a small persistent I_{evoked}. The exact nature of I_{basal} in native cells is usually not well defined, and the proportion of GPCR-dependent and -independent activity is expected to vary in distinct experimental settings (e.g., isolated cultured cells vs. brain slices vs. intact brain). Furthermore, studies in expression systems suggest that the GPCR-independent I_{basal} depends on subunit composition and contains Gβγ-dependent and Gβγ-independent components. Because in previous literature reviews I_{basal} was less explored than neurotransmitter-induced currents, we will extend the discussion on its physiological role and then discuss the molecular mechanisms.

4.1 Physiology of I_{basal}

4.1.1 Central Nervous System

The role of GIRK channels in the central nervous system (CNS) is reviewed in chapters "Localization and Targeting of GIRK Channels in Mammalian Central Neurons" by Luján and Aguado; "GIRK Channel Plasticity and Implications for Drug Addiction" by de Velasco et al.; "GIRK Channels: A Potential Link Between Learning and Addiction" by Tipps and Buck; and "Behavioral and Genetic Evidence for GIRK Channels in the CNS: Role in Physiology, Pathophysiology, and Drug Addiction" by Mayfield et al. Here, we provide only a brief note on the role of GIRK's I_{basal} in neurons. GIRK1, 2, and 3 are predominant in the brain, with a specific and variable distribution in brain regions as well as between different compartments of neurons (chapter "Localization and Targeting of GIRK Channels in Mammalian Central Neurons" by Luján and Aguado). Knockout of single or several

GIRK subunits usually causes a decrease in the expression of the others (Signorini, Liao, Duncan, Jan, & Stoffel, 1997; Torrecilla et al., 2002) and depolarizes neurons by up to 20 mV (locus coeruleus; Torrecilla et al., 2002; see chapters "GIRK Channel Plasticity and Implications for Drug Addiction" by de Velasco et al.; "GIRK Channels: A Potential Link Between Learning and Addiction" by Tipps and Buck; and "Behavioral and Genetic Evidence for GIRK Channels in the CNS: Role in Physiology, Pathophysiology, and Drug Addiction" by Mayfield et al.). Similarly, block of GIRKs by tertiapin substantially depolarizes neurons increases the resting excitability (e.g., Wiser et al., 2006). Thus, neurons may have a significant I_{basal}. The extent of GPCR- and Gβγ-dependence of this activity is not clear in most cases. For instance, I_{basal} recorded in cell-attached patches in dendrites of CA1 hippocampal neurons appeared to be GPCR-independent (Chen & Johnston, 2005), whereas whole-cell recordings in dorsal CA1 neurons suggested that most of I_{basal} was elicited by ambient tonic A1 adenosine receptor activation (Kim & Johnston, 2015). Whatever the nature of I_{basal} in the brain, it certainly carries important physiological functions. GIRK's basal activity and the balance between I_{basal} and I_{evoked} emerge as important determinants of the level of excitability and resting membrane potential in neurons (Chen & Johnston, 2005; Luscher et al., 1997; Torrecilla, Fernandez-Aedo, Arrue, Zumarraga, & Ugedo, 2013; Torrecilla et al., 2002; Wiser et al., 2006), bistability of some neuronal networks (Sanders, Berends, Major, Goldman, & Lisman, 2013), neuronal plasticity through depotentiation of long-term potentiation (Chung, Ge, et al., 2009; Chung, Qian, Ehlers, Jan, & Jan, 2009; Cooper et al., 2012), depressive behavior (Llamosas, Bruzos-Cidón, Rodríguez, Ugedo, & Torrecilla, 2015), dendritic integration (Makara & Magee, 2013), and possibly in working memory (Sanders et al., 2013), and have been proposed to be related to effects of Li^+, a drug used in the treatment of bipolar disorder (Farhy Tselnicker et al., 2014).

4.1.2 Heart

In the heart, acetylcholine (ACh) activates GIRK via m2R and eliciting the K^+ current, I_{KACh}, which mediates the negative chronotropic effect of the vagus (Drici, Diochot, Terrenoire, Romey, & Lazdunski, 2000; Noma & Trautwein, 1978; Tamargo, Caballero, Gomez, Valenzuela, & Delpon, 2004; Voigt et al., 2014; Walsh, 2015; Yamada et al., 1998) and regulates pacemaking (Kovoor et al., 2001; Mesirca et al., 2014, 2013). Heart expresses GIRK1 and GIRK4 in sinoatrial node, atrium, and Purkinje fibers, and much less in the ventricle (Beckmann et al., 2008; Calloe, Goodrow, Olesen, Antzelevitch, & Cordeiro, 2013; Gaborit et al., 2007; Karschin et al., 1994; Yamada et al., 1998). Nevertheless, GIRK1's RNA is as abundant as the major

G protein-independent inward rectifier, IRK1 (Kir2.1) in the ventricle (Marionneau et al., 2005). Accordingly, recent studies indicate a role for GIRK in shaping the action potential (AP) both in atrium and ventricle (Beckmann et al., 2008; Coote, 2013; Liang et al., 2014; Yang et al., 2010). The majority of cardiac GIRK channels are GIRK1/GIRK4 (GIRK1/4) heterotetramers (Bettahi, Marker, Roman, & Wickman, 2002; Nikolov & Ivanova-Nikolova, 2004a, 2004b; Yamada et al., 1998). GIRK4 homotetramers are functional but have low open probability (P_o) and normally contribute little to the total macroscopic GIRK current in the atrium (Bender et al., 2001; Bettahi et al., 2002; Chan et al., 1997; Chan, Langan, et al., 1996; Nikolov & Ivanova-Nikolova, 2004a, 2004b). However, the exact stoichiometry of $GIRK1_x/GIRK4_y$ heterotetramers and the contribution of GIRK4 homotetramers are unclear.

In healthy adult mammalian heart, GIRK's I_{basal} is low compared to I_{evoked}, and is insensitive to internal Na^+ and Gβγ-scavengers (Ito, Ono, & Noma, 1994; Kienitz, Mintert-Jancke, Hertel, & Pott, 2014; Mintert et al., 2007; Sakmann, Noma, & Trautwein, 1983). However, I_{basal} (often called constitutive I_{KACh}) is greatly increased, and plays a crucial role, in atrial fibrillation (AF) and remodeling. The increased I_{basal} correlates with action potential (AP) shortening (Cha et al., 2006; Dobrev et al., 2005; Ehrlich et al., 2004; Voigt et al., 2008) and involves complex changes in Na^+ regulation of IRK1 and of I_{KACh} (Voigt et al., 2014, 2013). It has been proposed that changes in Na^+ regulation in AF reflect a prevalence of GIRK4-dominated GIRK1/4, or homotetrameric GIRK4, which have high Na^+- and TPN-sensitivity (Mintert et al., 2007). However, heterologously expressed homomeric GIRK4 has very low I_{basal} (Chan, Langan, et al., 1996; Krapivinsky, Gordon, et al., 1995; Spauschus et al., 1996). Moreover, in human AF atria, the relative content of GIRK4 protein decreased compared to GIRK1 (Voigt et al., 2013). In all, variations of subunit stoichiometry within the GIRK1/4 heterotetramer, confirmed in a heterologous system (Grasser, Steinecker, Ahammer, & Schreibmayer, 2008), almost certainly occur in AF, but it is unclear which stoichiometry gives the high I_{basal}. Besides GIRK stoichiometry, potential factors that may underlie changes in I_{basal} in AF include balance of Gα and Gβγ; enhanced NDPK activity; changes in Na^+/PIP_2 basal regulation; phosphorylation by protein kinases (Voigt et al., 2014).

Interestingly, association of GIRK with IRK1 has been proposed (Ishihara, Yamamoto, & Kubo, 2009; Kokunai et al., 2014). If functional IRK1/GIRK4 channels exist and GIRK4 may regulate IRK1's current, our understanding of cardiac K^+ conductances and some pathologies will require considerable revision. GIRK4–IRK1 hypothesis could explain both cardiac and skeletal muscle Andersen–Tawil syndrome-like symptoms caused by a G378R mutation of GIRK4 (Kokunai et al., 2014); Andersen–Tawil syndrome is a known channelopathy of IRK1 (Donaldson, Yoon, Fu, & Ptacek, 2004; Tristani-Firouzi & Etheridge, 2010). The underlying mechanism has been proposed to involve a reduction in I_{basal} of IRK1/GIRK4$_{G387R}$ heterotetramers compared to wild-type IRK1/GIRK4 channels (Kokunai et al., 2014). However, coexpression of GIRK4 with IRK1 does not increase IRK1 currents nor yield any Gβγ-sensitive currents (Ishihara et al., 2009) and it is possible that GIRK4 regulates IRK1 biogenesis rather than producing functional heterotetramers. In all, formation of IRK1/GIRK4 channels and their contribution to I_{basal} needs further scrutiny.

4.1.3 Adrenal Gland

Recent studies suggest that somatic or germline mutations in *KCNJ5* gene, encoding the GIRK4 subunit, underlie 40–70% of all cases of primary aldosteronism, accounting for up to 10% of hypertension in humans (Funder, 2014; Williams et al., 2015). The most common mutations are within or around the channel's pore, in or close to the selectivity filter (Adachi et al., 2014; Charmandari et al., 2012; Choi et al., 2011; Kuppusamy et al., 2014; Monticone et al., 2013). Other mutations are outside the pore, within the CT cytosolic domain (Murthy et al., 2014). The current leading concept is that most mutations reduce GIRK selectivity and rectification, which triggers Na^+ influx and depolarization followed by Ca^{2+} influx and aldosterone secretion (Murthy et al., 2014; Williams et al., 2015). However, for non-pore mutants the mechanism could be different. Remarkably, GIRK4 expression is actually decreased in aldosterone-producing adenomas, for some mutations rather severely (Cheng et al., 2015; Fernandes-Rosa et al., 2015). Also, even with pore mutants of GIRK4, no severe cardiac phenotype is observed (as it would be expected from changes in K^+/Na^+ selectivity); elevated risk of cardiac complications in patients with PA is indirect, due to hypertension (Adachi et al., 2014; Al-Salameh et al., 2014). Therefore, changes in basal GIRK activity have been recently implicated as another mechanism in the disease etiology (Al-Salameh et al., 2014).

How changes in GIRK's I_{basal} can regulate aldosterone secretion? Basic mechanisms of GIRK function in adrenal gland are much less characterized than in the heart, but recent works suggest high I_{basal}. In adrenocortical NCI-H295R cells, GIRK blockade by tertiapin strongly reduces basal conductance (Tauber et al., 2014) and depolarizes the cells by >15 mV (Cheng et al., 2015), indicating a dramatic contribution of GIRK's I_{basal} to resting K^+ conductance. Accordingly, it has been proposed that in adrenal cortex the

GIRK channel is composed of GIRK1/4, and its high I_{basal} helps maintaining the negative V_m, preventing spontaneous depolarization and aldosterone secretion (Al-Salameh et al., 2014). However, mechanisms of I_{basal}, as well as GIRK subunit composition in adrenal gland, are unknown. No GIRK1 protein has been detected by proteomic screening (Kim et al., 2014), and the abundance of its RNA (*KCNJ3*) is ~30-fold lower than GIRK4 (*KCNJ5*) (Choi et al., 2011). Unraveling the mechanisms of high I_{basal} in adrenocortical cells may provide new clues to understanding and treating aldosteronism and the related hypertension.

4.2 Heterologous Expression Studies: GPCR-Independent I_{basal} and the Role of Gβγ

The mechanisms and subunit-dependence of I_{basal} have been widely addressed in heterologous expression systems, mainly *Xenopus* oocytes and HEK293 cells. The best-studied subunit compositions are GIRK1/4 and GIRK1/2, less extensive data are available on homotetramers of GIRK2a, GIRK4, and GIRK1*. I_{basal} was studied in the absence of added neurotransmitters and often without coexpressed GPCRs. Several lines of evidence suggest that there are little or no spontaneously activated endogenous G proteins under these conditions in these cells (discussed in Rubinstein et al., 2009). Therefore, the basal activity monitored in these cells is classified as GPCR-independent.

Most studies suggest that GPCR-independent I_{basal} depends on subunit composition and contains Gβγ-dependent and Gβγ-independent components (Chan et al., 1997; Leaney & Tinker, 2000; Peleg et al., 2002; Rishal et al., 2005). Considerable GPCR-independent I_{basal} has been reported for GIRK1/4, GIRK1/2, and GIRK1* (Chan et al., 1997; He et al., 2002; Kahanovitch et al., 2014; Leaney & Tinker, 2000; Peleg et al., 2002; Rishal et al., 2005; Rubinstein et al., 2007; Wydeven, Fernandez de Velasco, et al., 2014; Wydeven, Posokhova, Xia, Martemyanov, & Wickman, 2014). The involvement of Gβγ in I_{basal} has been studied by coexpressing proteins that are expected to "steal" the available Gβγ from GIRK, termed Gβγ scavengers: C-terminus of β-adrenergic kinase, phosducin (see Rishal et al., 2005), and an "inactive" mutant of $G\alpha_{i3}$, $G\alpha_{i3\text{-}G203A}$, which strongly binds Gβγ and is deficient in GDP–GTP exchange (Ogier-Denis, Houri, Bauvy, & Codogno, 1996). Overexpression of Gβγ scavengers or wild-type $G\alpha_{i/o}$ reduces I_{basal} of GIRK1/2 or GIRK1* channels by up to 80–90% in *Xenopus* oocytes (Rishal et al., 2005; Rubinstein et al., 2007) and HEK293 cells (Leaney, Milligan, & Tinker,

2000). Thus, for GIRK1/2, up to 80–90% of I_{basal} is Gβγ-dependent. Accordingly, 10–20% of GIRK1/2 I_{basal} are Gβγ-independent (Rishal et al., 2005). The nature of the constitutive Gβγ-independent component of I_{basal} is not known. Potential regulators are intracellular Na^+ and PIP_2, but other unidentified factors may be involved.

In contrast to GIRK1-containing channels, GIRK2a homotetramers have low basal activity compared to GIRK1* or GIRK1/2, at similar surface levels (densities) of the expressed channels, both in *Xenopus* oocytes (Kahanovitch et al., 2014; Rubinstein et al., 2009) and in HEK293 cells (Wydeven et al., 2012). Furthermore, I_{basal} of the homomeric GIRK2a is almost completely insensitive to coexpression of Gβγ scavengers and $G\alpha_{i/o}$, suggesting that most of I_{basal} of GIRK2a is Gβγ-insensitive (Rubinstein et al., 2009). Low basal activity was reported for homotetrameric GIRK4 and $GIRK4_{S143T}$ (a pore mutant with higher currents), but the effect of Gβγ scavengers or Gα was not studied in detail (Vivaudou et al., 1997). (There are no detailed studies as yet of I_{basal} GIRK1/3 and GIRK2/3, two important neuronal GIRK channels.)

In all, the available data suggest that the size and Gβγ-dependence of I_{basal} crucially depend on channel subunit composition. Heterologously expressed GIRK1-containing subunits have a large Gβγ-dependent component of I_{basal}, whereas homomeric GIRK2a has low and largely Gβγ-independent I_{basal}. I_{basal} observed in hippocampal neurons, where GIRK1/2 is the most prominent channel composition (Luscher & Slesinger, 2010), is also quite sizeable (Chen & Johnston, 2005; Farhy Tselnicker et al., 2014; Torrecilla et al., 2013; Wiser et al., 2006), though its relative magnitude compared to I_{evoked} is smaller than in *Xenopus* oocytes (I. Farhy Tselnicker & N. D., unpublished results). Thus, heterologous studies on GIRK1/2 appear to be relevant to neuronal cells. In contrast, in rat atrial cardiomyocytes coexpression of the Gβγ scavenger phosducin or $G\alpha_{i1}$ did not significantly reduce I_{basal}, and it has been proposed that GIRK1/4 in native cells does not contain a detectable Gβγ-dependent component (Kienitz et al., 2014). It is not known whether the difference with expression systems reflects a different cellular environment in the latter, differences in channel or Gβγ-sequestering proteins' expression levels, or GIRK1/4 subunit stoichiometries.

4.3 I_{basal} and I_{evoked} are Interrelated: The Roles of $G\alpha^{GDP}$ and Gβγ

Heterologous expression of wild-type $G\alpha_{i/o}$ reduces I_{basal} of GIRK1/4, GIRK1/2, and GIRK1*. However, unlike the Gβγ scavengers, $G\alpha_{i/o}$ does not reduce I_{evoked} but rather potentiates it to the extent that total GIRK current is at least as large as before the expression of $G\alpha_{i/o}$ (Leaney et al., 2000; Leaney & Tinker, 2000; Peleg et al., 2002; Rishal et al., 2005; Vivaudou et al., 1997). The obvious interpretation is that $G\alpha_{i/o}{}^{GDP}$ sequesters the "free" Gβγ that contributes to I_{basal}, and then releases it upon GPCR activation. Moreover, at expression levels that greatly reduce I_{basal}, $G\alpha_{i3}$ does not impair activation of GIRK1/2 by coexpressed Gβγ, also unlike Gβγ scavengers (Rubinstein et al., 2007). Thus, $G\alpha_i{}^{GDP}$ regulates GIRK gating, keeping I_{basal} low and preparing the channel for robust activation by agonist and Gβγ (Peleg et al., 2002; Rishal et al., 2005; Rubinstein et al., 2007; Vivaudou et al., 1997). We dubbed this effect of $G\alpha^{GDP}$ "priming," which is distinct from simple sequestration of Gα, possibly because heterotrimeric $G\alpha_{i/o}\beta\gamma$ remains close to the channel (maintaining a signaling complex), ready to release Gβγ when GPCR is activated (Rubinstein et al., 2007).

Titrated expression of GIRK1/2 in *Xenopus* oocytes revealed a distinctive relation between I_{basal} and I_{evoked}: a gradual incremental increase in the expression of GIRK1/2 (by titrating the amounts of injected RNA) was accompanied by a large increase in I_{basal} but much less change in I_{evoked}. The activation index R_a defined as I_{total}/I_{basal}, which essentially signifies the ratio between I_{evoked} and I_{basal}, was strongly reduced as the surface density of channels and I_{basal} increased (Peleg et al., 2002; Rishal et al., 2005; Rubinstein et al., 2007). The inverse relation between I_{evoked} (or R_a) and I_{basal} of GIRK1/2 appears to be ubiquitous, as it is also observed in HEK293 cells expressing GIRK1/2 (Rubinstein et al., 2009) and in cultured hippocampal neurons (I. Farhy Tselnicker & N. D., unpublished results; Fig. 4). Notably, in the oocytes, coexpression of $G\alpha_{i3}$ not only increased R_a but also eliminated the inverse I_{evoked}/I_{basal} relation, yielding a high, steady level of agonist-induced activation that did not depend anymore on the level of channel expression (Peleg et al., 2002). The high Gβγ-dependent component of I_{basal} in GIRK1/2 suggested that Gβγ available for channel activation is in excess of over $G\alpha_{i/o}$ (Rubinstein et al., 2007). In other words, there is more Gβγ than $G\alpha_{i/o}$ in channel's microenvironment, or associated with GIRK1/2 (see later and Fig. 5).

A single mechanism, which we call Gβγ recruitment, may explain the high I_{basal}, the excess of Gβγ, and the inverse I_{evoked}/I_{basal} relation of the GIRK1/2 channel. Several works suggested that GIRK1-containing channels themselves recruit Gβγ, but not Gα, to their own environment. Expression of GIRK1-containing channels, but not GIRK2a, increases the amount of Gβγ in the plasma membrane of *Xenopus* oocytes, whereas the amount of

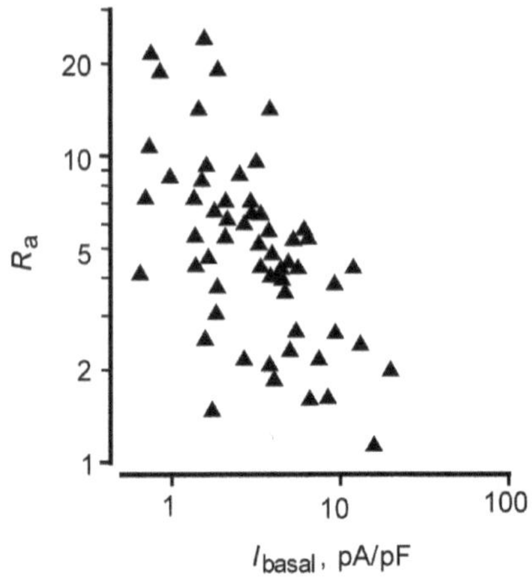

Figure 4 ***Basal and agonist-evoked GIRK currents in neuron are inversely related.*** The GIRK currents were measured in cultured hippocampal neurons in a high-K^+ external solution at holding potential of −70 mV, as described by Farhy Tselnicker et al. (2014). Note the logarithmic scale on both axes. The evoked current, I_{evoked}, was elicited by the application of a saturating concentration of baclofen, a GAB_B agonist. The basal current was measured by applying the GIRK channel blocker tertiapin-Q, TPNQ; the TPNQ-blocked current in the absence of agonists is the basal current, I_{basal}. The activation index (R_a) is defined as total current (basal + evoked) normalized to I_{basal} of the same cell. The inverse correlation between R_a and I_{basal} is highly significant, $p = 0.000000028$ by Spearman correlation test.

G$\alpha_{i/o}$ is increased less or not at all (Kahanovitch et al., 2014; Rishal et al., 2005). This means that as more GIRK1 channels are expressed in the plasma membrane, there is more available free Gβγ for each channel, relative to Gαβγ heterotrimers; thus higher I_{basal} and lower I_{evoked}. Coexpressed Gα restores the balance between free and Gα-bound Gβγ, reducing I_{basal} and increasing R_a.

Structure–function and mutagenesis studies demonstrated that the high Gβγ-dependent component of I_{basal} of GIRK1-containing channels depends on the unique distal CT domain of GIRK1 (Chan et al., 1997; Kahanovitch et al., 2014; Rubinstein et al., 2009; Wydeven, Fernandez de Velasco, et al., 2014; Wydeven, Posokhova, et al., 2014). Deleting distal CT from GIRK1*, or replacing it with the shorter distal CT of GIRK2a, reduces Gβγ binding to the cytosolic domain of GIRK1, eliminates Gβγ recruitment, and strongly reduces I_{basal} (although not to a level as low as that of GIRK2a). Lack of Gβγ recruitment explains the low, Gβγ-independent I_{basal} of GIRK2a and its greater relative extent of activation by coexpressed Gβγ compared to GIRK1* (Kahanovitch et al., 2014; Rubinstein et al.,

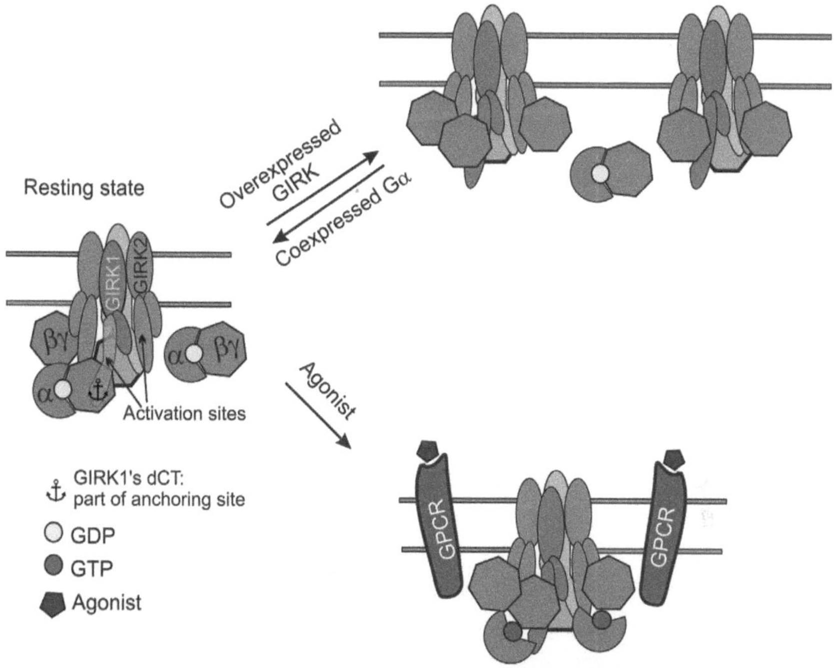

Figure 5 ***Cooperativity and stoichiometry are important factors in regulation of GIRK1/2 by Gβγ and Gα.*** Schematic representation of the GPCR–G protein–GIRK1/2 system based on studies in *Xenopus* oocytes. GIRK1/2 channel is a heterotetramer of two GIRK1 (gray) and two GIRK2 (green (dark gray in the print version)) subunits. The channel recruits Gβγ, but not Gα, to its immediate vicinity, in part due to the high-affinity association of Gβγ with the "anchoring" site in GIRK1, which comprises distal CT and likely other parts of the cytoplasmic domain. Our calculations suggest that, at any given time, there are up to four Gβγ available for GIRK1/2 but no more than 2 $G\alpha_{i/o}$. In the resting state, in the absence of an activated GPCR, $G\alpha^{GDP}$ is associated most of the time with Gβγ due to the very high-affinity interaction between the two (GDP is shown by a yellow (light gray in the print version) circle). The Gβγ molecules which are not associated with $G\alpha^{GDP}$ are free to bind at the interaction sites in the core of the cytoplasmic domain of the channel. As a consequence, in the resting (basal) state there are one to three free Gβγ molecules interacting with the channel, which gives a Gβγ-dependent basal activity that is 1–26% of $P_{o,max}$ (see Fig. 2). Activation of G proteins by an agonist (gray pentagon) via a GPCR (magenta (dark gray in the print version)) leads to the exchange of GDP to GTP (red (dark gray in the print version) circle) on Gα molecules, and to the subsequent dissociation of the Gαβγ heterotrimer, liberating additional Gβγ for channel activation; the maximal activation ($P_{o,max}$) is reached when all four Gβγ-binding sites are saturated (see Fig. 2). In *Xenopus* oocytes, and presumably in other cells, overexpression of GIRK1/2 is accompanied by preferential recruitment of Gβγ over Gα and decrease in GIRK:Gα ratio. This, in turn, leads to further increase in "basal" open probability and correspondingly smaller extent of GPCR-induced activation of the channel. The opposite process occurs upon overexpression of $G\alpha_{i/o}$, leading to a decrease in free Gβγ available for channel activation at rest and thus a reduction in I_{basal}.

2009). Thus, the distal CT of GIRK1 participates in, and is essential for, Gβγ recruitment, and we propose that it participates in Gβγ anchoring to GIRKs (Kahanovitch et al., 2014). BRET between GIRK and Gβγ is observed when the channel is still in the endoplasmic reticulum (Rebois et al., 2006), suggesting that GIRKs and Gβγ associate during biogenesis and cotrafficked to the plasma membrane (Zylbergold et al., 2014). Enrichment of regulatory proteins in the microenvironment of membrane protein can also be driven kinetically by high-affinity interactions, such as, for example, in the case of Ca^{2+} channels and calmodulin (Mori, Erickson, & Yue, 2004).

In summary, studies in heterologous expression systems suggest that $G\alpha^{GDP}$ and Gβγ are acting in concert, setting the level of I_{basal} and I_{evoked}. The inverse relation between I_{basal} and I_{evoked} (Fig. 4) indicates that this mechanism may be operating in neurons as well. I_{basal} and I_{evoked} are sensitively regulated by the availability of $G\alpha^{GDP}$, whereas recruitment of Gβγ ensures high level of activity of the channel under variable circumstances. The substantial level of I_{basal} of GIRK1/2 renders this channel with the property of bidirectional regulator which has been hypothesized to sensitively regulate neuronal excitability (Rubinstein et al., 2007): a decrease in I_{basal} (e.g., by activation of G_q-coupled GPCR) would increase the excitability, and channel activation by G_i-coupled GPCRs or drugs such as lithium (Farhy Tselnicker et al., 2014) or ethanol (Bodhinathan & Slesinger, 2014) will have an opposite effect. In comparison, homomeric GIRK2 channels, with their low I_{basal} and high sensitivity to Gβγ, are more suitable to serve as sensitive but mostly unidirectional, low-noise transduces of inhibitory neurotransmitter actions (Rubinstein et al., 2009).

5. MULTIMOLECULAR SIGNALING COMPLEX AND THE MECHANISM OF GIRK ACTIVATION

5.1 Preformed (Permanent) Signaling Complex in the GPCR–GIRK Cascade: Pro and Contra

The idea of strong association of proteins participating in the activation of GIRKs is central for the interpretation of much of the findings on GPCR–GIRK cascade over the last two decades. Fast kinetics (Hille, 1992), specificity and restricted localization of signaling cascades (Neubig, 1994) suggested close apposition and colocalization of the participants of the GPCR-coupled cascades. The evolving general concept of precoupling in "preformed," permanent signaling complexes that include GPCRs, G protein heterotrimers, effectors, and possibly other proteins, has been proposed for many

GPCR-coupled pathways (Davare et al., 2001; Golebiewska & Scarlata, 2008; Kamp & Hell, 2000), including the GPCR–GIRK cascade (see Doupnik, 2008; Luján et al., 2014; Luscher & Slesinger, 2010; Nagi & Pineyro, 2014; Padgett & Slesinger, 2010; Raveh et al., 2009, for recent reviews). Such complexes are formed before the activation of the GPCR by agonist and do not dissociate but rather undergo rearrangements upon activation; this ensures specificity and speed of signaling. Besides kinetic considerations, for the GPCR–GIRK cascade the concept is supported by the strong binding between the partners (e.g., GIRKs and $G\beta\gamma$; some GPCRs and GIRKs), and protein proximity and mobility studies by biophysical and optical methods (Berlin et al., 2010, 2011; Bunemann et al., 2003; Frank, Thumer, Lohse, & Bunemann, 2005; Gales et al., 2006; Huang et al., 1995; Ivanina et al., 2003; Lambert, 2008; Lober, Pereira, & Lambert, 2006; Rebois et al., 2006; Richard-Lalonde et al., 2013; Riven et al., 2006).

Nevertheless, many issues remains debated. First, the mode of coupling of the GPCR to G protein (catalytic collision coupling vs. permanent complexes) raises debate. Permanent FRET/BRET among heterologously coexpressed proteins has been a major argument in favor of precoupling (Gales et al., 2006; Nobles, Benians, & Tinker, 2005; Rebois & Hebert, 2003); however, robust overexpression of proteins may lead to "crowding," when FRET/BRET constantly occurs among reversibly interacting (rather than precoupled) proteins (Falkenburger et al., 2010; James, Oliveira, Carmo, Iaboni, & Davis, 2006; Lohse, Hein, et al., 2008; Lohse, Nikolaev, et al., 2008). It is becoming clear that mode of coupling is pathway-specific (for example, m3R-G_q are precoupled but m4R-G_o are not; Qin, Dong, Wu, & Lambert, 2011) and subcellular compartment—specific (Schwarzer, Nobles, & Tinker, 2010; van Hemert, Lazova, Snaar-Jagaska, & Schmidt, 2010). In the GPCR–GIRK cascade, there is a large body of evidence of precoupling in cases when the GPCR interacts not only with the G protein but also with the channel. The list includes several opioid, $GABA_B$, and dopaminergic D_2 and D_4 receptors (Luján et al., 2014; Nagi & Pineyro, 2014; Padgett & Slesinger, 2010). For other GPCRs such as the adenosine A1, α_{2A}-adrenergic receptors, m2R, and several others, kinetic and FRET analysis indicate rapid or restricted collision coupling, rather than precoupling, between these receptors and G proteins, both in heterologous model systems (Hein et al., 2005; Qin, Sethi, & Lambert, 2008; Vorobiov et al., 2000) and in native cells (Kienitz et al., 2014; Wellner-Kienitz et al., 2000). We note that additional regulatory and scaffolding proteins, most notably of RGS class, may participate in organization, localization, and specific functioning of the GPCR–G protein—effector complexes (Benians, Nobles, Hosny, & Tinker, 2005; Fajardo-Serrano et al., 2013; Labouebe et al., 2007; Schwenk et al., 2010). Nevertheless, the strong coupling within the GPCR–GIRK complexes observed in heterologous systems suggests that the basic components of the GPCR–GIRK cascade, especially the channels and the $G_{i/o}$ protein, are able to form a strong signaling complex. Reconstitution of precoupling of GPCR in such systems (see Luján et al., 2014; Nagi & Pineyro, 2014; Padgett & Slesinger, 2010) seems to require that the GPCR interacts with GIRK, indicating that the complex may be centered around the channel or the GPCR.

Several groups proposed a preformed complex of the whole heterotrimeric G protein (as $G\alpha_{i/o}\beta\gamma$) with GIRK, with or without the GPCR (Clancy et al., 2005; Huang et al., 1995; Peleg et al., 2002; Raveh et al., 2009; Rebois et al., 2006; Riven et al., 2006). The evidence for precoupling between GIRK1-containing channels and $G\beta\gamma$ is quite strong, at least for the best-studied cases of GIRK1/2 and GIRK1/4. Strong binding (Finley et al., 2004; He et al., 2002; Huang et al., 1995; Ivanina et al., 2003; Krapivinsky et al., 1998; Krapivinsky, Krapivinsky, Wickman, & Clapham, 1995; Mirshahi, Mittal, Zhang, Linder, & Logothetis, 2002) and FRET/BRET (Berlin et al., 2011; Rebois et al., 2006; Richard-Lalonde et al., 2013; Riven et al., 2006) between GIRK subunits and $G\beta\gamma$ are well established. The selective recruitment of $G\beta\gamma$ by GIRK1-containing channels also supports precoupling with $G\beta\gamma$ and underscores the importance of specific structural elements (distal CT), but also indicates that GIRK2 homotetramers may not precouple to $G\beta\gamma$ (Kahanovitch et al., 2014). Mackinnon and collaborators (Wang et al., 2014) pointed out that, at least for GIRK2, the affinity of $G\beta\gamma$ to the channel is rather low and local enrichment in the lipid phase, rather than direct high-affinity protein binding, contributes to the fast activation of the channel. Indeed, $G\beta\gamma$ deficient in lipid insertion does not activate GIRKs in excised patches of oocytes and neurons and in artificial membranes (Nakajima, Nakajima, & Kozasa, 1996; Schreibmayer et al., 1996; Wang et al., 2014). In fact, this argument may apply to all GIRKs. However, there may be additional mechanisms that underlie the inefficient activation of some GIRKs by nonlipidated $G\beta\gamma$. For GIRK1, the inhibitory element (lock) at the end of the distal CT might prevent activation by improperly located, nonmembrane attached $G\beta\gamma$ (Kahanovitch et al., 2014; Luchian et al., 1997). Another mechanism may involve $G\gamma$. There seems to be a requirement for $G\gamma$ subunit for full GIRK activation (Kawano et al., 1999) which, theoretically, might require prenylation of Gγ. Overall, the evidence for GIRK1–Gβγ association in living cells stands strong, although the complex may be dynamic rather than nondissociating (see below) and might require additional cellular elements such as cytoskeleton.

In comparison to Gβγ, precoupling between $G\alpha_{i/o}$ and GIRK is uncertain: (1) No FRET between $G\alpha_{i/o}$ and GIRKs is observed in expression systems (Berlin et al., 2011; Fowler, Aryal, Suen, & Slesinger, 2006), even within a strong complex with the δ-opioid receptor (Richard-Lalonde et al., 2013); (2) GIRK1 binds $G\alpha_i$ much less well than Gβγ (Berlin et al., 2011; Mase et al., 2012); (3) GIRK1/2 and GIRK2 do not recruit Gα to the membrane (Kahanovitch et al., 2014); (4) kinetics of Gβγ sequestration by $G\alpha_{i1}{}^{GDP}$ "away" from GIRK2 in lipid bilayer suggest that Gα does not stay bound to the channel (Wang et al., 2014). Together with functional data indicating lack of Gα associated with GIRK1/2 at high expression levels in heterologous systems (see above), these findings suggest that $G\alpha_{i/o}$ is a nonobligatory and potentially regulated part of the complex.

5.2 GIRK Channel and G Protein Subunits: Unique Association and Stoichiometry Rules

The preferential association of GIRK1-containing channels with Gβγ over Gα, and the potential absence of preassociation in homomeric GIRK2, calls for a revision of the popular model in which GIRKs are permanently precoupled with the "inactive" Gαβγ heterotrimer. We propose that GIRK1/2, and possibly other GIRK1-containing channels, operates within dynamic complexes with variable stoichiometries of Gαβγ and Gβγ (Kahanovitch et al., 2014). The organization of the basic signaling complex (without auxiliary proteins such as RGS, or a precoupled GPCR) and its regulation by channel overexpression and by activation of GPCR are presented schematically in Fig. 5. The scheme is related to GIRK1/2 and is based largely on heterologous expression studies, but we believe that the main principles of regulation by Gβγ and Gα are maintained in native cells.

The model incorporates the two-site hypothesis (He et al., 1999; Rubinstein et al., 2009), which proposes the existence of a high-affinity Gβγ anchoring site, and a lower-affinity site for Gβγ activation of the channel. The anchoring site comprises part of the distal CT of GIRK1 and other unidentified parts of the cytosolic domain (Kahanovitch et al., 2014), whereas the activation sites are located within the core of the cytosolic domain of all GIRK subunits, as suggested by structural and modeling studies (Mahajan et al., 2013; Whorton & MacKinnon, 2013; Yokogawa et al., 2011). In this basic "minimal complex," GIRK1/2 is strongly but reversibly (dynamically) associated with Gβγ, whereas $G\alpha^{GDP}$ is dispensable. The anchoring site may harbor free Gβγ or, potentially, the Gαβγ heterotrimer. It is plausible that Gα forms additional contacts with the channel and participates in anchoring (Clancy et al., 2005). However, because of the reasons presented in Section 5.1, among them the absence of FRET between Gα and GIRK, we believe that anchoring occurs mainly via GIRK contacts with Gβγ (Berlin et al., 2011; Kahanovitch et al., 2014). Since the binding of Gβγ to the anchoring site is reversible, the docked Gβγ molecules that are not associated with $G\alpha^{GDP}$ can dissociate, reassociate, and/or shift from the anchoring to the activation site (especially if the two sites partly overlap), promoting channel opening. When present, $G\alpha^{GDP}$ primes the channel for Gβγ activation by offsetting the excess of channel-associated Gβγ and reducing I_{basal} (Rubinstein et al., 2007).

To assess the amount of Gβγ and Gα available for GIRK1/2, we measured macroscopic and single-channel currents and surface densities of GIRK1/2 channels and Gβγ subunits in *Xenopus* oocytes, and employed mathematical modeling based on a previously published model (Yakubovich, Rishal, & Dascal, 2005), with modifications (D. Yakubovich, S. Berlin, U. K., & N. D., unpublished results). Our data and calculations suggest that there are at least three Gβγ molecules available for GIRK1/2 activation under all conditions tested. In contrast, the amount of available $G\alpha_{i/o}$ is at most two molecules per channel, and it further declines as the amount of expressed channel is increased. The estimate of two Gα molecules interacting with a tetrameric GIRK is in line with the model based on NMR study (Mase et al., 2012) discussed previously. This unique stoichiometry, and its channel density dependence, underlies the high I_{basal} and the inverse I_{evoked}/I_{basal} relation seen upon titrated expression of this channel (see Fig. 5).

The cooperative gating by Gβγ is another important factor underlying the complex regulation of channel function by both $G\alpha^{GDP}$ and Gβγ. Given the excess of Gβγ over $G\alpha^{GDP}$ in the resting state, there may be between 1 and 3 "free" Gβγ molecules available to activate channel. This gives a Gβγ-dependent I_{basal} that is 1–26% of $P_{o,max}$ (Ivanova-Nikolova et al., 1998; Sadja et al., 2002) (see Fig. 2). Activation by agonist or coexpression of Gβγ can further activate the channel, up to full $P_{o,max}$. We stress an important consequence of the constitutive preactivation and graded contribution by Gβγ: Gβγ biding to just one additional activation site, on top of one or two already bound Gβγ molecules, already activates the channel several fold (e.g., fourfold when going from two to three Gβγ-occupied activation sites). This makes the activation sensitive to modest changes in Gβγ levels even below saturation and could also contribute to speed of activation (no need to bind all four Gβγ). Anchoring imparts high local concentration of Gβγ and further ensures fast activation kinetics. Finally, the shift of Gβγ from the docking to the activation site may be very fast, especially if anchoring and activation sites partially overlap. This concurs with the notion that dissociation of Gβγ from Gα, rather than Gβγ binding, is the rate-limiting step in GIRK1/2 or GIRK1/4 activation (see above Lohse, Hein, et al., 2008; Lohse, Nikolaev, et al., 2008). In summary, we propose that the unique, unequal association of GIRK1/2 with G protein subunits, and the cooperative nature of GIRK gating by Gβγ, underlies the complex pattern of basal and agonist-evoked activities and allow GIRK1/2 to act as a sensitive bidirectional detector of both Gβγ and $G\alpha^{GDP}$.

5.3 Specificity of GPCR–GIRK Signaling: Relative Roles of Gα and Gβγ

Specificity is among the most complicated issues in GIRK physiology. It is still not clear how only certain GPCRs, via certain G proteins, activate GIRK. The factors that determine specificity of signaling may lay at all levels of complexity, from single amino acids in specific binding sites, to specific signaling complexes.

In native cells only GPCRs coupled to $G_{i/o}$ activate GIRK, thus Gα is essential for specific signaling (Bender, Wellner-Kienitz, Meyer, & Pott, 1998; Digby, Sethi, & Lambert, 2008; Leaney et al., 2000; Leaney & Tinker, 2000; Rusinova et al., 2007). Gα distal C-terminal domain swapping, to revert GPCR coupling from $G\alpha_q$ or $G\alpha_s$ to $G\alpha_{i/o}$, showed that the identity of the GPCR is essentially unimportant as long as it activates $G\alpha_{i/o}$ (Leaney et al., 2000; Rusinova et al., 2007). All members of the $G\alpha_{i/o}$ class, including the PTX-insensitive $G\alpha_z$ (Jeong & Ikeda, 1998) are able to donate Gβγ for GIRK activation (reviewed in Dascal, 2001). On the other hand, nonnegligible differences have been detected in binding and coupling of different Gα's to GIRKs, and in Gβγ and Gα binding to different GIRK subunits (Digby et al., 2008; Ivanina et al., 2003, 2004; Krapivinsky et al., 1998). Moderate variations in Gβγ and RGS sensitivity of GIRK2/3 versus GIRK1/3 or GIRK2 confer major differences in effect of GABA and of a drug of abuse (GHB), via $GABA_B$ receptor, in the ventral tegmental area (Cruz et al., 2004; Labouebe et al., 2007). Thus, diversity and specificity in GPCR–GIRK cascade depend on interactions at several interfaces: GPCR–Gαβγ, Gα–GIRK, Gβγ–GIRK, and GPCR–GIRK; additionally, auxiliary and scaffolding proteins may play a major role in determining signaling selectivity in some GPCR–GIRK cascades.

Most combinations of Gβ and Gγ can activate GIRK with similar potency (Wickman et al., 1994), with the exclusion of $G\beta_5$. The latter, when associated with Gγ subunits, has been found to bind GIRK and compete with GIRK-activating Gβ ($G\beta_1$ through $G\beta_4$), inhibiting GIRKs (Lei et al., 2003, 2000). Nevertheless, recent works suggest that, when paired with its natural partners of the RGS R7 subfamily (Slepak, 2009) and often with the scaffolding R7-binding protein, $G\beta_5$ is a genuine partner in the GPCR–GIRK cascade. $G\beta_5$/R7 complexes do not activate GIRKs but regulate the deactivation of neurotransmitter-evoked GIRK currents (I_{evoked}) in cardiac and neuronal cells (Fajardo-Serrano et al., 2013; Ostrovskaya et al., 2014; Posokhova, Wydeven, Allen, Wickman, & Martemyanov, 2010; Xie et al., 2010; Zhou et al., 2012).

The PTX-insensitive G proteins $G\alpha_s$ and $G\alpha_q$ normally do not donate Gβγ for the activation of GIRKs. However, G_s can activate GIRK, via the standard Gβγ-mediated mechanism, under conditions of overexpression of $G\alpha_s$ or the G_s-coupled β-adrenergic receptor, both in model cells and cardiomyocytes (Digby et al., 2008; Hatcher-Solis et al., 2014; Lim et al., 1995; Wellner-Kienitz et al., 2001). This activation is not due to a promiscuous β-adrenergic receptor coupling to $G\alpha_{i/o}$ (Leaney et al., 2000), because it is not suppressed by PTX (Digby et al., 2008; Hatcher-Solis et al., 2014). The proposed reasons for the "reluctant" activation via G_s are absence of interaction of $G\alpha_s$ with GIRK (Clancy et al., 2005) (only the CT of GIRK2 was examined), distinct localization of G_s but not GIRKs to membrane caveolae/rafts (Schwarzer et al., 2010) (but see Cui, Ho, Kim, & Cho, 2010), or the slow kinetics of dissociation of $G\alpha_s$ from Gβγ compared to $G\alpha_{i/o}$ (Digby et al., 2008). It can be that a combination of all factors is what prevents from $G\alpha_s$ to be a natural donor of Gβγ.

GPCRs coupled to G_q never activate GIRK (Ruiz-Velasco & Ikeda, 1998; Rusinova et al., 2007; Sharon et al., 1997); rather, they inhibit GIRKs, mainly through the two pathways mentioned before (see Fig. 3). There may be additional reasons why Gβγ derived from G_q does not activate GIRK. $G\alpha_q{}^{GDP}$ and AlF_3-activated $G\alpha_q$ (a mimic to $G\alpha^{GTP}$) bind to NT but not CT of GIRK2 or GIRK4 (Clancy et al., 2005; Koike-Tani et al., 2005). Based on $G\alpha_q$-NT binding, and the fast kinetics of GPCR-induced GIRK inhibition of neuronal GIRK and recombinant heterologously expressed

GIRK1/2, Koike-Tani et al. (2005) proposed that $G\alpha_q$ inhibits GIRK by a direct interaction with the channel. Replacing the helical domain of $G\alpha_q$ with that of a $G\alpha_i$ conferred to G_q chimera binding to the CT of GIRK4, and to G_q the ability to activate GIRK (Rusinova et al., 2007). Thus, the inability of $G\alpha_q$ to directly interact with the CT of GIRKs may be related to $G\alpha$ specificity; by inference, the binding of $G\alpha_{i/o}$ to GIRKs' CT may be important for specific signaling (Clancy et al., 2005; Rusinova et al., 2007).

Remarkable selectivity of GPCR–GIRK coupling is observed *within* the $G\alpha_{i/o}$ family. Antisense knockdown of specific mRNAs or $G\alpha$-specific antibodies suggested that $G\alpha_{i2}$ and $G\alpha_{i3}$ were the best donors of $G\beta\gamma$ for GIRK activation by somatostatin, with an exclusive role for $G\alpha_{i3}$ in AtT-20 pituitary cell line but mainly $G\alpha_{i2}$ in locus coeruleus neurons (Takano, Yasufuku-Takano, Kozasa, Nakajima, & Nakajima, 1997). Experiments in atrial cells and in cardiomyocytes derived from embryonic stem cells utilized targeted inactivation or knockout of $G\alpha$ and RGS subunits. These studies suggested that GIRK1/4 activation by the m2R receptor is mediated by $G\alpha_{i2}$ or $G\alpha_{i3}$ (Fu et al., 2006; Sowell et al., 1997) but not via $G\alpha_o$, despite the abundance of the latter (Luetje, Tietje, Christian, & Nathanson, 1988) and despite its specific participation in m2R coupling to voltage-dependent Ca^{2+} channels in the heart (Valenzuela et al., 1997; Ye, Sowell, Vassilev, Milstone, & Mortensen, 1999). The selective role of $G\alpha_{i2}$ but not $G\alpha_{i1}$ or $G\alpha_o$ in the muscarinic activation of cardiac GIRK has been confirmed in $G\alpha$ knockout mice (Zuberi, Birnbaumer, & Tinker, 2008). The exquisite selectivity observed in the heart cannot be easily reconstituted in heterologous systems, where all $G\alpha_{i/o}$ species can donate $G\beta\gamma$ (Leaney et al., 2000) to activate GIRKs. Nevertheless, moderate but significant selectivity in m2R-coupled GPCR–GIRK cascade was reconstituted in *Xenopus* oocytes through a titrated expression approach: $G\alpha_{i3}$ produced stronger and faster agonist-induced GIRK1/2 responses than $G\alpha_{i1}$, in correlation with $G\alpha_{i3}$'s better binding to N- and C-terminal domains of GIRK1 and GIRK2 (Ivanina et al., 2004). Thus, differential interactions between $G\alpha_{i/o}$ species with GIRKs may contribute to selectivity. However, it appears that scaffolding and regulation by auxiliary proteins such as RGS and R7 RGS/$G\beta_5$ pairs substantially contributes to specificity in $G\alpha_{i/o}$–GIRK coupling in different cells (Fu et al., 2006; Posokhova et al., 2013, 2010; Wydeven, Fernandez de Velasco, et al., 2014; Wydeven, Posokhova, et al., 2014). The study of signaling specificity in the GPCR–GIRK cascade is far from being over, and a better understanding of the underlying complex mechanisms will undoubtedly yield application relevant for the development of specific remedies for GIRK-related disorders.

ACKNOWLEDGMENTS

During the preparation of this chapter, the authors were supported by grants from the US-Israel Binational Science Foundation (grant 2013230) and the Israel Science Fund (grant 237/13). The authors thank Dr. Daniel Yakubovich and Dr. Paul Slesinger for insightful comments.

REFERENCES

Adachi, M., Muroya, K., Asakura, Y., Sugiyama, K., Homma, K., & Hasegawa, T. (2014). Discordant genotype-phenotype correlation in familial hyperaldosteronism type III with KCNJ5 gene mutation: A patient report and review of the literature. *Hormone Research in Pædiatrics*, *82*, 138–142. http://dx.doi.org/10.1159/000358197.

Albsoul-Younes, A. M., Sternweis, P. M., Zhao, P., Nakata, H., Nakajima, S., Nakajima, Y., et al. (2001). Interaction sites of the G protein β subunit with brain G protein-coupled inward rectifier K^+ channel. *Journal of Biological Chemistry*, *276*, 12712–12717.

Al-Salameh, A., Cohen, R., & Desailloud, R. (2014). Overview of the genetic determinants of primary aldosteronism. *The Application of Clinical Genetics*, 7, 67–79. http://dx.doi.org/10.2147/TACG.S45620.

Ambrosio, M., & Lohse, M. J. (2012). Nonequilibrium activation of a G-protein-coupled receptor. *Molecular Pharmacology*, *81*(6), 770–777. http://dx.doi.org/10.1124/mol.112.077693.

Beckmann, C., Rinne, A., Littwitz, C., Mintert, E., Bosche, L. I., Kienitz, M. C., et al. (2008). G protein-activated (GIRK) current in rat ventricular myocytes is masked by constitutive inward rectifier current (IK1). *Cellular Physiology and Biochemistry*, *21*(4), 259–268. http://dx.doi.org/10.1159/000129381000129381 [pii].

Bender, K., Nasrollahzadeh, P., Timpert, M., Liu, B., Pott, L., & Kienitz, M. C. (2008). A role for RGS10 in β-adrenergic modulation of G-protein-activated K^+ (GIRK) channel current in rat atrial myocytes. *Journal of Physiology*, *586*(8), 2049–2060.

Bender, K., Wellner-Kienitz, M. C., Bosche, L. I., Rinne, A., Beckmann, C., & Pott, L. (2004). Acute desensitization of GIRK current in rat atrial myocytes is related to K^+ current flow. *Journal of Physiology*, *561*(Pt. 2), 471–483.

Bender, K., Wellner-Kienitz, M. C., Inanobe, A., Meyer, T., Kurachi, Y., & Pott, L. (2001). Overexpression of monomeric and multimeric GIRK4 subunits in rat atrial myocytes removes fast desensitization and reduces inward rectification of muscarinic K+ current ($I_{K(ACh)}$). Evidence for functional homomeric GIRK4 channels. *Journal of Biological Chemistry*, *276*(31), 28873–28880.

Bender, K., Wellner-Kienitz, M. C., Meyer, T., & Pott, L. (1998). Activation of muscarinic K^+ current by β-adrenergic receptors in cultured atrial myocytes transfected with β1 subunit of heterotrimeric G proteins. *FEBS Letters*, *439*(1-2), 115–120.

Benians, A., Leaney, J. L., Milligan, G., & Tinker, A. (2003). The dynamics of formation and action of the ternary complex revealed in living cells using a G-protein gated K^+ channel as a biosensor. *Journal of Biological Chemistry*, *278*, 10851–10858.

Benians, A., Nobles, M., Hosny, S., & Tinker, A. (2005). Regulators of G-protein signaling form a quaternary complex with the agonist, receptor, and G-protein. A novel explanation for the acceleration of signaling activation kinetics. *Journal of Biological Chemistry*, *280*(14), 13383–13394.

Berlin, S., Keren-Raifman, T., Castel, R., Rubinstein, M., Dessauer, C. W., Ivanina, T., et al. (2010). $G\alpha_i$ and Gβγ jointly regulate the conformations of a Gβγ effector, the the neuronal G-protein activated K^+ channel (GIRK). *Journal of Biological Chemistry*, *285*, 6179–6185.

Berlin, S., Tsemakhovich, V. A., Castel, R., Ivanina, T., Dessauer, C. W., Keren-Raifman, T., et al. (2011). Two distinct aspects of coupling between $G\alpha_i$ and G protein-activated K^+ channel (GIRK) revealed by fluorescently-labeled $G\alpha_{i3}$ subunits. *Journal of Biological Chemistry, 286*, 33223–33235. http://dx.doi.org/10.1074/jbc.M111.271056.

Bettahi, I., Marker, C. L., Roman, I. M., & Wickman, K. (2002). Contribution of the Kir3.1 subunit to the muscarinic-gated atrial potassium channel I_{KACh}. *Journal of Biological Chemistry, 277*, 48282–48288.

Bodhinathan, K., & Slesinger, P. A. (2014). Alcohol modulation of G-protein-gated inwardly rectifying potassium channels: From binding to therapeutics. *Frontiers in Physiology, 5*, 76. http://dx.doi.org/10.3389/fphys.2014.00076.

Bondar, A., & Lazar, J. (2013). Dissociated GαGTP and Gβγ subunits are the major activated form of heterotrimeric $G_{i/o}$ proteins. *Journal of Biological Chemistry, 289*, 1271–1281. http://dx.doi.org/10.1074/jbc.M113.493643.

Breitwieser, G. E., & Szabo, G. (1985). Uncoupling of cardiac muscarinic and beta-adrenergic receptors from ion channels by a guanine nucleotide analogue. *Nature, 317*(6037), 538–540.

Breitwieser, G. E., & Szabo, G. (1988). Mechanism of muscarinic receptor-induced K+ channel activation as revealed by hydrolysis-resistant GTP analogues. *Journal of General Physiology, 91*(4), 469–493.

Bunemann, M., Bucheler, M. M., Philipp, M., Lohse, M. J., & Hein, L. (2001). Activation and deactivation kinetics of α2A- and α2C-adrenergic receptor-activated G protein-activated inwardly rectifying K^+ channel currents. *Journal of Biological Chemistry, 276*(50), 47512–47517. http://dx.doi.org/10.1074/jbc.M108652200.

Bunemann, M., Frank, M., & Lohse, M. J. (2003). Gi protein activation in intact cells involves subunit rearrangement rather than dissociation. *Proceedings of the National Academy of Sciences of the United States of America, 100*(26), 16077–16082.

Calloe, K., Goodrow, R., Olesen, S. P., Antzelevitch, C., & Cordeiro, J. M. (2013). Tissue-specific effects of acetylcholine in the canine heart. *American Journal of Physiology Heart and Circulatory Physiology, 305*(1), H66–H75. http://dx.doi.org/10.1152/ajpheart.00029.2013.

Cha, T. J., Ehrlich, J. R., Chartier, D., Qi, X. Y., Xiao, L., & Nattel, S. (2006). Kir3-based inward rectifier potassium current: Potential role in atrial tachycardia remodeling effects on atrial repolarization and arrhythmias. *Circulation, 113*(14), 1730–1737.

Chachisvilis, M., Zhang, Y.-L., & Frangos, J. A. (2006). G protein-coupled receptors sense fluid shear stress in endothelial cells. *Proceedings of the National Academy of Sciences of the United States of America, 103*(42), 15463–15468. http://dx.doi.org/10.1073/pnas.0607224103.

Chan, K. W., Langan, M. N., Sui, J. L., Kozak, J. A., Pabon, A., Ladias, J. A., et al. (1996). A recombinant inwardly rectifying potassium channel coupled to GTP-binding proteins. *Journal of General Physiology, 107*(3), 381–397.

Chan, K. W., Sui, J. L., Vivaudou, M., & Logothetis, D. E. (1996). Control of channel activity through a unique amino acid residue of a G protein-gated inwardly rectifying K+ channel subunit. *Proceedings of the National Academy of Sciences of the United States of America, 93*(24), 14193–14198.

Chan, K. W., Sui, J. L., Vivaudou, M., & Logothetis, D. E. (1997). Specific regions of heteromeric subunits involved in enhancement of G protein-gated K^+ channel activity. *Journal of Biological Chemistry, 272*(10), 6548–6555.

Charmandari, E., Sertedaki, A., Kino, T., Merakou, C., Hoffman, D. A., Hatch, M. M., et al. (2012). A novel point mutation in the KCNJ5 gene causing primary hyperaldosteronism and early-onset autosomal dominant hypertension. *The Journal of Clinical Endocrinology and Metabolism, 97*(8), E1532–E1539. http://dx.doi.org/10.1210/jc.2012-1334.

Chen, X., & Johnston, D. (2005). Constitutively active G-protein-gated inwardly rectifying K^+ channels in dendrites of hippocampal CA1 pyramidal neurons. *Journal of Neuroscience, 25*(15), 3787–3792.

Cheng, C. J., Sung, C. C., Wu, S. T., Lin, Y. C., Sytwu, H. K., Huang, C. L., et al. (2015). Novel KCNJ5 mutations in sporadic aldosterone-producing adenoma reduce Kir3.4 membrane abundance. *The Journal of Clinical Endocrinology and Metabolism, 100*(1), E155–E163. http://dx.doi.org/10.1210/jc.2014-3009.

Choi, M., Scholl, U. I., Yue, P., Bjorklund, P., Zhao, B., Nelson-Williams, C., et al. (2011). K^+ channel mutations in adrenal aldosterone-producing adenomas and hereditary hypertension. (Research Support, N.I.H., Extramural Research Support, Non-U.S. Gov't)- *Science, 331*(6018), 768–772. http://dx.doi.org/10.1126/science.1198785.

Chuang, H. H., & Chuang, A. Y. (2012). RGS proteins maintain robustness of GPCR-GIRK coupling by selective stimulation of the G protein subunit $G\alpha_o$. *Science Signaling, 5*, ra15.

Chuang, H. H., Yu, M., Jan, Y. N., & Jan, L. Y. (1998). Evidence that the nucleotide exchange and hydrolysis cycle of G proteins causes acute desensitization of G-protein gated inward rectifier K+ channels. *Proceedings of the National Academy of Sciences of the United States of America, 95*(20), 11727–11732.

Chung, H. J., Ge, W. P., Qian, X., Wiser, O., Jan, Y. N., & Jan, L. Y. (2009). G protein-activated inwardly rectifying potassium channels mediate depotentiation of long-term potentiation. *Proceedings of the National Academy of Sciences of the United States of America, 106*(2), 635–640.

Chung, H. J., Qian, X., Ehlers, M., Jan, Y. N., & Jan, L. Y. (2009). Neuronal activity regulates phosphorylation-dependent surface delivery of G protein-activated inwardly rectifying potassium channels. *Proceedings of the National Academy of Sciences of the United States of America, 106*(2), 629–634.

Clancy, S. M., Boyer, S. B., & Slesinger, P. A. (2007). Coregulation of natively expressed pertussis toxin-sensitive muscarinic receptors with G-protein-activated potassium channels. *Journal of Neuroscience, 27*(24), 6388–6399.

Clancy, S. M., Fowler, C. E., Finley, M., Suen, K. F., Arrabit, C., Berton, F., et al. (2005). Pertussis-toxin-sensitive $G\alpha$ subunits selectively bind to C-terminal domain of neuronal GIRK channels: Evidence for a heterotrimeric G-protein-channel complex. *Molecular and Cellular Neuroscience, 28*(2), 375–389.

Clayton, C. C., Xu, M., & Chavkin, C. (2009). Tyrosine phosphorylation of Kir3 following kappa-opioid receptor activation of p38 MAPK causes heterologous desensitization. *Journal of Biological Chemistry, 284*(46), 31872–31881. http://dx.doi.org/10.1074/jbc.M109.053793.

Cooper, A., Grigoryan, G., Guy-David, L., Tsoory, M. M., Chen, A., & Reuveny, E. (2012). Trisomy of the G protein-coupled K^+ channel gene, Kcnj6, affects reward mechanisms, cognitive functions, and synaptic plasticity in mice. *Proceedings of the National Academy of Sciences of the United States of America, 109*(7), 2642–2647. http://dx.doi.org/10.1073/pnas.1109099109.

Coote, J. H. (2013). Myths and realities of the cardiac vagus. *The Journal of Physiology, 591*, 4073–4085. http://dx.doi.org/10.1113/jphysiol.2013.257758.

Corey, S., & Clapham, D. E. (2001). The stoichiometry of $G\beta\gamma$ binding to G-protein-regulated inwardly rectifying K+ channels (GIRKs). *Journal of Biological Chemistry, 276*(14), 11409–11413.

Corey, S., Krapivinsky, G., Krapivinsky, L., & Clapham, D. E. (1998). Number and stoichiometry of subunits in the native atrial G-protein-gated K+ channel, IKACh. *Journal of Biological Chemistry, 273*(9), 5271–5278.

Cruz, H. G., Ivanova, T., Lunn, M. L., Stoffel, M., Slesinger, P. A., & Luscher, C. (2004). Bi-directional effects of $GABA_B$ receptor agonists on the mesolimbic dopamine system. *Nature Neuroscience, 7*(2), 153–159.

Cui, S., Ho, W. K., Kim, S. T., & Cho, H. (2010). Agonist-induced localization of Gq-coupled receptors and G protein-gated inwardly rectifying K^+ (GIRK) channels to caveolae determines receptor specificity of phosphatidylinositol 4,5-bisphosphate signaling. (Research Support, Non-U.S. Gov't). *Journal of Biological Chemistry*, *285*(53), 41732–41739. http://dx.doi.org/10.1074/jbc.M110.153312.

Dascal, N. (1997). Signalling via the G protein-activated K^+ channels. *Cellular Signalling*, *9*(8), 551–573.

Dascal, N. (2001). Ion-channel regulation by G proteins. *Trends in Endocrinology and Metabolism*, *12*(9), 391–398.

Davare, M. A., Avdonin, V., Hall, D. D., Peden, E. M., Burette, A., Weinberg, R. J., et al. (2001). A β2 adrenergic receptor signaling complex assembled with the Ca^{2+} channel $Ca_v1.2$. *Science*, *293*(5527), 98–101.

Digby, G. J., Sethi, P. R., & Lambert, N. A. (2008). Differential dissociation of G protein heterotrimers. *Journal of Physiology*, *586*(14), 3325–3335.

Dobrev, D., Friedrich, A., Voigt, N., Jost, N., Wettwer, E., Christ, T., et al. (2005). The G protein-gated potassium current IK, ACh is constitutively active in patients with chronic atrial fibrillation. *Circulation*, *112*(24), 3697–3706.

Donaldson, M. R., Yoon, G., Fu, Y. H., & Ptacek, L. J. (2004). Andersen-Tawil syndrome: A model of clinical variability, pleiotropy, and genetic heterogeneity. *Annals of Medicine*, *36*(Suppl. 1), 92–97.

Doupnik, C. A. (2008). GPCR-Kir channel signaling complexes: Defining rules of engagement. *Journal of Receptor and Signal Transduction Research*, *28*(1-2), 83–91. http://dx.doi.org/10.1080/10799890801941970.

Doupnik, C. A., Davidson, N., Lester, H. A., & Kofuji, P. (1997). RGS proteins reconstitute the rapid gating kinetics of Gβγ-activated inwardly rectifying K^+ channels. *Proceedings of the National Academy of Sciences of the United States of America*, *94*(19), 10461–10466.

Drici, M.-D., Diochot, S., Terrenoire, C., Romey, G., & Lazdunski, M. (2000). The bee venom peptide tertiapin underlines the role of IKACh in acetylcholine-induced atrioventricular blocks. *British Journal of Pharmacology*, *131*(3), 569–577. http://dx.doi.org/10.1038/sj.bjp.0703611.

Ehrlich, J. R., Cha, T. J., Zhang, L., Chartier, D., Villeneuve, L., Hebert, T. E., et al. (2004). Characterization of a hyperpolarization-activated time-dependent potassium current in canine cardiomyocytes from pulmonary vein myocardial sleeves and left atrium. *Journal of Physiology*, *557*(Pt. 2), 583–597.

Fajardo-Serrano, A., Wydeven, N., Young, D., Watanabe, M., Shigemoto, R., Martemyanov, K. A., et al. (2013). Association of Rgs7/Gβ5 complexes with Girk channels and $GABA_B$ receptors in hippocampal CA1 pyramidal neurons. *Hippocampus*, *23*(12), 1231–1245. http://dx.doi.org/10.1002/hipo.22161.

Falkenburger, B. H., Jensen, J. B., & Hille, B. (2010). Kinetics of M1 muscarinic receptor and G protein signaling to phospholipase C in living cells. *Journal of General Physiology*, *135*(2), 81–97. http://dx.doi.org/10.1085/jgp.200910344jgp.200910344 [pii].

Farhy Tselnicker, I., Tsemakhovich, V., Rishal, I., Kahanovitch, U., Dessauer, C. W., & Dascal, N. (2014). Dual regulation of G proteins and the G protein-activated K^+ channels by lithium. *Proceedings of the National Academy of Sciences of the United States of America*, *111*(13), 5018–5023. http://dx.doi.org/10.1073/pnas.1316425111.

Ferguson, S. S. (2001). Evolving concepts in G protein-coupled receptor endocytosis: The role in receptor desensitization and signaling. *Pharmacological Reviews*, *53*(1), 1–24.

Fernandes-Rosa, F. L., Amar, L., Tissier, F., Bertherat, J., Meatchi, T., Zennaro, M.-C., et al. (2015). Functional histopathological markers of aldosterone producing adenoma and somatic KCNJ5 mutations. *Molecular and Cellular Endocrinology*, *408*, 220–226. http://dx.doi.org/10.1016/j.mce.2015.01.020.

Fields, T. A., & Casey, P. J. (1997). Signalling functions and biochemical properties of pertussis toxin-resistant G-proteins. *Biochemical Journal, 321*(Pt. 3), 561–571.

Finley, M., Arrabit, C., Fowler, C., Suen, K. F., & Slesinger, P. A. (2004). βL-βM loop in the C-terminal domain of G protein-activated inwardly rectifying K^+ channels is important for Gβγ subunit activation. *Journal of Physiology, 555*(Pt. 3), 643–657.

Ford, C. E., Skiba, N. P., Bae, H., Daaka, Y., Reuveny, E., Shekter, L. R., et al. (1998). Molecular basis for interactions of G protein betagamma subunits with effectors. *Science, 280*(5367), 1271–1274.

Fowler, C. E., Aryal, P., Suen, K. F., & Slesinger, P. A. (2006). Evidence for association of $GABA_B$ receptors with Kir3 channels and RGS4 proteins. *Journal of Physiology (London), 580*, 51–65.

Frank, M., Thumer, L., Lohse, M. J., & Bunemann, M. (2005). G Protein activation without subunit dissociation depends on a Gαi-specific region. *Journal of Biological Chemistry, 280*(26), 24584–24590.

Fu, Y., Huang, X., Zhong, H., Mortensen, R. M., D'Alecy, L. G., & Neubig, R. R. (2006). Endogenous RGS proteins and Gα subtypes differentially control muscarinic and adenosine-mediated chronotropic effects. *Circulation Research, 98*(5), 659–666.

Funder, J. W. (2014). Genetics of primary aldosteronism. *Frontiers of Hormone Research, 43*, 70–78. http://dx.doi.org/10.1159/000360870.

Gaborit, N., Le Bouter, S., Szuts, V., Varro, A., Escande, D., Nattel, S., et al. (2007). Regional and tissue specific transcript signatures of ion channel genes in the non-diseased human heart. *Journal of Physiology, 582*(Pt. 2), 675–693. http://dx.doi.org/10.1113/jphysiol.2006.126714.

Gales, C., Rebois, R. V., Hogue, M., Trieu, P., Breit, A., Hebert, T. E., et al. (2005). Real-time monitoring of receptor and G-protein interactions in living cells. *Nature Methods, 2*(3), 177–184.

Gales, C., Van Durm, J. J. J., Schaak, S., Pontier, S., Percherancier, Y., Audet, M., et al. (2006). Probing the activation-promoted structural rearrangements in preassembled receptor-G protein complexes. *Nature Structural and Molecular Biology, 13*(9), 778–786. http://dx.doi.org/10.1038/nsmb1134.

Gibson, S. K., & Gilman, A. G. (2006). Giα and Gβ subunits both define selectivity of G protein activation by α2-adrenergic receptors. *Proceedings of the National Academy of Sciences of the United States of America, 103*(1), 212–217.

Gilman, A. G. (1987). G proteins: Transducers of receptor-generated signals. *Annual Review of Biochemistry, 56*, 615–649.

Golebiewska, U., & Scarlata, S. (2008). $G\alpha_q$ binds two effectors separately in cells: Evidence for predetermined signaling pathways. *Biophysical Journal, 95*(5), 2575–2582.

Grasser, E., Steinecker, B., Ahammer, H., & Schreibmayer, W. (2008). Subunit stoichiometry of heterologously expressed G-protein activated inwardly rectifying potassium channels analysed by fluorescence intensity ratio measurement. (Research Support, Non-U.S. Gov't). *Pflügers Archiv, 455*(6), 1017–1024. http://dx.doi.org/10.1007/s00424-007-0358-4.

Guyon, A. (2014). CXCL12 chemokine and its receptors as major players in the interactions between immune and nervous systems. *Frontiers in Cellular Neuroscience, 8*, 65. http://dx.doi.org/10.3389/fncel.2014.00065.

Hansen, S. B. (2015). Lipid agonism: The PIP_2 paradigm of ligand-gated ion channels. *Biochimica et Biophysica Acta (BBA)—Molecular and Cell Biology of Lipids, 1851*(5), 620–628. http://dx.doi.org/10.1016/j.bbalip.2015.01.011.

Hatcher-Solis, C., Fribourg, M., Spyridaki, K., Younkin, J., Ellaithy, A., Xiang, G., et al. (2014). G protein-coupled receptor signaling to Kir channels in *Xenopus* oocytes. *Current Pharmaceutical Biotechnology, 15*(10), 987–995.

He, C., Yang, X., Zhang, H., Mirshahi, T., Jin, T., Huang, A., et al. (2002). Identification of critical residues controlling GIRK channel activity through interactions with the βγ subunits of G proteins. *Journal of Biological Chemistry*, *277*, 6088–6096.

He, C., Zhang, H., Mirshahi, T., & Logothetis, D. E. (1999). Identification of a potassium channel site that interacts with G protein βγ subunits to mediate agonist-induced signaling. *Journal of Biological Chemistry*, *274*(18), 12517–12524.

Hein, P., & Bunemann, M. (2009). Coupling mode of receptors and G proteins. *Naunyn-Schmiedeberg's Archives of Pharmacology*, *379*(5), 435–443.

Hein, P., Frank, M., Hoffmann, C., Lohse, M. J., & Bunemann, M. (2005). Dynamics of receptor/G protein coupling in living cells. *EMBO Journal*, *24*(23), 4106–4114.

Hibino, H., Inanobe, A., Furutani, K., Murakami, S., Findlay, I., & Kurachi, Y. (2010). Inwardly rectifying potassium channels: Their structure, function, and physiological roles. *Physiological Reviews*, *90*(1), 291–366.

Hill, J. J., & Peralta, E. G. (2001). Inhibition of a Gi-activated potassium channel (GIRK1/4) by the Gq-coupled m1 muscarinic acetylcholine receptor. *Journal of Biological Chemistry*, *276*(8), 5505–5510.

Hille, B. (1992). G protein-coupled mechanisms and nervous signaling. *Neuron*, *9*(2), 187–195.

Hille, B. (2002). *Ion channels of excitable membranes*. Sunderland: Sinauer.

Hille, B., Dickson, E., Kruse, M., & Falkenburger, B. (2014). Dynamic metabolic control of an ion channel. *Progress in Molecular Biology and Translational Science*, *123*, 219–247.

Ho, I. H., & Murrell-Lagnado, R. D. (1999). Molecular mechanism for sodium-dependent activation of G protein-gated K^+ channels. *Journal of Physiology (London)*, *520*(Pt. 3), 645–651.

Hoffmann, C., Gaietta, G., Bunemann, M., Adams, S. R., Oberdorff-Maass, S., Behr, B., et al. (2005). A FlAsH-based FRET approach to determine G protein-coupled receptor activation in living cells. *Nature Methods*, *2*(3), 171–176.

Hofmann, K. P., Scheerer, P., Hildebrand, P. W., Choe, H. W., Park, J. H., Heck, M., et al. (2009). A G protein-coupled receptor at work: The rhodopsin model. *Trends in Biochemical Sciences*, *34*(11), 540–552. http://dx.doi.org/10.1016/j.tibs.2009.07.005.

Hollinger, S., & Hepler, J. R. (2002). Cellular regulation of RGS proteins: Modulators and integrators of G protein signaling. *Pharmacological Reviews*, *54*(3), 527–559.

Hosoya, Y., Yamada, M., Ito, H., & Kurachi, Y. (1996). A functional model for G protein activation of the muscarinic K^+ channel in guinea pig atrial myocytes. Spectral analysis of the effect of GTP on single-channel kinetics. *Journal of General Physiology*, *108*(6), 485–495.

Huang, C. L., Feng, S. Y., & Hilgemann, D. W. (1998). Direct activation of inward rectifier potassium channels by PIP2 and its stabilization by Gβγ. *Nature*, *391*(6669), 803–806.

Huang, C. L., Jan, Y. N., & Jan, L. Y. (1997). Binding of the G protein βγ subunit to multiple regions of G protein-gated inward-rectifying K^+ channels. *FEBS Letters*, *405*(3), 291–298.

Huang, C. L., Slesinger, P. A., Casey, P. J., Jan, Y. N., & Jan, L. Y. (1995). Evidence that direct binding of Gβγ to the GIRK1 G protein-gated inwardly rectifying K^+ channel is important for channel activation. *Neuron*, *15*(5), 1133–1143.

Inanobe, A., & Kurachi, Y. (2014). Membrane channels as integrators of G-protein-mediated signaling. *Biochimica et Biophysica Acta (BBA)—Biomembranes*, *1838*(2), 521–531. http://dx.doi.org/10.1016/j.bbamem.2013.08.018.

Inanobe, A., Matsuura, T., Nakagawa, A., & Kurachi, Y. (2007). Structural diversity in the cytoplasmic region of G protein-gated inward rectifier K^+ channels. *Channels*, *1*(1), 39–45.

Ippolito, D. L., Temkin, P. A., Rogalski, S. L., & Chavkin, C. (2002). N-terminal tyrosine residues within the potassium channel Kir3 modulate GTPase activity of $G\alpha_i$. *Journal of Biological Chemistry*, *277*(36), 32692–32696. http://dx.doi.org/10.1074/jbc.M204407200.

Ishihara, K., Yamamoto, T., & Kubo, Y. (2009). Heteromeric assembly of inward rectifier channel subunit Kir2.1 with Kir3.1 and with Kir3.4. *Biochemical and Biophysical Research Communications*, *380*(4), 832–837.

Ito, H., Ono, K., & Noma, A. (1994). Background conductance attributable to spontaneous opening of muscarinic K+ channels in rabbit sino-atrial node cells. *Journal of Physiology (London)*, *476*(1), 55–68.

Ito, H., Sugimoto, T., Kobayashi, I., Takahashi, K., Katada, T., Ui, M., et al. (1991). On the mechanism of basal and agonist-induced activation of the G protein-gated muscarinic K^+ channel in atrial myocytes of guinea pig heart. *Journal of General Physiology*, *98*(3), 517–533.

Ito, H., Tung, R. T., Sugimoto, T., Kobayashi, I., Takahashi, K., Katada, T., et al. (1992). On the mechanism of G protein βγ subunit activation of the muscarinic K^+ channel in guinea pig atrial cell membrane. Comparison with the ATP-sensitive K^+ channel. *Journal of General Physiology*, *99*(6), 961–983.

Ivanina, T., Rishal, I., Varon, D., Mullner, C., Frohnwieser-Steinecke, B., Schreibmayer, W., et al. (2003). Mapping the Gβγ-binding sites in GIRK1 and GIRK2 subunits of the G protein-activated K^+ channel. *Journal of Biological Chemistry*, *278*(31), 29174–29183.

Ivanina, T., Varon, D., Peleg, S., Rishal, I., Porozov, Y., Dessauer, C. W., et al. (2004). $G\alpha_{i1}$ and $G\alpha_{i3}$ differentially interact with, and regulate, the G protein-activated K^+ channel. *Journal of Biological Chemistry*, *279*(17), 17260–17268.

Ivanova-Nikolova, T. T., & Breitwieser, G. E. (1997). Effector contributions to Gβγ-mediated signaling as revealed by muscarinic potassium channel gating. *Journal of General Physiology*, *109*(2), 245–253.

Ivanova-Nikolova, T. T., Nikolov, E. N., Hansen, C., & Robishaw, J. D. (1998). Muscarinic K^+ channel in the heart. Modal regulation by G protein βγ subunits. *Journal of General Physiology*, *112*(2), 199–210.

James, J. R., Oliveira, M. I., Carmo, A. M., Iaboni, A., & Davis, S. J. (2006). A rigorous experimental framework for detecting protein oligomerization using bioluminescence resonance energy transfer. *Nature Methods*, *3*(12), 1001–1006.

Jensen, J. B., Lyssand, J. S., Hague, C., & Hille, B. (2009). Fluorescence changes reveal kinetic steps of muscarinic receptor-mediated modulation of phosphoinositides and $K_v7.2/7.3$ K^+ channels. *Journal of General Physiology*, *133*(4), 347–359. http://dx.doi.org/10.1085/jgp.200810075jgp.200810075 [pii].

Jeong, S. W., & Ikeda, S. R. (1998). G protein alpha subunit $G\alpha_z$ couples neurotransmitter receptors to ion channels in sympathetic neurons. *Neuron*, *21*(5), 1201–1212.

Jeong, S. W., & Ikeda, S. R. (2001). Differential regulation of G protein-gated inwardly rectifying K^+ channel kinetics by distinct domains of RGS8. *Journal of Physiology*, *535*(Pt. 2), 335–347.

Kahanovitch, U., Tsemakhovich, V., Berlin, S., Rubinstein, M., Styr, B., Castel, R., et al. (2014). Recruitment of Gβγ controls the basal activity of GIRK channels: Crucial role of distal C-terminus of GIRK1. *Journal of Physiology*, *592*(24), 5373–5390. http://dx.doi.org/10.1113/jphysiol.2014.283218.

Kamp, T. J., & Hell, J. W. (2000). Regulation of cardiac L-type calcium channels by protein kinase A and protein kinase C. *Circulation Research*, *87*(12), 1095–1102.

Karschin, C., Schreibmayer, W., Dascal, N., Lester, H., Davidson, N., & Karschin, A. (1994). Distribution and localization of a G protein-coupled inwardly rectifying K^+ channel in the rat. *FEBS Letters*, *348*(2), 139–144.

Kawano, T., Chen, L., Watanabe, S. Y., Yamauchi, J., Kaziro, Y., Nakajima, Y., et al. (1999). Importance of the G protein γ subunit in activating G protein-coupled inward rectifier K^+ channels. *FEBS Letters*, *463*(3), 355–359.

Kienitz, M.-C., Mintert-Jancke, E., Hertel, F., & Pott, L. (2014). Differential effects of genetically-encoded Gβγ scavengers on receptor-activated and basal Kir3.1/Kir3.4 channel current in rat atrial myocytes. *Cellular Signalling*, *26*(6), 1182–1192.

Kim, C. S., & Johnston, D. (2015). A1 adenosine receptor-mediated GIRK channels contributes to the resting conductance of CA1 neurons in the dorsal hippocampus. *Journal of Neurophysiology*, *113*(7), 2511–2523.

Kim, M.-S., Pinto, S. M., Getnet, D., Nirujogi, R. S., Manda, S. S., Chaerkady, R., et al. (2014). A draft map of the human proteome. *Nature*, *509*(7502), 575–581. http://dx.doi.org/10.1038/nature13302. http://www.nature.com/nature/journal/v509/n7502/abs/nature13302.html#supplementary-information.

Kobayashi, T., & Ikeda, K. (2006). G protein-activated inwardly rectifying potassium channels as potential therapeutic targets. *Current Pharmaceutical Design*, *12*(34), 4513–4523.

Kobayashi, T., Ikeda, K., Kojima, H., Niki, H., Yano, R., Yoshioka, T., et al. (1999). Ethanol opens G-protein-activated inwardly rectifying K+ channels. *Nature Neuroscience*, *2*(12), 1091–1097.

Kobrinsky, E., Mirshahi, T., Zhang, H., Jin, T., & Logothetis, D. E. (2000). Receptor-mediated hydrolysis of plasma membrane messenger PIP_2 leads to K^+-current desensitization. *Nature Cell Biology*, *2*(8), 507–514.

Koike-Tani, M., Collins, J. M., Kawano, T., Zhao, P., Zhao, Q., Kozasa, T., et al. (2005). Signal transduction pathway for the substance P-induced inhibition of Kir3 (GIRK) channel. *Journal of Physiology*, *564*, 489–500.

Kokunai, Y., Nakata, T., Furuta, M., Sakata, S., Kimura, H., Aiba, T., et al. (2014). A Kir3.4 mutation causes Andersen-Tawil syndrome by an inhibitory effect on Kir2.1. *Neurology*, *82*(12), 1058–1064. http://dx.doi.org/10.1212/wnl.0000000000000239.

Kovoor, P., Wickman, K., Maguire, C. T., Pu, W., Gehrmann, J., Berul, C. I., et al. (2001). Evaluation of the role of I_{KACh} in atrial fibrillation using a mouse knockout model. *Journal of the American College of Cardiology*, *37*(8), 2136–2143.

Krapivinsky, G., Gordon, E. A., Wickman, K., Velimirovic, B., Krapivinsky, L., & Clapham, D. E. (1995). The G-protein-gated atrial K^+ channel I_{KACh} is a heteromultimer of two inwardly rectifying K^+-channel proteins. *Nature*, *374*(6518), 135–141.

Krapivinsky, G., Kennedy, M. E., Nemec, J., Medina, I., Krapivinsky, L., & Clapham, D. E. (1998). Gβ binding to GIRK4 subunit is critical for G protein-gated K^+ channel activation. *Journal of Biological Chemistry*, *273*(27), 16946–16952.

Krapivinsky, G., Krapivinsky, L., Wickman, K., & Clapham, D. E. (1995). Gβγ binds directly to the G protein-gated K^+ channel, I_{KACh}. *Journal of Biological Chemistry*, *270*(49), 29059–29062.

Kuppusamy, M., Caroccia, B., Stindl, J., Bandulik, S., Lenzini, L., Gioco, F., et al. (2014). A novel KCNJ5-insT149 somatic mutation close to, but outside, the selectivity filter causes resistant hypertension by loss of selectivity for potassium. *The Journal of Clinical Endocrinology and Metabolism*, *99*(9), E1765–E1773. http://dx.doi.org/10.1210/jc.2014-1927.

Kurachi, Y., Ito, H., & Sugimoto, T. (1990). Positive cooperativity in activation of the cardiac muscarinic K+ channel by intracellular GTP. *Pflügers Archiv*, *416*(1–2), 216–218.

Kurachi, Y., Nakajima, T., & Sugimoto, T. (1986). On the mechanism of activation of muscarinic K+ channels by adenosine in isolated atrial cells: Involvement of GTP-binding proteins. *Pflügers Archiv*, *407*(3), 264–274.

Labouebe, G., Lomazzi, M., Cruz, H. G., Creton, C., Lujan, R., Li, M., et al. (2007). RGS2 modulates coupling between $GABA_B$ receptors and GIRK channels in dopamine neurons of the ventral tegmental area. *Nature Neuroscience*, *10*(12), 1559–1568.

Lambert, N. A. (2008). Dissociation of heterotrimeric G proteins in cells. *Science Signaling*, *1*(25), re5. http://dx.doi.org/10.1126/scisignal.125re5.

Leal-Pinto, E., Gomez-Llorente, Y., Sundaram, S., Tang, Q.-Y., Ivanova-Nikolova, T., Mahajan, R., et al. (2010). Gating of a G protein-sensitive mammalian Kir3.1—Prokaryotic Kir channel chimera in planar lipid bilayers. *Journal of Biological Chemistry*, *285*, 39790–39800. http://dx.doi.org/10.1074/jbc.M110.151373.

Leaney, J. L., Benians, A., Brown, S., Nobles, M., Kelly, D., & Tinker, A. (2004). Rapid desensitization of G protein-gated inwardly rectifying K^+ currents is determined by G protein cycle. *American Journal of Physiology Cell Physiology*, *287*(1), 10.

Leaney, J. L., Milligan, G., & Tinker, A. (2000). The G protein α subunit has a key role in determining the specificity of coupling to, but not the activation of, G protein-gated inwardly rectifying K^+ channels. *Journal of Biological Chemistry*, *275*(2), 921–929.

Leaney, J. L., & Tinker, A. (2000). The role of members of the pertussis toxin-sensitive family of G proteins in coupling receptors to the activation of the G protein-gated inwardly rectifying potassium channel. *Proceedings of the National Academy of Sciences of the United States of America*, *97*(10), 5651–5656.

Lei, Q., Jones, M. B., Talley, E. M., Garrison, J. C., & Bayliss, D. A. (2003). Molecular mechanisms mediating inhibition of G protein-coupled inwardly-rectifying K^+ channels. *Molecules and Cells*, *15*(1), 1–9.

Lei, Q., Jones, M. B., Talley, E. M., Schrier, A. D., McIntire, W. E., Garrison, J. C., et al. (2000). Activation and inhibition of G protein-coupled inwardly rectifying potassium (Kir3) channels by G protein βγ subunits. *Proceedings of the National Academy of Sciences of the United States of America*, *97*(17), 9771–9776.

Lei, Q., Talley, E. M., & Bayliss, D. A. (2001). Receptor-mediated inhibition of GIRK channels involves $G\alpha_q$ family subunits phospholipase C and a readily diffusible messenger. *Journal of Biological Chemistry*, *276*(20), 16720–16730.

Lewohl, J. M., Wilson, W. R., Mayfield, R. D., Brozowski, S. J., Morrisett, R. A., & Harris, R. A. (1999). G-protein-coupled inwardly rectifying potassium channels are targets of alcohol action. *Nature Neuroscience*, *2*(12), 1084–1090. http://dx.doi.org/10.1038/16012.

Liang, B., Nissen, J. D., Laursen, M., Wang, X., Skibsbye, L., Hearing, M. C., et al. (2014). G-protein coupled inward rectifier potassium current contributes to ventricular repolarization. *Cardiovascular Research*, *101*, 175–184. http://dx.doi.org/10.1093/cvr/cvt240.

Lim, N. F., Dascal, N., Labarca, C., Davidson, N., & Lester, H. A. (1995). A G protein-gated K channel is activated via β2-adrenergic receptors and Gβγ subunits in *Xenopus* oocytes. *Journal of General Physiology*, *105*(3), 421–439.

Linder, M. E., Ewald, D. A., Miller, R. J., & Gilman, A. G. (1990). Purification and characterization of Go alpha and three types of Gi alpha after expression in Escherichia coli. *Journal of Biological Chemistry*, *265*(14), 8243–8251.

Llamosas, N., Bruzos-Cidón, C., Rodríguez, J. J., Ugedo, L., & Torrecilla, M. (2015). Deletion of GIRK2 subunit of GIRK channels alters the 5-HT1A receptor-mediated signaling and results in a depression-resistant behavior. *The International Journal of Neuropsychopharmacology*. in press, http://dx.doi.org/10.1093/ijnp/pyv051.

Lober, R. M., Pereira, M. A., & Lambert, N. A. (2006). Rapid activation of inwardly rectifying potassium channels by immobile G-protein-coupled receptors. *Journal of Neuroscience*, *26*(48), 12602–12608.

Logothetis, D. E., Jin, T., Lupyan, D., & Rosenhouse-Dantsker, A. (2007). Phosphoinositide-mediated gating of inwardly rectifying K^+ channels. *Pflügers Archiv*, *455*, 83–95.

Logothetis, D. E., Kurachi, Y., Galper, J., Neer, E. J., & Clapham, D. E. (1987). The βγ subunits of GTP-binding proteins activate the muscarinic K^+ channel in heart. *Nature*, *325*(6102), 321–326.

Logothetis, D. E., Petrou, V. I., Zhang, M., Mahajan, R., Meng, X. Y., Adney, S. K., et al. (2015). Phosphoinositide control of membrane protein function: A frontier led by studies on ion channels. *Annual Review of Physiology*, 77, 81–104. http://dx.doi.org/10.1146/annurev-physiol-021113-170358.

Lohse, M. J., Hein, P., Hoffmann, C., Nikolaev, V. O., Vilardaga, J. P., & Bunemann, M. (2008). Kinetics of G-protein-coupled receptor signals in intact cells. *British Journal of Pharmacology*, *153*(Suppl. 1), S125–S132.

Lohse, M. J., Maiellaro, I., & Calebiro, D. (2014). Kinetics and mechanism of G protein-coupled receptor activation. *Current Opinion in Cell Biology*, *27*, 87–93. http://dx.doi.org/10.1016/j.ceb.2013.11.009.

Lohse, M. J., Nikolaev, V. O., Hein, P., Hoffmann, C., Vilardaga, J. P., & Bunemann, M. (2008). Optical techniques to analyze real-time activation and signaling of G-protein-coupled receptors. *Trends in Pharmacological Sciences*, *29*(3), 159–165. http://dx.doi.org/10.1016/j.tips.2007.12.002.

Luchian, T., Dascal, N., Dessauer, C., Platzer, D., Davidson, N., Lester, H. A., et al. (1997). A C-terminal peptide of the GIRK1 subunit directly blocks the G protein- activated K^+ channel (GIRK) expressed in Xenopus oocytes. *Journal of Physiology (London)*, *505*(Pt. 1), 13–22.

Luetje, C., Tietje, K., Christian, J., & Nathanson, N. (1988). Differential tissue expression and developmental regulation of guanine nucleotide binding regulatory proteins and their messenger RNAs in rat heart. *Journal of Biological Chemistry*, *263*(26), 13357–13365.

Luján, R., Marron Fernandez de Velasco, E., Aguado, C., & Wickman, K. (2014). New insights into the therapeutic potential of Girk channels. *Trends in Neurosciences*, *37*(1), 20–29. http://dx.doi.org/10.1016/j.tins.2013.10.006.

Lujan, R., Maylie, J., & Adelman, J. P. (2009). New sites of action for GIRK and SK channels. *Nature Reviews Neuroscience*, *10*(7), 475–480.

Luscher, C., Jan, L. Y., Stoffel, M., Malenka, R. C., & Nicoll, R. A. (1997). G protein-coupled inwardly rectifying K^+ channels (GIRKs) mediate postsynaptic but not presynaptic transmitter actions in hippocampal neurons. *Neuron*, *19*(3), 687–695.

Luscher, C., & Slesinger, P. A. (2010). Emerging roles for G protein-gated inwardly rectifying potassium (GIRK) channels in health and disease. *Nature Reviews Neuroscience*, *11*(5), 301–315.

Ma, D., Zerangue, N., Raab-Graham, K., Fried, S. R., Jan, Y. N., & Jan, L. Y. (2002). Diverse trafficking patterns due to multiple traffic motifs in G protein-activated inwardly rectifying potassium channels from brain and heart. *Neuron*, *33*(5), 715–729.

Mahajan, R., Ha, J., Zhang, M., Kawano, T., Kozasa, T., & Logothetis, D. E. (2013). A computational model predicts that Gβγ acts at a cleft between channel subunits to activate GIRK1 channels. *Science Signaling*, *6*(288), ra69. http://dx.doi.org/10.1126/scisignal.2004075.

Makara, J. K., & Magee, J. C. (2013). Variable dendritic integration in hippocampal CA3 pyramidal neurons. *Neuron*, *80*(6), 1438–1450. http://dx.doi.org/10.1016/j.neuron.2013.10.033.

Makary, S., Voigt, N., Maguy, A., Wakili, R., Nishida, K., Harada, M., et al. (2011). Differential protein kinase C isoform regulation and increased constitutive activity of acetylcholine-regulated potassium channels in atrial remodeling. *Circulation Research*, *109*(9), 1031–1043. http://dx.doi.org/10.1161/circresaha.111.253120.

Mao, J., Wang, X., Chen, F., Wang, R., Rojas, A., Shi, Y., et al. (2004). Molecular basis for the inhibition of G protein-coupled inward rectifier K^+ channels by protein kinase C.

Proceedings of the National Academy of Sciences of the United States of America, 101(4), 1087–1092.
Marcaggi, P., Mutoh, H., Dimitrov, D., Beato, M., & Knöpfel, T. (2009). Optical measurement of mGluR1 conformational changes reveals fast activation, slow deactivation, and sensitization. *Proceedings of the National Academy of Sciences of the United States of America, 106*(27), 11388–11393. http://dx.doi.org/10.1073/pnas.0901290106.
Marionneau, C., Couette, B., Liu, J., Li, H., Mangoni, M. E., Nargeot, J., et al. (2005). Specific pattern of ionic channel gene expression associated with pacemaker activity in the mouse heart. *Journal of Physiology, 562*(Pt. 1), 223–234. http://dx.doi.org/10.1113/jphysiol.2004.074047.
Mark, M. D., & Herlitze, S. (2000). G-protein mediated gating of inward-rectifier K^+ channels. *European Journal of Biochemistry, 267*(19), 5830–5836.
Mark, M. D., Ruppersberg, J. P., & Herlitze, S. (2000). Regulation of GIRK channel deactivation by Galpha(q) and Galpha(i/o) pathways. *Neuropharmacology, 39*(12), 2360–2373.
Mase, Y., Yokogawa, M., Osawa, M., & Shimada, I. (2012). Structural basis for the modulation of the gating property of G protein-gated inwardly rectifying potassium ion channel (GIRK) by the i/o-family G protein α subunit ($G\alpha_{i/o}$). *Journal of Biological Chemistry, 287*(23), 19537–19549. http://dx.doi.org/10.1074/jbc.M112.353888.
Medina, I., Krapivinsky, G., Arnold, S., Kovoor, P., Krapivinsky, L., & Clapham, D. E. (2000). A switch mechanism for Gβγ activation of I_{KACh}. *Journal of Biological Chemistry, 275*(38), 29709–29716.
Mesirca, P., Alig, J., Torrente, A. G., Muller, J. C., Marger, L., Rollin, A., et al. (2014). Cardiac arrhythmia induced by genetic silencing of 'funny' (f) channels is rescued by GIRK4 inactivation. *Nature Communications, 5*, 4664. http://dx.doi.org/10.1038/ncomms5664.
Mesirca, P., Marger, L., Toyoda, F., Rizzetto, R., Audoubert, M., Dubel, S., et al. (2013). The G-protein-gated K^+ channel, I_{KACh}, is required for regulation of pacemaker activity and recovery of resting heart rate after sympathetic stimulation. *Journal of General Physiology, 142*(2), 113–126. http://dx.doi.org/10.1085/jgp.201310996.
Meyer, T., Wellner-Kienitz, M. C., Biewald, A., Bender, K., Eickel, A., & Pott, L. (2000). Depletion of phosphatidylinositol 4,5 bisphosphate by activation of PLC-coupled receptors causes slow inhibition but not desensitisation of GIRK current in atrial myocytes. *Journal of Biological Chemistry, 276*, 5650–5658.
Mintert, E., Bösche, L. I., Rinne, A., Timpert, M., Kienitz, M.-C., Pott, L., et al. (2007). Generation of a constitutive Na^+-dependent inward-rectifier current in rat adult atrial myocytes by overexpression of Kir3.4. *Journal of Physiology, 585*(1), 3–13. http://dx.doi.org/10.1113/jphysiol.2007.140772.
Mirshahi, T., Mittal, V., Zhang, H., Linder, M. E., & Logothetis, D. E. (2002). Distinct sites on G protein βγ subunits regulate different effector functions. *Journal of Biological Chemistry, 277*, 36345–36350.
Mirshahi, T., Robillard, L., Zhang, H., Hebert, T. E., & Logothetis, D. E. (2002). Gβ residues that do not interact with Gα underlie agonist-independent activity of K^+ channels. *Journal of Biological Chemistry, 277*(9), 7348–7355.
Monticone, S., Hattangady, N. G., Penton, D., Isales, C., Edwards, M. A., Williams, T. A., et al. (2013). A novel Y152C KCNJ5 mutation responsible for familial hyperaldosteronism type III. *The Journal of Clinical Endocrinology and Metabolism, 98*, E1861–E1865. http://dx.doi.org/10.1210/jc.2013-2428.
Mori, M. X., Erickson, M. G., & Yue, D. T. (2004). Functional stoichiometry and local enrichment of calmodulin interacting with Ca^{2+} channels. *Science, 304*(5669), 432–435.
Mullner, C., Steinecker, B., Gorischek, A., & Schreibmayer, W. (2009). Identification of the structural determinant responsible for the phosphorylation of G-protein activated potassium channel 1 by cAMP-dependent protein kinase. *The FEBS Journal, 276*(21), 6218–6226.

Mullner, C., Vorobiov, D., Bera, A. K., Uezono, Y., Yakubovich, D., Frohnwieser, B., et al. (2000). Heterologous facilitation of G-protein-activated K^+ channels by β-adrenergic stimulation via cAMP dependent protein kinase. *Journal of General Physiology, 115*, 627–633.

Mullner, C., Yakubovich, D., Dessauer, C. W., Platzer, D., & Schreibmayer, W. (2003). Single channel analysis of the regulation of GIRK1/GIRK4 channels by protein phosphorylation. *Biophysical Journal, 84*(2), 1399–1409.

Murakami, S., Inanobe, A., & Kurachi, Y. (2013). Short-term desensitization of muscarinic K^+ current in the heart. *Biophysical Journal, 105*(6), 1515–1525.

Murakami, S., Suzuki, S., Ishii, M., Inanobe, A., & Kurachi, Y. (2010). Cellular modelling: Experiments and simulation to develop a physiological model of G-protein control of muscarinic K^+ channels in mammalian atrial cells (Research Support, Non-U.S. Gov't). *Philosophical Transactions Series A, Mathematical, Physical, and Engineering Sciences, 368*, 2983–3000. http://dx.doi.org/10.1098/rsta.2010.0093.

Murthy, M., Xu, S., Massimo, G., Wolley, M., Gordon, R. D., Stowasser, M., et al. (2014). Role for germline mutations and a rare coding single nucleotide polymorphism within the KCNJ5 potassium channel in a large cohort of sporadic cases of primary aldosteronism. *Hypertension, 63*(4), 783–789. http://dx.doi.org/10.1161/hypertensionaha.113.02234.

Nagi, K., & Pineyro, G. (2014). Kir3 channel signaling complexes: Focus on opioid receptor signaling. *Frontiers in Cellular Neuroscience, 8*, 186. http://dx.doi.org/10.3389/fncel.2014.00186.

Nakajima, Y., Nakajima, S., & Kozasa, T. (1996). Activation of G protein-coupled inward rectifier K^+ channels in brain neurons requires association of G protein βγ subunits with cell membrane. *FEBS Letters, 390*(2), 217–220.

Neubig, R. R. (1994). Membrane organization in G-protein mechanisms (Research Support, Non-U.S. Gov't Research Support, U.S. Gov't, P.H.S. Review). *FASEB Journal, 8*(12), 939–946.

Nikolov, E. N., & Ivanova-Nikolova, T. (2004a). Functional characterization of a small conductance GIRK channel in rat atrial cells. *Biophysical Journal, 87*, 3122–3136.

Nikolov, E. N., & Ivanova-Nikolova, T. T. (2004b). Coordination of membrane excitability through a GIRK1 signaling complex in the atria. *Journal of Biological Chemistry, 279*(22), 23630–23636.

Nishida, M., Cadene, M., Chait, B. T., & Mackinnon, R. (2007). Crystal structure of a Kir3.1-prokaryotic Kir channel chimera. *EMBO Journal, 26*(17), 4005–4015.

Nishida, M., & MacKinnon, R. (2002). Structural basis of inward rectification. Cytoplasmic pore of the G protein-gated inward rectifier GIRK1 at 1.8 A resolution. *Cell, 111*(7), 957–965.

Nobles, M., Benians, A., & Tinker, A. (2005). Heterotrimeric G proteins precouple with G protein-coupled receptors in living cells. *Proceedings of the National Academy of Sciences of the United States of America, 102*(51), 18706–18711.

Noma, A., & Trautwein, W. (1978). Relaxation of the ACh-induced potassium current in the rabbit sinoatrial node cell. *Pflügers Archiv, 377*(3), 193–200.

Nygaard, R., Zou, Y., Dror, R. O., Mildorf, T. J., Arlow, D. H., Manglik, A., et al. (2013). The dynamic process of β2-adrenergic receptor activation. *Cell, 152*(3), 532–542. http://dx.doi.org/10.1016/j.cell.2013.01.008.

Ogier-Denis, E., Houri, J.-J., Bauvy, C., & Codogno, P. (1996). Guanine nucleotide exchange on heterotrimeric Gi3 protein controls autophagic sequestration in HT-29 cells. *Journal of Biological Chemistry, 271*(45), 28593–28600.

Ostrovskaya, O., Xie, K., Masuho, I., Fajardo-Serrano, A., Lujan, R., Wickman, K., et al. (2014). RGS7/Gβ5/R7BP complex regulates synaptic plasticity and memory by modulating hippocampal $GABA_BR$-GIRK signaling. *eLife, 3*, e02053. http://dx.doi.org/10.7554/eLife.02053.

Padgett, C. L., & Slesinger, P. A. (2010). GABAB receptor coupling to G-proteins and ion channels. *Advances in Pharmacology, 58*, 123–147. http://dx.doi.org/10.1016/s1054-3589(10)58006-2.

Pegan, S., Arrabit, C., Zhou, W., Kwiatkowski, W., Collins, A., Slesinger, P. A., et al. (2005). Cytoplasmic domain structures of Kir2.1 and Kir3.1 show sites for modulating gating and rectification. *Nature Neuroscience, 8*(3), 279–287.

Peleg, S., Varon, D., Ivanina, T., Dessauer, C. W., & Dascal, N. (2002). $G\alpha_i$ controls the gating of the G-protein-activated K^+ channel, GIRK. *Neuron, 33*, 87–99.

Pfaffinger, P. J., Martin, J. M., Hunter, D. D., Nathanson, N. M., & Hille, B. (1985). GTP-binding proteins couple cardiac muscarinic receptors to a K channel. *Nature, 317*(6037), 536–538.

Posokhova, E., Ng, D., Opel, A., Masuho, I., Tinker, A., Biesecker, L. G., et al. (2013). Essential role of the m2R-RGS6-I_{KACh} pathway in controlling intrinsic heart rate variability. *PLoS One, 8*(10), e76973. http://dx.doi.org/10.1371/journal.pone.0076973.

Posokhova, E., Wydeven, N., Allen, K. L., Wickman, K., & Martemyanov, K. A. (2010). RGS6/Gβ5 complex accelerates I_{KACh} gating kinetics in atrial myocytes and modulates parasympathetic regulation of heart rate. *Circulation Research, 107*(11), 1350–1354. http://dx.doi.org/10.1161/circresaha.110.224212.

Pott, L. (1979). On the time course of the acetylcholine-induced hyperpolarization in quiescent guinea-pig atria. *Pflügers Archiv, 380*(1), 71–77.

Qin, K., Dong, C., Wu, G., & Lambert, N. A. (2011). Inactive-state preassembly of G_q-coupled receptors and G_q heterotrimers. *Nature Chemical Biology*, 7(10), 740–747. http://dx.doi.org/10.1038/nchembio.642.

Qin, K., Sethi, P. R., & Lambert, N. A. (2008). Abundance and stability of complexes containing inactive G protein-coupled receptors and G proteins (Research Support, N.I.H., Extramural Research Support, U.S. Gov't, Non-P.H.S.). *FASEB Journal, 22*(8), 2920–2927. http://dx.doi.org/10.1096/fj.08-105775.

Raveh, A., Cooper, A., Guy-David, L., & Reuveny, E. (2010). Nonenzymatic rapid control of GIRK channel function by a G protein-coupled receptor kinase. *Cell, 143*(5), 750–760. http://dx.doi.org/10.1016/j.cell.2010.10.018.

Raveh, A., Riven, I., & Reuveny, E. (2009). Elucidating the gating of the GIRK channel using spectroscopic approach. *Journal of Physiology, 587*, 5331–5335.

Rebois, R. V., & Hebert, T. E. (2003). Protein complexes involved in heptahelical receptor-mediated signal transduction. *Receptors & Channels, 9*(3), 169–194.

Rebois, R. V., Robitaille, M., Gales, C., Dupre, D. J., Baragli, A., Trieu, P., et al. (2006). Heterotrimeric G proteins form stable complexes with adenylyl cyclase and Kir3.1 channels in living cells. *Journal of Cell Science, 119*(Pt. 13), 2807–2818.

Reuveny, E., Slesinger, P. A., Inglese, J., Morales, J. M., Iniguez-Lluhi, J. A., Lefkowitz, R. J., et al. (1994). Activation of the cloned muscarinic potassium channel by G protein βγ subunits. *Nature, 370*(6485), 143–146.

Richard-Lalonde, M., Nagi, K., Audet, N., Sleno, R., Amraei, M., Hogue, M., et al. (2013). Conformational dynamics of Kir3.1/Kir3.2 channel activation via δ-opioid receptors. *Molecular Pharmacology, 83*(2), 416–428. http://dx.doi.org/10.1124/mol.112.081950mol.112.081950 [pii].

Rishal, I., Porozov, Y., Yakubovich, D., Varon, D., & Dascal, N. (2005). Gβγ-dependent and Gβγ-independent basal activity of G protein-activated K^+ channels. *Journal of Biological Chemistry, 280*(17), 16685–16694.

Riven, I., Iwanir, S., & Reuveny, E. (2006). GIRK channel activation involves a local rearrangement of a preformed G protein channel complex. *Neuron, 51*(5), 561–573.

Rogalski, S. L., Appleyard, S. M., Pattillo, A., Terman, G. W., & Chavkin, C. (2000). TrkB activation by brain-derived neurotrophic factor inhibits the G protein-gated inward rectifier Kir3 by tyrosine phosphorylation of the channel. *Journal of Biological Chemistry, 275*(33), 25082–25088.

Rogalski, S. L., & Chavkin, C. (2001). Eicosanoids inhibit the G-protein-gated inwardly rectifying potassium channel (Kir3) at the Na^+/PIP_2 gating site. *Journal of Biological Chemistry, 276*(18), 14855–14860.

Ross, E. M. (2008). Coordinating speed and amplitude in G-protein signaling. *Current Biology, 18*(17), R777–R783. http://dx.doi.org/10.1016/j.cub.2008.07.035.

Rubinstein, M., Peleg, S., Berlin, S., Brass, D., & Dascal, N. (2007). $G\alpha_{i3}$ primes the G protein-activated K^+ channels for activation by coexpressed Gβγ in intact *Xenopus* oocytes. *Journal of Physiology, 581*, 17–32.

Rubinstein, M., Peleg, S., Berlin, S., Brass, D., Keren-Raifman, T., Dessauer, C. W., et al. (2009). Divergent regulation of GIRK1 and GIRK2 subunits of the neuronal G protein gated K^+ channel by $G\alpha_i$GDP and Gβγ. *Journal of Physiology, 587*(14), 3473–3491.

Ruiz-Velasco, V., & Ikeda, S. R. (1998). Heterologous expression and coupling of G protein-gated inwardly rectifying K+ channels in adult rat sympathetic neurons. *Journal of Physiology (London), 513*(Pt. 3), 761–773.

Rusinova, R., Mirshahi, T., & Logothetis, D. E. (2007). Specificity of Gβγ signaling to Kir3 channels depends on the helical domain of pertussis toxin-sensitive Gα subunits. *Journal of Biological Chemistry, 282*, 34019–34030.

Rusinova, R., Shen, Y. M., Dolios, G., Padovan, J., Yang, H., Kirchberger, M., et al. (2009). Mass spectrometric analysis reveals a functionally important PKA phosphorylation site in a Kir3 channel subunit. *Pflügers Archiv, 458*(2), 303–314.

Sadana, R., & Dessauer, C. W. (2009). Physiological roles for G protein-regulated adenylyl cyclase isoforms: Insights from knockout and overexpression studies. *Neuro-Signals, 17*(1), 5–22. http://dx.doi.org/10.1159/000166277.

Sadja, R., Alagem, N., & Reuveny, E. (2002). Graded contribution of the Gβγ binding domains to GIRK channel activation. *Proceedings of the National Academy of Sciences of the United States of America, 99*, 10783–10788.

Sadja, R., & Reuveny, E. (2009). Activation gating kinetics of GIRK channels are mediated by cytoplasmic residues adjacent to transmembrane domains. *Channels (Austin), 3*(3), 205–214.

Saitoh, O., Kubo, Y., Miyatani, Y., Asano, T., & Nakata, H. (1997). RGS8 accelerates G-protein-mediated modulation of K+ currents. *Nature, 390*(6659), 525–529.

Sakmann, B., Noma, A., & Trautwein, W. (1983). Acetylcholine activation of single muscarinic K+ channels in isolated pacemaker cells of the mammalian heart. *Nature, 303*(5914), 250–253.

Sanders, H., Berends, M., Major, G., Goldman, M. S., & Lisman, J. E. (2013). NMDA and $GABA_B$ (KIR) conductances: The "perfect couple" for bistability (Research Support, N.I.H., Extramural Research Support, Non-U.S. Gov't Research Support, U.S. Gov't, Non-P.H.S.). *The Journal of Neuroscience, 33*(2), 424–429. http://dx.doi.org/10.1523/JNEUROSCI.1854-12.2013.

Schreibmayer, W., Dessauer, C. W., Vorobiov, D., Gilman, A. G., Lester, H. A., Davidson, N., et al. (1996). Inhibition of an inwardly rectifying K channel by G-protein α-subunits. *Nature, 380*(6575), 624–627.

Schwarzer, S., Nobles, M., & Tinker, A. (2010). Do caveolae have a role in the fidelity and dynamics of receptor activation of G-protein-gated inwardly rectifying potassium channels? *Journal of Biological Chemistry, 285*(36), 27817–27826. http://dx.doi.org/10.1074/jbc.M110.103598M110.103598 [pii].

Schwenk, J., Metz, M., Zolles, G., Turecek, R., Fritzius, T., Bildl, W., et al. (2010). Native $GABA_B$ receptors are heteromultimers with a family of auxiliary subunits. *Nature, 465*(7295), 231–235.

Sharon, D., Vorobiov, D., & Dascal, N. (1997). Positive and negative coupling of the metabotropic glutamate receptors to a G protein-activated K^+ channel, GIRK, in *Xenopus* oocytes. *Journal of General Physiology, 109*(4), 477–490.

Shukla, A. K., Xiao, K., & Lefkowitz, R. J. (2011). Emerging paradigms of β-arrestin-dependent seven transmembrane receptor signaling. *Trends in Biochemical Sciences, 36*(9), 457–469. http://dx.doi.org/10.1016/j.tibs.2011.06.003.

Sickmann, T., & Alzheimer, C. (2003). Short-term desensitization of G-protein-activated, inwardly rectifying K^+ (GIRK) currents in pyramidal neurons of rat neocortex. *Journal of Neurophysiology, 90*(4), 2494–2503.

Signorini, S., Liao, Y. J., Duncan, S. A., Jan, L. Y., & Stoffel, M. (1997). Normal cerebellar development but susceptibility to seizures in mice lacking G protein-coupled, inwardly rectifying K+ channel GIRK2. *Proceedings of the National Academy of Sciences of the United States of America, 94*(3), 923–927.

Slepak, V. Z. (2009). Structure, function, and localization of Gβ5-RGS complexes. *Progress in Molecular Biology and Translational Science, 86*, 157–203. http://dx.doi.org/10.1016/s1877-1173(09)86006-7.

Slesinger, P. A., Reuveny, E., Jan, Y. N., & Jan, L. Y. (1995). Identification of structural elements involved in G protein gating of the GIRK1 potassium channel. *Neuron, 15*(5), 1145–1156.

Sodickson, D. L., & Bean, B. P. (1996). $GABA_B$ receptor-activated inwardly rectifying potassium current in dissociated hippocampal CA3 neurons. *Journal of Neuroscience, 16*(20), 6374–6385.

Sodickson, D. L., & Bean, B. P. (1998). Neurotransmitter activation of inwardly rectifying potassium current in dissociated hippocampal CA3 neurons: Interactions among multiple receptors. *Journal of Neuroscience, 18*, 8153–8162.

Soejima, M., & Noma, A. (1984). Mode of regulation of the ACh-sensitive K-channel by the muscarinic receptor in rabbit atrial cells. *Pflügers Archiv, 400*(4), 424–431.

Sowell, M. O., Ye, C., Ricupero, D. A., Hansen, S., Quinn, S. J., Vassilev, P. M., et al. (1997). Targeted inactivation of αi2 or α i3 disrupts activation of the cardiac muscarinic K^+ channel, I_{KACh}, in intact cells. *Proceedings of the National Academy of Sciences of the United States of America, 94*(15), 7921–7926.

Spauschus, A., Lentes, K. U., Wischmeyer, E., Dissmann, E., Karschin, C., & Karschin, A. (1996). A G-protein-activated inwardly rectifying K+ channel (GIRK4) from human hippocampus associates with other GIRK channels. *Journal of Neuroscience, 16*(3), 930–938.

Stanfield, P. R., Nakajima, S., & Nakajima, Y. (2003). Constitutively active and G-protein coupled inward rectifier K^+ channels: Kir2.0 and Kir3.0. *Reviews of Physiology, Biochemistry and Pharmacology, 145*, 47–179.

Stevens, E. B., Shah, B. S., Pinnock, R. D., & Lee, K. (1999). Bombesin receptors inhibit G protein-coupled inwardly rectifying K+ channels expressed in Xenopus oocytes through a protein kinase C-dependent pathway. *Molecular Pharmacology, 55*(6), 1020–1027.

Sui, J. L., Chan, K. W., & Logothetis, D. E. (1996). Na^+ activation of the muscarinic K^+ channel by a G-protein-independent mechanism. *Journal of General Physiology, 108*(5), 381–391.

Sui, J. L., Petit-Jacques, J., & Logothetis, D. E. (1998). Activation of the atrial KACh channel by the betagamma subunits of G proteins or intracellular Na+ ions depends on the presence of phosphatidylinositol phosphates. *Proceedings of the National Academy of Sciences of the United States of America, 95*(3), 1307–1312.

Takano, K., Yasufuku-Takano, J., Kozasa, T., Nakajima, S., & Nakajima, Y. (1997). Different G proteins mediate somatostatin-induced inward rectifier K^+ currents in murine brain and endocrine cells. *Journal of Physiology, 502*(Pt. 3), 559–567.

Tamargo, J., Caballero, R., Gomez, R., Valenzuela, C., & Delpon, E. (2004). Pharmacology of cardiac potassium channels. *Cardiovascular Research, 62*(1), 9–33. http://dx.doi.org/10.1016/j.cardiores.2003.12.026.

Tauber, P., Penton, D., Stindl, J., Humberg, E., Tegtmeier, I., Sterner, C., et al. (2014). Pharmacology and pathophysiology of mutated KCNJ5 found in adrenal aldosterone-producing adenomas. *Endocrinology, 155*(4), 1353–1362. http://dx.doi.org/10.1210/en.2013-1944.

Torrecilla, M., Fernandez-Aedo, I., Arrue, A., Zumarraga, M., & Ugedo, L. (2013). Role of GIRK channels on the noradrenergic transmission in vivo: An electrophysiological and neurochemical study on GIRK2 mutant mice. *The International Journal of Neuropsychopharmacology, 16*(5), 1093–1104. http://dx.doi.org/10.1017/s1461145712000971.

Torrecilla, M., Marker, C. L., Cintora, S. C., Stoffel, M., Williams, J. T., & Wickman, K. (2002). G-protein-gated potassium channels containing Kir3.2 and Kir3.3 subunits mediate the acute inhibitory effects of opioids on locus ceruleus neurons. *Journal of Neuroscience, 22*(11), 4328–4334.

Treiber, F., Rosker, C., Keren-Raifman, T., Steinecker, B., Gorischek, A., Dascal, N., et al. (2013). Molecular basis of the facilitation of the heterooligomeric GIRK1/GIRK4 complex by cAMP dependent protein kinase. *Biochimica et Biophysica Acta, 1828*(4), 1214–1221. http://dx.doi.org/10.1016/j.bbamem.2012.12.016.

Tristani-Firouzi, M., & Etheridge, S. P. (2010). Kir 2.1 channelopathies: The Andersen-Tawil syndrome. *Pflügers Archiv, 460*(2), 289–294. http://dx.doi.org/10.1007/s00424-010-0820-6.

Valenzuela, D., Han, X., Mende, U., Fankhauser, C., Mashimo, H., Huang, P., et al. (1997). $G\alpha_o$ is necessary for muscarinic regulation of Ca^{2+} channels in mouse heart. *Proceedings of the National Academy of Sciences of the United States of America, 94*(5), 1727–1732.

van Hemert, F., Lazova, M. D., Snaar-Jagaska, B. E., & Schmidt, T. (2010). Mobility of G proteins is heterogeneous and polarized during chemotaxis (Research Support, Non-U.S. Gov't.) *Journal of Cell Science, 123*(Pt. 17), 2922–2930. http://dx.doi.org/10.1242/jcs.063990.

Vilardaga, J. P., Bunemann, M., Krasel, C., Castro, M., & Lohse, M. J. (2003). Measurement of the millisecond activation switch of G protein-coupled receptors in living cells. *Nature Biotechnology, 21*(7), 807–812.

Vivaudou, M., Chan, K. W., Sui, J. L., Jan, L. Y., Reuveny, E., & Logothetis, D. E. (1997). Probing the G-protein regulation of GIRK1 and GIRK4, the two subunits of the K_{ACh} channel, using functional homomeric mutants. *Journal of Biological Chemistry, 272*(50), 31553–31560.

Voigt, N., Abu-Taha, I., Heijman, J., & Dobrev, D. (2014). Chapter thirteen—Constitutive activity of the acetylcholine-activated potassium current IK, ACh in cardiomyocytes. In T. Ya-Xiong (Ed.), *Advances in Pharmacology, 70*, pp. 393–409.

Voigt, N., Friedrich, A., Bock, M., Wettwer, E., Christ, T., Knaut, M., et al. (2007). Differential phosphorylation-dependent regulation of constitutively active and muscarinic receptor-activated IK, ACh channels in patients with chronic atrial fibrillation. *Cardiovascular Research, 74*(3), 426–437.

Voigt, N., Heijman, J., Trausch, A., Mintert-Jancke, E., Pott, L., Ravens, U., et al. (2013). Impaired Na^+-dependent regulation of acetylcholine-activated inward-rectifier K^+ current modulates action potential rate dependence in patients with chronic atrial fibrillation. *Journal of Molecular and Cellular Cardiology, 61*, 142–152. http://dx.doi.org/10.1016/j.yjmcc.2013.03.011.

Voigt, N., Maguy, A., Yeh, Y.-H., Qi, X., Ravens, U., Dobrev, D., et al. (2008). Changes in IK, ACh single-channel activity with atrial tachycardia remodelling in canine atrial cardiomyocytes. *Cardiovascular Research*, 77(1), 35–43. http://dx.doi.org/10.1093/cvr/cvm051.

Vorobiov, D., Bera, A. K., Keren-Raifman, T., Barzilai, R., & Dascal, N. (2000). Coupling of the muscarinic m2 receptor to G protein-activated K^+ channels via $G\alpha_z$ and a receptor-$G\alpha_z$ fusion protein. Fusion between the receptor and $G\alpha_z$ eliminates catalytic (collision) coupling. *Journal of Biological Chemistry*, *275*, 4166–4170.

Walsh, K. B. (2015). Targeting cardiac potassium channels for state-of-the-art drug discovery. *Expert Opinion on Drug Discovery*, *10*(2), 157–169. http://dx.doi.org/10.1517/17460441.2015.983471.

Wang, W., Whorton, M. R., & MacKinnon, R. (2014). Quantitative analysis of mammalian GIRK2 channel regulation by G proteins, PIP_2 and Na^+ in a reconstituted system. *eLife*, *3*, e03671. http://dx.doi.org/10.7554/eLife.03671.

Weiss, J. N. (1997). The Hill equation revisited: Uses and misuses. [Research Support, U.S. Gov't, P.H.S.]*FASEB Journal*, *11*(11), 835–841.

Wellner-Kienitz, M. C., Bender, K., Meyer, T., Bunemann, M., & Pott, L. (2000). Overexpressed A1 adenosine receptors reduce activation of acetylcholine-sensitive K^+ current by native muscarinic M2 receptors in rat atrial myocytes. *Circulation Research*, *86*(6), 643–648.

Wellner-Kienitz, M. C., Bender, K., & Pott, L. (2001). Overexpression of β1 and β2 adrenergic receptors in rat atrial myocytes. Differential coupling to G protein-gated inward rectifier K^+ channels via G_s and $G_{i/o}$. *Journal of Biological Chemistry*, *276*(40), 37347–37354.

Whorton, M. R., & MacKinnon, R. (2011). Crystal structure of the mammalian GIRK2 K^+ channel and gating regulation by G proteins, PIP_2, and sodium. *Cell*, *147*(1), 199–208. http://dx.doi.org/10.1016/j.cell.2011.07.046.

Whorton, M. R., & MacKinnon, R. (2013). X-ray structure of the mammalian GIRK2-βγ G-protein complex. *Nature*, *498*, 190–197. http://dx.doi.org/10.1038/nature12241.

Wickman, K., & Clapham, D. E. (1995). Ion channel regulation by G proteins. *Physiological Reviews*, *75*(4), 865–885.

Wickman, K. D., Iniguez-Lluhl, J. A., Davenport, P. A., Taussig, R., Krapivinsky, G. B., Linder, M. E., et al. (1994). Recombinant G-protein βγsubunits activate the muscarinic-gated atrial potassium channel. *Nature*, *368*(6468), 255–257.

Williams, T. A., Monticone, S., & Mulatero, P. (2015). KCNJ5 mutations are the most frequent genetic alteration in primary aldosteronism. *Hypertension*, *65*, 507–509. http://dx.doi.org/10.1161/hypertensionaha.114.04636.

Wiser, O., Qian, X., Ehlers, M., Ja, W. W., Roberts, R. W., Reuveny, E., et al. (2006). Modulation of basal and receptor-induced GIRK potassium channel activity and neuronal excitability by the mammalian PINS homolog LGN. *Neuron*, *50*(4), 561–573.

Wydeven, N., Fernandez de Velasco, E. M., Du, Y., Benneyworth, M. A., Hearing, M. C., Fischer, R. A., et al. (2014). Mechanisms underlying the activation of G-protein-gated inwardly rectifying K^+ (GIRK) channels by the novel anxiolytic drug, ML297. *Proceedings of the National Academy of Sciences of the United States of America*, *111*, 10755–10760. http://dx.doi.org/10.1073/pnas.1405190111.

Wydeven, N., Posokhova, E., Xia, Z., Martemyanov, K. A., & Wickman, K. (2014). RGS6, but not RGS4, is the dominant regulator of G protein signaling (RGS) modulator of the parasympathetic regulation of mouse heart rate. *Journal of Biological Chemistry*, *289*(4), 2440–2449. http://dx.doi.org/10.1074/jbc.M113.520742.

Wydeven, N., Young, D., Mirkovic, K., & Wickman, K. (2012). Structural elements in the Girk1 subunit that potentiate G protein-gated potassium channel activity. *Proceedings of the National Academy of Sciences of the United States of America*, *109*(52), 21492–21497. http://dx.doi.org/10.1073/pnas.1212019110.

Xie, K., Allen, K. L., Kourrich, S., Colon-Saez, J., Thomas, M. J., Wickman, K., et al. (2010). Gβ5 recruits R7 RGS proteins to GIRK channels to regulate the timing of neuronal inhibitory signaling. *Nature Neuroscience*, *13*(6), 661–663.

Yakubovich, D., Pastushenko, V., Bitler, A., Dessauer, C. W., & Dascal, N. (2000). Slow modal gating of single G protein-activated K^+ channels expressed in *Xenopus* oocytes. *Journal of Physiology*, *524*(3), 737–755.

Yakubovich, D., Rishal, I., & Dascal, N. (2005). Kinetic modeling of Na^+-induced, Gβγ-dependent activation of G-protein-gated K^+ channels. *Journal of Molecular Neuroscience*, *25*(1), 7–20.

Yamada, M., Inanobe, A., & Kurachi, Y. (1998). G protein regulation of potassium ion channels. *Pharmacological Reviews*, *50*(4), 723–760.

Yamada, M., Jahangir, A., Hosoya, Y., Inanobe, A., Katada, T., & Kurachi, Y. (1993). GK* and brain Gβγ activate muscarinic K^+ channel through the same mechanism. *Journal of Biological Chemistry*, *268*(33), 24551–24554.

Yang, Y., Liang, B., Liu, J., Li, J., Grunnet, M., Olesen, S. P., et al. (2010). Identification of a Kir3.4 mutation in congenital long QT syndrome. *The American Journal of Human Genetics*, *86*(6), 872–880.

Ye, C., Sowell, M. O., Vassilev, P. M., Milstone, D. S., & Mortensen, R. M. (1999). Gαi2, Gαi3 and Gαo are all required for normal muscarinic inhibition of the cardiac calcium channels in nodal/atrial-like cultured cardiocytes. *Journal of Molecular and Cellular Cardiology*, *31*(9), 1771–1781.

Yokogawa, M., Osawa, M., Takeuchi, K., Mase, Y., & Shimada, I. (2011). NMR analyses of the Gβγ binding and conformational rearrangements of the cytoplasmic pore of G protein-activated inwardly rectifying potassium channel 1 (GIRK1) (Research Support, Non-U.S. Gov't). *Journal of Biological Chemistry*, *286*(3), 2215–2223. http://dx.doi.org/10.1074/jbc.M110.160754.

Zhang, Q., Dickson, A., & Doupnik, C. A. (2004). GIRK channel activation kinetics via Gαi and Gαo-coupled receptors are determined by isoform-specific interdomain interactions that affect GDP release rates. *Journal of Biological Chemistry*, *279*, 29787–29796.

Zhao, Q., Kawano, T., Nakata, H., Nakajima, Y., Nakajima, S., & Kozasa, T. (2003). Interaction of G protein β subunit with inward rectifier K^+ channel Kir3. *Molecular Pharmacology*, *64*(5), 1085–1091. http://dx.doi.org/10.1124/mol.64.5.1085.

Zhong, H., Wade, S. M., Woolf, P. J., Linderman, J. J., Traynor, J. R., & Neubig, R. R. (2003). A spatial focusing model for G protein signals. Regulator of G protein signaling (RGS) protein-mediated kinetic scaffolding. *Journal of Biological Chemistry*, *278*(9), 7278–7284.

Zhou, H., Chisari, M., Raehal, K. M., Kaltenbronn, K. M., Bohn, L. M., Mennerick, S. J., et al. (2012). GIRK channel modulation by assembly with allosterically regulated RGS proteins. *Proceedings of the National Academy of Sciences of the United States of America*, *109*(49), 19977–19982. http://dx.doi.org/10.1073/pnas.1214337109.

Zuberi, Z., Birnbaumer, L., & Tinker, A. (2008). The role of inhibitory heterotrimeric G proteins in the control of in vivo heart rate dynamics (Research Support, N.I.H., Intramural Research Support, Non-U.S. Gov't). *American Journal of Physiology Regulatory, Integrative and Comparative Physiology*, *295*(6), R1822–R1830. http://dx.doi.org/10.1152/ajpregu.90625.2008.

Zylbergold, P., Ramakrishnan, N., & Hebert, T. (2010). The role of G proteins in assembly and function of Kir3 inwardly rectifying potassium channels. *Channels*, *4*(5), 411–421. http://dx.doi.org/10.4161/chan.4.5.13327.

Zylbergold, P., Sleno, R., Khan, S. M., Jacobi, A. M., Belhke, M. A., & Hebert, T. E. (2014). Kir3 channel ontogeny—The role of Gβγ subunits in channel assembly and trafficking. *Frontiers in Cellular Neuroscience*, *8*, 108. http://dx.doi.org/10.3389/fncel.2014.00108.

CHAPTER THREE

RGS Redundancy and Implications in GPCR–GIRK Signaling

Craig A. Doupnik[1]
Department of Molecular Pharmacology & Physiology, University of South Florida College of Medicine, Tampa, Florida, USA
[1]Corresponding author: e-mail address: cdoupnik@health.usf.edu

Contents

1. RGS Proteins—Background and Significance 88
2. Direct "Membrane-Delimited" GPCR-to-Ion Channel Signaling 89
3. RGS Actions on GPCR–GIRK Channel Gating Kinetics 90
4. RGS Functional Redundancy—Emerging Evidence from Retinal Neurons 94
5. RGS Functional Redundancy in Neuronal GPCR–GIRK Channel Signaling 96
6. RGS Functional Redundancy in Cardiac GPCR-GIRK Channel Signaling 100
7. RGS Actions on G_q-Mediated Changes in GIRK Channel Activity 101
8. Conclusions 102
References 105

Abstract

Regulators of G protein signaling (RGS proteins) are key components of GPCR complexes, interacting directly with G protein α-subunits to enhance their intrinsic GTPase activity. The functional consequence is an accelerated termination of G protein effectors including certain ion channels. RGS proteins have a profound impact on the membrane-delimited gating behavior of G-protein-activated inwardly rectifying K^+ (GIRK) channels as demonstrated in reconstitution assays and recent RGS knockout mice studies. Akin to GPCRs and G protein αβγ subunits, multiple RGS isoforms are expressed within single GIRK-expressing neurons, suggesting functional redundancy and/or specificity in GPCR-GIRK channel signaling. The extent and impact of RGS redundancy in neuronal GPCR–GIRK channel signaling is currently not fully appreciated; however, recent studies from RGS knockout mice are providing important new clues on the impact of individual endogenous RGS proteins and the extent of RGS functional redundancy. Incorporating "tools" such as engineered RGS-resistant Gαi/o subunits provide an important assessment method for determining the impact of all endogenous RGS proteins on a given GPCR response and an accounting benchmark to assess the impact of individual RGS knockouts on overall RGS redundancy within a given neuron. Elucidating the degree of regulation attributable to specific RGS proteins in GIRK channel function will aid in the assessment of individual RGS proteins as viable therapeutic targets in epilepsy, ataxia's, memory disorders, and a growing list of neurological disorders.

International Review of Neurobiology, Volume 123
ISSN 0074-7742
http://dx.doi.org/10.1016/bs.irn.2015.05.010

1. RGS PROTEINS—BACKGROUND AND SIGNIFICANCE

Regulators of G protein signaling (RGSs) represent a large family of proteins that function allosterically as GTPase-activating proteins (GAPs) for heterotrimeric G proteins (De Vries, Zheng, Fischer, Elenko, & Farquhar, 2000; Hollinger & Hepler, 2002; Ross & Wilkie, 2000). They are defined by a conserved ~110 amino acid helical bundle called an "RGS domain" (or "RGS-like domain"), which directly interacts with the GTP-binding α subunit to accelerate intrinsic GTP hydrolysis (Berman, Wilkie, & Gilman, 1996; De Vries, Mousli, Wurmser, & Farquhar, 1995; Hunt, Fields, Casey, & Peralta, 1996; Nance et al., 2013; Tesmer, 2009; Tesmer, Berman, Gilman, & Sprang, 1997; Watson, Linder, Druey, Kehrl, & Blumer, 1996). The human genome contains more than 30 genes encoding RGS or RGS-like domains and cluster into at least six distinct subfamilies based on amino acid sequence similarity (Hollinger & Hepler, 2002; Willars, 2006). Multiple RGSs are expressed in the mammalian nervous system with several demonstrating high expression in specific regional areas of the brain (Gold, Ni, Dohlman, & Nestler, 1997; Grafstein-Dunn, Young, Cockett, & Khawaja, 2001; Larminie et al., 2004). Additionally, several RGS transcripts undergo extensive alternative splicing that adds to an overall rich diversity of RGS protein expression within the human nervous system and individual neurons (Xie & Palmer, 2007).

The discovery of RGS proteins, now two decades past, resolved a discrepancy between the "slow" biochemical rate of GTP hydrolysis measured from purified G proteins *in vitro* and the more rapid effector kinetics measured in living cells (Doupnik, Jaen, & Zhang, 2004; Ross & Wilkie, 2000). The resulting paradigm shift generally accepted today is a default involvement of RGS proteins in GPCR signaling where they effectively function as a "brake" (i.e., negative regulator) on signal transduction via Gi/o-, Gq-, and G11/12-coupled receptors by enhancing intrinsic GTPase activity and signal termination.

The encoded full-length RGS proteins are highly modular in overall structure, containing several protein–lipid and protein–protein interaction domains that participate in the assembly of macromolecular signaling complexes and subcellular and membrane targeting (Hollinger & Hepler, 2002; Willars, 2006). The RGS domains from different RGS proteins have been shown to have varying degrees of GAP specificity for different Gα subunits, with some showing high specificity for Gαq (i.e., RGS2) (Heximer,

Watson, Linder, Blumer, & Hepler, 1997), high specificity for Gαo (R7 subfamily members) (Hooks et al., 2003), and others being relatively nonspecific among Gαi/o/q subunits (i.e., RGS3 and RGS4) (Hepler, Berman, Gilman, & Kozasa, 1997; Heximer et al., 1997; Scheschonka et al., 2000). Gαs GAP activity is generally not affected by RGS proteins (Berman et al., 1996) with the exception of a sorting nexin (SNX13) involved in vesicular trafficking and endocytosis (Zheng et al., 2001).

The RGS GAP specificity suggests that different GPCRs known to couple to specific classes of G proteins (i.e., adrenergic, serotonergic, muscarinic, etc.) may also couple selectively to specific RGS proteins, invoking the notion of distinct signaling complexes comprised of a unique receptor, G protein αβγ heterotrimer, and RGS protein. Validating this model has proved difficult in heterologous expression systems and reconstitution assays that assemble various combinations and stoichiometries of GPCRs, RGSs, and effectors. Several RGSs tested in these high-expression assays generally display a promiscuous capacity to regulate different GPCR–effector signaling events that occur through "collision-coupling" versus a selective precoupled mechanism (Jaen & Doupnik, 2006; cf. Table 1).

To better address questions of selective RGS regulation of GPCR–effector signaling events in native cells, several genetically engineered RGS "knockout" ($RGS^{-/-}$) mice are beginning to provide new insight regarding the role of individual RGS proteins in well-established GPCR signaling events. Moreover, the cross breeding of different $RGS^{-/-}$ mice to generate double RGS knockout strains is providing compelling evidence for limited RGS functional redundancy in selected GPCR signaling events (Chen et al., 2010; Mojumder, Qian, & Wensel, 2009; Shim et al., 2012; Wydeven, Posokhova, Xia, Martemyanov, &Wickman, 2014; Zhang et al., 2010).

2. DIRECT "MEMBRANE-DELIMITED" GPCR-TO-ION CHANNEL SIGNALING

As described in detail in earlier chapters, the G-protein-activated inwardly rectifying K^+ (GIRK) channel is a classic G protein effector where direct protein–protein interaction of Gβγ dimers causes channel activation (increased channel open probability) (Logothetis, Kurachi, Galper, Neer, & Clapham, 1987). GTP-bound Gα(i/o) subunits then deactivate the channel via resequestration of the Gβγ dimers following GTP hydrolysis. The Gαi/o subunits may also "dock" to the GIRK channel protein as part of a larger

GPCR–GIRK channel macromolecular complex (Berlin et al., 2010; Clancy et al., 2005; Wang, Whorton, & MacKinnon, 2014).

This signaling mechanism has been called "membrane-delimited" since it does not require soluble cytosolic second messengers (i.e., cyclic AMP) and is confined to a two-dimensional membrane microdomain where GPCR and GIRK channels reside in either close proximity or as a quasistable signaling complex (Breitwieser, 1991; Brown, 1990). Similar membrane-delimited signaling has been described for pre- and postsynaptic GPCR inhibition of voltage-gated calcium channels (i.e., N- and P/Q-type Cav2 calcium channels), where Gβγ dimers interact directly and inhibit Ca^{2+} channel activity in a voltage-dependent manner (Dolphin, 2003; Ikeda, 1996; Tedford & Zamponi, 2006). More recently, the TrpM1 channel expressed in retinal on-bipolar neurons (OBNs) has also been shown to be inhibited directly by Gβγ dimers, and is thought to mediate TrpM1 channel inhibition during mGluR6 receptor activation by glutamate released from dark-adapted photoreceptors (Koike et al., 2010; Shen, Rampino, Carroll, & Nawy, 2012). Membrane-delimited GPCR regulation of ion channel effectors provides a means for high-fidelity electrical signaling, which is necessary for the temporal resolution demands required in GPCR-mediated synaptic transmission and visual scene processing.

3. RGS ACTIONS ON GPCR–GIRK CHANNEL GATING KINETICS

Heterologous expression of most RGS proteins belonging to the R4 and R7 families cause multiple changes in GIRK channel function (Chuang et al., 1998; Doupnik et al., 1997; Herlitze et al., 1999; Kovoor et al., 2000; Saitoh et al., 1997, 1999; Zhang et al., 2002). First, receptor-independent "basal" GIRK channel activity is significantly reduced, consistent with RGS-enhanced GAP activity promoting the sequestration of "free" Gβγ dimers via increased resident time of GDP-bound Gα subunits (Doupnik et al., 1997; Zhang et al., 2002). Second, the kinetics for receptor agonist-mediated GIRK channel activation and deactivation are dramatically accelerated with moderate changes in steady-state peak current amplitude (Doupnik et al., 1997; Saitoh et al., 1997). And lastly, the time course for GIRK channel "rapid" desensitization during continuous receptor agonist activation is also significantly accelerated (Chuang et al., 1998; Doupnik et al., 1997; Mutneja, Berton, Suen, Luscher, & Slesinger, 2005; Saitoh et al., 1997). A summary list of reconstituted RGS protein–GIRK channel interactions is provided in Table 1.

Table 1 RGS-Accelerated GIRK Channel Gating Assessed in Heterologous Expression Systems

RGS Protein	Heterologous Expression System(s)	Coexpressed GPCR (Gi/o-coupled)	Accelerated GIRK Channel Gating Kinetics?	Comments	References
R4 subfamily					
RGS1	*Xenopus oocytes*	5-HT1A Muscarinic m2	Yes	Endogenous expression mostly in lymphocytes	Doupnik, Davidson, Lester, and Kofuji (1997)
RGS2	*X. oocytes*	5-HT1A Muscarinic m2	No (Yes)	High expression in the nervous system. Positive Gi/o effects observed at high ratiometric expression levels. Mostly Gq specific	Doupnik et al. (1997); Herlitze, Ruppersberg, and Mark (1999)
RGS3	*X. oocytes*, CHO-K1 cells	5-HT1A Muscarinic m2	Yes	Broad tissue expression. The short splice RGS3 isoform effects occur via "collision-coupling"	Doupnik et al. (1997); Jaen and Doupnik (2005, 2006)
RGS4	*X. oocytes*, CHO-K1 cells	5-HT1A Muscarinic m2 Muscarinic m4 μ-opioid receptor $GABA_BR1/R2$ Adenosine A1 Dopamine D2	Yes	High expression in the nervous system. Directly associates with multiple GPCRs.	Chuang, Yu, Jan, and Jan (1998); Doupnik et al. (1997); Fowler, Aryal, Suen, and Slesinger (2007); Jaen and Doupnik (2005, 2006)
RGS5	*X. oocytes*	Muscarinic m2	Yes	High expression in pericytes and smooth muscle cells	Herlitze et al. (1999)

Continued

Table 1 RGS-Accelerated GIRK Channel Gating Assessed in Heterologous Expression Systems—cont'd

RGS Protein	Heterologous Expression System(s)	Coexpressed GPCR (Gi/o-coupled)	Accelerated GIRK Channel Gating Kinetics?	Comments	References
RGS8	*X. oocytes*, HEK293 cells	Muscarinic m2 Muscarinic m4 Dopamine D2 Adrenergic α2A Adenosine A1 $GABA_BR1/R2$	Yes	Precouples with multiple GPCRs, High expression in nervous system	Benians, Leaney, and Tinker (2003); Saitoh, Kubo, Miyatani, Asano, and Nakata (1997)
RGS13	ND			Endogenous expression in lymphocytes	Druey (2009); Kehrl (2004)
RGS16	ND			Endogenous expression in lung, lymphocytes	Druey (2009); Kehrl (2004)
RGS18	ND			Endogenous expression in platelets	Louwette, Van Geet, and Freson (2012)
RGS21	ND			Endogenous expression in taste receptors and airway epithelia	Cohen et al. (2012); von Buchholtz et al. (2004)
R7 subfamily					
RGS6	ND		Yes	Region-specific expression in the nervous system. Positive effects confirmed from $RGS6^{-/-}$ mice	See Table 2
RGS7	*X. oocytes*	Muscarinic m2 $GABA_BR1/R2$ Dopamine D2 Muscarinic m2	Yes	Region-specific expression in the nervous system. Augmented by Gβ5 and R7BP	Drenan et al. (2005); Drenan et al. (2006); Kovoor et al. (2000); Zhang, Pacheco, & Doupnik (2002)

RGS9	*X. oocytes* CHO cells	Muscarinic m2 Dopamine D2	Yes	Region-specific expression in the nervous system. Augmented by Gβ5 and R7BP/R9AP	Granneman et al. (1998); Kovoor et al. (2000)
RGS11	*X. oocytes*	mGluR6	Yes	Region-specific expression in the nervous system. Augmented by Gβ5 and R9AP	Masuho, Celver, Kovoor, and Martemyanov (2010)
RZ subfamily					
RGS17	ND				
RGS19	ND				
RGS20	*X. oocytes*	mGluR6	Yes	High expression in retinal ON-bipolar neurons	Dhingra, Faurobert, Dascal, Sterling, and Vardi (2004)
R12 subfamily					
RGS10	Rat atrial myocytes (adenovirus)	Muscarinic m2 Adenosine A1 (endogenous)	Yes		Bender et al. (2008)
RGS12	ND				
RGS14	ND				

ND, not done.

All of the RGS-mediated kinetic changes in GIRK channel gating have been attributed to the RGS-accelerated GAP activity of Gαi/o subunits (Lambert et al., 2010; Qin, Sethi, & Lambert, 2008). Moreover, the kinetic effects produced by RGS proteins do not require a requisite preassembled GPCR-GIRK channel macromolecular complex as originally postulated (Lambert et al., 2010; Qin et al., 2008). Nevertheless, GIRK channel co-assembly with GPCRs (Fowler et al., 2007; Jaen & Doupnik, 2006; Lavine et al., 2002) and RGS co-assembly with GPCRs suggest the presence of "primed" precoupled complexes that may engender an initial level of molecular GPCR–RGS–GIRK channel signaling specificity.

Convergent activation of GIRK channels by multiple GPCRs in acutely isolated hippocampal neurons reveals a predominant nonadditive effect that is more consistent with a model whereby individual GIRK channels respond to a pool of G proteins that can be activated by multiple receptors (Andrade, Malenka, & Nicoll, 1986; Sodickson & Bean, 1998). Therefore preassembled GPCR–RGS–GIRK channel complexes (nonactivated) must necessarily be "quasistable," where agonist-mediated receptor activation would enable G proteins to dissociate from the channel complex and activate neighboring channels (Doupnik, 2008; Riven, Iwanir, & Reuveny, 2006; Sodickson & Bean, 1998). These molecular assemblies are key in determining the time course for inhibitory postsynaptic potentials mediated by neuronal GIRK channels that have wide ranging influence on neuronal signaling and implicated in several neuronal disorders (D'Adamo, Catacuzzeno, Di Giovanni, Franciolini, & Pessia, 2013; Kobayashi & Ikeda, 2006; Lujan, Fernandez, de Velasco, Aguado, & Wickman, 2014; Luscher, Jan, Stoffel, Malenka, & Nicoll, 1997; Luscher & Slesinger, 2010). RGS-mediated suppression of "basal" GIRK channel activity may also play a key role in this process (Chen & Johnston, 2005; Kim & Johnston, 2015).

4. RGS FUNCTIONAL REDUNDANCY—EMERGING EVIDENCE FROM RETINAL NEURONS

Functional redundancy is a well-established design principle in engineered systems and is found in several biological systems including signal transduction networks (Citri & Yarden, 2006). Functional redundancy at the cellular level improves overall network-level "robustness" at a certain energetic cost, where the loss of function of an individual signaling component does not compromise the function of an entire signaling circuit because of the availability of redundant components (Kitano, 2004; Stelling, Sauer, Szallasi, Doyle, & Doyle, 2004).

The number of redundant RGS proteins affecting specific GPCR-regulated effector pathways in neurons is currently not understood, but recent studies in knockout mice are beginning to provide important new clues for RGS functional redundancy in retinal neurons (Chen et al., 2010; Mojumder et al., 2009; Zhang et al., 2010). These findings are in contrast to initial studies in photoreceptors, where a single RGS protein (RGS9-1) was found to be the predominant GAP for transducin (Gαt) and responsible for accelerating phototransduction signaling kinetics (Chen et al., 2000; Cowan, Fariss, Sokal, Palczewski, & Wensel, 1998; He, Cowan, & Wensel, 1998; Lyubarsky et al., 2001; Makino, Handy, Li, & Arshavsky, 1999). Humans carrying recessive RGS9 mutations display a slowed photoresponse phenotype with reduced motion detection that correlate with the biochemical and physiological actions of RGS9-1 (Nishiguchi et al., 2004).

The R7 proteins selective for Gαo (RGS6, RGS7, RGS9, and RGS11) are distinct in overall structure and unique binding partners that include the Gβ5 subunit (Cabrera, de Freitas, Satpaev, & Slepak, 1998; Slepak, 2009). Gβ5 binds to a Gγ-like (GGL) domain found in all four R7 RGS isoforms and stabilizes the native protein complex (Snow et al., 1998). As a result, all four R7 proteins are severely degraded in $G\beta5^{-/-}$ mice and effectively yield mice lacking all four R7 proteins; RGS6, RGS7, RGS9, and RGS11 (Chen et al., 2003). Utilization of $G\beta5^{-/-}$ mice have proved to be extremely valuable in phenotypic screening and identification of physiological roles of R7 proteins.

Electroretinography (ERG) of $G\beta5^{-/-}$ mice reveals an absence of light-evoked b-waves that are normally produced by retinal OBNs and suggests the involvement of R7 proteins in OBN signaling (Rao, Dallman, Henderson, & Chen, 2007). OBNs play a critical role in the initial stage of signal processing from the rod and cone photoreceptors. Glutamate released from dark-adapted photoreceptors bind to Gαo-coupled type-6 metabotropic glutamate receptors (mGluR6) on OBN dendrites, causing inhibition of TrpM1 ion channels. With light stimulation, the decrease in photoreceptor glutamate release results in the disinhibition (i.e., opening) of TrpM1 channels, causing membrane depolarization and synaptic activation of innervated retinal ganglion neurons. The mechanism for mGluR6 inhibition of TrpM1 channels in OBN dendrites is membrane-delimited and involves Gβγ dimers (Shen et al., 2012).

Both RGS7 and RGS11 proteins have been immunohistochemically identified at the dendritic tips of OBNs, where RGS6 and RGS9 were not detected (Morgans et al., 2007). ERG recordings from individual R7 knockout mice ($RGS7^{-/-}$ or $RGS11^{-/-}$) reveal delays in b-waves, where the b-wave is eliminated in the double knockout ($RGS7^{-/-}$ + $RGS11^{-/-}$) (Chen et al., 2010; Mojumder et al., 2009; Shim et al., 2012; Zhang et al., 2010). These findings provide strong evidence for functional redundancy involving RGS7 and RGS11 in the mGluR6 regulation of OBN TrpM1 channels.

In a fascinating series of related studies, RGS7 and RGS11 have been found to assemble with previously unidentified orphan GPCRs (GPR156 and GPR178) that target RGS7 and RGS11 to the OBN dendrites and are necessary for the functional effects produced by these GAP proteins (Orlandi, Cao, & Martemyanov, 2013; Orlandi et al., 2012; Ray et al., 2014). Moreover, autosomal-recessive mutations in GPR179 result in complete congenital stationary night blindness (Audo et al., 2012; Bijveld et al., 2013; Peachey et al., 2012).

Although the details are still not clear yet on how R7-coupled GPR156/178 receptors form heterooligomeric complexes with mGluR6 and TrpM1 channels (Orlandi et al., 2013), this area of investigation is likely to continue to inspire future work on this retinal synapse and many other neuronal GPCR signaling circuits in the coming years.

5. RGS FUNCTIONAL REDUNDANCY IN NEURONAL GPCR–GIRK CHANNEL SIGNALING

The functional and kinetic features of hippocampal GABA-activated GIRK channels have been well characterized and essentially dictate the magnitude and kinetics of slow IPSCs within the hippocampal synaptic circuitry (Nicoll, 1988; Otis, De Koninck, & Mody, 1993; Sodickson & Bean, 1996). To begin to decipher which neuronal R7 proteins might accelerate the membrane-delimited signaling between GPCRs and GIRK channels, the $GABA_B$ receptor-activated GIRK channels in hippocampal neurons from $G\beta5^{-/-}$ mice were compared to those recorded from wild-type mice (Xie et al., 2010). In $G\beta5^{-/-}$ mice, both the receptor activation and deactivation kinetics of GABA-evoked GIRK channels were significantly slower than wild-type responses, indicating the involvement of R7 GAP regulation in $GABA_B$ receptor-activated GIRK channels. The slower GIRK channel gating was also reflected in significantly slower IPSCs recorded from CA1 pyramidal neurons in $G\beta5^{-/-}$ hippocampal slices, resulting in a two- to threefold enhancement in inhibitory K^+ current. Co-immunoprecipitation of RGS7 with GIRK channel subunits from hippocampal tissue supports the participation of RGS7 in $GABA_B$ receptor-GIRK channel signaling (Xie et al., 2010).

Follow-up studies in $RGS7^{-/-}$ mice confirm that RGS7 ablation results in slower deactivation kinetics of GABA-activated GIRK channels in hippocampal neurons (Fajardo-Serrano et al., 2013; Ostrovskaya et al., 2014). Moreover, the GIRK deactivation time constants were roughly equivalent in both $G\beta5^{-/-}$ and $RGS7^{-/-}$ hippocampal neurons (~3 s compared to ~1 s in wild-type), suggesting that RGS7 is the predominant R7 isoform involved (Ostrovskaya et al., 2014; Xie et al., 2010). Parallel experiments in $RGS6^{-/-}$ mice found no significant influence of RGS6 ablation on hippocampal GIRK channel gating kinetics (Ostrovskaya et al., 2014), although responses from $RGS9^{-/-}$ and $RGS11^{-/-}$ mice were not tested. Earlier studies of $G\alpha o^{-/-}$ mice also reported a similar slowing of hippocampal GIRK current kinetics (Greif, Sodickson, Bean, Neer, & Mende, 2000), which may reflect the loss of RGS7-accelerated GIRK channel gating given R7 selectivity for Gαo subunits (Hooks et al., 2003; Lan, Zhong, Nanamori, & Neubig, 2000; Saitoh et al., 1999; Zhang et al., 2002).

Although RGS6 ablation did not impact hippocampal $GABA_B$ receptor-activated GIRK channels (Ostrovskaya et al., 2014), Maity and coworkers found that the deactivation kinetics of GABA-activated GIRK channels from cerebellar granule neurons (CGNs) were slower in $RGS6^{-/-}$ mice compared to wild-type CGNs (Maity et al., 2012). A greater role of RGS6 in the cerebellum is supported by the impaired gait and locomotor phenotype in the $RGS6^{-/-}$ mice (Maity et al., 2012). The impact of RGS6 ablation on the GIRK channel deactivation kinetics in CGNs was less robust (time constant ~2 s) than that described earlier with RGS7 ablation in hippocampal neurons (time constant ~3 s). In addition, both RGS6 and RGS7 did not coimmunoprecipitate with GIRK channel subunits from cerebellar or cerebral tissues in contrast to the hippocampal study (Maity et al., 2012). Given the lesser effect of RGS6 on GABA-activated GIRK channel deactivation kinetics in CGNs, the likelihood of R7 functional redundancy in CGNs seems worthy of testing in $G\beta5^{-/-}$ mice.

Another key contributor to neuronal R7 modulation is the subcellular targeting role of the R7-binding protein (R7BP) (Drenan et al., 2005; Hepler, 2005; Jayaraman, Zhou, Jia, Cain, & Blumer, 2009; Liapis et al.,

2012; Martemyanov, Yoo, Skiba, & Arshavsky, 2005). R7BP is reversibly palmitoylated and functions as a membrane anchoring protein through protein–protein interactions with a DEP domain (Dishevelled, Egl-10, and Pleckstrin domain) located at the N-terminus of all R7 proteins (Drenan et al., 2005; Jia, Linder, & Blumer, 2011; Martemyanov et al., 2005; Song, Waataja, & Martemyanov, 2006). In addition to membrane localization, R7BP protects R7-Gβ5 complexes from proteolytic degradation, thereby increasing R7-Gβ5 steady-state cellular concentrations (Anderson et al., 2007).

Heterologous expression of R7BP has been shown to dramatically enhance RGS7–Gβ5 acceleration of GPCR–GIRK channel gating kinetics in a palmitoylation-dependent manner (Drenan et al., 2005, 2006). Additionally, R7BP is necessary for R7–Gβ5 association with the GIRK channel subunits, where the Gβ5 subunit does not compete for channel sites necessary for Gβγ dimer-mediated channel activation (Zhou et al., 2012).

Whether RGS redundancy for neuronal GPCR-activated GIRK channels involves RGS proteins other than R7 proteins (i.e., R4 proteins) remains to be determined but seems likely. For example, RGS4 is highly expressed in brain and co-assembles with multiple GPCRs and GIRK channels in heterologous expression systems to accelerate GIRK channel gating kinetics (Chuang & Chuang, 2012; Fowler et al., 2007; Jaen & Doupnik, 2006). RGS4 also colocalizes and directly associates with $GABA_B$ receptors immunoprecipitated from brain (Kim et al., 2014). RGS8, which also is highly expressed in brain and structurally very similar to RGS4, is another likely candidate (Gold et al., 1997; Saitoh et al., 1997). Altogether, the recently emerging model is one where different populations of GPCR–R4 complexes and GPCR–R7/Gβ5 complexes can each couple to GIRK channels for kinetic acceleration of channel gating. Whether coupling to specific RGS proteins confers some other functional or subcellular trafficking features of channel regulation (i.e., specific membrane targeting or additional regulation) also remains to be elucidated.

Quantitative comparison of GIRK deactivation time constants from the $G\beta5^{-/-}$ and $RGS7^{-/-}$ hippocampal neurons (~3 s) strongly suggests other "redundant" RGS proteins contribute to the GIRK deactivation kinetics. Heterologous expression of an engineered RGS-resistant Gαo subunit (G184S) in rat sympathetic neurons yield GIRK deactivation times markedly slower (>20 s) and provide an estimate of total endogenous RGS influence on GIRK gating kinetics (Jeong & Ikeda, 2001). In other heterologous expression systems where "background" endogenous RGS protein levels are low, the time constant for GIRK deactivation is considerably larger (i.e., slower time course) (~10 s in CHO-K1 cells; >10 s in *Xenopus* oocytes) (Doupnik et al., 2004; Jaen & Doupnik, 2005, 2006; Mutneja et al., 2005; Zhang et al., 2002). These quantitative considerations are likely to reflect the presence of additional RGS proteins in hippocampal and cerebellar neurons that contribute toward the accelerated RGS-dependent GIRK channel gating kinetics in a redundant manner (see Fig. 1). Determining the relative contribution of each RGS protein in GPCR–GIRK channel signaling will no doubt be challenging, but cross-breeding $RGS^{-/-}$ mice may help to pave the way in this endeavor (Chen et al., 2010; Mojumder et al., 2009; Shim et al., 2012; Wydeven et al., 2014; Zhang et al., 2010).

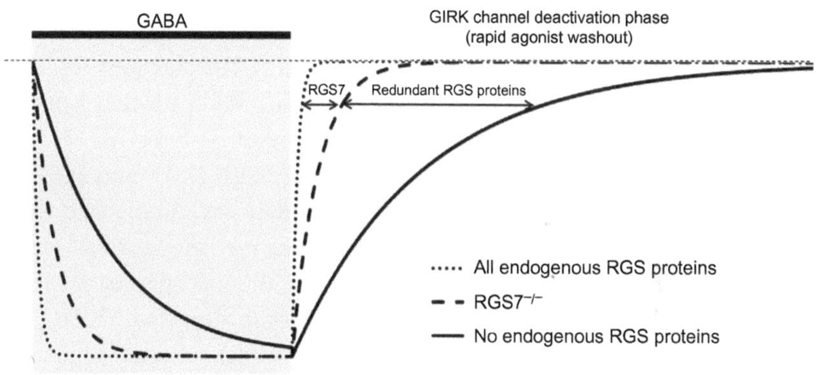

Figure 1 Simulation of $GABA_B$ receptor-activated hippocampal GIRK currents under varying extents of RGS regulation. GABA application (bar) activates inward (downward) potassium current under experimental recording conditions that returns to baseline upon agonist removal (GIRK deactivation phase). The kinetic influence of all endogenous RGS proteins is indicated by the green trace (dotted line in the print version), and reflects neuronal GIRK channel activation and deactivation gating characteristics (Ostrovskaya et al., 2014, Sodickson & Bean, 1996). RGS7 ablation as seen in $RGS7^{-/-}$ hippocampal neurons is depicted by the blue trace (dashed line in the print version), and reflects the influence of all endogenous RGS proteins minus RGS7 (Ostrovskaya et al., 2014). The red trace (solid line in the print version) simulates the activation and deactivation GIRK current time course in the absence of RGS modulation, as revealed with receptor coupling to engineered RGS-resistant Gαo(G184S) subunits (Jeong & Ikeda, 2001). All the currents were normalized to maximal amplitudes to highlight the different kinetic features associated with varying degrees of RGS modulation.

6. RGS FUNCTIONAL REDUNDANCY IN CARDIAC GPCR-GIRK CHANNEL SIGNALING

Autonomic regulation of the heartbeat involves sympathetic and parasympathetic regulation of cardiac ion channels via distinct GPCR signaling pathways (Harvey & Belevych, 2003; Robinson, Epstein, Beiser, & Braunwald, 1966). Key in vagal parasympathetic slowing of the heartbeat is acetylcholine activation of muscarinic m2 receptor-coupled GIRK channels expressed in sinoatrial and atrioventricular node pacemaker cells, and atrial cardiomyocytes (Hibino et al., 2010; Yamada, Inanobe, & Kurachi, 1998). Acetylcholine activation of cardiac GIRK channels slow pacemaker action potential firing by increasing the cell membrane K^+ conductance and hyperpolarizing the cell membrane potential.

Initial studies exploring the role of RGS proteins in the heart found expression of multiple RGS genes much like that described for the brain (Kardestuncer, Wu, Lim, & Neer, 1998; Tamirisa, Blumer, & Muslin, 1999; Zhang et al., 1998). Within individual GIRK-expressing atrial myocytes, the expression of at least seven distinct RGS protein isoforms were found representing multiple RGS subfamilies; R4 (RGS2, RGS3, and RGS4), R7 (RGS6), R10 (RGS10), and RZ (RGS17 and RGS19) (Doupnik, Xu, & Shinaman, 2001). Several studies have implicated functional roles for RGS2 and RGS4 in regulating Gq-mediated cardiac signaling (Anger et al., 2007; Hao et al., 2006; Muslin, 2006; Rogers et al., 1999, 2001; Tamirisa et al., 1999; Zhang et al., 1998, 2006; Zou et al., 2006), and the role of endogenous RGS proteins in cardiac GIRK channel regulation by Gi/o-mediated receptor signaling is supported by "knock-in" mice expressing engineered RGS-resistant Gαi/o subunits (Fu, Huang, Piao, Lopatin, & Neubig, 2007; Fu et al., 2006).

In 2008, studies utilizing $RGS4^{-/-}$ mice found evidence for RGS4 in muscarinic control of the sinoatrial GIRK channel current, I_{KACh} (Cifelli et al., 2008; Neubig, 2008). The reported cardiac phenotype for RGS4 null mice was consistent with a role of RGS4 in parasympathetic control of heart rate, showing increased bradycardia and heightened sensitivity to cholinergic agonists. Moreover, RGS4 ablation significantly slowed the native muscarinic m2 receptor-activated GIRK channel kinetics in sinoatrial pacemaker cells (Cifelli et al., 2008).

Subsequent studies in $RGS6^{-/-}$ mice similarly found a significant role for RGS6 in parasympathetic heart rate control (Posokhova, Wydeven,

Allen, Wickman, & Martemyanov, 2010; Posokhova et al., 2013; Yang et al., 2010). Cardiomyocytes from RGS6$^{-/-}$ mice also display slowed I_{KACh} kinetics consistent with RGS6/Gβ5 complexes regulating muscarinic m2 receptor controlled GIRK channel gating (Posokhova et al., 2010; Yang et al., 2010). Since R7BP is not expressed in the heart, the cardiac RGS6/Gβ5 findings raise the possibility of unidentified cardiac R7 trafficking protein(s), analogous to the orphan receptors GPR156 and GPR178 in OBNs proteins (Orlandi et al., 2013, 2012; Ray et al., 2014).

Thus, RGS4 and RGS6 both appear to provide functional redundancy for cardiac muscarinic m2 receptor-gated GIRK channel signaling (Ang, Opel, & Tinker, 2012; Jones, Tuomi, & Chidiac, 2012; Stewart, Huang, & Fisher, 2012). More recently, however, the effects of RGS4 and RGS6 were compared in parallel experiments from RGS4$^{-/-}$ and RGS6$^{-/-}$ mice and found a predominant role of RGS6 (GIRK deactivation time constant ~5 s) with little to no effect of RGS4 ablation on I_{KACh} kinetics (RGS4$^{-/-}$ GIRK deactivation time constant <1 s, and not different from wild-type) (Wydeven et al., 2014). Interestingly, I_{KACh} kinetics from Gβ5$^{-/-}$ cardiomyocytes indicated a much more prominent slowing in GIRK kinetics (GIRK deactivation time constant ~8 s) compared to RGS6$^{-/-}$ cardiomyocytes, suggesting that additional cardiac-expressed R7 proteins provide some degree of functional redundancy greater than RGS4 (Wydeven et al., 2014). Lastly, despite the lack of a prominent phenotype for the RGS4$^{-/-}$ cardiomyocytes, results from double knockout mice (RGS4$^{-/-}$+RGS6$^{-/-}$ mice) indicated some level of interaction between RGS4 and RGS6, where RGS4 may compete and suppress the influence of another cardiac R7 protein. The explanation for the two opposing RGS4$^{-/-}$ findings is currently not clear, but may reflect specific experimental conditions that affect RGS functional redundancy by diminishing or accentuating the role of RGS4 in cardiac GIRK channel regulation (Cifelli et al., 2008; Wydeven et al., 2014).

7. RGS ACTIONS ON G_q-MEDIATED CHANGES IN GIRK CHANNEL ACTIVITY

In contrast to receptors coupled to Gαi/o subunits, neuronal GIRK channels are downregulated by receptors coupled to Gαq/11 subunits that activate phospholipase C and alter membrane phosphatidylinositol lipid composition. As described by Logothetis and colleagues in chapter "Unifying Mechanism of Controlling Kir3 Channel Activity by G Proteins and Phosphoinositides", phosphatidylinositol 4, 5-bisphosphate (PIP2) directly binds to GIRK channels and is necessary for channel gating by Gβγ dimers (Huang, Feng, & Hilgemann, 1998). A reduction in local PIP2 concentrations during Gαq activation of phospholipase C effectively inhibits GIRK channel activity. Neurotransmitters that activate Gαq/11-coupled receptors promote neuronal excitability via the associated "slow" depolarization of the membrane potential resulting from the GIRK channel inhibition (Bajic, Koike, Albsoul-Younes, Nakajima, & Nakajima, 2002; Nakajima, Nakajima, & Inoue, 1988; Stanfield, Nakajima, & Nakajima, 2002; Takano et al., 1996).

These Gq-mediated GIRK channel inhibitory responses are reduced by RGS GAPs selective for Gq subunits such as RGS2 (Herlitze et al., 1999; Lei, Jones, Talley, Garrison, & Bayliss, 2003). There is also intriguing evidence for cross-talk between Gq- and Gi/o-coupled receptors converging on neuronal GIRK channels that likely involve multiple RGS proteins (Labouebe et al., 2007). Genetic silencing of RGS2 increased $GABA_B$ (Gi/o) receptor-GIRK channel-coupling efficiency (i.e., agonist sensitivity) in dopamine neurons of the ventral tegmental area of $RGS2^{-/-}$ mice, resulting in altered neuronal excitability (Labouebe et al., 2007). These findings and others described in additional chapters of this volume point toward RGS dysregulation in the mechanistic neuropharmacological underpinnings of drug addiction, with GIRK channels being just one of many effectors involved.

The extent that individual RGS proteins alter "basal" agonist-independent GIRK channel activity, either via enhanced Gβγ sequestration or reduced PIP2 concentrations, remains an important and yet largely unexplored area of the RGS-dependent GIRK channel regulation in $RGS^{-/-}$ mice. Notably, it is now recognized that increased basal GIRK activity in atrial myocytes is associated with atrial fibrillation (AF)-induced remodeling and promotes re-entrant arrhythmias (Bingen et al., 2013; Cha et al., 2006; Sarmast et al., 2003; Voigt, Abu-Taha, Heijman, & Dobrev, 2014; Voigt et al., 2008). Whether RGS remodeling in neurological excitability disorders, analogous to AF, contributes to an increased basal GIRK channel activity remains an open and tantalizing hypothesis (Lomazzi, Slesinger, & Luscher, 2008; Luscher & Slesinger, 2010).

8. CONCLUSIONS

Evidence to date indicate robust functional redundancy in RGS protein regulation of neuronal GIRK channels (see Tables 1 and 2). The total influence of endogenous RGS proteins on GIRK channel gating, as revealed

Table 2 Genetically Engineered Mice Effecting RGS-GIRK Interactions

Targeted Gene	Native GIRK-Expressing Cell/Tissue Type	Endogenous GPCR	GIRK Function Effected	References
Gα subunit "knock-outs"				
$G\alpha o^{-/-}$	Hippocampal CA3 neurons	Muscarinic m2 receptors	Slowed deactivation kinetics	Greif et al. (2000)
	Hippocampal CA3 neurons	Adenosine A1 receptors	Slowed deactivation kinetics	Greif et al. (2000)
Gα subunit RGS-resistant "knockin" mutations				
Gαo (G184S)+/+	Whole heart, agonist mediated bradycardia	Muscarinic m2 receptors	Decreased receptor-GIRK channel-coupling efficiency	Fu et al. (2006)
	Whole heart, agonist mediated bradycardia	Adenosine A1 receptors	No effect	Fu et al. (2006)
Gαi2 (G184S)+/+	Whole heart, agonist mediated bradycardia	Muscarinic m2 receptors	Decreased receptor-GIRK channel-coupling efficiency	Fu et al. (2006)
R4 subfamily "knockouts"				
$RGS2^{-/-}$	Ventral tegmental area dopamine neurons	$GABA_B$ receptors	Increased receptor-GIRK channel-coupling efficiency	Labouebe et al. (2007)
$RGS4^{-/-}$	Sinoatrial nodel cells	Muscarinic m2 receptors	Slowed deactivation kinetics	Cifelli et al. (2008)
	Sinoatrial nodel cells Atrial myocytes in primary culture	Muscarinic m2 receptors	No effect	Wydevenet al. (2014)

Continued

Table 2 Genetically Engineered Mice Effecting RGS-GIRK Interactions—cont'd

Targeted Gene	Native GIRK-Expressing Cell/Tissue Type	Endogenous GPCR	GIRK Function Effected	References
R7 subfamily "knockouts"				
$RGS6^{-/-}$	Sinoatrial nodel cells Atrial myocytes in primary culture	Muscarinic m2 receptors	Slowed deactivation kinetics	Posokhova et al. (2010); Yang et al. (2010)
	Hippocampal neurons in primary culture	$GABA_B$ receptors	No effect	Ostrovskaya et al. (2014)
	Cerebellar granule neurons in primary culture	$GABA_B$ receptors	Slowed deactivation kinetics	Maity et al. (2012)
$RGS7^{-/-}$	Hippocampal neurons in primary culture	$GABA_B$ receptors	Slowed deactivation kinetics	Ostrovskaya et al. (2014)
R7-associated protein "knockouts"				
$G\beta5^{-/-}$	Hippocampal neurons in primary culture	$GABA_B$ receptors	Slowed deactivation kinetics	Xie et al. (2010)
	Sinoatrial nodel cells atrial myocytes in primary culture	Muscarinic m2 receptors	Slowed deactivation kinetics	Wydeven et al. (2014)
$R7BP^{-/-}$	Hippocampal neurons in primary culture	$GABA_B$ receptors	Slowed deactivation kinetics	Ostrovskaya et al. (2014); Zhou et al. (2012)

in the GIRK deactivation time course when receptors are coupled to RGS-resistant $G\alpha i/o$ subunits (Jeong & Ikeda, 2001; Mutneja et al., 2005), indicate single RGS knockouts (i.e., RGS2, RGS4, RGS6, and RGS7) affect only a small subset of the total endogenous RGS influence on GIRK channel gating (Cifelli et al., 2008; Ostrovskaya et al., 2014; Posokhova et al., 2010). The relative contribution of each RGS protein toward accelerating the GIRK channel gating kinetics appears to be cell-specific, determined by different RGS isoform expression levels within specific neurons of the nervous system, which may occur through either collision-coupled or precoupled GPCR–effector mechanisms.

Despite the observed RGS redundancy at the cellular level, the broader systemic phenotypes associated with $RGS^{-/-}$ mice represent the combined RGS influence on a multitude of other signaling pathways beyond GIRK channel regulation and predict phenotypic expression in genetically afflicted individuals carrying single RGS gene mutations that may cause epilepsy, ataxia, memory disorders, and other neurological disorders (Terzi, Stergiou, King, & Zachariou, 2009). Bradyopsia associated with RGS9 gene mutations provide a context for RGS dysfunction in humans (Nishiguchi et al., 2004). Hypertension associated with RGS2 gene mutations that heighten global Gq/11 signaling provides yet another (Yang et al., 2005).

RGS knockout mice will continue to be an important animal model providing insight on the implications and role of specific RGS isoforms on complex neurological functions (Table 2), with heterologous expression studies providing mechanistic insight (Table 1). A comprehensive understanding of the impact of individual RGS proteins on GPCR-regulated effectors in specific neuronal cell types will aid current drug development efforts targeting specific RGS protein function within the nervous system (Kimple, Bosch, Giguere, & Siderovski, 2011; Neubig, 2002; Sjogren, Blazer, & Neubig, 2010; Sjogren & Neubig, 2010; Storaska et al., 2013; Zhong & Neubig, 2001).

REFERENCES

Anderson, G. R., Lujan, R., Semenov, A., Pravetoni, M., Posokhova, E. N., Song, J. H., et al. (2007). Expression and localization of RGS9-2/G β5/R7BP complex *in vivo* is set by dynamic control of its constitutive degradation by cellular cysteine proteases. *The Journal of Neuroscience*, *27*(51), 14117–14127. http://dx.doi.org/10.1523/JNEUROSCI.3884-07.2007.

Andrade, R., Malenka, R. C., & Nicoll, R. A. (1986). A G protein couples serotonin and GABAB receptors to the same channels in hippocampus. *Science*, *234*(4781), 1261–1265.

Ang, R., Opel, A., & Tinker, A. (2012). The role of inhibitory G proteins and regulators of G protein signaling in the *in vivo* control of heart rate and predisposition to cardiac arrhythmias. *Frontiers in Physiology*, *3*, 96. http://dx.doi.org/10.3389/fphys.2012.00096.

Anger, T., Klintworth, N., Stumpf, C., Daniel, W. G., Mende, U., & Garlichs, C. D. (2007). RGS protein specificity towards Gq- and Gi/o-mediated ERK 1/2 and Akt activation, *in vitro*. *Journal of Biochemistry and Molecular Biology*, *40*(6), 899–910.

Audo, I., Bujakowska, K., Orhan, E., Poloschek, C. M., Defoort-Dhellemmes, S., Drumare, I., et al. (2012). Whole-exome sequencing identifies mutations in GPR179 leading to autosomal-recessive complete congenital stationary night blindness. *American Journal of Human Genetics*, *90*(2), 321–330. http://dx.doi.org/10.1016/j.ajhg.2011.12.007.

Bajic, D., Koike, M., Albsoul-Younes, A. M., Nakajima, S., & Nakajima, Y. (2002). Two different inward rectifier K+ channels are effectors for transmitter-induced slow excitation in brain neurons. *Proceedings of the National Academy of Sciences of the United States of America*, *99*(22), 14494–14499. http://dx.doi.org/10.1073/pnas.222379999.

Bender, K., Nasrollahzadeh, P., Timpert, M., Liu, B., Pott, L., & Kienitz, M. C. (2008). A role for RGS10 in beta-adrenergic modulation of G-protein-activated K+ (GIRK) channel current in rat atrial myocytes. *The Journal of Physiology*, *586*(8), 2049–2060. http://dx.doi.org/10.1113/jphysiol.2007.148346.

Benians, A., Leaney, J. L., & Tinker, A. (2003). Agonist unbinding from receptor dictates the nature of deactivation kinetics of G protein-gated K+ channels. *Proceedings of the National Academy of Sciences of the United States of America*, *100*(10), 6239–6244. http://dx.doi.org/10.1073/pnas.1037595100.

Berlin, S., Keren-Raifman, T., Castel, R., Rubinstein, M., Dessauer, C. W., Ivanina, T., et al. (2010). G α(i) and G βγ jointly regulate the conformations of a G βγ effector, the neuronal G protein-activated K+ channel (GIRK). *The Journal of Biological Chemistry*, *285*(9), 6179–6185. http://dx.doi.org/10.1074/jbc.M109.085944.

Berman, D. M., Wilkie, T. M., & Gilman, A. G. (1996). GAIP and RGS4 are GTPase-activating proteins for the Gi subfamily of G protein α subunits. *Cell*, *86*(3), 445–452.

Bijveld, M. M., Florijn, R. J., Bergen, A. A., van den Born, L. I., Kamermans, M., Prick, L., et al. (2013). Genotype and phenotype of 101 dutch patients with congenital stationary night blindness. *Ophthalmology*, *120*(10), 2072–2081. http://dx.doi.org/10.1016/j.ophtha.2013.03.002.

Bingen, B. O., Neshati, Z., Askar, S. F., Kazbanov, I. V., Ypey, D. L., Panfilov, A. V., et al. (2013). Atrium-specific Kir3.x determines inducibility, dynamics, and termination of fibrillation by regulating restitution-driven alternans. *Circulation*, *128*(25), 2732–2744. http://dx.doi.org/10.1161/CIRCULATIONAHA.113.005019.

Breitwieser, G. E. (1991). G protein-mediated ion channel activation. *Hypertension*, *17*(5), 684–692.

Brown, A. M. (1990). Regulation of heartbeat by G protein-coupled ion channels. *The American Journal of Physiology*, *259*(6 Pt. 2), H1621–H1628.

Cabrera, J. L., de Freitas, F., Satpaev, D. K., & Slepak, V. Z. (1998). Identification of the Gβ5-RGS7 protein complex in the retina. *Biochemical and Biophysical Research Communications*, *249*(3), 898–902. http://dx.doi.org/10.1006/bbrc.1998.9218.

Cha, T. J., Ehrlich, J. R., Chartier, D., Qi, X. Y., Xiao, L., & Nattel, S. (2006). Kir3-based inward rectifier potassium current: Potential role in atrial tachycardia remodeling effects on atrial repolarization and arrhythmias. *Circulation*, *113*(14), 1730–1737. http://dx.doi.org/10.1161/CIRCULATIONAHA.105.561738.

Chen, C. K., Burns, M. E., He, W., Wensel, T. G., Baylor, D. A., & Simon, M. I. (2000). Slowed recovery of rod photoresponse in mice lacking the GTPase accelerating protein RGS9-1. *Nature*, *403*(6769), 557–560. http://dx.doi.org/10.1038/35000601.

Chen, C. K., Eversole-Cire, P., Zhang, H., Mancino, V., Chen, Y. J., He, W., et al. (2003). Instability of GGL domain-containing RGS proteins in mice lacking the G protein beta-subunit Gβ5. *Proceedings of the National Academy of Sciences of the United States of America*, *100*(11), 6604–6609. http://dx.doi.org/10.1073/pnas.0631825100.

Chen, X., & Johnston, D. (2005). Constitutively active G-protein-gated inwardly rectifying K+ channels in dendrites of hippocampal CA1 pyramidal neurons. *The Journal of Neuroscience*, *25*(15), 3787–3792. http://dx.doi.org/10.1523/JNEUROSCI.5312-04.2005.

Chen, F. S., Shim, H., Morhardt, D., Dallman, R., Krahn, E., McWhinney, L., et al. (2010). Functional redundancy of R7 RGS proteins in ON-bipolar cell dendrites. *Investigative Ophthalmology & Visual Science, 51*(2), 686–693. http://dx.doi.org/10.1167/iovs.09-4084.
Chuang, H. H., & Chuang, A. Y. (2012). RGS proteins maintain robustness of GPCR-GIRK coupling by selective stimulation of the G protein subunit Gαo. *Science Signaling, 5*(212), ra15. http://dx.doi.org/10.1126/scisignal.2002202.
Chuang, H. H., Yu, M., Jan, Y. N., & Jan, L. Y. (1998). Evidence that the nucleotide exchange and hydrolysis cycle of G proteins causes acute desensitization of G-protein gated inward rectifier K+ channels. *Proceedings of the National Academy of Sciences of the United States of America, 95*(20), 11727–11732.
Cifelli, C., Rose, R. A., Zhang, H., Voigtlaender-Bolz, J., Bolz, S. S., Backx, P. H., et al. (2008). RGS4 regulates parasympathetic signaling and heart rate control in the sinoatrial node. *Circulation Research, 103*(5), 527–535. http://dx.doi.org/10.1161/CIRCRESAHA.108.180984.
Citri, A., & Yarden, Y. (2006). EGF-ERBB signalling: Towards the systems level. *Nature Reviews. Molecular Cell Biology, 7*(7), 505–516. http://dx.doi.org/10.1038/nrm1962.
Clancy, S. M., Fowler, C. E., Finley, M., Suen, K. F., Arrabit, C., Berton, F., et al. (2005). Pertussis-toxin-sensitive Gα subunits selectively bind to C-terminal domain of neuronal GIRK channels: Evidence for a heterotrimeric G-protein-channel complex. *Molecular and Cellular Neurosciences, 28*(2), 375–389. http://dx.doi.org/10.1016/j.mcn.2004.10.009.
Cohen, S. P., Buckley, B. K., Kosloff, M., Garland, A. L., Bosch, D. E., Cheng, G., Jr., et al. (2012). Regulator of G-protein signaling-21 (RGS21) is an inhibitor of bitter gustatory signaling found in lingual and airway epithelia. *The Journal of Biological Chemistry, 287*(50), 41706–41719. http://dx.doi.org/10.1074/jbc.M112.423806.
Cowan, C. W., Fariss, R. N., Sokal, I., Palczewski, K., & Wensel, T. G. (1998). High expression levels in cones of RGS9, the predominant GTPase accelerating protein of rods. *Proceedings of the National Academy of Sciences of the United States of America, 95*(9), 5351–5356.
D'Adamo, M. C., Catacuzzeno, L., Di Giovanni, G., Franciolini, F., & Pessia, M. (2013). K(+) channelepsy: Progress in the neurobiology of potassium channels and epilepsy. *Frontiers in Cellular Neuroscience, 7*, 134. http://dx.doi.org/10.3389/fncel.2013.00134.
De Vries, L., Mousli, M., Wurmser, A., & Farquhar, M. G. (1995). GAIP, a protein that specifically interacts with the trimeric G protein Gαi3, is a member of a protein family with a highly conserved core domain. *Proceedings of the National Academy of Sciences of the United States of America, 92*(25), 11916–11920.
De Vries, L., Zheng, B., Fischer, T., Elenko, E., & Farquhar, M. G. (2000). The regulator of G protein signaling family. *Annual Review of Pharmacology and Toxicology, 40*, 235–271. http://dx.doi.org/10.1146/annurev.pharmtox.40.1.235.
Dhingra, A., Faurobert, E., Dascal, N., Sterling, P., & Vardi, N. (2004). A retinal-specific regulator of G-protein signaling interacts with Gα(o) and accelerates an expressed metabotropic glutamate receptor 6 cascade. *The Journal of Neuroscience, 24*(25), 5684–5693. http://dx.doi.org/10.1523/JNEUROSCI.0492-04.2004.
Dolphin, A. C. (2003). G protein modulation of voltage-gated calcium channels. *Pharmacological Reviews, 55*(4), 607–627. http://dx.doi.org/10.1124/pr.55.4.3.
Doupnik, C. A. (2008). GPCR-Kir channel signaling complexes: Defining rules of engagement. *Journal of Receptor and Signal Transduction Research, 28*(1–2), 83–91. http://dx.doi.org/10.1080/10799890801941970.
Doupnik, C. A., Davidson, N., Lester, H. A., & Kofuji, P. (1997). RGS proteins reconstitute the rapid gating kinetics of Gβγ-activated inwardly rectifying K+ channels. *Proceedings of the National Academy of Sciences of the United States of America, 94*(19), 10461–10466.

Doupnik, C. A., Jaen, C., & Zhang, Q. (2004). Measuring the modulatory effects of RGS proteins on GIRK channels. *Methods in Enzymology, 389*, 131–154. http://dx.doi.org/10.1016/S0076-6879(04)89009-8.

Doupnik, C. A., Xu, T., & Shinaman, J. M. (2001). Profile of RGS expression in single rat atrial myocytes. *Biochimica et Biophysica Acta, 1522*(2), 97–107.

Drenan, R. M., Doupnik, C. A., Boyle, M. P., Muglia, L. J., Huettner, J. E., Linder, M. E., et al. (2005). Palmitoylation regulates plasma membrane-nuclear shuttling of R7BP, a novel membrane anchor for the RGS7 family. *The Journal of Cell Biology, 169*(4), 623–633. http://dx.doi.org/10.1083/jcb.200502007.

Drenan, R. M., Doupnik, C. A., Jayaraman, M., Buchwalter, A. L., Kaltenbronn, K. M., Huettner, J. E., et al. (2006). R7BP augments the function of RGS7*Gβ5 complexes by a plasma membrane-targeting mechanism. *The Journal of Biological Chemistry, 281*(38), 28222–28231. http://dx.doi.org/10.1074/jbc.M604428200.

Druey, K. M. (2009). Regulation of G-protein-coupled signaling pathways in allergic inflammation. *Immunologic Research, 43*(1–3), 62–76. http://dx.doi.org/10.1007/s12026-008-8050-0.

Fajardo-Serrano, A., Wydeven, N., Young, D., Watanabe, M., Shigemoto, R., Martemyanov, K. A., et al. (2013). Association of Rgs7/Gβ5 complexes with Girk channels and GABAB receptors in hippocampal CA1 pyramidal neurons. *Hippocampus, 23*(12), 1231–1245. http://dx.doi.org/10.1002/hipo.22161.

Fowler, C. E., Aryal, P., Suen, K. F., & Slesinger, P. A. (2007). Evidence for association of GABA(B) receptors with Kir3 channels and regulators of G protein signalling (RGS4) proteins. *The Journal of Physiology, 580*(Pt. 1), 51–65. http://dx.doi.org/10.1113/jphysiol.2006.123216.

Fu, Y., Huang, X., Piao, L., Lopatin, A. N., & Neubig, R. R. (2007). Endogenous RGS proteins modulate SA and AV nodal functions in isolated heart: Implications for sick sinus syndrome and AV block. *American Journal of Physiology. Heart and Circulatory Physiology, 292*(5), H2532–H2539. http://dx.doi.org/10.1152/ajpheart.01391.2006.

Fu, Y., Huang, X., Zhong, H., Mortensen, R. M., D'Alecy, L. G., & Neubig, R. R. (2006). Endogenous RGS proteins and Gα subtypes differentially control muscarinic and adenosine-mediated chronotropic effects. *Circulation Research, 98*(5), 659–666. http://dx.doi.org/10.1161/01.RES.0000207497.50477.60.

Gold, S. J., Ni, Y. G., Dohlman, H. G., & Nestler, E. J. (1997). Regulators of G-protein signaling (RGS) proteins: Region-specific expression of nine subtypes in rat brain. *The Journal of Neuroscience, 17*(20), 8024–8037.

Grafstein-Dunn, E., Young, K. H., Cockett, M. I., & Khawaja, X. Z. (2001). Regional distribution of regulators of G-protein signaling (RGS) 1, 2, 13, 14, 16, and GAIP messenger ribonucleic acids by in situ hybridization in rat brain. *Brain Research. Molecular Brain Research, 88*(1–2), 113–123.

Granneman, J. G., Zhai, Y., Zhu, Z., Bannon, M. J., Burchett, S. A., Schmidt, C. J., et al. (1998). Molecular characterization of human and rat RGS 9 L, a novel splice variant enriched in dopamine target regions, and chromosomal localization of the RGS 9 gene. *Molecular Pharmacology, 54*(4), 687–694.

Greif, G. J., Sodickson, D. L., Bean, B. P., Neer, E. J., & Mende, U. (2000). Altered regulation of potassium and calcium channels by GABA(B) and adenosine receptors in hippocampal neurons from mice lacking Gα(o). *Journal of Neurophysiology, 83*(2), 1010–1018.

Hao, J., Michalek, C., Zhang, W., Zhu, M., Xu, X., & Mende, U. (2006). Regulation of cardiomyocyte signaling by RGS proteins: Differential selectivity towards G proteins and susceptibility to regulation. *Journal of Molecular and Cellular Cardiology, 41*(1), 51–61. http://dx.doi.org/10.1016/j.yjmcc.2006.04.003.

Harvey, R. D., & Belevych, A. E. (2003). Muscarinic regulation of cardiac ion channels. *British Journal of Pharmacology, 139*(6), 1074–1084. http://dx.doi.org/10.1038/sj.bjp.0705338.

He, W., Cowan, C. W., & Wensel, T. G. (1998). RGS9, a GTPase accelerator for phototransduction. *Neuron, 20*(1), 95–102.

Hepler, J. R. (2005). R7BP: A surprising new link between G proteins, RGS proteins, and nuclear signaling in the brain. *Science's STKE, 2005*(294), pe38. http://dx.doi.org/10.1126/stke.2942005pe38.

Hepler, J. R., Berman, D. M., Gilman, A. G., & Kozasa, T. (1997). RGS4 and GAIP are GTPase-activating proteins for Gqα and block activation of phospholipase C beta by gamma-thio-GTP-Gqα. *Proceedings of the National Academy of Sciences of the United States of America, 94*(2), 428–432.

Herlitze, S., Ruppersberg, J. P., & Mark, M. D. (1999). New roles for RGS2, 5 and 8 on the ratio-dependent modulation of recombinant GIRK channels expressed in Xenopus oocytes. *The Journal of Physiology, 517*(Pt 2), 341–352.

Heximer, S. P., Watson, N., Linder, M. E., Blumer, K. J., & Hepler, J. R. (1997). RGS2/G0S8 is a selective inhibitor of Gqα function. *Proceedings of the National Academy of Sciences of the United States of America, 94*(26), 14389–14393.

Hibino, H., Inanobe, A., Furutani, K., Murakami, S., Findlay, I., & Kurachi, Y. (2010). Inwardly rectifying potassium channels: Their structure, function, and physiological roles. *Physiological Reviews, 90*(1), 291–366. http://dx.doi.org/10.1152/physrev.00021.2009.

Hollinger, S., & Hepler, J. R. (2002). Cellular regulation of RGS proteins: Modulators and integrators of G protein signaling. *Pharmacological Reviews, 54*(3), 527–559.

Hooks, S. B., Waldo, G. L., Corbitt, J., Bodor, E. T., Krumins, A. M., & Harden, T. K. (2003). RGS6, RGS7, RGS9, and RGS11 stimulate GTPase activity of Gi family G-proteins with differential selectivity and maximal activity. *The Journal of Biological Chemistry, 278*(12), 10087–10093. http://dx.doi.org/10.1074/jbc.M211382200.

Huang, C. L., Feng, S., & Hilgemann, D. W. (1998). Direct activation of inward rectifier potassium channels by PIP2 and its stabilization by Gβγ. *Nature, 391*(6669), 803–806. http://dx.doi.org/10.1038/35882.

Hunt, T. W., Fields, T. A., Casey, P. J., & Peralta, E. G. (1996). RGS10 is a selective activator of Gαi GTPase activity. *Nature, 383*(6596), 175–177. http://dx.doi.org/10.1038/383175a0.

Ikeda, S. R. (1996). Voltage-dependent modulation of N-type calcium channels by G-protein βγ subunits. *Nature, 380*(6571), 255–258. http://dx.doi.org/10.1038/380255a0.

Jaen, C., & Doupnik, C. A. (2005). Neuronal Kir3.1/Kir3.2a channels coupled to serotonin 1A and muscarinic m2 receptors are differentially modulated by the "short" RGS3 isoform. *Neuropharmacology, 49*(4), 465–476. http://dx.doi.org/10.1016/j.neuropharm.2005.04.010.

Jaen, C., & Doupnik, C. A. (2006). RGS3 and RGS4 differentially associate with G protein-coupled receptor-Kir3 channel signaling complexes revealing two modes of RGS modulation. Precoupling and collision coupling. *The Journal of Biological Chemistry, 281*(45), 34549–34560. http://dx.doi.org/10.1074/jbc.M603177200.

Jayaraman, M., Zhou, H., Jia, L., Cain, M. D., & Blumer, K. J. (2009). R9AP and R7BP: Traffic cops for the RGS7 family in phototransduction and neuronal GPCR signaling. *Trends in Pharmacological Sciences, 30*(1), 17–24. http://dx.doi.org/10.1016/j.tips.2008.10.002.

Jeong, S. W., & Ikeda, S. R. (2001). Differential regulation of G protein-gated inwardly rectifying K(+) channel kinetics by distinct domains of RGS8. *The Journal of Physiology, 535*(Pt 2), 335–347.

Jia, L., Linder, M. E., & Blumer, K. J. (2011). Gi/o signaling and the palmitoyltransferase DHHC2 regulate palmitate cycling and shuttling of RGS7 family-binding protein. *The Journal of Biological Chemistry, 286*(15), 13695–13703. http://dx.doi.org/10.1074/jbc.M110.193763.

Jones, D. L., Tuomi, J. M., & Chidiac, P. (2012). Role of cholinergic innervation and RGS2 in atrial arrhythmia. *Frontiers in Physiology*, *3*, 239. http://dx.doi.org/10.3389/fphys.2012.00239.

Kardestuncer, T., Wu, H., Lim, A. L., & Neer, E. J. (1998). Cardiac myocytes express mRNA for ten RGS proteins: Changes in RGS mRNA expression in ventricular myocytes and cultured atria. *FEBS Letters*, *438*(3), 285–288.

Kehrl, J. H. (2004). G-protein-coupled receptor signaling, RGS proteins, and lymphocyte function. *Critical Reviews in Immunology*, *24*(6), 409–423.

Kim, C. S., & Johnston, D. (2015). A1 adenosine receptor-mediated GIRK channels contribute to the resting conductance of CA1 neurons in the dorsal hippocampus. *Journal of Neurophysiology*, *113*, 2511–2523. http://dx.doi.org/10.1152/jn.00951.2014.

Kim, G., Jung, S., Son, H., Kim, S., Choi, J., Lee, D. H., et al. (2014). The GABAB receptor associates with regulators of G-protein signaling 4 protein in the mouse prefrontal cortex and hypothalamus. *BMB Reports*, *47*(6), 324–329.

Kimple, A. J., Bosch, D. E., Giguere, P. M., & Siderovski, D. P. (2011). Regulators of G-protein signaling and their Gα substrates: Promises and challenges in their use as drug discovery targets. *Pharmacological Reviews*, *63*(3), 728–749. http://dx.doi.org/10.1124/pr.110.003038.

Kitano, H. (2004). Biological robustness. *Nature Reviews. Genetics*, *5*(11), 826–837. http://dx.doi.org/10.1038/nrg1471.

Kobayashi, T., & Ikeda, K. (2006). G protein-activated inwardly rectifying potassium channels as potential therapeutic targets. *Current Pharmaceutical Design*, *12*(34), 4513–4523.

Koike, C., Obara, T., Uriu, Y., Numata, T., Sanuki, R., Miyata, K., et al. (2010). TRPM1 is a component of the retinal ON bipolar cell transduction channel in the mGluR6 cascade. *Proceedings of the National Academy of Sciences of the United States of America*, *107*(1), 332–337. http://dx.doi.org/10.1073/pnas.0912730107.

Kovoor, A., Chen, C. K., He, W., Wensel, T. G., Simon, M. I., & Lester, H. A. (2000). Co-expression of Gβ5 enhances the function of two Ggamma subunit-like domain-containing regulators of G protein signaling proteins. *The Journal of Biological Chemistry*, *275*(5), 3397–3402.

Labouebe, G., Lomazzi, M., Cruz, H. G., Creton, C., Lujan, R., Li, M., et al. (2007). RGS2 modulates coupling between GABAB receptors and GIRK channels in dopamine neurons of the ventral tegmental area. *Nature Neuroscience*, *10*(12), 1559–1568. http://dx.doi.org/10.1038/nn2006.

Lambert, N. A., Johnston, C. A., Cappell, S. D., Kuravi, S., Kimple, A. J., Willard, F. S., et al. (2010). Regulators of G-protein signaling accelerate GPCR signaling kinetics and govern sensitivity solely by accelerating GTPase activity. *Proceedings of the National Academy of Sciences of the United States of America*, *107*(15), 7066–7071. http://dx.doi.org/10.1073/pnas.0912934107.

Lan, K. L., Zhong, H., Nanamori, M., & Neubig, R. R. (2000). Rapid kinetics of regulator of G-protein signaling (RGS)-mediated Gαi and Gαo deactivation. Gα specificity of RGS4 and RGS7. *The Journal of Biological Chemistry*, *275*(43), 33497–33503. http://dx.doi.org/10.1074/jbc.M005785200.

Larminie, C., Murdock, P., Walhin, J. P., Duckworth, M., Blumer, K. J., Scheideler, M. A., et al. (2004). Selective expression of regulators of G-protein signaling (RGS) in the human central nervous system. *Brain Research. Molecular Brain Research*, *122*(1), 24–34. http://dx.doi.org/10.1016/j.molbrainres.2003.11.014.

Lavine, N., Ethier, N., Oak, J. N., Pei, L., Liu, F., Trieu, P., et al. (2002). G protein-coupled receptors form stable complexes with inwardly rectifying potassium channels and adenylyl cyclase. *The Journal of Biological Chemistry*, *277*(48), 46010–46019. http://dx.doi.org/10.1074/jbc.M205035200.

Lei, Q., Jones, M. B., Talley, E. M., Garrison, J. C., & Bayliss, D. A. (2003). Molecular mechanisms mediating inhibition of G protein-coupled inwardly-rectifying K+ channels. *Molecules and Cells*, *15*(1), 1–9.

Liapis, E., Sandiford, S., Wang, Q., Gaidosh, G., Motti, D., Levay, K., et al. (2012). Subcellular localization of regulator of G protein signaling RGS7 complex in neurons and transfected cells. *Journal of Neurochemistry*, *122*(3), 568–581. http://dx.doi.org/10.1111/j.1471-4159.2012.07811.x.

Logothetis, D. E., Kurachi, Y., Galper, J., Neer, E. J., & Clapham, D. E. (1987). The βγ subunits of GTP-binding proteins activate the muscarinic K+ channel in heart. *Nature*, *325*(6102), 321–326. http://dx.doi.org/10.1038/325321a0.

Lomazzi, M., Slesinger, P. A., & Luscher, C. (2008). Addictive drugs modulate GIRK-channel signaling by regulating RGS proteins. *Trends in Pharmacological Sciences*, *29*(11), 544–549. http://dx.doi.org/10.1016/j.tips.2008.07.011.

Louwette, S., Van Geet, C., & Freson, K. (2012). Regulators of G protein signaling: Role in hematopoiesis, megakaryopoiesis and platelet function. *Journal of Thrombosis and Haemostasis*, *10*(11), 2215–2222. http://dx.doi.org/10.1111/j.1538-7836.2012.04903.x.

Lujan, R., Fernandez, M., de Velasco, E., Aguado, C., & Wickman, K. (2014). New insights into the therapeutic potential of Girk channels. *Trends in Neurosciences*, *37*(1), 20–29. http://dx.doi.org/10.1016/j.tins.2013.10.006.

Luscher, C., Jan, L. Y., Stoffel, M., Malenka, R. C., & Nicoll, R. A. (1997). G protein-coupled inwardly rectifying K+ channels (GIRKs) mediate postsynaptic but not presynaptic transmitter actions in hippocampal neurons. *Neuron*, *19*(3), 687–695.

Luscher, C., & Slesinger, P. A. (2010). Emerging roles for G protein-gated inwardly rectifying potassium (GIRK) channels in health and disease. *Nature Reviews. Neuroscience*, *11*(5), 301–315. http://dx.doi.org/10.1038/nrn2834.

Lyubarsky, A. L., Naarendorp, F., Zhang, X., Wensel, T., Simon, M. I., & Pugh, E. N., Jr. (2001). RGS9-1 is required for normal inactivation of mouse cone phototransduction. *Molecular Vision*, 7, 71–78.

Maity, B., Stewart, A., Yang, J., Loo, L., Sheff, D., Shepherd, A. J., et al. (2012). Regulator of G protein signaling 6 (RGS6) protein ensures coordination of motor movement by modulating GABAB receptor signaling. *The Journal of Biological Chemistry*, *287*(7), 4972–4981. http://dx.doi.org/10.1074/jbc.M111.297218.

Makino, E. R., Handy, J. W., Li, T., & Arshavsky, V. Y. (1999). The GTPase activating factor for transducin in rod photoreceptors is the complex between RGS9 and type 5 G protein beta subunit. *Proceedings of the National Academy of Sciences of the United States of America*, *96*(5), 1947–1952.

Martemyanov, K. A., Yoo, P. J., Skiba, N. P., & Arshavsky, V. Y. (2005). R7BP, a novel neuronal protein interacting with RGS proteins of the R7 family. *The Journal of Biological Chemistry*, *280*(7), 5133–5136. http://dx.doi.org/10.1074/jbc.C400596200.

Masuho, I., Celver, J., Kovoor, A., & Martemyanov, K. A. (2010). Membrane anchor R9AP potentiates GTPase-accelerating protein activity of RGS11-Gβ5 complex and accelerates inactivation of the mGluR6-G(o) signaling. *The Journal of Biological Chemistry*, *285*(7), 4781–4787. http://dx.doi.org/10.1074/jbc.M109.058511.

Mojumder, D. K., Qian, Y., & Wensel, T. G. (2009). Two R7 regulator of G-protein signaling proteins shape retinal bipolar cell signaling. *The Journal of Neuroscience*, *29*(24), 7753–7765. http://dx.doi.org/10.1523/JNEUROSCI.1794-09.2009.

Morgans, C. W., Weiwei, L., Wensel, T. G., Brown, R. L., Perez-Leon, J. A., Bearnot, B., et al. (2007). Gβ5-RGS complexes co-localize with mGluR6 in retinal ON-bipolar cells. *The European Journal of Neuroscience*, *26*(10), 2899–2905. http://dx.doi.org/10.1111/j.1460-9568.2007.05867.x.

Muslin, A. J. (2006). Tuning cardiomyocyte Gq signaling with RGS2. *Journal of Molecular and Cellular Cardiology*, *41*(1), 14–16. http://dx.doi.org/10.1016/j.yjmcc.2006.04.004.

Mutneja, M., Berton, F., Suen, K. F., Luscher, C., & Slesinger, P. A. (2005). Endogenous RGS proteins enhance acute desensitization of GABA(B) receptor-activated GIRK currents in HEK-293 T cells. *Pflügers Archiv*, *450*(1), 61–73. http://dx.doi.org/10.1007/s00424-004-1367-1.

Nakajima, Y., Nakajima, S., & Inoue, M. (1988). Pertussis toxin-insensitive G protein mediates substance P-induced inhibition of potassium channels in brain neurons. *Proceedings of the National Academy of Sciences of the United States of America*, *85*(10), 3643–3647.

Nance, M. R., Kreutz, B., Tesmer, V. M., Sterne-Marr, R., Kozasa, T., & Tesmer, J. J. (2013). Structural and functional analysis of the regulator of G protein signaling 2-Gαq complex. *Structure*, *21*(3), 438–448. http://dx.doi.org/10.1016/j.str.2012.12.016.

Neubig, R. R. (2002). Regulators of G protein signaling (RGS proteins): Novel central nervous system drug targets. *The Journal of Peptide Research*, *60*(6), 312–316.

Neubig, R. R. (2008). And the winner is ... RGS4! *Circulation Research*, *103*(5), 444–446. http://dx.doi.org/10.1161/CIRCRESAHA.108.183384.

Nicoll, R. A. (1988). The coupling of neurotransmitter receptors to ion channels in the brain. *Science*, *241*(4865), 545–551.

Nishiguchi, K. M., Sandberg, M. A., Kooijman, A. C., Martemyanov, K. A., Pott, J. W., Hagstrom, S. A., et al. (2004). Defects in RGS9 or its anchor protein R9AP in patients with slow photoreceptor deactivation. *Nature*, *427*(6969), 75–78. http://dx.doi.org/10.1038/nature02170.

Orlandi, C., Cao, Y., & Martemyanov, K. A. (2013). Orphan receptor GPR179 forms macromolecular complexes with components of metabotropic signaling cascade in retina ON-bipolar neurons. *Investigative Ophthalmology & Visual Science*, *54*(10), 7153–7161. http://dx.doi.org/10.1167/iovs.13-12907.

Orlandi, C., Posokhova, E., Masuho, I., Ray, T. A., Hasan, N., Gregg, R. G., et al. (2012). GPR158/179 regulate G protein signaling by controlling localization and activity of the RGS7 complexes. *The Journal of Cell Biology*, *197*(6), 711–719. http://dx.doi.org/10.1083/jcb.201202123.

Ostrovskaya, O., Xie, K., Masuho, I., Fajardo-Serrano, A., Lujan, R., Wickman, K., et al. (2014). RGS7/Gβ5/R7BP complex regulates synaptic plasticity and memory by modulating hippocampal GABABR-GIRK signaling. *Elife*, *3*, e02053. http://dx.doi.org/10.7554/eLife.02053.

Otis, T. S., De Koninck, Y., & Mody, I. (1993). Characterization of synaptically elicited GABAB responses using patch-clamp recordings in rat hippocampal slices. *The Journal of Physiology*, *463*, 391–407.

Peachey, N. S., Ray, T. A., Florijn, R., Rowe, L. B., Sjoerdsma, T., Contreras-Alcantara, S., et al. (2012). GPR179 is required for depolarizing bipolar cell function and is mutated in autosomal-recessive complete congenital stationary night blindness. *American Journal of Human Genetics*, *90*(2), 331–339. http://dx.doi.org/10.1016/j.ajhg.2011.12.006.

Posokhova, E., Ng, D., Opel, A., Masuho, I., Tinker, A., Biesecker, L. G., et al. (2013). Essential role of the m2R-RGS6-IKACh pathway in controlling intrinsic heart rate variability. *PLoS One*, *8*(10), e76973. http://dx.doi.org/10.1371/journal.pone.0076973.

Posokhova, E., Wydeven, N., Allen, K. L., Wickman, K., & Martemyanov, K. A. (2010). RGS6/Gβ5 complex accelerates IKACh gating kinetics in atrial myocytes and modulates parasympathetic regulation of heart rate. *Circulation Research*, *107*(11), 1350–1354. http://dx.doi.org/10.1161/CIRCRESAHA.110.224212.

Qin, K., Sethi, P. R., & Lambert, N. A. (2008). Abundance and stability of complexes containing inactive G protein-coupled receptors and G proteins. *The FASEB Journal*, *22*(8), 2920–2927. http://dx.doi.org/10.1096/fj.08-105775.

Rao, A., Dallman, R., Henderson, S., & Chen, C. K. (2007). Gβ5 is required for normal light responses and morphology of retinal ON-bipolar cells. *The Journal of Neuroscience*, *27*(51), 14199–14204. http://dx.doi.org/10.1523/JNEUROSCI.4934-07.2007.

Ray, T. A., Heath, K. M., Hasan, N., Noel, J. M., Samuels, I. S., Martemyanov, K. A., et al. (2014). GPR179 is required for high sensitivity of the mGluR6 signaling cascade in depolarizing bipolar cells. *The Journal of Neuroscience*, *34*(18), 6334–6343. http://dx.doi.org/10.1523/JNEUROSCI.4044-13.2014.

Riven, I., Iwanir, S., & Reuveny, E. (2006). GIRK channel activation involves a local rearrangement of a preformed G protein channel complex. *Neuron*, *51*(5), 561–573. http://dx.doi.org/10.1016/j.neuron.2006.08.017.

Robinson, B. F., Epstein, S. E., Beiser, G. D., & Braunwald, E. (1966). Control of heart rate by the autonomic nervous system. Studies in man on the interrelation between baroreceptor mechanisms and exercise. *Circulation Research*, *19*(2), 400–411.

Rogers, J. H., Tamirisa, P., Kovacs, A., Weinheimer, C., Courtois, M., Blumer, K. J., et al. (1999). RGS4 causes increased mortality and reduced cardiac hypertrophy in response to pressure overload. *The Journal of Clinical Investigation*, *104*(5), 567–576. http://dx.doi.org/10.1172/JCI6713.

Rogers, J. H., Tsirka, A., Kovacs, A., Blumer, K. J., Dorn, G. W., 2nd., & Muslin, A. J. (2001). RGS4 reduces contractile dysfunction and hypertrophic gene induction in Gαq overexpressing mice. *Journal of Molecular and Cellular Cardiology*, *33*(2), 209–218. http://dx.doi.org/10.1006/jmcc.2000.1307.

Ross, E. M., & Wilkie, T. M. (2000). GTPase-activating proteins for heterotrimeric G proteins: Regulators of G protein signaling (RGS) and RGS-like proteins. *Annual Review of Biochemistry*, *69*, 795–827. http://dx.doi.org/10.1146/annurev.biochem.69.1.795.

Saitoh, O., Kubo, Y., Miyatani, Y., Asano, T., & Nakata, H. (1997). RGS8 accelerates G-protein-mediated modulation of K+ currents. *Nature*, *390*(6659), 525–529. http://dx.doi.org/10.1038/37385.

Saitoh, O., Kubo, Y., Odagiri, M., Ichikawa, M., Yamagata, K., & Sekine, T. (1999). RGS7 and RGS8 differentially accelerate G protein-mediated modulation of K+ currents. *The Journal of Biological Chemistry*, *274*(14), 9899–9904.

Sarmast, F., Kolli, A., Zaitsev, A., Parisian, K., Dhamoon, A. S., Guha, P. K., et al. (2003). Cholinergic atrial fibrillation: I(K, ACh) gradients determine unequal left/right atrial frequencies and rotor dynamics. *Cardiovascular Research*, *59*(4), 863–873.

Scheschonka, A., Dessauer, C. W., Sinnarajah, S., Chidiac, P., Shi, C. S., & Kehrl, J. H. (2000). RGS3 is a GTPase-activating protein for G(iα) and G(qα) and a potent inhibitor of signaling by GTPase-deficient forms of G(qα) and G(11α). *Molecular Pharmacology*, *58*(4), 719–728.

Shen, Y., Rampino, M. A., Carroll, R. C., & Nawy, S. (2012). G-protein-mediated inhibition of the Trp channel TRPM1 requires the Gβγ dimer. *Proceedings of the National Academy of Sciences of the United States of America*, *109*(22), 8752–8757. http://dx.doi.org/10.1073/pnas.1117433109.

Shim, H., Wang, C. T., Chen, Y. L., Chau, V. Q., Fu, K. G., Yang, J., et al. (2012). Defective retinal depolarizing bipolar cells in regulators of G protein signaling (RGS) 7 and 11 double null mice. *The Journal of Biological Chemistry*, *287*(18), 14873–14879. http://dx.doi.org/10.1074/jbc.M112.345751.

Sjogren, B., Blazer, L. L., & Neubig, R. R. (2010). Regulators of G protein signaling proteins as targets for drug discovery. *Progress in Molecular Biology and Translational Science*, *91*, 81–119. http://dx.doi.org/10.1016/S1877-1173(10)91004-1.

Sjogren, B., & Neubig, R. R. (2010). Thinking outside of the "RGS box": New approaches to therapeutic targeting of regulators of G protein signaling. *Molecular Pharmacology*, *78*(4), 550–557. http://dx.doi.org/10.1124/mol.110.065219.

Slepak, V. Z. (2009). Structure, function, and localization of Gβ5-RGS complexes. *Progress in Molecular Biology and Translational Science*, *86*, 157–203. http://dx.doi.org/10.1016/S1877-1173(09)86006-7.
Snow, B. E., Krumins, A. M., Brothers, G. M., Lee, S. F., Wall, M. A., Chung, S., et al. (1998). A G protein γ subunit-like domain shared between RGS11 and other RGS proteins specifies binding to Gβ5 subunits. *Proceedings of the National Academy of Sciences of the United States of America*, *95*(22), 13307–13312.
Sodickson, D. L., & Bean, B. P. (1996). GABAB receptor-activated inwardly rectifying potassium current in dissociated hippocampal CA3 neurons. *The Journal of Neuroscience*, *16*(20), 6374–6385.
Sodickson, D. L., & Bean, B. P. (1998). Neurotransmitter activation of inwardly rectifying potassium current in dissociated hippocampal CA3 neurons: Interactions among multiple receptors. *The Journal of Neuroscience*, *18*(20), 8153–8162.
Song, J. H., Waataja, J. J., & Martemyanov, K. A. (2006). Subcellular targeting of RGS9-2 is controlled by multiple molecular determinants on its membrane anchor, R7BP. *The Journal of Biological Chemistry*, *281*(22), 15361–15369. http://dx.doi.org/10.1074/jbc.M600749200.
Stanfield, P. R., Nakajima, S., & Nakajima, Y. (2002). Constitutively active and G-protein coupled inward rectifier K+ channels: Kir2.0 and Kir3.0. *Reviews of Physiology, Biochemistry and Pharmacology*, *145*, 47–179.
Stelling, J., Sauer, U., Szallasi, Z., Doyle, F. J., 3rd., & Doyle, J. (2004). Robustness of cellular functions. *Cell*, *118*(6), 675–685. http://dx.doi.org/10.1016/j.cell.2004.09.008.
Stewart, A., Huang, J., & Fisher, R. A. (2012). RGS proteins in heart: Brakes on the vagus. *Frontiers in Physiology*, *3*, 95. http://dx.doi.org/10.3389/fphys.2012.00095.
Storaska, A. J., Mei, J. P., Wu, M., Li, M., Wade, S. M., Blazer, L. L., et al. (2013). Reversible inhibitors of regulators of G-protein signaling identified in a high-throughput cell-based calcium signaling assay. *Cellular Signalling*, *25*(12), 2848–2855. http://dx.doi.org/10.1016/j.cellsig.2013.09.007.
Takano, K., Yasufuku-Takano, J., Kozasa, T., Singer, W. D., Nakajima, S., & Nakajima, Y. (1996). Gq/11 and PLC-β1 mediate the substance P-induced inhibition of an inward rectifier K+ channel in brain neurons. *Journal of Neurophysiology*, *76*(3), 2131–2136.
Tamirisa, P., Blumer, K. J., & Muslin, A. J. (1999). RGS4 inhibits G-protein signaling in cardiomyocytes. *Circulation*, *99*(3), 441–447.
Tedford, H. W., & Zamponi, G. W. (2006). Direct G protein modulation of Cav2 calcium channels. *Pharmacological Reviews*, *58*(4), 837–862. http://dx.doi.org/10.1124/pr.58.4.11.
Terzi, D., Stergiou, E., King, S. L., & Zachariou, V. (2009). Regulators of G protein signaling in neuropsychiatric disorders. *Progress in Molecular Biology and Translational Science*, *86*, 299–333. http://dx.doi.org/10.1016/S1877-1173(09)86010-9.
Tesmer, J. J. (2009). Structure and function of regulator of G protein signaling homology domains. *Progress in Molecular Biology and Translational Science*, *86*, 75–113. http://dx.doi.org/10.1016/S1877-1173(09)86004-3.
Tesmer, J. J., Berman, D. M., Gilman, A. G., & Sprang, S. R. (1997). Structure of RGS4 bound to AlF4–activated Giα1: Stabilization of the transition state for GTP hydrolysis. *Cell*, *89*(2), 251–261.
Voigt, N., Abu-Taha, I., Heijman, J., & Dobrev, D. (2014). Constitutive activity of the acetylcholine-activated potassium current IK,ACh in cardiomyocytes. *Advances in Pharmacology*, *70*, 393–409. http://dx.doi.org/10.1016/B978-0-12-417197-8.00013-4.
Voigt, N., Maguy, A., Yeh, Y. H., Qi, X., Ravens, U., Dobrev, D., et al. (2008). Changes in I K,ACh single-channel activity with atrial tachycardia remodelling in canine atrial cardiomyocytes. *Cardiovascular Research*, 77(1), 35–43. http://dx.doi.org/10.1093/cvr/cvm051.

von Buchholtz, L., Elischer, A., Tareilus, E., Gouka, R., Kaiser, C., Breer, H., et al. (2004). RGS21 is a novel regulator of G protein signalling selectively expressed in subpopulations of taste bud cells. *The European Journal of Neuroscience*, *19*(6), 1535–1544. http://dx.doi.org/10.1111/j.1460-9568.2004.03257.x.

Wang, W., Whorton, M. R., & MacKinnon, R. (2014). Quantitative analysis of mammalian GIRK2 channel regulation by G proteins, the signaling lipid PIP2 and Na + in a reconstituted system. *Elife*, *3*, e03671. http://dx.doi.org/10.7554/eLife.03671.

Watson, N., Linder, M. E., Druey, K. M., Kehrl, J. H., & Blumer, K. J. (1996). RGS family members: GTPase-activating proteins for heterotrimeric G-protein α-subunits. *Nature*, *383*(6596), 172–175. http://dx.doi.org/10.1038/383172a0.

Willars, G. B. (2006). Mammalian RGS proteins: Multifunctional regulators of cellular signalling. *Seminars in Cell & Developmental Biology*, *17*(3), 363–376. http://dx.doi.org/10.1016/j.semcdb.2006.03.005.

Wydeven, N., Posokhova, E., Xia, Z., Martemyanov, K. A., & Wickman, K. (2014). RGS6, but not RGS4, is the dominant regulator of G protein signaling (RGS) modulator of the parasympathetic regulation of mouse heart rate. *The Journal of Biological Chemistry*, *289*(4), 2440–2449. http://dx.doi.org/10.1074/jbc.M113.520742.

Xie, K., Allen, K. L., Kourrich, S., Colon-Saez, J., Thomas, M. J., Wickman, K., et al. (2010). Gβ5 recruits R7 RGS proteins to GIRK channels to regulate the timing of neuronal inhibitory signaling. *Nature Neuroscience*, *13*(6), 661–663. http://dx.doi.org/10.1038/nn.2549.

Xie, G. X., & Palmer, P. P. (2007). How regulators of G protein signaling achieve selective regulation. *Journal of Molecular Biology*, *366*(2), 349–365. http://dx.doi.org/10.1016/j.jmb.2006.11.045.

Yamada, M., Inanobe, A., & Kurachi, Y. (1998). G protein regulation of potassium ion channels. *Pharmacological Reviews*, *50*(4), 723–760.

Yang, J., Huang, J., Maity, B., Gao, Z., Lorca, R. A., Gudmundsson, H., et al. (2010). RGS6, a modulator of parasympathetic activation in heart. *Circulation Research*, *107*(11), 1345–1349. http://dx.doi.org/10.1161/CIRCRESAHA.110.224220.

Yang, J., Kamide, K., Kokubo, Y., Takiuchi, S., Tanaka, C., Banno, M., et al. (2005). Genetic variations of regulator of G-protein signaling 2 in hypertensive patients and in the general population. *Journal of Hypertension*, *23*(8), 1497–1505.

Zhang, W., Anger, T., Su, J., Hao, J., Xu, X., Zhu, M., et al. (2006). Selective loss of fine tuning of Gq/11 signaling by RGS2 protein exacerbates cardiomyocyte hypertrophy. *The Journal of Biological Chemistry*, *281*(9), 5811–5820. http://dx.doi.org/10.1074/jbc.M507871200.

Zhang, J., Jeffrey, B. G., Morgans, C. W., Burke, N. S., Haley, T. L., Duvoisin, R. M., et al. (2010). RGS7 and -11 complexes accelerate the ON-bipolar cell light response. *Investigative Ophthalmology & Visual Science*, *51*(2), 1121–1129. http://dx.doi.org/10.1167/iovs.09-4163.

Zhang, Q., Pacheco, M. A., & Doupnik, C. A. (2002). Gating properties of GIRK channels activated by Gαo- and Gαi-coupled muscarinic m2 receptors in Xenopus oocytes: The role of receptor precoupling in RGS modulation. *The Journal of Physiology*, *545*(Pt 2), 355–373.

Zhang, S., Watson, N., Zahner, J., Rottman, J. N., Blumer, K. J., & Muslin, A. J. (1998). RGS3 and RGS4 are GTPase activating proteins in the heart. *Journal of Molecular and Cellular Cardiology*, *30*(2), 269–276. http://dx.doi.org/10.1006/jmcc.1997.0591.

Zheng, B., Ma, Y. C., Ostrom, R. S., Lavoie, C., Gill, G. N., Insel, P. A., et al. (2001). RGS-PX1, a GAP for Gαs and sorting nexin in vesicular trafficking. *Science*, *294*(5548), 1939–1942. http://dx.doi.org/10.1126/science.1064757.

Zhong, H., & Neubig, R. R. (2001). Regulator of G protein signaling proteins: Novel multifunctional drug targets. *The Journal of Pharmacology and Experimental Therapeutics*, *297*(3), 837–845.

Zhou, H., Chisari, M., Raehal, K. M., Kaltenbronn, K. M., Bohn, L. M., Mennerick, S. J., et al. (2012). GIRK channel modulation by assembly with allosterically regulated RGS proteins. *Proceedings of the National Academy of Sciences of the United States of America*, *109*(49), 19977–19982. http://dx.doi.org/10.1073/pnas.1214337109.

Zou, M. X., Roy, A. A., Zhao, Q., Kirshenbaum, L. A., Karmazyn, M., & Chidiac, P. (2006). RGS2 is upregulated by and attenuates the hypertrophic effect of α1-adrenergic activation in cultured ventricular myocytes. *Cellular Signalling*, *18*(10), 1655–1663. http://dx.doi.org/10.1016/j.cellsig.2006.01.012.

CHAPTER FOUR

Structural Insights into GIRK Channel Function

Ian W. Glaaser, Paul A. Slesinger[1]
Department of Neuroscience, Friedman Brain Institute, Icahn School of Medicine at Mount Sinai, New York, New York, USA
[1]Corresponding author: e-mail address: paul.slesinger@mssm.edu

Contents

1. Introduction 118
2. Selectivity and Permeation 121
3. Rectification 128
4. Gating 130
   4.1 Gating at the Bundle Crossing 130
   4.2 Gating at the G Loop 131
   4.3 Activation 132
5. Conclusions 144
Acknowledgments 145
References 145

Abstract

G protein-gated inwardly rectifying potassium (GIRK; Kir3) channels, which are members of the large family of inwardly rectifying potassium channels (Kir1–Kir7), regulate excitability in the heart and brain. GIRK channels are activated following stimulation of G protein-coupled receptors that couple to the $G_{i/o}$ (pertussis toxin-sensitive) G proteins. GIRK channels, like all other Kir channels, possess an extrinsic mechanism of inward rectification involving intracellular Mg^{2+} and polyamines that occlude the conduction pathway at membrane potentials positive to E_K. In the past 17 years, more than 20 high-resolution atomic structures containing GIRK channel cytoplasmic domains and transmembrane domains have been solved. These structures have provided valuable insights into the structural determinants of many of the properties common to all inward rectifiers, such as permeation and rectification, as well as revealing the structural bases for GIRK channel gating. In this chapter, we describe advances in our understanding of GIRK channel function based on recent high-resolution atomic structures of inwardly rectifying K^+ channels discussed in the context of classical structure–function experiments.

International Review of Neurobiology, Volume 123
ISSN 0074-7742
http://dx.doi.org/10.1016/bs.irn.2015.05.014

1. INTRODUCTION

In 1949, Bernard Katz observed that muscle fibers in high external potassium (K^+) displayed an inward K^+ conductance that increased at membrane potentials negative to the Nernst potential for potassium (E_K) and decreased upon membrane depolarization, contrary to the prototypical delayed outward K^+ conductance. Surprisingly, the gating of inward current shifted with different extracellular K^+ concentrations, instead of occurring over a fixed range of membrane potentials (Katz, 1949). This "anomalous rectification," subsequently referred to as inward rectification, was later observed in a wide variety of tissues including brain (Andrade, Malenka, & Nicoll, 1986; Andrade & Nicoll, 1987; Inoue, Nakajima, & Nakajima, 1988; North, Williams, Surprenant, & Christie, 1987; Penington, Kelly, & Fox, 1993), heart (Breitwieser & Szabo, 1985; Kurachi, 1985; Pfaffinger, Martin, Hunter, Nathanson, & Hille, 1985; Vereecke, Isenberg, & Carmeliet, 1980), skeletal muscle (Barrett-Jolley, Dart, & Standen, 1999; Beam & Donaldson, 1983; Duval & Leoty, 1980; Stanfield, Nakajima, & Nakajima, 2002), kidney (Hebert, 1995; Ho et al., 1993), and pancreas (Cook & Hales, 1984; Findlay, Dunne, & Petersen, 1985; Isomoto et al., 1996; Iwanir & Reuveny, 2008; Yoshimoto et al., 1999), in addition to blood cells (Lewis, Ikeda, Aryee, & Joho, 1991; McKinney & Gallin, 1988) and oocytes (Hagiwara, Miyazaki, & Rosenthal, 1976; Hagiwara & Takahashi, 1974). In general, inwardly rectifying K^+ channels are critical in maintaining the resting membrane potential because the voltage range over which the channel passes outward current is near the cell's resting membrane potential (V_m) (Armstrong & Binstock, 1965; Noble & Tsien, 1968). The family of inwardly rectifying K^+ channels (Kir) now includes seven subfamilies (Kir1–Kir7) (see IUPHAR database, 2015). This chapter focuses on the subfamily of G protein-gated inwardly rectifying K^+ (GIRK) channels. In the Kir nomenclature, GIRK channels are also referred to as Kir3 channels, with the human gene name of KCNJ. We refer the reader to other reviews on the family of Kir channels (Bichet, Haass, & Jan, 2003; Hibino et al., 2010).

One of the first GIRK channels described was in atrial myocytes (Hutter & Noble, 1960; Hutter & Trautwein, 1955a, 1955b). Acetylcholine (ACh) was found to slow the heart rate by inducing a K^+ current (I_{KACh}), which was distinguished from a background inwardly rectifying K^+ current (I_{K1}) (Sakmann, Noma, & Trautwein, 1983). Activation of the muscarinic-type cholinergic receptors was shown to induce I_{KACh} via pertussis toxin (PTX)-sensitive G proteins (Breitwieser & Szabo, 1985, 1988; Pfaffinger et al., 1985). In the brain, GIRK channels are activated by various neurotransmitters, including GABA, serotonin, dopamine, opioids, and somatostatin, via their respective G protein-coupled receptors (GPCRs) (Andrade et al., 1986; Andrade & Nicoll, 1987; Christie & North, 1988; Ikeda et al., 1996, 1997; Ikeda, Kobayashi, Ichikawa, Usui, & Kumanishi, 1995; Inoue et al., 1988). All GIRK channels are activated following stimulation of GPCRs that couple to $G_{i/o}$ (PTX-sensitive) G proteins. GIRK channels, like all other Kir channels, possess a mechanism of inward rectification involving intracellular Mg^{2+} and polyamines (e.g., spermine, spermidine, and putriscein), which occlude the conduction pathway at membrane potentials positive to E_K (Lopatin, Makhina, & Nichols, 1994; Matsuda, Saigusa, & Irisawa, 1987; Taglialatela, Wible, Caporaso, & Brown, 1994; Vandenberg, 1987).

After identification of the cDNA clones for ROMK1 (Kir1.1) (Ho et al., 1993) and IRK1 (Kir2.1) (Kubo, Baldwin, Jan, & Jan, 1993), the GIRK1 (Kir3.1) cDNA was first isolated (Dascal et al., 1993; Kubo, Reuveny, Slesinger, Jan, & Jan, 1993), followed by the cloning of three additional GIRK subunits (GIRK2–GIRK4) (Kofuji, Davidson, & Lester, 1995; Krapivinsky, Gordon, et al., 1995; Lesage et al., 1994). Toxin-binding studies using chimeras, mutants, and concatemers, as well as protein biochemistry, demonstrated that GIRK channels (as well as other Kir channels) assemble into tetramers (Glowatzki et al., 1995; Inanobe, Ito, Ito, Hosoya, & Kurachi, 1995; MacKinnon, 1991; Tinker, Jan, & Jan, 1996; Yang, Jan, & Jan, 1995b). Using heterologous expression systems and biochemical measurements, the I_{KACh} channel was determined to be a heterotetramer composed of GIRK1 and GIRK4 subunits (Krapivinsky, Gordon, et al., 1995), while brain GIRK channels were heterotetramers of GIRK1 and GIRK2, GIRK1 and GIRK3, or GIRK2 and GIRK3 (Jelacic, Kennedy, Wickman, & Clapham, 2000; Jelacic, Sims, & Clapham, 1999; Krapivinsky, Gordon, et al., 1995). Only GIRK2 and GIRK4 subunits also form functional homotetramers in the brain and heart, respectively (Duprat et al., 1995; Jelacic et al., 2000; Karschin, Dissmann, Stuhmer, & Karschin, 1996; Kobayashi et al., 1995; Kofuji et al., 1995; Lesage et al., 1994; Liao, Jan, & Jan, 1996). For GIRK2, there are four splice variants, termed GIRK2a–d (Inanobe et al., 1999; Isomoto et al., 1996; Lesage et al., 1994, 1995; Wei et al., 1998; Wickman, Pu, & Clapham, 2002), of which GIRK2a–c are expressed in the brain and GIRK2d in the testes. The brain isoforms differ in the length and composition of their respective carboxy-terminal domains, yielding subunits that are 414 (GIRK2a), 327 (GIRK2b), and 425 (GIRK2c) amino acids long; only GIRK2c contains a PDZ-binding motif (see below).

GIRK channels share significant structural homology with other Kir channels. Each monomer contains two integral membrane-spanning regions (M1 and M2), a reentrant P-loop, which controls permeation and ion selectivity, and cytoplasmic amino- and carboxy-termini (Doupnik, Davidson, & Lester, 1995; Ho et al., 1993; Kubo, Baldwin, et al., 1993; Kubo, Reuveny, et al., 1993).

Patch-clamp recordings from atrial myocytes revealed that acetylcholine (ACh) activates GIRK channels (I_{KACh}) through a membrane-delimited mechanism; that is, inclusion of ACh in the patch pipette of cell-attached patch recordings leads to robust activation of GIRK channels, whereas ACh added to the bath, outside the patch electrode, does not activate the channel (Soejima & Noma, 1984). These findings also raised the possibility that G proteins directly activate GIRK channels. In the late 1980s, a disagreement arose over whether the activated Gα-GTPγS subunit or Gβγ dimer was responsible for activating GIRK channels (Clapham & Neer, 1993; Codina, Yatani, Grenet, Brown, & Birnbaumer, 1987; Logothetis, Kim, Northup, Neer & Clapham, 1988; Logothetis, Kurachi, Galper, Neer, & Clapham, 1987; Neer & Clapham, 1988; Okabe et al., 1990; Yatani, Codina, Brown, & Birnbaumer, 1987; Yatani et al., 1988). This was subsequently resolved, with the preponderance of evidence demonstrating that Gβγ activates GIRK channels through direct membrane delimited and biochemical interactions (Huang, Slesinger, Casey, Jan, & Jan, 1995; Krapivinsky, Krapivinsky, Wickman, & Clapham, 1995; Reuveny et al., 1994).

The identification of cDNAs for different families of Kir channel and the development of molecular biology techniques for making site-specific point mutations spawned an era of so-called structure–function experiments. Here, the "structure" was inferred from a rigorous analysis of functional and biochemical data studying the effect of mutations on the channel properties. Over the course of the past ~17 years, significant progress has been made in determining the high-resolution atomic structure of ion channels. Techniques have been developed such that sufficient yields of mammalian membrane proteins can be screened and produced (Hattori, Hibbs, & Gouaux, 2012; Junge et al., 2008; Kawate & Gouaux, 2006), high-throughput crystallization steps are faster (Caffrey & Cherezov, 2009;

Fogg & Wilkinson, 2008), and high-energy beam lines (Smith, Fischetti, & Yamamoto, 2012) have significantly improved the resolution. Currently, there are over 20 GIRK channel structures deposited in the Protein Data Bank (Balana et al., 2011; Inanobe, Matsuura, Nakagawa, & Kurachi, 2011; Inanobe, Nakagawa, & Kurachi, 2011, 2013; Inanobe, Nakagawa, Matsuura, & Kurachi, 2010; Nishida, Cadene, Chait, & MacKinnon, 2007; Nishida & MacKinnon, 2002; Pegan et al., 2005; Whorton & MacKinnon, 2011, 2013; Xu, Shin, Szep, & Lu, 2009) that contain part of or most of the GIRK channel (Table 1). In this chapter, we describe advances in the understanding of GIRK channel function based on results of structure–function experiments discussed in the context of recent high-resolution atomic structures of inwardly rectifying K^+ channels.

2. SELECTIVITY AND PERMEATION

The pore of K^+ channels was characterized through experiments designed to determine the extent to which other ions interfere with K^+ conductance through K^+ channels, and how well other cations permeated through K^+ channels (Adelman & Senft, 1968; Bezanilla & Armstrong, 1972). These studies found that the pore of the K^+ channel was perfectly evolved for selectively coordinating K^+ because larger ions (Cs^+ and TEA) blocked the channel from the inner side, while ions smaller than K^+, such as Na^+, were too small to be coordinated in the pore without high energetic cost. From these studies, the diameter of the selectivity filter was predicted to be 2.6–3.0 Å, and the diameter of the inner mouth of the channel was predicted to be 8 Å (Bezanilla & Armstrong, 1972).

With the advent of cloning K^+ channel subunits (Tempel, Jan, & Jan, 1988; Tempel, Papazian, Schwarz, Jan, & Jan, 1987), similar experiments could be performed to test the effects of mutations on conductance and selectivity. Additionally, the sequence information revealed a conserved region among the K^+ channels that is not found in other channels. Mutagenesis of residues within this region, in both inwardly rectifying and voltage-gated K^+ channels, resulted in changes in ion selectivity, toxin-binding, or channel block consistent with the P-loop forming the selectivity filter (Anderson, MacKinnon, Smith, & Miller, 1988; Brown et al., 1993; Hartmann et al., 1991; Heginbotham, Abramson, & MacKinnon, 1992; Heginbotham, Lu, Abramson, & MacKinnon, 1994; Heginbotham & MacKinnon, 1993; Kavanaugh et al., 1991; Kirsch et al., 1992; MacKinnon, Heginbotham, & Abramson, 1990; MacKinnon & Miller,

Table 1 Kir Structures

Kir	PDB ID	Resolution (Å)	R-Value	R-Free	Space Group	References
GIRK1/Kir3.1 CTD	1N9P	1.80	0.232 (obs.)	0.253	$P\ 4\ 2_1\ 2$	Nishida and MacKinnon (2002)
GIRK1/Kir3.1 CTD	1U4E	2.09	0.196 (obs.)	0.235	$P\ 4\ 2_1\ 2$	Pegan et al. (2005)
GIRK2/Kir3.2 CTD	2E4F	2.30	0.244 (work)	0.279	$P\ 4\ 2_1\ 2$	Inanobe, Matsuura, Nakagawa, and Kurachi (2007)
Kir2.1 CTD	1U4F	2.41	0.225 (work)	0.281	$C\ 2\ 2\ 2_1$	Pegan et al. (2005)
Kir2.1 CTD R218Q/T309K + MPD	2GIX	2.02	0.191 (obs.)	0.229	C 2 (C 1 2 1)	Pegan, Arrabit, Slesinger, and Choe (2006)
KirBac1.3/GIRK1	2QKS	2.20	0.232 (obs.)	0.256	P 4	Nishida et al. (2007)
Kir2.2	3JYC	3.11	0.242 (obs.)	0.270	I 4	Tao, Avalos, Chen, and MacKinnon (2009)
Kir2.2 +PIP2	3SPI	3.31	0.237 (obs.)	0.278	I 4	Hansen, Tao, and MacKinnon (2011)
GIRK2/Kir3.2						
+Na	3SYO	3.54	0.261 (obs.)	0.273	$P\ 4\ 2_1\ 2$	Whorton and MacKinnon (2011)
+ PIP_2/Na	3SYA	2.98	0.242 (obs.)	0.269	$P\ 4\ 2_1\ 2$	
D228N	3SYC	3.41	0.252 (obs.)	0.271	$P\ 4\ 2_1\ 2$	
R201A	3SYP	3.12	0.265 (obs.)	0.296	$P\ 4\ 2_1\ 2$	
R201A + PIP2	3SYQ	3.44	0.301 (obs.)	0.323	$C\ 2\ 2\ 2_1$	
GIRK2/Kir3.2 CTD S225E	3K6N	2.00	0.254 (work)	0.266	$P\ 4\ 2_1\ 2$	Xu et al. (2009)
GIRK2/Kir3.2-Gβγ + PIP_2	4KFM	3.45	0.230 (obs.)	0.265	I 4 2 2	Whorton and MacKinnon (2013)

Table 1 Kir Structures—cont'd

Kir	PDB ID	Resolution (Å)	R-Value	R-Free	Space Group	References
GIRK2/Kir3.2 CTD-Na	3AGW	2.20	0.200 (obs.)	0.236	I 4 2 2	Inanobe et al. (2010)
GIRK2/Kir3.2 CTD + Cd^{2+}	3AUW	3.56	0.287 (obs.)	0.331	P 4 2_1 2	Inanobe, Matsuura, et al. (2011)
GIRK2/Kir3.2 CTD E236R	3VSQ	2.00	0.216 (obs.)	0.259	P 4 2_1 2	Inanobe et al. (2013)
GIRK3/Kir3.3						
SNX27-PDZ fused to GGESESKV	3QDO	1.88	0.147 (obs.)	0.203	I 4_1	Balana et al. (2011)
SNX27-PDZ fused to ESESKV	3QE1	1.68	0.161 (obs.)	0.223	H 3	
SNX27-PDZ + ESESKV peptide	3QGL	3.31	0.235 (obs.)	0.254	P 1 2_1 1	

1988, 1989; Slesinger et al., 1996; Yool & Schwarz, 1991). These studies identified a "signature sequence," T-X-G-(F/Y)-G, for determining selectivity in all K^+ channels, including both inward-rectifiers and voltage-gated K^+ channels.

In 1998, Rod Mackinnon and colleagues determined the structure of a bacterial K^+ channel, KcsA, at 3.2 Å (Doyle et al., 1998). Although KcsA does not share significant sequence homology with eukaryotic inward-rectifiers, it has the same general structural architecture (2 TM domains and P-loop) and shares many of its signature properties. The crystal structure exquisitely demonstrated the basis for K^+ selectivity. In the selectivity filter, unhydrated K^+ ions are coordinated by main chain carbonyl oxygens of the T-X-G-(F/Y)-G signature sequence in an environment that geometrically and energetically mimics the hydration by water molecules in the inner and outer vestibule (Morais-Cabral, Zhou, & MacKinnon, 2001; Roux & MacKinnon, 1999; Zhou, Morais-Cabral, Kaufman, & MacKinnon, 2001). Residues surrounding the signature sequence induce structural constraints such that the carbonyl oxygens coordinate K^+ but not the smaller Na^+. The energy landscape was such that the environment for K^+ was changed little from entering the vestibule as a hydrated ion to coordination by the carbonyl backbones in the absence of water (Aqvist & Luzhkov, 2000; Berneche & Roux, 2001; Morais-Cabral et al., 2001; Zhou, Morais-Cabral, et al., 2001). The subsequent structures of voltage-gated K^+ channels further illustrated the absolute conservation of the selectivity filter among all K^+ channels (Jiang et al., 2003; Long, Campbell, & Mackinnon, 2005; Fig. 1).

In addition to the signature sequence residues, there are several residues outside the pore region in GIRK channels that can affect selectivity and permeation. For example, the *weaver* mouse, a mutant mouse with neuronal and motor behavioral deficits, is caused by a mutation (G156S) in the signature sequence of GIRK2 (Patil et al., 1995; Fig. 1). Characterization of this mutation in heterologous expression systems demonstrated constitutively open channels, which were no longer selective for K^+ (Kofuji, Hofer, et al., 1996; Navarro et al., 1996; Slesinger et al., 1996; Surmeier, Mermelstein, & Goldowitz, 1996). In the related Kir2.1 channel, mutation of Glutamate (Kir2.1 E138) and Arginine (Kir2.1 R149) residues located adjacent to the selectivity filter resulted in significant losses in selectivity (Yang, Yu, Jan, & Jan, 1997). These charged amino acids are conserved in GIRK channels (GIRK1 E139/R149; GIRK2 E150/R160) and likely stabilize the pore through the formation of salt bridges. Taken together these

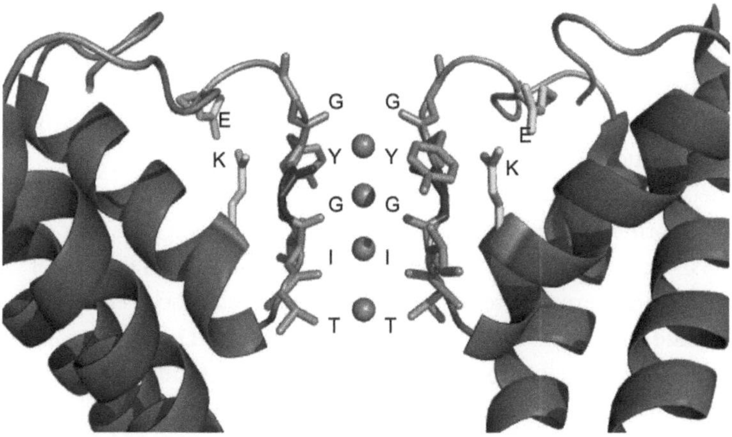

Figure 1 Structure of ion selectivity filter. View of the ion selectivity filter in GIRK2 (PDB: 3SYO). Front and back monomers have been removed for clarity. The residues lining the selectivity filter, TIGYG (signature sequence), from bottom to top are labeled in single-letter amino acid code. Both side chain and main chain sticks are represented. Four possible K^+-binding sites are represented with orange (gray in the print version) spheres. The locus of the weaver mouse mutation (G156) is represented in dark blue (black in the print version). Additional residues implicated in selectivity in Kir2.1 channels, glutamate and lysine, are represented by cyan (light gray in the print version) side chains, indicating their relative proximity.

data demonstrate that in addition to the signature sequence residues, adjacent residues that stabilize the local structure play a significant role in selectivity by providing structural stability in the pore region.

Although functional data provided evidence of a role for the cytoplasmic domains (CTDs) in permeation (Yang, Jan, & Jan, 1995a), additional insights were provided by additional atomic structures. The GIRK1 CTD, solved at 1.8 Å resolution, was the first structure containing part of a GIRK channel (PDB ID: 1N9P, see Table 1; Nishida & MacKinnon, 2002). This GIRK1 structure contains a truncated portion of the GIRK1 amino terminus (residues 41–63) that is fused directly to a truncated portion of the carboxy terminus (residues 190–371), bypassing the transmembrane (TM) domains. The fused CTDs were expressed in *Escherichia coli* bacteria and the purified protein formed tetramers in solution (Nishida & MacKinnon, 2002). In addition to the GIRK1 CTD, the structure of the Kir2.1 CTD was also solved (2.0 Å; PDB ID: 1U4F) (Pegan et al., 2005). Subsequently, structures of the GIRK1 and GIRK2 CTD under various conditions were similarly determined (see Table 1; Inanobe et al., 2007,

2013, 2010; Inanobe, Matsuura, et al., 2011; Inanobe, Nakagawa, et al., 2011; Pegan et al., 2006; Xu et al., 2009). All of the CTDs have the same basic structural elements, containing 14 beta strands (βA-βN) and two alpha helices (αA) and (αB), with βD-βE, βL-βM, and βH-βI loops forming significant functional units (see later) (Fig. 2).

The structures of the Kir2.1, GIRK1, and GIRK2 CTDs were striking in that all formed a long cytoplasmic pore (Fig. 3), which is distinct from that of the voltage-gated K^+ channels, and not seen in the KcsA channel where the intracellular termini are removed. Moreover, a unique region was identified in the structures of the Kir2.1 and GIRK1 CTDs, which formed the narrowest part of the cytoplasmic structure, and was termed the cytoplasmic "G loop" gate (Pegan et al., 2005; Fig. 2). The opening formed by the G loops (βH-βI) was too small to pass a hydrated K^+ (for details, see later).

The interior face of the cytoplasmic pore is lined with many polar residues, contributing to negative electrostatic potential; this has the effect of concentrating K^+ as well as impermeant cations and cationic compounds such as those for producing rectification. Residues linked to rectification, which would bind Mg^{2+} and polyamines and face the aqueous cytoplasmic

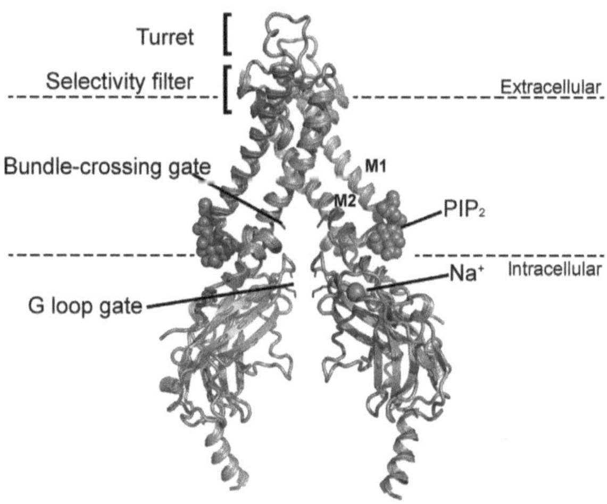

Figure 2 Structure of the GIRK2 channel. Side view of GIRK2 (green (light gray in the print version)) channel. The front and back subunits have been removed for clarity. Na^+ ions are shown as orange (gray in the print version) spheres. PIP_2 molecules (C8-PIP_2) are shown as space-filling molecules (only the head group is resolved). Two gates exist, the bundle-crossing gate (F192, blue (dark gray in the print version) stick side chains) and G loop gate in the βH-βI loop (M313 side chains and M319 main chain carbonyl oxygens are highlighted, red sticks (dark gray in the print version)). All Kir channels contain M1 and M2 transmembrane helices, a selectivity filter and two gates. GIRK2 also has a turret for toxin binding. Schematic is based on PDB ID: 3SYA.

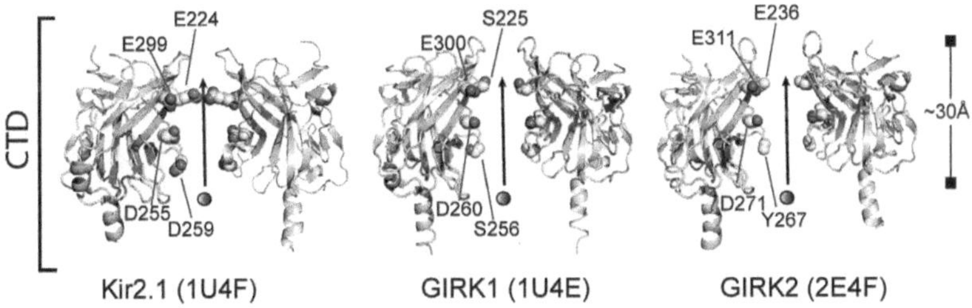

Figure 3 Structural determinants of permeation and rectification in the CTDs. Side views of the cytoplasmic domains of Kir2.1 (PDB ID: 1U4F), GIRK1 (PDB ID: 1U4E), and GIRK2 (PDB ID: 2E4F) CTDs, highlighting residues, implicated in permeation and rectification. The front and back subunits have been removed for clarity. The dark gray ball represents a K^+ and the arrow indicates location of the central cytoplasmic pore. There is no side-chain density for E236 or Y267 in the GIRK2 CTD structure. Indicated residues are important for establishing the negative electrostatic potential of the cytoplasmic pore.

pore, would be expected to affect permeation. Several studies illustrate this particularly well in the context of the CTD structures. Structure–function experiments initially identified two primary sites in determining rectification in the strong inward-rectifier Kir2.1, an aspartate in the M2 TM domain and several acidic residues in the cytoplasmic pore. One of the acidic residues in the CTD, E224 in the βC–βD loop of the Kir2.1, was shown to affect not only rectification but also permeation. Mutation of Kir2.1 E224 to glycine, which is found in the weak rectifier Kir1.1, also altered permeation (Yang et al., 1995a). Mutation of E224 to Lysine or Arginine revealed blocker-independent (i.e., absence of Mg^{2+} and polyamines) inward rectification in cell-excised patches (Taglialatela, Ficker, Wible, & Brown, 1995; Yang et al., 1995a). A structure of the GIRK1 CTD with a S225E mutation solved at 2.0 Å shows four hydrated Na^+ (used as a surrogate for K^+) coordinated by polar residues along its permeation pathway (PDB ID:3K6N, Table 1), providing further evidence for a long, cytoplasmic pore lined by acidic residues (Xu et al., 2009).

To provide more information on the relationship between the TM domains and the CTDs, a full-length structure was solved at 2.2 Å for a chimeric channel in which the TM domain was largely that of the bacterial inward-rectifier KirBac3.1 and the CTDs and the lower one-third of the TM2 domain was from GIRK1 in Nishida et al. (2007) (PDB ID:2QKS; Table 1). Two different conformations were observed in which the channel was either opened or closed at the cytoplasmic G loop gate (Fig. 2). Importantly, the width of the cytoplasmic pore is narrower when the G loop gate is open, promoting permeation by coordinating hydrated K^+ ions. Conversely, the CTD with a constricted G loop gate is wider through the cytoplasmic pore (Nishida et al., 2007), a conformation which was shown to be more amenable to divalent cation binding that blocks permeation (Inanobe, Nakagawa, et al., 2011).

Whorton and MacKinnon (2011) determined the structure of a full-length eukaryotic GIRK2 homomeric channel, which contained two TM domains, the pore and truncated amino- and carboxy-termini (same as in CTD structures) (Fig. 2) by itself and cocrystallized with channel modifiers such as Na^+ and PIP_2 (see Table 1). These structures confirm the presence of a cytoplasmic pore in inward rectifiers. The functional data and high-resolution structures appear to indicate that specific cytoplasmic residues play a significant role in permeation through the cytoplasmic pore, in addition to the selectivity filter.

3. RECTIFICATION

The degree of rectification for Kir channel subfamilies varies from weak (Kir1) to strong, where Kir2 > Kir3 ≫ Kir1 (Bichet et al., 2003; Hibino et al., 2010). Initially, the blocking particle was identified as internal Mg^{2+} (Matsuda et al., 1987; Vandenberg, 1987). However, when some rectification was observed even in the absence of internal Mg^{2+} ions, other rectifying factors were investigated. Polyamines, such as spermine, spermidine, and putriscein, which are present in high intracellular concentrations, occlude the conduction pathway at membrane potentials positive to E_K, thereby inducing rectification (Lopatin et al., 1994). Structure–function experiments implicated two regions in Kir channels in determining rectification. A polar residue (aspartate or asparagine) in the M2 TM domain forming the intracellular side of the central cavity (Kir 2.1 D171/GIRK1 D173/GIRK2 N184) contributes to the extent of rectification (Lu & MacKinnon, 1994; Stanfield et al., 1994; Wible, Taglialatela, Ficker, & Brown, 1994) and several acidic residues in the carboxy terminus (Kubo & Murata, 2001; Taglialatela et al., 1995; Yang et al., 1995a). With regard to the M2 domain site, inwardly rectifying channels with an aspartate exhibited strong rectification while those containing an asparagine tended to have weak rectification. However, the relatively strong inward rectifier GIRK2 contains an asparagine at this M2 locus and Kir7.1 has an aspartate but is a weak rectifier, suggesting that other sites contribute to rectification (Doring et al., 1998; Krapivinsky et al., 1998).

Mutagenesis studies identified several residues in the carboxy-terminus that change blocker affinity, suggesting that they might be part of a conduction pathway. The aforementioned Kir2.1 E224 (Taglialatela et al., 1995; Yang et al., 1995a), as well Kir2.1 E299 (Kubo & Murata, 2001), were identified as critical residues for rectification affinity and their mutation to non-acidic residues resulted in decreased affinity for block by Mg^{2+} and polyamines (Fig. 3). Several of the residues identified from Kir2 functional studies (Kir2.1 E224, D255, D259, and E299) and their equivalent residues in GIRK1 (S244, S256, D260, and E300) and GIRK2 (E236, Y267, E271, E311) are shown in the structure to be protruding from beta sheets lining the interior of the protein, inside the cytoplasmic pore (Fig. 3). Interestingly, in addition to the acidic side chains, many hydrophobic residues were implicated in increasing the affinity of polyamine binding (Guo & Lu, 2003). The arrangement of acidic side chains with hydrophobic residues interspersed is ideal for binding to polyamines with positive amino group coordinated by acidic side chains and alkane spacers associating with hydrophobic side chains of the channel.

Mutation of a diaspartate (D255/D259) cluster in Kir2.1 reduced inward rectification (Pegan et al., 2005). In the homologous region of GIRK1, D255 is a serine, a tyrosine in GIRK2, and an aspartate in GIRK4. Kir2.1 D259 is conserved in all GIRK subunits (Kurata, Cheng, Arrabit, Slesinger, & Nichols, 2007; Pegan et al., 2005). Computational studies of the relative impact of these residues using the structural data predict that residues in the central cytoplasmic pore (Kir2.1 E224, D259, and E299) have long-range electrostatic effects through the channel and that Kir2.1 E224 residue in particular is critical not only for rectification but is also important in establishing the negative electrostatic potential of the cytoplasmic pore (Robertson, Palmer, & Roux, 2008). The M2 locus for inward rectification (Kir2.1 D171/GIRK2 N184) can also be visualized in the TM domain containing GIRK2 channel structures (GIRK2 and the KirBac/GIRK1 chimera) between the selectivity filter and inner helix bundle crossing, where the aperture is narrow enough that binding of Mg^{2+} or polyamines would act as a blocking particle preventing hydrated K^{+} ions from approaching the pore.

To better understand the mechanism of inward rectification, a GIRK structure solved with Mg^{2+} or polyamines would be needed. All of the structures determined thus far are at least partially closed and structural determination in the presence of Mg^{2+} or polyamines may require a fully open but blocked channel. Nevertheless, the functional data maps well onto the structures determined thus far.

4. GATING

In addition to rectification, two other regions of Kir channels have been identified as important molecular determinants of gating: the inner helix (or bundle crossing) formed by the M2 TM domain, and the G loop in the CTD. Kir channel opening is postulated to require opening of both gates.

4.1 Gating at the Bundle Crossing

The first gate was originally described in the bacterial KcsA channel, where the opening around the inner helix bundle crossing was too small to allow for the passage of K^+ ions, suggesting a closed channel (Doyle et al., 1998; Perozo, Cortes, & Cuello, 1999). Additional structures of KcsA and MthK, the Ca^{2+}-activated K^+ channel from the archaeon *Methanobacterium thermoautotrophicum* coupled with functional studies suggested that activation occurs via bending and rotation of the second TM domain (M2) through a highly conserved glycine (Jiang et al., 2002b; Jin et al., 2002; Liu, Sompornpisut, & Perozo, 2001; Sadja, Smadja, Alagem, & Reuveny, 2001). This glycine acts as a "hinge," which allows dilation of the inner helix bundle crossing, permitting flow of K^+. A random mutagenesis of GIRK1 channels demonstrated that a proline substitution near this glycine introduced a structural kink in the M2 domain that constitutively opened the channel (Sadja et al., 2001; Yi, Lin, Jan, & Jan, 2001). Interestingly, this mutant also rendered the channel insensitive to activation by Gβγ subunits, suggesting that this conformation change underlies channel activation by G proteins. A kink in the M2 domain opens the helix bundle crossing sufficiently to allow the passage of hydrated K^+ (Jiang et al., 2002a, 2002b). The conservation of the glycine as the hinge as well as the phenylalanine forming the narrowest part of the helix bundle crossing was demonstrated in GIRK4 channels (Jin et al., 2002). A proline scan of the M2 domain revealed that proline substitutions above the bundle crossing rendered the channels constitutively activated and Gβγ insensitive, supporting the glycine hinge-model for activation.

Currently, there are no high-resolution structures of a fully open GIRK channel. A R201A mutation was found to produce constitutively open channels, which lacked additional receptor-dependent activation

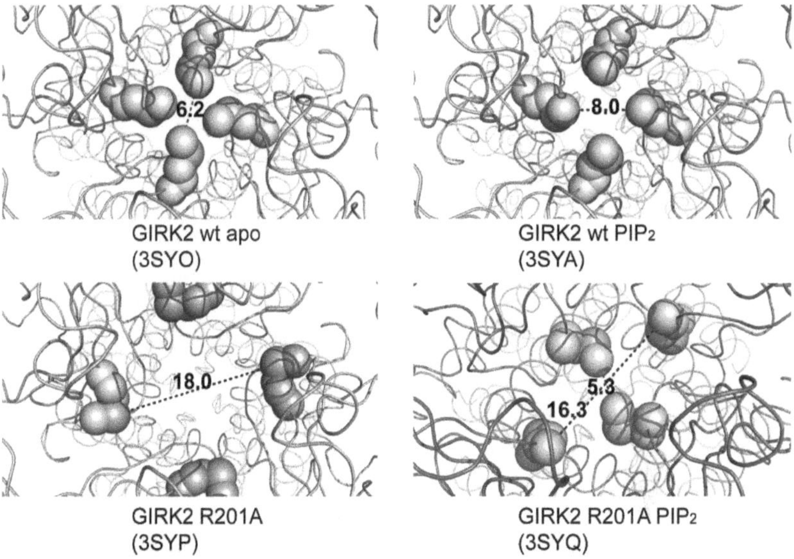

Figure 4 The G loop of GIRK2 in different conformations. Space-filling side chains at M313 in various conformations. Distances measured are 6.3 Å in wild-type (wt) apo, 8.0 Å in wt PIP_2, and 18.0 Å in R201A. The structure containing R201A and PIP_2 (3SYQ) has symmetrical subunits with G loops closed and open, with distances 5.3 and 16.3 Å, respectively.

(Whorton & MacKinnon, 2011) (see Table 1). This mutation was proposed to represent a G protein-activated state, although this has yet to be confirmed. The structure of GIRK2-R201A in the presence of PIP_2 was solved (3.44 Å) and revealed two of four subunits in an open conformation, where the phenylalanine in the M2 domain and the G loop have moved sufficiently to open the channel gates (Fig. 4). Modeling the entire inner helix this way would result in an opening of 11 Å, sufficient to pass a hydrated K^+ (Whorton & MacKinnon, 2011). Interestingly, this opening is not as large as that observed in voltage-gated K^+ channel and prokaryotic Ca^{2+}-gated K^+ channels (Doyle et al., 1998; Jiang et al., 2002a, 2002b).

4.2 Gating at the G Loop

A second gate was identified based on structures of the Kir2.1 and GIRK1 CTDs (see Table 1) (Pegan et al., 2005). A loop formed by βH and βI strands, termed "G loop," formed the narrowest part of the cytoplasmic structure (Fig. 4). The G loop includes G300, M301, A306, M307, and T308 in Kir2.1 channels. Conformational difference between the two isoforms suggested intrinsic flexibility in this region. Furthermore, the opening formed by the G loops in the Kir2.1 CTD structure is too small to pass a hydrated K^+. In support of the G loop as a gating structure, mutations associated with genetically inherited diseases in Kir2.1 and Kir6.2 channels were localized to this region. A mutation at residue I296 (I296L) in Kir6.2, which corresponds to M308 in GIRK1, results in developmental delay, epilepsy, and neonatal diabetes syndrome (Proks et al., 2005). Characterization of this mutation demonstrated resistance to SUR1 activation and an increase in open probability. Andersen–Tawil syndrome is a genetic disorder characterized by developmental abnormalities and cardiac arrhythmias caused by mutations in Kir2.1 (Plaster et al., 2001). Several mutations associated with this disease are located in the G loop of Kir2.1 channels (e.g., G300V, V302M, and E303K) and functionally exhibit dominant-negative current inhibition, which appears to be independent of trafficking defects (Pegan et al., 2006).

The structure of a chimeric GIRK1-prokaryotic Kirbac channel was solved in open and closed conformations, with the greatest movement in the G loop (Nishida et al., 2007). In these structures, the constricted or closed structure is occluded by side chains of G loop residues (T306–T309), whereas in the open state, the loop is dilated and contains two K^+ coordinated by local side chains. The R201A mutation in GIRK2 also results in dilation at the G loop (Fig. 4). The pore is dilated in the "open" subunits of the crystal structure of the GIRK2 channel containing this mutant in the presence of PIP_2 (3SYQ, Table 1) relative to the wild-type channel without PIP_2. Taken together, these structures indicate that the G loop is open, or dilated, in the constitutively open Kir2.1 CTD structure, and closed, or constricted, in the GIRK1 CTD structure. The data further support a model in the GIRK channel where the G loop opens in the presence of PIP_2 and a channel activator such as Gβγ, alcohol, or the R201A mutation (Fig. 4).

4.3 Activation

4.3.1 G Protein Activation

The Gβγ dimer of heterotrimeric G proteins directly activates GIRK channels, originally demonstrated for the atrial GIRK channel (I_{KACh}), but subsequently for other GIRK channels as well (Kofuji et al., 1995; Krapivinsky, Krapivinsky, et al., 1995; Logothetis et al., 1987; Reuveny et al., 1994). Direct physical association between Gβγ and the amino- and carboxy-termini was initially demonstrated using various biochemical methods including coimmunoprecipitation, yeast two-hybrid, and protein–protein-binding assays (He, Zhang, Mirshahi, & Logothetis, 1999; Huang, Jan, & Jan, 1997; Huang et al., 1995; Kunkel & Peralta, 1995). The specific region or amino acids involved in Gβγ binding was further mapped to residues in the N-terminus and the βD-βE and βL-βM sheets of the C-terminus (Finley, Arrabit, Fowler, Suen, & Slesinger, 2004; He et al., 2002; Ivanina et al., 2003). In particular, a leucine (GIRK1 L333, GIRK2 L344, and GIRK4 L339) in the βL-βM sheet of the carboxy-terminal domain was particularly critical in supporting G protein-mediated activation; mutation of this residue to glutamate ablated G protein activation. Other residues implicated in Gβγ activation include an N-terminal histidine (GIRK1 H57, GIRK2 H69, and GIRK4 H64), a leucine in the carboxy-terminal βE-βF loop (GIRK1 L262, GIRK2 L273, and GIRK4 L268), and glycine in the βL-βM loop (GIRK1 G336, GIRK2 G347, and GIRK4 G342) (Clancy et al., 2005; Finley et al., 2004; He et al., 1999; Slesinger, Reuveny, Jan, & Jan, 1995; Zhao et al., 2003, 2006). The initial structures of the cytoplasmic regions confirmed the outer surface clustering of some of the residues (e.g., GIRK1 L333) previously implicated, but not others (e.g., GIRK1 L262), which is buried in the protein structure (Nishida & MacKinnon, 2002). GIRK channels interact with Gβγ proteins via the Gβ subunit (Ford et al., 1998). The crystal structures of the Gαβγ heterotrimer (Lambright et al., 1996; Wall et al., 1995) highlighted two regions on the Gβ subunit in mediating protein–protein interactions: the switch interface and the amino-terminal interface. Functional studies implicated residues on the amino-terminal interface (Gβ L55, K78, I80, K89, W99) in mediating the GIRK–Gβ interaction (Ford et al., 1998). Further studies identified additional residues on Gβ: S67 and T128, based on mutations that reduced the agonist-induced current (Mirshahi, Robillard, Zhang, Hebert, & Logothetis, 2002).

NMR analysis of chemical shifts induced by Gβγ binding to GIRK1 CTD confirmed the βD-βE and βL-βM sheets as primary sites of Gβγ–GIRK1 interactions (Yokogawa, Osawa, Takeuchi, Mase, & Shimada, 2011). Among the GIRK1 residues with the largest intensity reductions in the presence of Gβγ were the previously identified critical residues, L333 and G336. This study also predicted several residues on the Gβ subunit that were important in coordinating the GIRK1 binding, including L55, K78, I80, K89, W99, D228, D246, W332, most of which are located in the Gα-binding site. Finally, two of the G loop residues (GIRK1 G307 and M308) display significant chemical shift perturbations in the presence of Gβγ, which could represent Gβγ-dependent opening of this gate.

A computational study of the GIRK1/GIRK4 heterotetramer used the previously generated functional data and mapped the Gβγ binding to a cleft between the subunits (Mahajan et al., 2013). This model predicted that the leucine at position 55 of the Gβ subunit would occupy a cleft between the critical GIRK1 L333 and F243 residues. The model predicts and the experiments support the critical role of Gβ L55 and K89 in coordinating GIRK–Gβ interactions. The model further predicts that the βD-βE and βL-βM loops are in close contact in the closed state, but that the βL-βM loop adopts a raised conformation upon Gβγ subunit binding (Mahajan et al., 2013). Mutation of key residues on GIRK1 and the Gβ subunit supports the model predictions and suggests that the "raised" βL-βM loop propagates its signal such that the βC-βD loop, coordinated by PIP_2, induces a conformational change that dilates the G loop, effectively opening this gate (Mahajan et al., 2013).

Further structural insight was provided by a study in which fluorescence energy transfer (FRET) measurements of CFP- and YFP-tagged GIRK1 and GIRK4 subunits of GIRK1/GIRK4 heterotetramers were examined in the presence of Gβγ subunits (Riven, Kalmanzon, Segev, & Reuveny, 2003). Gβγ decreased the FRET, indicating an increase in the distance between tagged GIRK subunits. Furthermore, FRET measurements with variations in how the amino- and carboxy-termini of GIRK1 and GIRK4 subunits were tagged, increased the resolution with which the distance could be calculated even in the presence of such large fluorescent tags. Based on the change in FRET efficiency, the authors calculated an ~10 Å rotation of the CTDs (Riven et al., 2003).

The GIRK2–Gβγ crystal structure demonstrates interactions between the GIRK2 CTD and Gβ subunit through the βK, βL, βM, and βN sheets from one GIRK2 subunit and the βD and βE from the adjacent subunit, as predicted by previous studies (Whorton & MacKinnon, 2013). The GIRK2 channel regions that mediate the Gβγ interactions are well conserved among GIRK subunits, notably GIRK2 Q248, F254, and L344. Q248 makes contact with three residues (Q75, S98, and W99) on Gβ, which have been shown to functionally affect Gβγ activation when mutated (Ford et al., 1998). The Gβ subunit L55 and K89, both identified as important residues from previous studies, interact with the critical GIRK2 L344 residue found in the βL-βM loop, confirming the importance of these residues in Gβγ activation (Mahajan et al., 2013; Whorton & MacKinnon, 2013; Yokogawa et al., 2011).

Interestingly, the structure of the GIRK2–Gβγ G protein complex does not adopt an open conformation at either of the gates. Although there is some splaying of the helices at the bundle crossing, it is not sufficient to pass a hydrated K^+ ion. Instead, it is postulated to be in a "pre-open" conformation in which the CTDs have undergone a rotation of ~4°. The CTD in the GIRK2 R201A and PIP_2 structure rotates by ~8°, and both the bundle crossing and G loop gates are open for the two subunits in this conformation. Interestingly, the aforementioned FRET study also predicted a rotation of ~10° for the CTD upon Gβγ binding, providing further evidence for this as the structural mechanism for channel opening. Taken together, these studies suggest a series of conformational changes in the carboxy-terminus result in the opening of the two gates with Gβγ but the complete opening of these gates requires the binding of PIP_2.

Although the Gα subunit does not activate GIRK channels, numerous studies have demonstrated that the Gα subunit interacts with the channel directly and can affect function (Clancy et al., 2005; Huang et al., 1995; Ivanina et al., 2004; Leal-Pinto et al., 2010; Peleg, Varon, Ivanina, Dessauer, & Dascal, 2002; Rubinstein et al., 2009; Schreibmayer et al., 1996). These studies have suggested that there is a macromolecular signaling complex in which the heterotrimeric G protein binds the GIRK channel and the GCPR, bringing together the appropriate GPCR with the GIRK channel and ensuring rapid deactivation by placing the Gα-GDP subunit in close proximity to the channel to reassociate with Gβγ and close the channel (Berlin et al., 2010, 2011; Peleg et al., 2002; Rubinstein, Peleg, Berlin, Brass, & Dascal, 2007; Rubinstein et al., 2009). The role of Gα in regulating the function of GIRK channels is discussed in more detail in chapter "The Roles of Gβγ and Gα in Gating and Regulation of GIRK Channels" by Dascal and colleagues.

$G\alpha_o$-GDP binds to both the amino- and carboxy-termini of GIRK2, with C321 providing a critical contact with PTX-sensitive Gα subunits (Clancy et al., 2005). In addition to these roles, it has been proposed that Gα regulates basal GIRK current through channel inhibition in the absence of GPCR activation (Kahanovitch et al., 2014; Peleg et al., 2002). The Gα subunit contains a GTPase domain and a helical domain. The helical domain has been proposed as the determinant of GIRK activation by specific $G_{i/o}$-coupled GPCRs (Rusinova, Mirshahi, & Logothetis, 2007). A recent study used NMR spectroscopy to examine interactions of GIRK1 CTD with $G\alpha_{i/o}$ (Mase, Yokogawa, Osawa, & Shimada, 2012). The GIRK1 residues implicated in interactions with Gα are primarily located on the αA helix along with a glutamate (E242) in the βE sheet of GIRK1. These GIRK1 residues are predicted to directly bind the α2/α3 site of Gα, with a single Gα subunit binding adjacent GIRK channel CTDs (Mase et al., 2012). The binding regions identified from protein biochemistry similarly contained a large part of the distal carboxy-terminus. However, there is significant heterogeneity in this region between GIRK channels, suggesting that the Gα affinity could vary between isoforms. A model is constructed for a ternary complex of the GIRK channel CTD, Gα, and Gβγ, which is consistent with previous biochemical data and places the Gα-GTP in close proximity to Gβγ, such that upon GTP hydrolysis, Gα-GDP can immediately reassociate with Gβγ resulting in closure of channel gates. There is currently no high-resolution structure of a GIRK channel or GIRK channel domain bound to both Gα and Gβγ.

4.3.1.1 PIP_2 Requirement

One common feature of all Kir channels is their regulation by the lipid molecule phosphotidylinositol-4,5-bisphosphate (PIP_2) (Baukrowitz et al., 1998; Hilgemann & Ball, 1996; Huang, Feng, & Hilgemann, 1998; Zhang, He, Yan, Mirshahi, & Logothetis, 1999). The PIP_2 requirement for channel opening was demonstrated in studies by depleting PIP_2 via anti-PIP_2 antibodies or polycations such as poly-lysine (Huang et al., 1998; Lopes et al., 2002) and observing a reduction in channel activity (Huang et al., 1998; Schulze, Krauter, Fritzenschaft, Soom, & Baukrowitz, 2003; Zeng, Li, Hilgemann, & Huang, 2003). Reintroducing PIP_2 restored channel activity. Thus far, all activators of GIRK channels (e.g., Gβγ, ethanol, and Na) require PIP_2 as a cofactor (Bodhinathan & Slesinger, 2013; Huang et al., 1998; Zhang et al., 1999). The role of PIP_2 in regulating the function of GIRK channels is discussed in more detail in chapter "Unifying Mechanism of Controlling Kir3 Channel Activity by G Proteins and Phosphoinositides" by Logothetis and colleagues.

The polycationic inhibition suggests that PIP_2 binds the channel via electrostatic interactions. PIP_2 is found primarily in the cytoplasmic leaflet of eukaryotic membranes (Nikawa, Kodaki, & Yamashita, 1987; Nikawa & Yamashita, 1982). The search for the PIP_2-binding site identified a cluster of basic residues located near the membrane interface that is relatively conserved between Kir channels. In GIRK2 channels, these residues include K64, K90, W91, K92, K194, K199, K200 (Baukrowitz et al., 1998; Donaldson et al., 2003; Lopes et al., 2002; Pegan et al., 2006; Rohacs, Chen, Prestwich, & Logothetis, 1999; Rohacs et al., 2003; Soom et al., 2001; Zeng et al., 2003; Zhang et al., 1999). Although the electrostatic interactions are necessary for the coordination of PIP_2, the structure of Kir2 channel with glycerol pyrophosphatidic acid, which is essentially a smaller version of the head group only and does not activate Kir2 channels, suggests that other interactions may be important for channel activation (Cheng, D'Avanzo, Doyle, & Nichols, 2011; Hansen et al., 2011). In fact, PIP_2 with longer acyl chains demonstrates a higher affinity interaction with GIRK channels and chimeras, implicating membrane interactions in PIP_2 gating effects (Leal-Pinto et al., 2010; Rohacs et al., 1999).

Several studies have shown many of the residues involved in PIP_2 binding are conserved among different Kir channels. However, the relative affinity differs between the channels (D'Avanzo, Cheng, Doyle, & Nichols, 2010; Gilman et al., 2002; Rohacs et al., 1999, 2003). Examination of the structures of Kir2.2 and GIRK2 channels indicates structural differences. The primary difference between the Kir2.2 and GIRK2 structures cocrystallized with PIP_2 is the interaction of the TM domains with the CTD, and the conformation of the G loop (Hansen et al., 2011; Whorton & MacKinnon, 2011). In the Kir2.2 structure, the CTD is extended below from the TM domains in the absence of PIP_2. The addition of PIP_2 brings the CTD up toward the TM domain, resulting in opening at the inner helix gate. A conformational change occurs in the opening at the G loop in the presence of PIP_2, ensuring that the G loop gate remains open in the presence of PIP_2 (Hansen et al., 2011). However, the PIP_2-bound Kir2.2 structure may not be fully open. By contrast, the TM and CTDs in the GIRK2 structure are in close proximity in the absence of PIP_2. While the addition of PIP_2 results in small conformational changes within the CTD, its importance is more pronounced in the presence of additional effectors of channel gating such as the $G\beta\gamma$ dimer, Na^+, or the R201A mutation. The R201A mutant by itself causes a 6° rotation of the CTD, but in the presence of PIP_2, this rotation increases to 8° and the G loop and helix bundle crossing are opened, though in only two of the four subunits (Whorton & MacKinnon, 2011, 2013).

4.3.1.2 The Role of Na^+

Investigations into the cytosolic components involved in GIRK channel gating revealed that native atrial GIRK channels, as well as recombinant GIRK1/GIRK4 channels, were activated by an increase in cytosolic Na^+ in addition to $G\beta\gamma$ and required the presence of PIP_2 (Sui, Petit-Jacques, & Logothetis, 1998). Increased intracellular $[Na^+]$ following an action potential would increase GIRK channel activation and cell inhibition. Further investigations revealed that Na^+-dependent gating was also a property of GIRK2-containing channels. Here, Na^+ binding was primarily mediated through a conserved aspartate residue in either GIRK4 (D229) or GIRK2 (D228) found in the βC-βD loop, of the carboxy terminus (Fig. 1; Ho & Murrell-Lagnado, 1999a, 1999b; Whorton & MacKinnon, 2011; Zhang et al., 1999). The equivalent residue is an asparagine (N217) in GIRK1, and mutation of this residue to an aspartate (N217D) conferred Na^+ sensitivity to these channels (Ho & Murrell-Lagnado, 1999a, 1999b). Comparisons of dose-response fits with Hill curves demonstrate that the GIRK1/GIRK2 and GIRK1/GIRK4 channels have a Hill coefficient of about two, whereas the GIRK2 and GIRK4 homomers have Hill coefficients of four, suggesting that these channels bind two and four sodium ions, respectively (Rosenhouse-Dantsker et al., 2008). The structure for the GIRK2 channel containing both TM domains and CTDs was solved for the wild-type channel as well as a channel containing the D228N mutant (Whorton & MacKinnon, 2011), which would not be predicted to coordinate a sodium ion. In the wild-type channel, a Na^+ is located in a pocket in the carboxy-terminus coordinated by the D228 side chain, as well as main chain carbonyl oxygen atoms of residues in the βC-βD loop (R230, N231, and S232), the loop between βF and βG sheets (L275, V276) and flanking histidines (H69 and H233). The structure also provides some clues as to how Na^+ binding might influence channel activation. The arginine (R201), which when mutated to alanine (R201A) results in constitutive G protein-independent activation (Whorton & MacKinnon, 2011), is hydrogen bonded to D228. Sodium ions binding to this position could weaken the bond between R201 and D228, lowering the energy barrier for G protein-induced conformational changes leading to the opening of the gates, similar to the R201A mutant. A recent study that quantitatively analyzed the effects of PIP_2, Na^+, and Gβγ found that PIP_2 and Gβγ were necessary for interaction and that Na^+ was not necessary for channel activation, but augmented the activation, suggesting interdependence of these ligands for maximal activation (Wang, Whorton, & MacKinnon, 2014).

4.3.2 Activation by Alcohol

Many different types of ion channels are modulated by ethanol. Some of the best evidence for direct ethanol activation of an ion channel comes from work on GIRK channels. In 1999, two groups demonstrated that GIRK channels could be activated by alcohol in heterologous expression systems and in neurons (Kobayashi et al., 1999; Lewohl et al., 1999). These studies provided evidence that alcohol activation was largely independent of Gβγ or other second messengers. Alcohol with carbon chain lengths of up to four carbons (methanol, ethanol, propanol, and butanol) activated GIRK1/GIRK2 channels, although butanol inhibited GIRK1/GIRK4 channels. However, all alcohols with carbon chain lengths longer than four (pentanol, hexanol, etc.) inhibited the channel, suggesting that there were spatial constraints on the interactions of alcohol with GIRK channels. Interestingly, all other Kir channels were blocked by alcohol, demonstrating that this mode of activation is unique to GIRK channels among the inward-rectifiers.

The original studies implicated the distal carboxy-terminus in alcohol activation (Hara, Lewohl, Yamakura, & Harris, 2001; Lewohl et al., 1999). The physical structure through which ethanol-activated GIRK channels, however, remained unknown until 2006. Slesinger and colleagues crystallized the CTD of a related inward-rectifier, Kir2.1, with the alcohol 2-methyl-2,4-pentanediol (MPD) bound to hydrophobic pockets at the interface of two adjacent subunits (Pegan et al., 2006) (see Table 1). MPD binding in this pocket was largely coordinated by hydrophobic side chains (F47, L232, L245, L330, and Y337). Additionally, the hydroxyl groups of MPD appear to make contacts with backbone carbonyls (P244) and side-chain hydroxyl groups of Y242 and Y337 (Pegan et al., 2006). Comparison of the hydrophobic pocket in Kir2.1 and GIRK2 (Fig. 5)

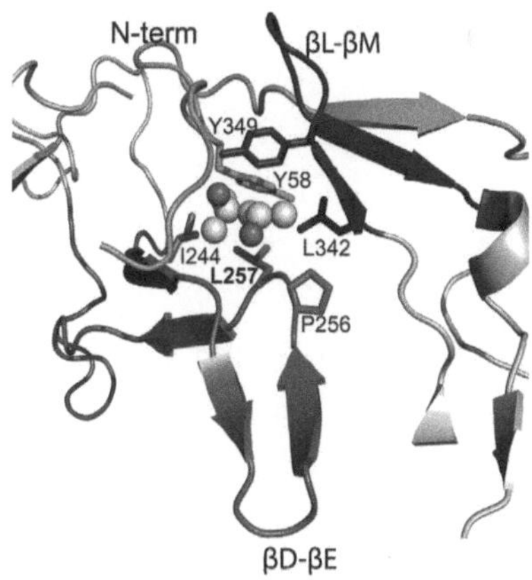

Figure 5 Structure of the alcohol pocket in GIRK2. A model of GIRK2 CTD in complex with the alcohol 2-methyl-2,4-pentanediol (MPD). The amino acids indicated form the hydrophobic alcohol pocket, coordinating MPD, at the interface of two adjacent subunits. Amino acid residues in stick format are colored according to the domain from which they originate, N-terminus (green (light gray in the print version)), βD-βE loop (purple (dark gray in the print version)), and βL-βM loop (blue (black in the print version)). MPD is shown as a space-filling molecule. L257 (red (dark gray in the print version)) was shown previously to be a critical determinant of alcohol-dependent activation of GIRK2 channels.

revealed substantial conservation of amino acids implicated in the alcohol-binding pocket (Aryal, Dvir, Choe, & Slesinger, 2009). Functional analysis demonstrated that MPD activated GIRK2 channels similar to other alcohols. Furthermore, coexpression of the Gβγ chelator phosducin did not affect alcohol-induced channel activation, further demonstrating that alcohol activation was G protein-independent. Subsequently, it was demonstrated that mutations in the homologous pocket of GIRK2, particularly at L257, altered alcohol activation in a way that correlated with the volume of the side chain (Aryal et al., 2009). Interestingly, mutations at L257 inhibited G protein-mediated activation in GIRK2 channels, suggesting that Gβγ activation converges on the same structural region of the channel. The recent structure of GIRK2 cocrystallized with the Gβγ dimer supports this conclusion (Whorton & MacKinnon, 2013), providing additional evidence that the alcohol pocket is critical for channel activation.

Through experiments chemically modifying L257C with reagents that mimic ethanol, it was shown that methane-thiol-sulfonate hydroxyethyl (MTS-HE) treatment was sufficient to constitutively activate the channel, analogous to irreversible ethanol activation (Bodhinathan & Slesinger, 2013). Use of other MTS-conjugated chemicals provided evidence that the hydrophobic component of the ligand is more critical for channel activation than the hydroxyl group. Both ethanol and chemical activation increased the relative affinity for PIP_2, leading to an increase in channel activity (Bodhinathan & Slesinger, 2013). Structures of the MTS- and ethanol-activated channels in the presence of PIP_2 are needed to better understand the structural changes that occur with these ligands.

4.3.3 Structural Determinants of Gating Modifiers

GIRK channels are attractive drug targets both as potential sites of inhibition for regulating heart rate and behavior and as potential sites of activation for ischemia following stroke or epilepsy, for example. Many existing drugs interact with GIRK channels (Kobayashi & Ikeda, 2006; Lujan, Marron Fernandez de Velasco, Aguado, & Wickman, 2014), although their mechanisms and sites of interaction are not well understood.

4.3.3.1 Local Anesthetics

Local anesthetics are common drugs used in treatment of cardiac arrhythmias and pain management (Butterworth & Strichartz, 1990). Local anesthetics have diverse effects on channels and receptors and they have been shown to inhibit voltage-gated K^+ channels (Gray, Zhorov, & Moczydlowski, 2013; Madeja, Steffen, Mesic, Garic, & Zhorov, 2010; Valenzuela, Delpon, Tamkun, Tamargo, & Snyders, 1995; Valenzuela, Snyders, Bennett, Tamargo, & Hondeghem, 1995). One study demonstrated inhibition of GIRK channels by the local anesthetics bupivacaine, lidocaine, and the permanently charged lidocaine derivative, QX-314 (Zhou, Arrabit, Choe, & Slesinger, 2001). The authors proposed that this occurred through modulation of channel–PIP_2 interactions because Kir1 and Kir2 were insensitive to local anesthetics and GIRK chimeras containing the Kir1- or Kir2 PIP_2-binding regions that were similarly insensitive to local anesthetics.

Cocaine also acts as a local anesthetic, in addition to its more commonly characterized use as a drug of abuse. Similar to the effects of other local anesthetics, cocaine inhibited GIRK1/GIRK2 and GIRK1/GIRK4 heteromultimers and GIRK2 homomers, while Kir1.1 and Kir2.1 were insensitive (Kobayashi, Nishizawa, Iwamura, & Ikeda, 2007). However, cocaine inhibition was weaker than that of other local anesthetics and did not show the small pH-dependent increase in efficacy seen for lidocaine and bupivacaine.

4.3.3.2 Volatile Anesthetics

In addition to local anesthetics, general anesthetics, such as the volatile anesthetic halothane, have also been shown to act on GIRK channels (Weigl & Schreibmayer, 2001; Yamakura, Lewohl, & Harris, 2001). GIRK1/GIRK4 channel activation by halothane appears to involve a Gβγ-dependent enhancement of basal current, while it also inhibits agonist-induced current (Weigl & Schreibmayer, 2001). The TM domains and carboxy-terminus were shown to mediate halothane interactions (Milovic, Steinecker-Frohnwieser, Schreibmayer, & Weigl, 2004; Styer et al., 2010; Yamakura et al., 2001). Chimeric studies probing the determinants of halothane sensitivity implicated the phenylalanine at the helix bundle crossing (GIRK2 F192; Fig. 2) and demonstrated that mutation of this residue to the equivalent residue in halothane-insensitive Kir2.1, a methionine, resulted in reversal of halothane inhibition. Similar results were also demonstrated for a mutation that inhibited Gβγ activation (GIRK2 L273I) (Styer et al., 2010). Several other anesthetics acted as agonist or antagonists toward GIRK channels. Isoflurane and F3 (1-chloro-1,2,2-trifluorocyclobutane) displayed similar profiles to halothane, although F3 was an even more potent inhibitor, especially of GIRK2-containing channels (Yamakura et al., 2001). In addition, nitrous oxide was shown to act as an activator of GIRK channels, but not other inward rectifiers (Yamakura et al., 2001).

4.3.3.3 GIRK Channel Drug Interactions

Because modulators of GIRK channel function can play an important role in regulation neuronal and cardiac excitability, the response of GIRK channels to the direct application of numerous drugs has been examined. Various antipsychotics and antidepressants, including Fluoxetine (Prozac) and tricyclic antidepressants (Kobayashi et al., 2009; Kobayashi, Washiyama, & Ikeda, 2003, 2004, 2006b, 2011), have been examined and almost all inhibit GIRK channels, but not Kir1 or Kir2 channels. A follow-up study of fluoxetine derivatives found an agonist and three antagonists (Nishizawa, Gajya, & Ikeda, 2011). The structural mechanism of this inhibition has not been examined. The α_1 adrenergic receptor antagonist Ifenprodil has been shown to have neuroprotective and anticonvulsant effects and is related to the antipsychotic haloperidol, which weekly inhibits GIRK channels. Ifenprodil is also an inhibitor of GIRK channels, as well as other inward rectifiers, all with similar affinity (Kobayashi, Washiyama, & Ikeda, 2006a).

A recent study reported that the antitussive drug, Tipepidine, also blocks GIRK channels (Hamasaki, Shirasaki, Soeda, & Takahama, 2013). Interestingly, this drug appears to have novel antidepressant-like effects that may be mediated through activation of mesolimbic dopamine neurons through GIRK channel inhibition (Hamao et al., 2015; Hamasaki et al., 2013; Kawahara et al., 2013). Two pilot studies suggest that tipepidine may improve adolescent depression and ADHD (Sasaki, Hashimoto, Tachibana, Kurata, Kimura, et al., 2014; Sasaki, Hashimoto, Tachibana, Kurata, Okawada, et al., 2014). These studies point to GIRK channels as an additional therapeutic target in depression and behavior. The molecular site for tipepidine inhibition is unknown.

4.3.3.4 GIRK Channel Agonists and Antagonists

GIRK channels are blocked by the small bee venom peptide tertiapin, which also has some inhibiting activity with Kir1.1 and the voltage-dependent Ca^{2+}-activated (BK) K^+ channels (Jin, Klem, Lewis, & Lu, 1999; Jin & Lu, 1998; Kanjhan, Coulson, Adams, & Bellingham, 2005). Tertiapin binds to the extracellular vestibule and has proven to be a useful molecular probe for examining the physiological functions and mechanisms of GIRK channels in the brain (Fig. 2). Comparison of Kir2.2 and GIRK2 structures reveals a more open turret on the extracellular side that could interact with tertiapin.

In addition to ethanol and halothane, Naringin, a bioflavonoid found in grapefruit, has been identified as a GIRK channel agonist (Yow et al., 2011). Studies suggest that Naringin might exert its agonist effects extracellularly, within a site containing aromatic tyrosine residues on the GIRK1 extracellular vestibule (GIRK1 Y148 and Y150), which are known to be important in tertiapin binding. Naringin and tertiapin may share a binding site (Yow et al., 2011), although there is no proposed mechanism as to how this might lead to channel activation.

Recently, several studies have focused on synthesizing and identifying high-affinity agonists and antagonists for GIRK channels. Initially, NIP-142, a benzopyrane derivative, was shown to suppress atrial fibrillation via inhibition of I_{KACh} block with an EC_{50} of 0.64 μM (Matsuda et al., 2006). Further optimization resulted in the synthesis of NTC-801, which blocked the GIRK1-GIRK4 channel with an IC_{50} of 0.7 nM, as opposed to GIRK1/2 inhibition at 24 nM (Machida et al., 2011; Yamamoto et al., 2014). Another GIRK channel modulator, ML297, has been characterized (Kaufmann et al., 2013). ML297 is a potent activator with an EC_{50} of 233 ± 38 nM, compared to affinities on the order of hundreds of micromolar for other agonists and antagonists. ML297 is also specific for GIRK1-containing heterotetramers, with no agonist activity on GIRK2 homomers (Kaufmann et al., 2013). A recent study identified two amino acids unique to GIRK1 (F137 in the pore helix and D173 in the M2 domain) that are necessary and sufficient for selective activation of GIRK1-containing channels by ML297; mutation of these residues in GIRK2, which is otherwise ML297 insensitive, confers ML297 sensitivity upon these channels (Wydeven et al., 2014). Interestingly, these residues have been implicated in isoform differences in selectivity and gating (F137) (Chan, Sui, Vivaudou, & Logothetis, 1996; Slesinger, 2001) and inward rectification (D173) (Kofuji, Doupnik, Davidson, & Lester, 1996).

4.3.4 Trafficking Motifs

4.3.4.1 SNX27 Regulation and GIRK3

Movement of GIRK channels is controlled by a number of intrinsic amino acid sequence motifs that control ER export, internalization from plasma membrane, lysosomal targeting, and association with trafficking proteins (Ma et al., 2002; Mirshahi & Logothetis, 2004). Both GIRK2c and GIRK3 channels have been shown to associate with sorting nexin 27 (SNX27) via a

PDZ (PSD95/Disc large/Zona occludens) domain interaction (Lunn et al., 2007). SNX27 is a member of a large family of endosomal trafficking proteins (Cullen, 2008) and recently implicated in intractable myoclonic epilepsy (Damseh et al., 2015). GIRK2c and GIRK3 contain a C terminal class I PDZ-binding motif, E-S-K-V, that associates directly with the PDZ domain of SNX27. The canonical consensus sequence for class I motif is X-Ser/Thr-X-Φ (where X is any residue and Φ is a hydrophobic amino acid) (Hung & Sheng, 2002; Songyang et al., 1997). The PDZ domain of SNX27 exhibits a unique level of specificity compared to other PDZ domains, binding to GIRK2c and GIRK3 channels but not to related inward rectifiers, such as Kir2.1, even though Kir2.1 also contain a class I PDZ motif (E-S-E-I). Using high-resolution structures and protein biochemistry (see Table 1), Balana et al. examined the mechanism of PDZ binding with GIRK3 (2011). The SNX27–PDZ binding specificity appeared to be determined by two distinct structural regions, the canonical-binding site involving a hydrophobic groove near the C-terminal residue (0) of the PDZ ligand and a second region containing a charged pocket that interacts with the −4 and −5 residues in the GIRK3 channel. The hydrophobic pocket is formed by an αB helix, βB strand, and the carboxylate-binding (GLGF) loop from the PDZ domain. Mutating the valine to isoleucine ('ESKV" to "ESKI") in the carboxy-terminal domain of GIRK3 eliminates binding to SNX27–PDZ (Lunn et al., 2007), likely because the isoleucine provides steric hindrance and is unfavorable in the 0 position. In addition, the second region involving the charged pocket of SNX27–PDZ preferentially associates with glutamate in GIRK2c and GIRK3 and is repulsed by arginine in Kir2.1. These differences are sufficient to explain the selective association of SNX27 with GIRK2c/3 channels but not the related Kir2.1. The role of SNX27 in regulating the function of GIRK channels in the brain is discussed in more detail in chapter "GIRK Channel Plasticity and Implications for Drug Addiction" by Wickman and colleagues.

5. CONCLUSIONS

High-resolution structures of GIRK channels have allowed for the visualization of the channel in many conformational states and in the presence of several modulatory compounds. These "snapshots" have expedited the interpretation of previous functional and biochemical experiments as well as the design of subsequent experiments to test dynamic properties predicted by the structures. Still, many challenges remain. Future structures should provide additional information about the mechanisms underlying channel gating, provide a picture of a fully open channel, and determine heterotetrameric interactions. Combining these new physiologically important structures with existing and future functional and biochemical data will provide valuable insights into GIRK channel physiology.

ACKNOWLEDGMENTS

We thank Drs. Peter Ung and Avner Schlessinger (Icahn School of Medicine at Mount Sinai) for the GIRK2 model with MPD. Work from the Slesinger lab cited in this chapter was supported by grants from the National Institute on Alcohol Abuse and Alcoholism (AA018734) and the National Institute on Drug Abuse (DA037170; DA025236; DA029706). We apologize to our colleagues whose works could not be included owing to space limitations.

REFERENCES

Adelman, W. J., & Senft, J. P. (1968). Dynamic asymmetries in the squid axon membrane. *The Journal of General Physiology*, *51*(5), 102–114.

Anderson, C. S., MacKinnon, R., Smith, C., & Miller, C. (1988). Charybdotoxin block of single Ca^{2+}–activated K+ channels. Effects of channel gating, voltage, and ionic strength. *The Journal of General Physiology*, *91*(3), 317–333.

Andrade, R., Malenka, R. C., & Nicoll, R. A. (1986). A G protein couples serotonin and GABAB receptors to the same channels in hippocampus. *Science*, *234*(4781), 1261–1265.

Andrade, R., & Nicoll, R. A. (1987). Pharmacologically distinct actions of serotonin on single pyramidal neurones of the rat hippocampus recorded in vitro. *The Journal of Physiology*, *394*, 99–124.

Aqvist, J., & Luzhkov, V. (2000). Ion permeation mechanism of the potassium channel. *Nature*, *404*(6780), 881–884. http://dx.doi.org/10.1038/35009114.

Armstrong, C. M., & Binstock, L. (1965). Anomalous rectification in the squid giant axon injected with tetraethylammonium chloride. *The Journal of General Physiology*, *48*, 859–872.

Aryal, P., Dvir, H., Choe, S., & Slesinger, P. A. (2009). A discrete alcohol pocket involved in GIRK channel activation. *Nature Neuroscience*, *12*(8), 988–995. http://dx.doi.org/10.1038/nn.2358, nn.2358 [pii].

Balana, B., Maslennikov, I., Kwiatkowski, W., Stern, K. M., Bahima, L., Choe, S., et al. (2011). Mechanism underlying selective regulation of G protein-gated inwardly rectifying potassium channels by the psychostimulant-sensitive sorting nexin 27. *Proceedings of the National Academy of Sciences of the United States of America*, *108*(14), 5831–5836. http://dx.doi.org/10.1073/pnas.1018645108.

Barrett-Jolley, R., Dart, C., & Standen, N. B. (1999). Direct block of native and cloned (Kir2.1) inward rectifier K+ channels by chloroethylclonidine. *British Journal of Pharmacology*, *128*(3), 760–766. http://dx.doi.org/10.1038/sj.bjp.0702819.

Baukrowitz, T., Schulte, U., Oliver, D., Herlitze, S., Krauter, T., Tucker, S. J., et al. (1998). PIP2 and PIP as determinants for ATP inhibition of KATP channels. *Science*, *282*(5391), 1141–1144.

Beam, K. G., & Donaldson, P. L. (1983). Slow components of potassium tail currents in rat skeletal muscle. *The Journal of General Physiology*, *81*(4), 513–530.

Berlin, S., Keren-Raifman, T., Castel, R., Rubinstein, M., Dessauer, C. W., Ivanina, T., et al. (2010). G alpha(i) and G betagamma jointly regulate the conformations of a

G betagamma effector, the neuronal G protein-activated K+ channel (GIRK). *The Journal of Biological Chemistry, 285*(9), 6179–6185. http://dx.doi.org/10.1074/jbc.M109.085944.

Berlin, S., Tsemakhovich, V. A., Castel, R., Ivanina, T., Dessauer, C. W., Keren-Raifman, T., et al. (2011). Two distinct aspects of coupling between galpha(i) protein and G protein-activated K+ channel (GIRK) revealed by fluorescently labeled galpha(i3) protein subunits. *The Journal of Biological Chemistry, 286*(38), 33223–33235. http://dx.doi.org/10.1074/jbc.M111.271056.

Berneche, S., & Roux, B. (2001). Energetics of ion conduction through the K+ channel. *Nature, 414*(6859), 73–77. http://dx.doi.org/10.1038/35102067.

Bezanilla, F., & Armstrong, C. M. (1972). Negative conductance caused by entry of sodium and cesium ions into the potassium channels of squid axons. *The Journal of General Physiology, 60*(5), 588–608.

Bichet, D., Haass, F. A., & Jan, L. Y. (2003). Merging functional studies with structures of inward-rectifier K(+) channels. *Nature Reviews. Neuroscience, 4*(12), 957–967. http://dx.doi.org/10.1038/nrn1244.

Bodhinathan, K., & Slesinger, P. A. (2013). Molecular mechanism underlying ethanol activation of G-protein-gated inwardly rectifying potassium channels. *Proceedings of the National Academy of Sciences of the United States of America, 110*(45), 18309–18314. http://dx.doi.org/10.1073/pnas.1311406110.

Breitwieser, G. E., & Szabo, G. (1985). Uncoupling of cardiac muscarinic and beta-adrenergic receptors from ion channels by a guanine nucleotide analogue. *Nature, 317*(6037), 538–540.

Breitwieser, G. E., & Szabo, G. (1988). Mechanism of muscarinic receptor-induced K+ channel activation as revealed by hydrolysis-resistant GTP analogues. *The Journal of General Physiology, 91*(4), 469–493.

Brown, A. M., Drewe, J. A., Hartmann, H. A., Taglialatela, M., De Biasi, M., Soman, K., et al. (1993). The potassium pore and its regulation. *Annals of the New York Academy of Sciences, 707*, 74–80.

Butterworth, J. F., IV, & Strichartz, G. R. (1990). Molecular mechanisms of local anesthesia: A review. *Anesthesiology, 72*(4), 711–734.

Caffrey, M., & Cherezov, V. (2009). Crystallizing membrane proteins using lipidic mesophases. *Nature Protocols, 4*(5), 706–731. http://dx.doi.org/10.1038/nprot.2009.31.

Chan, K. W., Sui, J. L., Vivaudou, M., & Logothetis, D. E. (1996). Control of channel activity through a unique amino acid residue of a G protein-gated inwardly rectifying K+ channel subunit. *Proceedings of the National Academy of Sciences of the United States of America, 93*(24), 14193–14198.

Cheng, W. W., D'Avanzo, N., Doyle, D. A., & Nichols, C. G. (2011). Dual-mode phospholipid regulation of human inward rectifying potassium channels. *Biophysical Journal, 100*(3), 620–628. http://dx.doi.org/10.1016/j.bpj.2010.12.3724.

Christie, M. J., & North, R. A. (1988). Agonists at mu-opioid, M2-muscarinic and GABAB-receptors increase the same potassium conductance in rat lateral parabrachial neurones. *British Journal of Pharmacology, 95*(3), 896–902.

Clancy, S. M., Fowler, C. E., Finley, M., Suen, K. F., Arrabit, C., Berton, F., et al. (2005). Pertussis-toxin-sensitive Galpha subunits selectively bind to C-terminal domain of neuronal GIRK channels: Evidence for a heterotrimeric G-protein-channel complex. *Molecular and Cellular Neurosciences, 28*(2), 375–389. http://dx.doi.org/10.1016/j.mcn.2004.10.009.

Clapham, D. E., & Neer, E. J. (1993). New roles for G-protein beta gamma-dimers in transmembrane signalling. *Nature, 365*(6445), 403–406. http://dx.doi.org/10.1038/365403a0.

Codina, J., Yatani, A., Grenet, D., Brown, A. M., & Birnbaumer, L. (1987). The alpha subunit of the GTP binding protein Gk opens atrial potassium channels. *Science*, *236*(4800), 442–445.

Cook, D. L., & Hales, C. N. (1984). Intracellular ATP directly blocks K+ channels in pancreatic B-cells. *Nature*, *311*(5983), 271–273.

Cullen, Peter J. (2008). Endosomal sorting and signalling: An emerging role for sorting nexins. *Nature Reviews. Molecular Cell Biology*, *9*(7), 574–582.

Damseh, N., Danson, C. M., Al-Ashhab, M., Abu-Libdeh, B., Gallon, M., Sharma, K., et al. (2015). A defect in the retromer accessory protein, SNX27, manifests by infantile myoclonic epilepsy and neurodegeneration. *Neurogenetics*. http://dx.doi.org/10.1007/s10048-015-0446-0, [Epub ahead of print].

Dascal, N., Lim, N. F., Schreibmayer, W., Wang, W., Davidson, N., & Lester, H. A. (1993). Expression of an atrial G-protein-activated potassium channel in Xenopus oocytes. *Proceedings of the National Academy of Sciences of the United States of America*, *90*(14), 6596–6600.

D'Avanzo, N., Cheng, W. W., Doyle, D. A., & Nichols, C. G. (2010). Direct and specific activation of human inward rectifier K+ channels by membrane phosphatidylinositol 4,5-bisphosphate. *The Journal of Biological Chemistry*, *285*(48), 37129–37132. http://dx.doi.org/10.1074/jbc.C110.186692.

Donaldson, M. R., Jensen, J. L., Tristani-Firouzi, M., Tawil, R., Bendahhou, S., Suarez, W. A., et al. (2003). PIP2 binding residues of Kir2.1 are common targets of mutations causing Andersen syndrome. *Neurology*, *60*(11), 1811–1816.

Doring, F., Derst, C., Wischmeyer, E., Karschin, C., Schneggenburger, R., Daut, J., et al. (1998). The epithelial inward rectifier channel Kir7.1 displays unusual K+ permeation properties. *The Journal of Neuroscience*, *18*(21), 8625–8636.

Doupnik, C. A., Davidson, N., & Lester, H. A. (1995). The inward rectifier potassium channel family. *Current Opinion in Neurobiology*, *5*(3), 268–277.

Doyle, D. A., Morais Cabral, J., Pfuetzner, R. A., Kuo, A., Gulbis, J. M., Cohen, S. L., et al. (1998). The structure of the potassium channel: Molecular basis of K+ conduction and selectivity. *Science*, *280*(5360), 69–77.

Duprat, F., Lesage, F., Guillemare, E., Fink, M., Hugnot, J. P., Bigay, J., et al. (1995). Heterologous multimeric assembly is essential for K+ channel activity of neuronal and cardiac G-protein-activated inward rectifiers. *Biochemical and Biophysical Research Communications*, *212*(2), 657–663. http://dx.doi.org/10.1006/bbrc.1995.2019.

Duval, A., & Leoty, C. (1980). Ionic currents in slow twitch skeletal muscle in the rat. *The Journal of Physiology*, *307*, 23–41.

Findlay, I., Dunne, M. J., & Petersen, O. H. (1985). ATP-sensitive inward rectifier and voltage- and calcium-activated K+ channels in cultured pancreatic islet cells. *The Journal of Membrane Biology*, *88*(2), 165–172.

Finley, M., Arrabit, C., Fowler, C., Suen, K. F., & Slesinger, P. A. (2004). betaL-betaM loop in the C-terminal domain of G protein-activated inwardly rectifying K(+) channels is important for G(betagamma) subunit activation. *The Journal of Physiology*, *555*(Pt 3), 643–657. http://dx.doi.org/10.1113/jphysiol.2003.056101.

Fogg, M. J., & Wilkinson, A. J. (2008). Higher-throughput approaches to crystallization and crystal structure determination. *Biochemical Society Transactions*, *36*(Pt. 4), 771–775. http://dx.doi.org/10.1042/BST0360771.

Ford, C. E., Skiba, N. P., Bae, H., Daaka, Y., Reuveny, E., Shekter, L. R., et al. (1998). Molecular basis for interactions of G protein betagamma subunits with effectors. *Science*, *280*(5367), 1271–1274.

Gilman, A. G., Simon, M. I., Bourne, H. R., Harris, B. A., Long, R., Ross, E. M., et al. (2002). Overview of the alliance for cellular signaling. *Nature*, *420*(6916), 703–706. http://dx.doi.org/10.1038/nature01304.

Glowatzki, E., Fakler, G., Brandle, U., Rexhausen, U., Zenner, H. P., Ruppersberg, J. P., et al. (1995). Subunit-dependent assembly of inward-rectifier K+ channels. *Proceedings of the Biological Sciences*, *261*(1361), 251–261. http://dx.doi.org/10.1098/rspb.1995.0145.

Gray, N. W., Zhorov, B. S., & Moczydlowski, E. G. (2013). Interaction of local anesthetics with the K (+) channel pore domain: KcsA as a model for drug-dependent tetramer stability. *Channels (Austin, Tex.)*, 7(3), 182–193. http://dx.doi.org/10.4161/chan.24455.

Guo, D., & Lu, Z. (2003). Interaction mechanisms between polyamines and IRK1 inward rectifier K+ channels. *The Journal of General Physiology*, *122*(5), 485–500. http://dx.doi.org/10.1085/jgp.200308890.

Hagiwara, S., Miyazaki, S., & Rosenthal, N. P. (1976). Potassium current and the effect of cesium on this current during anomalous rectification of the egg cell membrane of a starfish. *The Journal of General Physiology*, *67*(6), 621–638.

Hagiwara, S., & Takahashi, K. (1974). The anomalous rectification and cation selectivity of the membrane of a starfish egg cell. *The Journal of Membrane Biology*, *18*(1), 61–80.

Hamao, K., Kawaura, K., Soeda, F., Hamasaki, R., Shirasaki, T., & Takahama, K. (2015). Tipepidine increases dopamine level in the nucleus accumbens without methamphetamine-like behavioral sensitization. *Behavioural Brain Research*, *284*, 118–124. http://dx.doi.org/10.1016/j.bbr.2015.02.012.

Hamasaki, R., Shirasaki, T., Soeda, F., & Takahama, K. (2013). Tipepidine activates VTA dopamine neuron via inhibiting dopamine D(2) receptor-mediated inward rectifying K(+) current. *Neuroscience*, *252*, 24–34. http://dx.doi.org/10.1016/j.neuroscience.2013.07.044.

Hansen, S. B., Tao, X., & MacKinnon, R. (2011). Structural basis of PIP2 activation of the classical inward rectifier K+ channel Kir2.2. *Nature*, *477*(7365), 495–498. http://dx.doi.org/10.1038/nature10370.

Hara, K., Lewohl, J. M., Yamakura, T., & Harris, R. A. (2001). Mutational analysis of ethanol interactions with G-protein-coupled inwardly rectifying potassium channels. *Alcohol*, *24*(1), 5–8.

Hartmann, H. A., Kirsch, G. E., Drewe, J. A., Taglialatela, M., Joho, R. H., & Brown, A. M. (1991). Exchange of conduction pathways between two related K+ channels. *Science*, *251*(4996), 942–944.

Hattori, M., Hibbs, R. E., & Gouaux, E. (2012). A fluorescence-detection size-exclusion chromatography-based thermostability assay for membrane protein precrystallization screening. *Structure*, *20*(8), 1293–1299. http://dx.doi.org/10.1016/j.str.2012.06.009.

He, C., Yan, X., Zhang, H., Mirshahi, T., Jin, T., Huang, A., et al. (2002). Identification of critical residues controlling G protein-gated inwardly rectifying K(+) channel activity through interactions with the beta gamma subunits of G proteins. *The Journal of Biological Chemistry*, *277*(8), 6088–6096. http://dx.doi.org/10.1074/jbc.M104851200.

He, C., Zhang, H., Mirshahi, T., & Logothetis, D. E. (1999). Identification of a potassium channel site that interacts with G protein betagamma subunits to mediate agonist-induced signaling. *The Journal of Biological Chemistry*, *274*(18), 12517–12524.

Hebert, S. C. (1995). An ATP-regulated, inwardly rectifying potassium channel from rat kidney (ROMK). *Kidney International*, *48*(4), 1010–1016.

Heginbotham, L., Abramson, T., & MacKinnon, R. (1992). A functional connection between the pores of distantly related ion channels as revealed by mutant K^+ channels. *Science*, *258*(5085), 1152–1155.

Heginbotham, L., Lu, Z., Abramson, T., & MacKinnon, R. (1994). Mutations in the K+ channel signature sequence. *Biophysical Journal*, *66*(4), 1061–1067. http://dx.doi.org/10.1016/S0006-3495(94)80887-2.

Heginbotham, L., & MacKinnon, R. (1993). Conduction properties of the cloned Shaker K+ channel. *Biophysical Journal*, *65*(5), 2089–2096. http://dx.doi.org/10.1016/S0006-3495(93)81244-X.

Hibino, H., Inanobe, A., Furutani, K., Murakami, S., Findlay, I., & Kurachi, Y. (2010). Inwardly rectifying potassium channels: Their structure, function, and physiological roles. *Physiological Reviews*, *90*(1), 291–366. http://dx.doi.org/10.1152/physrev.00021.2009.

Hilgemann, D. W., & Ball, R. (1996). Regulation of cardiac Na+, Ca2+ exchange and KATP potassium channels by PIP2. *Science*, *273*(5277), 956–959.

Ho, I. H., & Murrell-Lagnado, R. D. (1999a). Molecular determinants for sodium-dependent activation of G protein-gated K+ channels. *The Journal of Biological Chemistry*, *274*(13), 8639–8648.

Ho, I. H., & Murrell-Lagnado, R. D. (1999b). Molecular mechanism for sodium-dependent activation of G protein-gated K+ channels. *The Journal of Physiology*, *520*(Pt. 3), 645–651.

Ho, K., Nichols, C. G., Lederer, W. J., Lytton, J., Vassilev, P. M., Kanazirska, M. V., et al. (1993). Cloning and expression of an inwardly rectifying ATP-regulated potassium channel. *Nature*, *362*(6415), 31–38. http://dx.doi.org/10.1038/362031a0.

Huang, C. L., Feng, S., & Hilgemann, D. W. (1998). Direct activation of inward rectifier potassium channels by PIP2 and its stabilization by Gbetagamma. *Nature*, *391*(6669), 803–806. http://dx.doi.org/10.1038/35882.

Huang, C. L., Jan, Y. N., & Jan, L. Y. (1997). Binding of the G protein betagamma subunit to multiple regions of G protein-gated inward-rectifying K+ channels. *FEBS Letters*, *405*(3), 291–298.

Huang, C. L., Slesinger, P. A., Casey, P. J., Jan, Y. N., & Jan, L. Y. (1995). Evidence that direct binding of G beta gamma to the GIRK1 G protein-gated inwardly rectifying K+ channel is important for channel activation. *Neuron*, *15*(5), 1133–1143.

Hung, A. Y., & Sheng, M. (2002). PDZ domains: Structural modules for protein complex assembly. *The Journal of Biological Chemistry*, *277*(8), 5699–5702.

Hutter, O. F., & Noble, D. (1960). Rectifying properties of heart muscle. *Nature*, *188*, 495.

Hutter, O. F., & Trautwein, W. (1955a). Effect of vagal stimulation on the sinus venosus of the frog's heart. *Nature*, *176*(4480), 512–513.

Hutter, O. F., & Trautwein, W. (1955b). Vagal effects on the sinus venosus of the frog's heart. *The Journal of Physiology*, *129*(3), 48P.

Ikeda, K., Kobayashi, T., Ichikawa, T., Usui, H., Abe, S., & Kumanishi, T. (1996). Comparison of the three mouse G-protein-activated K+ (GIRK) channels and functional couplings of the opioid receptors with the GIRK1 channel. *Annals of the New York Academy of Sciences*, *801*, 95–109.

Ikeda, K., Kobayashi, T., Ichikawa, T., Usui, H., & Kumanishi, T. (1995). Functional couplings of the delta- and the kappa-opioid receptors with the G-protein-activated K+ channel. *Biochemical and Biophysical Research Communications*, *208*(1), 302–308. http://dx.doi.org/10.1006/bbrc.1995.1338.

Ikeda, K., Kobayashi, K., Kobayashi, T., Ichikawa, T., Kumanishi, T., Kishida, H., et al. (1997). Functional coupling of the nociceptin/orphanin FQ receptor with the G-protein-activated K+ (GIRK) channel. *Brain Research. Molecular Brain Research*, *45*(1), 117–126.

Inanobe, A., Horio, Y., Fujita, A., Tanemoto, M., Hibino, H., Inageda, K., et al. (1999). Molecular cloning and characterization of a novel splicing variant of the Kir3.2 subunit predominantly expressed in mouse testis. *The Journal of Physiology*, *521*(Pt. 1), 19–30.

Inanobe, A., Ito, H., Ito, M., Hosoya, Y., & Kurachi, Y. (1995). Immunological and physical characterization of the brain G protein-gated muscarinic potassium channel. *Biochemical and Biophysical Research Communications*, *217*(3), 1238–1244. http://dx.doi.org/10.1006/bbrc.1995.2901.

Inanobe, A., Matsuura, T., Nakagawa, A., & Kurachi, Y. (2007). Structural diversity in the cytoplasmic region of G protein-gated inward rectifier K+ channels. *Channels (Austin, Tex.)*, *1*(1), 39–45.

Inanobe, A., Matsuura, T., Nakagawa, A., & Kurachi, Y. (2011). Inverse agonist-like action of cadmium on G-protein-gated inward-rectifier K(+) channels. *Biochemical and Biophysical Research Communications*, *407*(2), 366–371. http://dx.doi.org/10.1016/j.bbrc.2011.03.025.

Inanobe, A., Nakagawa, A., & Kurachi, Y. (2011). Interactions of cations with the cytoplasmic pores of inward rectifier K(+) channels in the closed state. *The Journal of Biological Chemistry*, *286*(48), 41801–41811. http://dx.doi.org/10.1074/jbc.M111.278531.

Inanobe, A., Nakagawa, A., & Kurachi, Y. (2013). Conformational changes underlying pore dilation in the cytoplasmic domain of mammalian inward rectifier K+ channels. *PLoS One*, *8*(11), e79844. http://dx.doi.org/10.1371/journal.pone.0079844.

Inanobe, A., Nakagawa, A., Matsuura, T., & Kurachi, Y. (2010). A structural determinant for the control of PIP2 sensitivity in G protein-gated inward rectifier K+ channels. *The Journal of Biological Chemistry*, *285*(49), 38517–38523. http://dx.doi.org/10.1074/jbc.M110.161703.

Inoue, M., Nakajima, S., & Nakajima, Y. (1988). Somatostatin induces an inward rectification in rat locus coeruleus neurones through a pertussis toxin-sensitive mechanism. *The Journal of Physiology*, *407*, 177–198.

Isomoto, S., Kondo, C., Takahashi, N., Matsumoto, S., Yamada, M., Takumi, T., et al. (1996). A novel ubiquitously distributed isoform of GIRK2 (GIRK2B) enhances GIRK1 expression of the G-protein-gated K+ current in Xenopus oocytes. *Biochemical and Biophysical Research Communications*, *218*(1), 286–291. http://dx.doi.org/10.1006/bbrc.1996.0050.

IUPHAR. (2015). IUPHAR/BPS guide to pharmacology.

Ivanina, T., Rishal, I., Varon, D., Mullner, C., Frohnwieser-Steinecke, B., Schreibmayer, W., et al. (2003). Mapping the Gbetagamma-binding sites in GIRK1 and GIRK2 subunits of the G protein-activated K+ channel. *The Journal of Biological Chemistry*, *278*(31), 29174–29183. http://dx.doi.org/10.1074/jbc.M304518200.

Ivanina, T., Varon, D., Peleg, S., Rishal, I., Porozov, Y., Dessauer, C. W., et al. (2004). Galphai1 and Galphai3 differentially interact with, and regulate, the G protein-activated K+ channel. *The Journal of Biological Chemistry*, *279*(17), 17260–17268. http://dx.doi.org/10.1074/jbc.M313425200.

Iwanir, S., & Reuveny, E. (2008). Adrenaline-induced hyperpolarization of mouse pancreatic islet cells is mediated by G protein-gated inwardly rectifying potassium (GIRK) channels. *Pflügers Archiv*, *456*(6), 1097–1108. http://dx.doi.org/10.1007/s00424-008-0479-4.

Jelacic, T. M., Kennedy, M. E., Wickman, K., & Clapham, D. E. (2000). Functional and biochemical evidence for G-protein-gated inwardly rectifying K^+ (GIRK) channels composed of GIRK2 and GIRK3. *Journal of Biological Chemistry*, *275*(46), 36211–36216.

Jelacic, T. M., Sims, S. M., & Clapham, D. E. (1999). Functional expression and characterization of G-protein-gated inwardly rectifying K^+ channels containing GIRK3. *The Journal of Membrane Biology*, *169*(2), 123–129.

Jiang, Y., Lee, A., Chen, J., Cadene, M., Chait, B. T., & MacKinnon, R. (2002a). Crystal structure and mechanism of a calcium-gated potassium channel. *Nature*, *417*(6888), 515–522.

Jiang, Y., Lee, A., Chen, J., Cadene, M., Chait, B. T., & MacKinnon, R. (2002b). The open pore conformation of potassium channels. *Nature*, *417*(6888), 523–526.

Jiang, Y., Lee, A., Chen, J., Ruta, V., Cadene, M., Chait, B. T., et al. (2003). X-ray structure of a voltage-dependent K+ channel. *Nature*, *423*(6935), 33–41. http://dx.doi.org/10.1038/nature01580.

Jin, W., Klem, A. M., Lewis, J. H., & Lu, Z. (1999). Mechanisms of inward-rectifier K+ channel inhibition by tertiapin-Q. *Biochemistry*, *38*(43), 14294–14301.

Jin, W., & Lu, Z. (1998). A novel high-affinity inhibitor for inward-rectifier K+ channels. *Biochemistry, 37*(38), 13291–13299. http://dx.doi.org/10.1021/bi981178p.

Jin, T., Peng, L., Mirshahi, T., Rohacs, T., Chan, K. W., Sanchez, R., et al. (2002). The (beta)gamma subunits of G proteins gate a K(+) channel by pivoted bending of a transmembrane segment. *Molecular Cell, 10*(3), 469–481.

Junge, F., Schneider, B., Reckel, S., Schwarz, D., Dotsch, V., & Bernhard, F. (2008). Large-scale production of functional membrane proteins. *Cellular and Molecular Life Sciences, 65*(11), 1729–1755. http://dx.doi.org/10.1007/s00018-008-8067-5.

Kahanovitch, U., Tsemakhovich, V., Berlin, S., Rubinstein, M., Styr, B., Castel, R., et al. (2014). Recruitment of Gbetagamma controls the basal activity of G-protein coupled inwardly rectifying potassium (GIRK) channels: Crucial role of distal C terminus of GIRK1. *The Journal of Physiology, 592*(Pt. 24), 5373–5390. http://dx.doi.org/10.1113/jphysiol.2014.283218.

Kanjhan, R., Coulson, E. J., Adams, D. J., & Bellingham, M. C. (2005). Tertiapin-Q blocks recombinant and native large conductance K+ channels in a use-dependent manner. *The Journal of Pharmacology and Experimental Therapeutics, 314*(3), 1353–1361. http://dx.doi.org/10.1124/jpet.105.085928, jpet.105.085928 [pii].

Karschin, C., Dissmann, E., Stuhmer, W., & Karschin, A. (1996). IRK(1–3) and GIRK(1–4) inwardly rectifying K+ channel mRNAs are differentially expressed in the adult rat brain. *The Journal of Neuroscience, 16*(11), 3559–3570.

Katz, B. (1949). Les constantes electriques de la membrane du muscle. *Archives des Sciences Physiologiques, 2*, 285–299.

Kaufmann, K., Romaine, I., Days, E., Pascual, C., Malik, A., Yang, L., et al. (2013). ML297 (VU0456810), the first potent and selective activator of the GIRK potassium channel, displays antiepileptic properties in mice. *ACS Chemical Neuroscience, 4*(9), 1278–1286. http://dx.doi.org/10.1021/cn400062a.

Kavanaugh, M. P., Varnum, M. D., Osborne, P. B., Christie, M. J., Busch, A. E., Adelman, J. P., et al. (1991). Interaction between tetraethylammonium and amino acid residues in the pore of cloned voltage-dependent potassium channels. *The Journal of Biological Chemistry, 266*(12), 7583–7587.

Kawahara, R., Soeda, F., Kawaura, K., Honda, S., Miki, R., Noguchi, T., et al. (2013). Effect of tipepidine with novel antidepressant-like action on c-fos-like protein expression in rat brain. *Brain Research, 1513*, 135–142. http://dx.doi.org/10.1016/j.brainres.2013.03.034.

Kawate, T., & Gouaux, E. (2006). Fluorescence-detection size-exclusion chromatography for precrystallization screening of integral membrane proteins. *Structure, 14*(4), 673–681. http://dx.doi.org/10.1016/j.str.2006.01.013.

Kirsch, G. E., Drewe, J. A., Hartmann, H. A., Taglialatela, M., de Biasi, M., Brown, A. M., et al. (1992). Differences between the deep pores of K+ channels determined by an interacting pair of nonpolar amino acids. *Neuron, 8*(3), 499–505.

Kobayashi, T., Hirai, H., Iino, M., Fuse, I., Mitsumura, K., Washiyama, K., et al. (2009). Inhibitory effects of the antiepileptic drug ethosuximide on G protein-activated inwardly rectifying K+ channels. *Neuropharmacology, 56*(2), 499–506. http://dx.doi.org/10.1016/j.neuropharm.2008.10.003.

Kobayashi, T., & Ikeda, K. (2006). G protein-activated inwardly rectifying potassium channels as potential therapeutic targets. *Current Pharmaceutical Design, 12*(34), 4513–4523.

Kobayashi, T., Ikeda, K., Ichikawa, T., Abe, S., Togashi, S., & Kumanishi, T. (1995). Molecular cloning of a mouse G-protein-activated K+ channel (mGIRK1) and distinct distributions of three GIRK (GIRK1, 2 and 3) mRNAs in mouse brain. *Biochemical and Biophysical Research Communications, 208*(3), 1166–1173.

Kobayashi, T., Ikeda, K., Kojima, H., Niki, H., Yano, R., Yoshioka, T., et al. (1999). Ethanol opens G-protein-activated inwardly rectifying K+ channels. *Nature Neuroscience*, *2*(12), 1091–1097. http://dx.doi.org/10.1038/16019.

Kobayashi, T., Nishizawa, D., Iwamura, T., & Ikeda, K. (2007). Inhibition by cocaine of G protein-activated inwardly rectifying K+ channels expressed in Xenopus oocytes. *Toxicology In Vitro*, *21*(4), 656–664. http://dx.doi.org/10.1016/j.tiv.2007.01.009.

Kobayashi, T., Washiyama, K., & Ikeda, K. (2003). Inhibition of G protein-activated inwardly rectifying K+ channels by fluoxetine (Prozac). *British Journal of Pharmacology*, *138*(6), 1119–1128. http://dx.doi.org/10.1038/sj.bjp.0705172.

Kobayashi, T., Washiyama, K., & Ikeda, K. (2004). Inhibition of G protein-activated inwardly rectifying K+ channels by various antidepressant drugs. *Neuropsychopharmacology*, *29*(10), 1841–1851. http://dx.doi.org/10.1038/sj.npp.1300484.

Kobayashi, T., Washiyama, K., & Ikeda, K. (2006a). Inhibition of G protein-activated inwardly rectifying K+ channels by ifenprodil. *Neuropsychopharmacology*, *31*(3), 516–524. http://dx.doi.org/10.1038/sj.npp.1300844.

Kobayashi, T., Washiyama, K., & Ikeda, K. (2006b). Inhibition of G protein-activated inwardly rectifying K+ channels by the antidepressant paroxetine. *Journal of Pharmacological Sciences*, *102*(3), 278–287.

Kobayashi, T., Washiyama, K., & Ikeda, K. (2011). Inhibition of G protein-activated inwardly rectifying K+ channels by different classes of antidepressants. *PloS One*, *6*(12), e28208. http://dx.doi.org/10.1371/journal.pone.0028208.

Kofuji, P., Davidson, N., & Lester, H. A. (1995). Evidence that neuronal G-protein-gated inwardly rectifying K+ channels are activated by G beta gamma subunits and function as heteromultimers. *Proceedings of the National Academy of Sciences of the United States of America*, *92*(14), 6542–6546.

Kofuji, P., Doupnik, C. A., Davidson, N., & Lester, H. A. (1996). A unique P-region residue is required for slow voltage-dependent gating of a G protein-activated inward rectifier K+ channel expressed in Xenopus oocytes. *The Journal of Physiology*, *490*(Pt 3), 633–645.

Kofuji, P., Hofer, M., Millen, K. J., Millonig, J. H., Davidson, N., Lester, H. A., et al. (1996). Functional analysis of the *weaver* mutant GIRK2 K^+ channel and rescue of *weaver* granule cells. *Neuron*, *16*, 941–952.

Krapivinsky, G., Gordon, E. A., Wickman, K., Velimirovic, B., Krapivinsky, L., & Clapham, D. E. (1995). The G-protein-gated atrial K+ channel IKACh is a heteromultimer of two inwardly rectifying K(+)-channel proteins. *Nature*, *374*(6518), 135–141. http://dx.doi.org/10.1038/374135a0.

Krapivinsky, G., Krapivinsky, L., Wickman, K., & Clapham, D. E. (1995). G beta gamma binds directly to the G protein-gated K+ channel, IKACh. *The Journal of Biological Chemistry*, *270*(49), 29059–29062.

Krapivinsky, G., Medina, I., Eng, L., Krapivinsky, L., Yang, Y., & Clapham, D. E. (1998). A novel inward rectifier K+ channel with unique pore properties. *Neuron*, *20*(5), 995–1005.

Kubo, Y., Baldwin, T. J., Jan, Y. N., & Jan, L. Y. (1993). Primary structure and functional expression of a mouse inward rectifier potassium channel. *Nature*, *362*(6416), 127–133. http://dx.doi.org/10.1038/362127a0.

Kubo, Y., & Murata, Y. (2001). Control of rectification and permeation by two distinct sites after the second transmembrane region in Kir2.1 K+ channel. *The Journal of Physiology*, *531*(Pt 3), 645–660.

Kubo, Y., Reuveny, E., Slesinger, P. A., Jan, Y. N., & Jan, L. Y. (1993). Primary structure and functional expression of a rat G-protein-coupled muscarinic potassium channel. *Nature*, *364*(6440), 802–806. http://dx.doi.org/10.1038/364802a0.

Kunkel, M. T., & Peralta, E. G. (1995). Identification of domains conferring G protein regulation on inward rectifier potassium channels. *Cell, 83*(3), 443–449.

Kurachi, Y. (1985). Voltage-dependent activation of the inward-rectifier potassium channel in the ventricular cell membrane of guinea-pig heart. *The Journal of Physiology, 366*, 365–385.

Kurata, H. T., Cheng, W. W., Arrabit, C., Slesinger, P. A., & Nichols, C. G. (2007). The role of the cytoplasmic pore in inward rectification of Kir2.1 channels. *The Journal of General Physiology, 130*(2), 145–155. http://dx.doi.org/10.1085/jgp.200709742.

Lambright, D. G., Sondek, J., Bohm, A., Skiba, N. P., Hamm, H. E., & Sigler, P. B. (1996). The 2.0 A crystal structure of a heterotrimeric G protein. *Nature, 379*(6563), 311–319. http://dx.doi.org/10.1038/379311a0.

Leal-Pinto, E., Gomez-Llorente, Y., Sundaram, S., Tang, Q. Y., Ivanova-Nikolova, T., Mahajan, R., et al. (2010). Gating of a G protein-sensitive mammalian Kir3.1 prokaryotic Kir channel chimera in planar lipid bilayers. *The Journal of Biological Chemistry, 285*(51), 39790–39800. http://dx.doi.org/10.1074/jbc.M110.151373.

Lesage, F., Duprat, F., Fink, M., Guillemare, E., Coppola, T., Lazdunski, M., et al. (1994). Cloning provides evidence for a family of inward rectifier and G-protein coupled K+ channels in the brain. *FEBS Letters, 353*(1), 37–42.

Lesage, F., Guillemare, E., Fink, M., Duprat, F., Heurteaux, C., Fosset, M., et al. (1995). Molecular properties of neuronal G-protein-activated inwardly rectifying K+ channels. *The Journal of Biological Chemistry, 270*(48), 28660–28667.

Lewis, D. L., Ikeda, S. R., Aryee, D., & Joho, R. H. (1991). Expression of an inwardly rectifying K+ channel from rat basophilic leukemia cell mRNA in Xenopus oocytes. *FEBS Letters, 290*(1–2), 17–21.

Lewohl, J. M., Wilson, W. R., Mayfield, R. D., Brozowski, S. J., Morrisett, R. A., & Harris, R. A. (1999). G-protein-coupled inwardly rectifying potassium channels are targets of alcohol action. *Nature Neuroscience, 2*(12), 1084–1090. http://dx.doi.org/10.1038/16012.

Liao, Y. J., Jan, Y. N., & Jan, L. Y. (1996). Heteromultimerization of G-protein-gated inwardly rectifying K+ channel proteins GIRK1 and GIRK2 and their altered expression in weaver brain. *The Journal of Neuroscience, 16*(22), 7137–7150.

Liu, Y. S., Sompornpisut, P., & Perozo, E. (2001). Structure of the KcsA channel intracellular gate in the open state. *Nature Structural Biology, 8*(10), 883–887. http://dx.doi.org/10.1038/nsb1001-883.

Logothetis, D. E., Kim, D. H., Northup, J. K., Neer, E. J., & Clapham, D. E. (1988). Specificity of action of guanine nucleotide-binding regulatory protein subunits on the cardiac muscarinic K+ channel. *Proceedings of the National Academy of Sciences of the United States of America, 85*(16), 5814–5818.

Logothetis, D. E., Kurachi, Y., Galper, J., Neer, E. J., & Clapham, D. E. (1987). The beta gamma subunits of GTP-binding proteins activate the muscarinic K+ channel in heart. *Nature, 325*(6102), 321–326. http://dx.doi.org/10.1038/325321a0.

Long, S. B., Campbell, E. B., & Mackinnon, R. (2005). Crystal structure of a mammalian voltage-dependent Shaker family K+ channel. *Science, 309*(5736), 897–903. http://dx.doi.org/10.1126/science.1116269.

Lopatin, A. N., Makhina, E. N., & Nichols, C. G. (1994). Potassium channel block by cytoplasmic polyamines as the mechanism of intrinsic rectification. *Nature, 372*(6504), 366–369. http://dx.doi.org/10.1038/372366a0.

Lopes, C. M., Zhang, H., Rohacs, T., Jin, T., Yang, J., & Logothetis, D. E. (2002). Alterations in conserved Kir channel-PIP2 interactions underlie channelopathies. *Neuron, 34*(6), 933–944.

Lu, Z., & MacKinnon, R. (1994). Electrostatic tuning of Mg^{2+} affinity in an inward-rectifier K^{+} channel. *Nature, 371*(6494), 243–246.

Lujan, R., Marron Fernandez de Velasco, E., Aguado, C., & Wickman, K. (2014). New insights into the therapeutic potential of Girk channels. *Trends in Neurosciences*, *37*(1), 20–29. http://dx.doi.org/10.1016/j.tins.2013.10.006.

Lunn, M. L., Nassirpour, R., Arrabit, C., Tan, J., McLeod, I., Arias, C. M., et al. (2007). A unique sorting nexin regulates trafficking of potassium channels via a PDZ domain interaction. *Nature Neuroscience*, *10*(10), 1249–1259. http://dx.doi.org/10.1038/nn1953.

Ma, D., Zerangue, N., Raab-Graham, K., Fried, S. R., Jan, Y. N., & Jan, L. Y. (2002). Diverse trafficking patterns due to multiple traffic motifs in G protein-activated inwardly rectifying potassium channels from brain and heart. *Neuron*, *33*(5), 715–729.

Machida, T., Hashimoto, N., Kuwahara, I., Ogino, Y., Matsuura, J., Yamamoto, W., et al. (2011). Effects of a highly selective acetylcholine-activated K+ channel blocker on experimental atrial fibrillation. *Circulation Arrhythmia and Electrophysiology*, *4*(1), 94–102. http://dx.doi.org/10.1161/CIRCEP.110.951608.

MacKinnon, R. (1991). Determination of the subunit stoichiometry of a voltage-activated potassium channel. *Nature*, *350*(6315), 232–235. http://dx.doi.org/10.1038/350232a0.

MacKinnon, R., Heginbotham, L., & Abramson, T. (1990). Mapping the receptor site for charybdotoxin, a pore-blocking potassium channel inhibitor. *Neuron*, *5*(6), 767–771.

MacKinnon, R., & Miller, C. (1988). Mechanism of charybdotoxin block of the high-conductance, Ca2+ — activated K+ channel. *The Journal of General Physiology*, *91*(3), 335–349.

MacKinnon, R., & Miller, C. (1989). Mutant potassium channels with altered binding of charybdotoxin, a pore-blocking peptide inhibitor. *Science*, *245*(4924), 1382–1385.

Madeja, M., Steffen, W., Mesic, I., Garic, B., & Zhorov, B. S. (2010). Overlapping binding sites of structurally different antiarrhythmics flecainide and propafenone in the subunit interface of potassium channel Kv2.1. *The Journal of Biological Chemistry*, *285*(44), 33898–33905. http://dx.doi.org/10.1074/jbc.M110.159897.

Mahajan, R., Ha, J., Zhang, M., Kawano, T., Kozasa, T., & Logothetis, D. E. (2013). A computational model predicts that Gbetagamma acts at a cleft between channel subunits to activate GIRK1 channels. *Science Signaling*, *6*(288), ra69. http://dx.doi.org/10.1126/scisignal.2004075.

Mase, Y., Yokogawa, M., Osawa, M., & Shimada, I. (2012). Structural basis for modulation of gating property of G protein-gated inwardly rectifying potassium ion channel (GIRK) by i/o-family G protein alpha subunit (Galphai/o). *The Journal of Biological Chemistry*, *287*(23), 19537–19549. http://dx.doi.org/10.1074/jbc.M112.353888.

Matsuda, T., Ito, M., Ishimaru, S., Tsuruoka, N., Saito, T., Iida-Tanaka, N., et al. (2006). Blockade by NIP-142, an antiarrhythmic agent, of carbachol-induced atrial action potential shortening and GIRK1/4 channel. *Journal of Pharmacological Sciences*, *101*(4), 303–310.

Matsuda, H., Saigusa, A., & Irisawa, H. (1987). Ohmic conductance through the inwardly rectifying K channel and blocking by internal Mg2+. *Nature*, *325*(7000), 156–159.

McKinney, L. C., & Gallin, E. K. (1988). Inwardly rectifying whole-cell and single-channel K currents in the murine macrophage cell line J774.1. *The Journal of Membrane Biology*, *103*(1), 41–53.

Milovic, S., Steinecker-Frohnwieser, B., Schreibmayer, W., & Weigl, L. G. (2004). The sensitivity of G protein-activated K+ channels toward halothane is essentially determined by the C terminus. *The Journal of Biological Chemistry*, *279*(33), 34240–34249. http://dx.doi.org/10.1074/jbc.M403448200.

Mirshahi, T., & Logothetis, D. E. (2004). Molecular determinants responsible for differential cellular distribution of G protein-gated inwardly rectifying K+ channels. *The Journal of Biological Chemistry*, *279*(12), 11890–11897. http://dx.doi.org/10.1074/jbc.M313322200.

Mirshahi, T., Robillard, L., Zhang, H., Hebert, T. E., & Logothetis, D. E. (2002). Gbeta residues that do not interact with Galpha underlie agonist-independent activity of K+ channels. *The Journal of Biological Chemistry*, *277*(9), 7348–7355. http://dx.doi.org/10.1074/jbc.M109999200.

Morais-Cabral, J. H., Zhou, Y., & MacKinnon, R. (2001). Energetic optimization of ion conduction rate by the K+ selectivity filter. *Nature*, *414*(6859), 37–42.

Navarro, B., Kennedy, M. E., Velimirovic, B., Bhat, D., Peterson, A. S., & Clapham, D. E. (1996). Nonselective and Gβγ-insensitive weaver K^+ channels. *Science*, *272*(5270), 1950–1953.

Neer, E. J., & Clapham, D. E. (1988). Roles of G protein subunits in transmembrane signalling. *Nature*, *333*(6169), 129–134. http://dx.doi.org/10.1038/333129a0.

Nikawa, J., Kodaki, T., & Yamashita, S. (1987). Primary structure and disruption of the phosphatidylinositol synthase gene of Saccharomyces cerevisiae. *The Journal of Biological Chemistry*, *262*(10), 4876–4881.

Nikawa, J., & Yamashita, S. (1982). Yeast mutant defective in synthesis of phosphatidylinositol. Isolation and characterization of a CDPdiacylglycerol–inositol 3-phosphatidyltransferase Km mutant. *European Journal of Biochemistry*, *125*(2), 445–451.

Nishida, M., Cadene, M., Chait, B. T., & MacKinnon, R. (2007). Crystal structure of a Kir3.1-prokaryotic Kir channel chimera. *The EMBO Journal*, *26*(17), 4005–4015.

Nishida, M., & MacKinnon, R. (2002). Structural basis of inward rectification: Cytoplasmic pore of the G protein-gated inward rectifier GIRK1 at 1.8 A resolution. *Cell*, *111*(7), 957–965.

Nishizawa, D., Gajya, N., & Ikeda, K. (2011). Identification of selective agonists and antagonists to g protein-activated inwardly rectifying potassium channels: Candidate medicines for drug dependence and pain. *Current Neuropharmacology*, *9*(1), 113–117. http://dx.doi.org/10.2174/157015911795017227.

Noble, D., & Tsien, R. W. (1968). The kinetics and rectifier properties of the slow potassium current in cardiac Purkinje fibres. *The Journal of Physiology*, *195*(1), 185–214.

North, R. A., Williams, J. T., Surprenant, A., & Christie, M. J. (1987). Mu and delta receptors belong to a family of receptors that are coupled to potassium channels. *Proceedings of the National Academy of Sciences of the United States of America*, *84*(15), 5487–5491.

Okabe, K., Yatani, A., Evans, T., Ho, Y. K., Codina, J., Birnbaumer, L., et al. (1990). Beta gamma dimers of G proteins inhibit atrial muscarinic K+ channels. *The Journal of Biological Chemistry*, *265*(22), 12854–12858.

Patil, N., Cox, D. R., Bhat, D., Faham, M., Myers, R. M., & Peterson, A. S. (1995). A potassium channel mutation in weaver mice implicates membrane excitability in granule cell differentiation. *Nature Genetics*, *11*(2), 126–129. http://dx.doi.org/10.1038/ng1095-126.

Pegan, S., Arrabit, C., Slesinger, P. A., & Choe, S. (2006). Andersen's syndrome mutation effects on the structure and assembly of the cytoplasmic domains of Kir2.1. *Biochemistry*, *45*(28), 8599–8606.

Pegan, S., Arrabit, C., Zhou, W., Kwiatkowski, W., Collins, A., Slesinger, P. A., et al. (2005). Cytoplasmic domain structures of Kir2.1 and Kir3.1 show sites for modulating gating and rectification. *Nature Neuroscience*, *8*(3), 279–287. http://dx.doi.org/10.1038/nn1411.

Peleg, S., Varon, D., Ivanina, T., Dessauer, C. W., & Dascal, N. (2002). G(alpha)(i) controls the gating of the G protein-activated K(+) channel, GIRK. *Neuron*, *33*(1), 87–99.

Penington, N. J., Kelly, J. S., & Fox, A. P. (1993). Unitary properties of potassium channels activated by 5-HT in acutely isolated rat dorsal raphe neurones. *The Journal of Physiology*, *469*, 407–426.

Perozo, E., Cortes, D. M., & Cuello, L. G. (1999). Structural rearrangements underlying K + – channel activation gating. *Science, 285*(5424), 73–78.

Pfaffinger, P. J., Martin, J. M., Hunter, D. D., Nathanson, N. M., & Hille, B. (1985). GTP-binding proteins couple cardiac muscarinic receptors to a K channel. *Nature, 317*(6037), 536–538.

Plaster, N. M., Tawil, R., Tristani-Firouzi, M., Canun, S., Bendahhou, S., Tsunoda, A., et al. (2001). Mutations in Kir2.1 cause the developmental and episodic electrical phenotypes of Andersen's syndrome. *Cell, 105*(4), 511–519.

Proks, P., Girard, C., Haider, S., Gloyn, A. L., Hattersley, A. T., Sansom, M. S., et al. (2005). A gating mutation at the internal mouth of the Kir6.2 pore is associated with DEND syndrome. *EMBO Reports, 6*(5), 470–475. http://dx.doi.org/10.1038/sj.embor.7400393.

Reuveny, E., Slesinger, P. A., Inglese, J., Morales, J. M., Iniguez-Lluhi, J. A., Lefkowitz, R. J., et al. (1994). Activation of the cloned muscarinic potassium channel by G protein beta gamma subunits. *Nature, 370*(6485), 143–146. http://dx.doi.org/10.1038/370143a0.

Riven, I., Kalmanzon, E., Segev, L., & Reuveny, E. (2003). Conformational rearrangements associated with the gating of the G protein-coupled potassium channel revealed by FRET microscopy. *Neuron, 38*(2), 225–235.

Robertson, J. L., Palmer, L. G., & Roux, B. (2008). Long-pore electrostatics in inward-rectifier potassium channels. *The Journal of General Physiology, 132*(6), 613–632. http://dx.doi.org/10.1085/jgp.200810068.

Rohacs, T., Chen, J., Prestwich, G. D., & Logothetis, D. E. (1999). Distinct specificities of inwardly rectifying K(+) channels for phosphoinositides. *The Journal of Biological Chemistry, 274*(51), 36065–36072.

Rohacs, T., Lopes, C. M., Jin, T., Ramdya, P. P., Molnar, Z., & Logothetis, D. E. (2003). Specificity of activation by phosphoinositides determines lipid regulation of Kir channels. *Proceedings of the National Academy of Sciences of the United States of America, 100*(2), 745–750. http://dx.doi.org/10.1073/pnas.0236364100.

Rosenhouse-Dantsker, A., Sui, J. L., Zhao, Q., Rusinova, R., Rodriguez-Menchaca, A. A., Zhang, Z., et al. (2008). A sodium-mediated structural switch that controls the sensitivity of Kir channels to PtdIns(4,5)P(2). *Nature Chemical Biology, 4*(10), 624–631. http://dx.doi.org/10.1038/nchembio.112, nchembio.112 [pii].

Roux, B., & MacKinnon, R. (1999). The cavity and pore helices in the KcsA K+ channel: Electrostatic stabilization of monovalent cations. *Science, 285*(5424), 100–102.

Rubinstein, M., Peleg, S., Berlin, S., Brass, D., & Dascal, N. (2007). Galphai3 primes the G protein-activated K+ channels for activation by coexpressed Gbetagamma in intact Xenopus oocytes. *The Journal of Physiology, 581*(Pt. 1), 17–32. http://dx.doi.org/10.1113/jphysiol.2006.125864.

Rubinstein, M., Peleg, S., Berlin, S., Brass, D., Keren-Raifman, T., Dessauer, C. W., et al. (2009). Divergent regulation of GIRK1 and GIRK2 subunits of the neuronal G protein gated K+ channel by GalphaiGDP and Gbetagamma. *The Journal of Physiology, 587*(Pt. 14), 3473–3491. http://dx.doi.org/10.1113/jphysiol.2009.173229.

Rusinova, R., Mirshahi, T., & Logothetis, D. E. (2007). Specificity of Gbetagamma signaling to Kir3 channels depends on the helical domain of pertussis toxin-sensitive Galpha subunits. *The Journal of Biological Chemistry, 282*(47), 34019–34030. http://dx.doi.org/10.1074/jbc.M704928200.

Sadja, R., Smadja, K., Alagem, N., & Reuveny, E. (2001). Coupling Gbetagamma-dependent activation to channel opening via pore elements in inwardly rectifying potassium channels. *Neuron, 29*(3), 669–680.

Sakmann, B., Noma, A., & Trautwein, W. (1983). Acetylcholine activation of single muscarinic K^+ channels in isolated pacemaker cells of the mammalian heart. *Nature, 303*(5914), 250–253.

Sasaki, T., Hashimoto, K., Tachibana, M., Kurata, T., Kimura, H., Komatsu, H., et al. (2014). Tipepidine in adolescent patients with depression: A 4 week, open-label, preliminary study. *Neuropsychiatric Disease and Treatment, 10*, 719–722. http://dx.doi.org/10.2147/NDT.S63075.

Sasaki, T., Hashimoto, K., Tachibana, M., Kurata, T., Okawada, K., Ishikawa, M., et al. (2014). Tipepidine in children with attention deficit/hyperactivity disorder: A 4-week, open-label, preliminary study. *Neuropsychiatric Disease and Treatment, 10*, 147–151. http://dx.doi.org/10.2147/NDT.S58480.

Schreibmayer, W., Dessauer, C. W., Vorobiov, D., Gilman, A. G., Lester, H. A., Davidson, N., et al. (1996). Inhibition of an inwardly rectifying K+ channel by G-protein alpha-subunits. *Nature, 380*(6575), 624–627. http://dx.doi.org/10.1038/380624a0.

Schulze, D., Krauter, T., Fritzenschaft, H., Soom, M., & Baukrowitz, T. (2003). Phosphatidylinositol 4,5-bisphosphate (PIP2) modulation of ATP and pH sensitivity in Kir channels. A tale of an active and a silent PIP2 site in the N terminus. *The Journal of Biological Chemistry, 278*(12), 10500–10505. http://dx.doi.org/10.1074/jbc.M208413200.

Slesinger, P. A. (2001). Ion selectivity filter regulates local anesthetic inhibition of G-protein-gated inwardly rectifying K+ channels. *Biophysical Journal, 80*(2), 707–718. http://dx.doi.org/10.1016/S0006-3495(01)76050-X.

Slesinger, P. A., Patil, N., Liao, Y. J., Jan, Y. N., Jan, L. Y., & Cox, D. R. (1996). Functional effects of the mouse weaver mutation on G protein-gated inwardly rectifying K^+ channels. *Neuron, 16*(2), 321–331.

Slesinger, P. A., Reuveny, E., Jan, Y. N., & Jan, L. Y. (1995). Identification of structural elements involved in G protein gating of the GIRK1 potassium channel. *Neuron, 15*(5), 1145–1156.

Smith, J. L., Fischetti, R. F., & Yamamoto, M. (2012). Micro-crystallography comes of age. *Current Opinion in Structural Biology, 22*(5), 602–612. http://dx.doi.org/10.1016/j.sbi.2012.09.001.

Soejima, M., & Noma, A. (1984). Mode of regulation of the ACh-sensitive K-channel by the muscarinic receptor in rabbit atrial cells. *Pflügers Archiv, 400*(4), 424–431.

Songyang, Z., Fanning, A. S., Fu, C., Xu, J., Marfatia, S. M., Chishti, A. H., et al. (1997). Recognition of unique carboxyl-terminal motifs by distinct PDZ domains. *Science, 275*(5296), 73–77.

Soom, M., Schonherr, R., Kubo, Y., Kirsch, C., Klinger, R., & Heinemann, S. H. (2001). Multiple PIP2 binding sites in Kir2.1 inwardly rectifying potassium channels. *FEBS Letters, 490*(1–2), 49–53.

Stanfield, P. R., Davies, N. W., Shelton, P. A., Sutcliffe, M. J., Khan, I. A., Brammar, W. J., et al. (1994). A single aspartate residue is involved in both intrinsic gating and blockage by Mg2+ of the inward rectifier, IRK1. *The Journal of Physiology, 478*(Pt. 1), 1–6.

Stanfield, P. R., Nakajima, S., & Nakajima, Y. (2002). Constitutively active and G-protein coupled inward rectifier K+ channels: Kir2.0 and Kir3.0. *Reviews of Physiology, Biochemistry and Pharmacology, 145*, 47–179.

Styer, A. M., Mirshahi, U. L., Wang, C., Girard, L., Jin, T., Logothetis, D. E., et al. (2010). G protein {beta}{gamma} gating confers volatile anesthetic inhibition to Kir3 channels. *The Journal of Biological Chemistry, 285*(53), 41290–41299. http://dx.doi.org/10.1074/jbc.M110.178541.

Sui, J. L., Petit-Jacques, J., & Logothetis, D. E. (1998). Activation of the atrial KACh channel by the betagamma subunits of G proteins or intracellular Na+ions depends on the presence of phosphatidylinositol phosphates. *Proceedings of the National Academy of Sciences of the United States of America, 95*(3), 1307–1312.

Surmeier, D. J., Mermelstein, P. G., & Goldowitz, D. (1996). The weaver mutation of GIRK2 results in a loss of inwardly rectifying K^+ current in cerebellar granule cells. *Proceedings of the National Academy of Sciences of the United States of America, 93*(20), 11191–11195.

Taglialatela, M., Ficker, E., Wible, B. A., & Brown, A. M. (1995). C-terminus determinants for Mg2+ and polyamine block of the inward rectifier K+ channel IRK1. *The EMBO Journal, 14*(22), 5532–5541.

Taglialatela, M., Wible, B. A., Caporaso, R., & Brown, A. M. (1994). Specification of pore properties by the carboxyl terminus of inwardly rectifying K+ channels. *Science, 264*(5160), 844–847.

Tao, X., Avalos, J. L., Chen, J., & MacKinnon, R. (2009). Crystal structure of the eukaryotic strong inward-rectifier K^+ channel Kir2.2 at 3.1 Å resolution. *Science, 326*(5960), 1668–1674. http://dx.doi.org/10.1126/science.1180310, 326/5960/1668 [pii].

Tempel, B. L., Jan, Y. N., & Jan, L. Y. (1988). Cloning of a probable potassium channel gene from mouse brain. *Nature, 332*(6167), 837–839. http://dx.doi.org/10.1038/332837a0.

Tempel, B. L., Papazian, D. M., Schwarz, T. L., Jan, Y. N., & Jan, L. Y. (1987). Sequence of a probable potassium channel component encoded at Shaker locus of Drosophila. *Science, 237*(4816), 770–775.

Tinker, A., Jan, Y. N., & Jan, L. Y. (1996). Regions responsible for the assembly of inwardly rectifying potassium channels. *Cell, 87*(5), 857–868.

Valenzuela, C., Delpon, E., Tamkun, M. M., Tamargo, J., & Snyders, D. J. (1995). Stereoselective block of a human cardiac potassium channel (Kv1.5) by bupivacaine enantiomers. *Biophysical Journal, 69*(2), 418–427. http://dx.doi.org/10.1016/S0006-3495(95)79914-3.

Valenzuela, C., Snyders, D. J., Bennett, P. B., Tamargo, J., & Hondeghem, L. M. (1995). Stereoselective block of cardiac sodium channels by bupivacaine in guinea pig ventricular myocytes. *Circulation, 92*(10), 3014–3024.

Vandenberg, C. A. (1987). Inward rectification of a potassium channel in cardiac ventricular cells depends on internal magnesium ions. *Proceedings of the National Academy of Sciences of the United States of America, 84*(8), 2560–2564.

Vereecke, J., Isenberg, G., & Carmeliet, E. (1980). K efflux through inward rectifying K channels in voltage clamped Purkinje fibers. *Pflügers Archiv, 384*(3), 207–217.

Wall, M. A., Coleman, D. E., Lee, E., Iniguez-Lluhi, J. A., Posner, B. A., Gilman, A. G., et al. (1995). The structure of the G protein heterotrimer Gi alpha 1 beta 1 gamma 2. *Cell, 83*(6), 1047–1058.

Wang, W., Whorton, M. R., & MacKinnon, R. (2014). Quantitative analysis of mammalian GIRK2 channel regulation by G proteins, PIP2 and Na+ in a reconstituted system. *Elife (Cambridge), 3*, e03671. http://dx.doi.org/10.7554/eLife.03671.

Wei, J., Hodes, M. E., Piva, R., Feng, Y., Wang, Y., Ghetti, B., et al. (1998). Characterization of murine Girk2 transcript isoforms: Structure and differential expression. *Genomics, 51*(3), 379–390. http://dx.doi.org/10.1006/geno.1998.5369.

Weigl, L. G., & Schreibmayer, W. (2001). G protein-gated inwardly rectifying potassium channels are targets for volatile anesthetics. *Molecular Pharmacology, 60*(2), 282–289.

Whorton, M. R., & MacKinnon, R. (2011). Crystal structure of the mammalian GIRK2 K+ channel and gating regulation by G proteins, PIP2, and sodium. *Cell, 147*(1), 199–208. http://dx.doi.org/10.1016/j.cell.2011.07.046.

Whorton, M. R., & MacKinnon, R. (2013). X-ray structure of the mammalian GIRK2-betagamma G-protein complex. *Nature, 498*(7453), 190–197. http://dx.doi.org/10.1038/nature12241.

Wible, B. A., Taglialatela, M., Ficker, E., & Brown, A. M. (1994). Gating of inwardly rectifying K^+ channels localized to a single negatively charged residue. *Nature, 371*(6494), 246–249.

Wickman, K., Pu, W. T., & Clapham, D. E. (2002). Structural characterization of the mouse Girk genes. *Gene, 284*(1–2), 241–250.

Wydeven, N., Marron Fernandez de Velasco, E., Du, Y., Benneyworth, M. A., Hearing, M. C., et al. (2014). Mechanisms underlying the activation of G-protein-gated inwardly rectifying K+ (GIRK) channels by the novel anxiolytic drug, ML297. *Proceedings of the National Academy of Sciences of the United States of America, 111*(29), 10755–10760. http://dx.doi.org/10.1073/pnas.1405190111.

Xu, Y., Shin, H. G., Szep, S., & Lu, Z. (2009). Physical determinants of strong voltage sensitivity of K(+) channel block. *Nature Structural & Molecular Biology, 16*(12), 1252–1258. http://dx.doi.org/10.1038/nsmb.1717.

Yamakura, T., Lewohl, J. M., & Harris, R. A. (2001). Differential effects of general anesthetics on G protein-coupled inwardly rectifying and other potassium channels. *Anesthesiology, 95*(1), 144–153.

Yamamoto, W., Hashimoto, N., Matsuura, J., Machida, T., Ogino, Y., Kobayashi, T., et al. (2014). Effects of the selective KACh channel blocker NTC-801 on atrial fibrillation in a canine model of atrial tachypacing: Comparison with class Ic and III drugs. *Journal of Cardiovascular Pharmacology, 63*(5), 421–427. http://dx.doi.org/10.1097/FJC.0000000000000065.

Yang, J., Jan, Y. N., & Jan, L. Y. (1995a). Control of rectification and permeation by residues in two distinct domains in an inward rectifier K+ channel. *Neuron, 14*(5), 1047–1054.

Yang, J., Jan, Y. N., & Jan, L. Y. (1995b). Determination of the subunit stoichiometry of an inwardly rectifying potassium channel. *Neuron, 15*(6), 1441–1447.

Yang, J., Yu, M., Jan, Y. N., & Jan, L. Y. (1997). Stabilization of ion selectivity filter by pore loop ion pairs in an inwardly rectifying potassium channel. *Proceedings of the National Academy of Sciences of the United States of America, 94*(4), 1568–1572.

Yatani, A., Codina, J., Brown, A. M., & Birnbaumer, L. (1987). Direct activation of mammalian atrial muscarinic potassium channels by GTP regulatory protein Gk. *Science, 235*(4785), 207–211.

Yatani, A., Hamm, H., Codina, J., Mazzoni, M. R., Birnbaumer, L., & Brown, A. M. (1988). A monoclonal antibody to the alpha subunit of Gk blocks muscarinic activation of atrial K+ channels. *Science, 241*(4867), 828–831.

Yi, B. A., Lin, Y. F., Jan, Y. N., & Jan, L. Y. (2001). Yeast screen for constitutively active mutant G protein-activated potassium channels. *Neuron, 29*(3), 657–667.

Yokogawa, M., Osawa, M., Takeuchi, K., Mase, Y., & Shimada, I. (2011). NMR analyses of the Gbetagamma binding and conformational rearrangements of the cytoplasmic pore of G protein-activated inwardly rectifying potassium channel 1 (GIRK1). *The Journal of Biological Chemistry, 286*(3), 2215–2223. http://dx.doi.org/10.1074/jbc.M110.160754.

Yool, A. J., & Schwarz, T. L. (1991). Alteration of ionic selectivity of a K+ channel by mutation of the H5 region. *Nature, 349*(6311), 700–704. http://dx.doi.org/10.1038/349700a0.

Yoshimoto, Y., Fukuyama, Y., Horio, Y., Inanobe, A., Gotoh, M., & Kurachi, Y. (1999). Somatostatin induces hyperpolarization in pancreatic islet alpha cells by activating a G protein-gated K+ channel. *FEBS Letters, 444*(2–3), 265–269.

Yow, T. T., Pera, E., Absalom, N., Heblinski, M., Johnston, G. A., Hanrahan, J. R., et al. (2011). Naringin directly activates inwardly rectifying potassium channels at an overlapping binding site to tertiapin-Q. *British Journal of Pharmacology, 163*(5), 1017–1033. http://dx.doi.org/10.1111/j.1476-5381.2011.01315.x.

Zeng, W. Z., Li, X. J., Hilgemann, D. W., & Huang, C. L. (2003). Protein kinase C inhibits ROMK1 channel activity via a phosphatidylinositol 4,5-bisphosphate-dependent mechanism. *The Journal of Biological Chemistry, 278*(19), 16852–16856. http://dx.doi.org/10.1074/jbc.M300619200.

Zhang, H., He, C., Yan, X., Mirshahi, T., & Logothetis, D. E. (1999). Activation of inwardly rectifying K+ channels by distinct PtdIns(4,5)P2 interactions. *Nature Cell Biology, 1*(3), 183–188. http://dx.doi.org/10.1038/11103.

Zhao, Q., Albsoul-Younes, A. M., Zhao, P., Kozasa, T., Nakajima, Y., & Nakajima, S. (2006). Dominant negative effects of a Gbeta mutant on G-protein coupled inward rectifier K+ channel. *FEBS Letters*, *580*(16), 3879–3882. http://dx.doi.org/10.1016/j.febslet.2006.06.016.

Zhao, Q., Kawano, T., Nakata, H., Nakajima, Y., Nakajima, S., & Kozasa, T. (2003). Interaction of G protein beta subunit with inward rectifier K(+) channel Kir3. *Molecular Pharmacology*, *64*(5), 1085–1091. http://dx.doi.org/10.1124/mol.64.5.1085.

Zhou, W., Arrabit, C., Choe, S., & Slesinger, P. A. (2001). Mechanism underlying bupivacaine inhibition of G protein-gated inwardly rectifying K^+ channels. *Proceedings of the National Academy of Sciences of the United States of America*, *98*(11), 6482–6487.

Zhou, Y., Morais-Cabral, J. H., Kaufman, A., & MacKinnon, R. (2001). Chemistry of ion coordination and hydration revealed by a K+ channel-Fab complex at 2.0 A resolution. *Nature*, *414*(6859), 43–48. http://dx.doi.org/10.1038/35102009.

CHAPTER FIVE

Localization and Targeting of GIRK Channels in Mammalian Central Neurons

Rafael Luján[1], Carolina Aguado

Instituto de Investigación en Discapacidades Neurológicas (IDINE), Departamento de Ciencias Médicas, Facultad de Medicina, Universidad Castilla-La Mancha, Campus Biosanitario, Albacete, Spain

[1]Corresponding author: e-mail address: rafael.lujan@uclm.es

Contents

1. Introduction 162
2. Molecular Organization and Heterogeneity of GIRK Channels 163
3. Neuroanatomical Approaches to the Study of GIRK Channel Distribution 164
4. Regional Distribution of GIRK Channel Subunits in the CNS 169
   4.1 Distribution of GIRK1, GIRK2, and GIRK3 Subunits 170
   4.2 Distribution of GIRK4 Subunit 173
5. Cellular Distribution of GIRK Channel Subunits 175
   5.1 GIRK Channels in Neurochemically and Functionally Different Cell Populations 176
   5.2 Cell Type-Dependent Localization of GIRK Channel Subunits 177
6. Subcellular Distribution of GIRK Channel Subunits 179
   6.1 Compartment-Dependent Localization 180
   6.2 Localization of GIRK Channels Relative to Glutamate Release Sites 183
   6.3 Input-Dependent Localization 184
   6.4 Plasticity- and Drug-Dependent Changes in Distribution 184
   6.5 Presynaptic Localization of GIRK Channel Subunits 185
   6.6 Localization of GIRK-Associated Proteins 188
7. Developmental Aspects of GIRK Channel Distribution 190
   7.1 Distribution of GIRK Channels During Pre- and Postnatal Development 190
   7.2 Acquisition of the GIRK2 Neurochemical Phenotype 192
8. Conclusions and Future Perspectives 193

Acknowledgments 194

References 194

Abstract

G protein-gated inwardly rectifying K^+ ($GIRK/K_{ir}3$) channels are critical to brain function. They hyperpolarize neurons in response to activation of different G protein-coupled receptors, reducing cell excitability. Molecular cloning has revealed four distinct mammalian genes (GIRK1–4), which, with the exception of GIRK4, are broadly expressed in

International Review of Neurobiology, Volume 123
ISSN 0074-7742
http://dx.doi.org/10.1016/bs.irn.2015.05.009

the central nervous system (CNS) and have been implicated in a variety of neurological disorders. Although the molecular structure and composition of GIRK channels are key determinants of their biophysical properties, their cellular and subcellular localization patterns and densities on the neuronal surface are just as important to nerve function. Current data obtained with high-resolution quantitative localization techniques reveal complex, subcellular compartment-specific distribution patterns of GIRK channel subunits. Recent efforts have focused on determining the associated proteins that form macromolecular complexes with GIRK channels. Demonstration of the precise subcellular compartmentalization of GIRK channels and their associated proteins represents a crucial step in understanding the contribution of these channels to specific aspects of neuronal function under both physiological and pathological conditions. Here, we present an overview of studies aimed at determining the cellular and subcellular localization of GIRK channel subunits in mammalian brain neurons and discuss implications for neuronal physiology.

1. INTRODUCTION

Ion channels are classified by their gating properties and ion selectivity for Na^+, Ca^{2+}, Cl^-, and K^+ (Hille, 2001). Potassium (K^+)-selective channels are key determinants of membrane excitability and regulate a variety of cellular processes including membrane potential, signal transduction, hormone release, vascular tone, cell volume, and immune responses (Curran, 1998). Four different subfamilies of K^+ channels have been proposed based on their structural and phylogenetic relationship: voltage-gated K^+ (K_v) channels, Ca^{2+}-activated K^+ (K_{Ca}) channels, two-pore K^+ (K_{2P}) channels, and inwardly rectifying (K_{ir}) channels (Gutman et al., 2005).

K_{ir} channels are formed by 16 subunits falling into 7 subfamilies ($K_{ir}1$–$K_{ir}7$), including the $K_{ir}3$ subfamily, which is directly coupled to heterotrimeric G proteins and mediates the inhibitory effect of activated G protein-coupled receptors (GPCRs) on neuronal excitability (Dascal, 1997; Kubo et al., 2005; Luján, Marron Fernandez de Velasco, Aguado, & Wickman, 2014; Luján, Maylie, & Adelman, 2009; Yamada, Inanobe, & Kurachi, 1998). This subfamily of K_{ir} channels, also known as G protein-gated inwardly rectifying K^+ (GIRK) channels, is composed of four subunits, designated GIRK1–4 (Dascal, 1997; Luján et al., 2014; Lüscher & Slesinger, 2010).

The past decade has witnessed great progress in our understanding of the diverse functional roles that GIRK channels fulfill in different neuron populations and brain regions. Molecular cloning of GIRK channel genes has led to the design and synthesis of antibodies that, combined with several immunohistochemical techniques, facilitated studies on the distribution of GIRK channel subunits in the CNS. Indeed, immunohistochemical approaches at both the light and electron microscopic levels have yielded important insights into the distribution and precise subcellular localization of GIRK channels in neurons. In particular, the use of high-resolution immunohistochemical techniques at the electron microscopic level has allowed unparalleled precision, showing a highly regulated subcellular distribution patterns for GIRK channel subunits that are also region- and cell type-dependent (Luján et al., 2014, 2009). The demonstration of this precise subcellular compartmentalization provides a new understanding of the role of GIRK channels in information transfer and processing within neurons and neural networks under physiological and pathological conditions. In this chapter, we summarize current anatomical knowledge describing the regional and cellular distribution of the four GIRK channel subunits in the mammalian CNS. The precise subcellular distribution of GIRK channels at the surface of central neurons will be discussed in view of their relationship to neurotransmitter release sites and of possible functional implications.

2. MOLECULAR ORGANIZATION AND HETEROGENEITY OF GIRK CHANNELS

The structure and function of GIRK channels are reviewed in other chapters of this volume, and we refer the reader to those sections for further information. Here, we briefly describe basic molecular, biochemical, and physiological features of GIRK channels.

GIRK channels are composed of four different, but homologous, subunits (GIRK1–4) that are conserved in mouse, rat, and human (Hibino et al., 2010; Yamada et al., 1998). Each subunit contains hydrophilic amino- and carboxy-terminal domains located at the cytoplasmic side of the membrane, and two transmembrane segments (M1 and M2) flanking a hydrophobic (P) domain. Following coassembly, the P-domain of each GIRK subunit participates in the formation of a K^+-selective pore (Hibino et al., 2010). Functional GIRK channels are homo- or heterotetrameric complexes (Yang, Jan, & Jan, 1995) that are activated by direct interaction with $G\beta\gamma$ released from $G\alpha_{i/o}$ G proteins; $G\beta\gamma$ binding increases their probability of opening (Krapivinsky, Krapivinsky, Wickman, & Clapham, 1995; Logothetis, Kurachi, Galper, Neer, & Clapham, 1987; Wickman & Clapham, 1995). The GIRK2 and GIRK4 subunits are able to form homomeric channels, but when exogenously expressed, these homomeric channels produce brief and poorly resolved open-time kinetics that are different from those of native channels. GIRK1 and GIRK3 subunits are unable to independently form functional channels, and, furthermore, GIRK1 requires coexpression with another GIRK subunit to achieve membrane distribution (Hibino et al., 2010; Lüscher & Slesinger, 2010). Indeed. GIRK1 is retained in the rough endoplasmic reticulum (rER) when GIRK2 and GIRK3 are absent (Koyrakh et al., 2005). GIRK1 and GIRK2 subunits can coassemble with GIRK3 to form GIRK1/3 and GIRK2/3, and, interestingly, the latter have reduced Gβγ sensitivity compared to other heteromers containing GIRK1 subunits (Jelacic, Kennedy, Wickman, & Clapham, 2000).

The physiological role of GIRK channels is to maintain the resting membrane potential near the potassium equilibrium potential, to regulate cellular excitability, and to maintain the potassium homeostasis that lead to hyperpolarization and the reduction of membrane excitability (Lüscher & Slesinger, 2010). Thus, GIRK channels are important regulators of electrical excitability in neurons. Neuronal GIRK channels mediate the inhibitory effect of numerous neurotransmitters including opioids, acetylcholine, adenosine, dopamine, γ-aminobutyric acid (GABA), glutamate, serotonin, norepinephrine, somatostatin, and neuropeptide Y (Jaén & Doupnik, 2005; Lüscher et al., 1997; North, 1989; Sosulina, Schwesig, Seifert, & Pape, 2008; Witkowski, Szulczyk, Rola, & Szulczyk, 2008) by stimulating their cognate GPCRs, which couple to the $G_{i/o}$ family of G proteins. The opening of GIRK channels following GPCR activation decreases membrane excitability by hyperpolarizing the membrane potential and opposing membrane depolarization (Dascal, 1997). Loss of neuronal GIRK channels leads to hyperexcitability, seizures, hyperalgesia, decreased analgesic response to morphine administration, decreased anxiety, decreased baclofen-induced ataxia, increased operant responding for food, blunted behavioral responses to ethanol, and reduced cocaine self-administration (reviewed by Luján et al., 2014).

3. NEUROANATOMICAL APPROACHES TO THE STUDY OF GIRK CHANNEL DISTRIBUTION

The molecular definition of the mammalian GIRK family has led to the generation of subunit-specific molecular tools for investigating expression levels and cellular patterns of GIRK mRNAs in brain tissue with different techniques. Thus, northern blots, western blots, reverse transcription polymerase chain reaction (RT-PCR), *in situ* hybridization, histoblots, and immunohistochemical studies have been used to observe the regional, cellular, and subcellular distribution of GIRK channels in the CNS (Aguado et al., 2008; Chen, Ehrhard, Goldowitz, & Smeyne, 1997; Ciruela et al., 2010; Fernández-Alacid et al., 2009; Fernández-Alacid, Watanabe, Molnár, Wickman, & Luján, 2011; Inanobe et al., 1999; Karschin, Dissmann, Stuhmer, & Karschin, 1996; Kobayashi et al., 1995; Koyrakh et al., 2005; Liao, Jan, & Jan, 1996; Miyashita & Kubo, 1997; Ponce et al., 1996; Saenz del Burgo et al., 2008). Some of these techniques like northern blots, RT-PCR, and *in situ* hybridization have provided invaluable information regarding the relative expression and distribution patterns of GIRK channel subunits in different brain regions and in specific neuronal populations of the developing and adult brain, under both normal and pathological conditions. Although these techniques possess high specificity and are relatively simple and inexpensive to perform, they have a number of limitations (for instance, an inherently high false-positive rate, the difficulty in identifying targets that have low DNA and RNA copies, risk of mRNA degradation, high doses of radioactivity and formaldehyde are a risk for workers or the lack of subcellular resolution) that are only overcome using western blots, histoblots, and immunohistochemical techniques with selective subunit-specific antibodies.

Immunoelectron microscopy (iEM) is currently the method of choice for examining the subcellular localization of proteins. The resolution offered by this technique is unparalleled, yielding precise localization of ion channels, neurotransmitter receptors, and associated proteins in any compartment within neurons. iEM is a technically demanding procedure that provides quantitative analysis of target localization, e.g., membrane versus intracellular, pre- versus postsynaptic. The development of specific antibodies was a key for the success of these iEM studies. This issue is particularly critical, as antibody quality and specificity are absolutely fundamental to achieving reliable immunohistochemical results. There are now a number of subunit-specific antibodies that can be used to identify and map the cellular and subcellular localization patterns of GIRK channel subunits in the CNS. Some of these antibodies have been validated using GIRK knockout mice, confirming the specificity of the immunolabeling patterns using different techniques (Aguado et al., 2008, 2013; Booker et al., 2013; Ciruela et al., 2010; Fernández-Alacid et al., 2009, 2011; Kirizs, Kerti-Szigeti, Lorincz, & Nusser, 2014; Koyrakh et al., 2005; Kulik et al., 2006).

Four high-resolution immunohistochemical techniques have been developed and successfully used for the subcellular distribution of GIRK channels: (i) preembedding immunoperoxidase (Drake, Bausch, Milner, & Chavkin, 1997; Fernández-Alacid et al., 2009; Ponce et al., 1996), (ii) preembedding immunogold (Aguado et al., 2008; Booker et al., 2013; Ciruela et al., 2010; Fernández-Alacid et al., 2009, 2011; Koyrakh et al., 2005; Kulik et al., 2006; Labouebe et al., 2007; Marker, Luján, Loh, & Wickman, 2005), (iii) postembedding immunogold (Aguado et al., 2008; Fernández-Alacid et al., 2009, 2011; Koyrakh et al., 2005; Marker et al., 2005), and (iv) the sodium dodecyl sulfate (SDS)-digested freeze-fracture replica labeling (SDS-FRL; Kirizs et al., 2014; Kulik et al., 2006) methods, which have revealed that GIRK channel subunits are localized with extraordinary precision in defined subcellular compartments of neurons in the CNS (Fig. 1). Each of these four iEM methods provides different information due to their unique advantages and disadvantages (Luján, 2010), as we describe below.

The *preembedding immunoperoxidase method* uses enzymatic reactions to visualize antibody–antigen complexes (Fig. 1A). This method is highly sensitive and provides valuable information on the regional distribution and subcellular localization of proteins (Fig. 1A). However, owing to the diffusible nature of the peroxidase reaction-end-product, this technique is not suitable for localizing the channels at synaptic sites, or for quantitative evaluation of different antigenic sites. These limitations are overcome by the *preembedding immunogold method* (Fig. 1B). This technique uses secondary antibodies covalently conjugated to a gold particle of small size (0.8–1.4 nm), which is further enhanced using a silver kit, thus producing a nondiffusible label allowing target identification at the precise site of reaction (Luján, 2010). This method has been used to study GIRK channels at extrasynaptic and perisynaptic sites in single and serial sections and has been used successfully to quantify GIRK channels along the neuronal plasma membrane (Aguado et al., 2008; Ciruela et al., 2010; Fernández-Alacid et al., 2009, 2011; Koyrakh et al., 2005; Kulik et al., 2006; Labouebe et al., 2007; Marker et al., 2005; Fig. 1B). However, the main limitation of this technique is that the strength of labeling intensity decreases with tissue depth, and antibodies may penetrate unevenly into brain tissue. As a result, synaptic GIRK channels, or any other protein located along the postsynaptic density of glutamatergic synapses, cannot generally be detected using this approach (Luján, 2010).

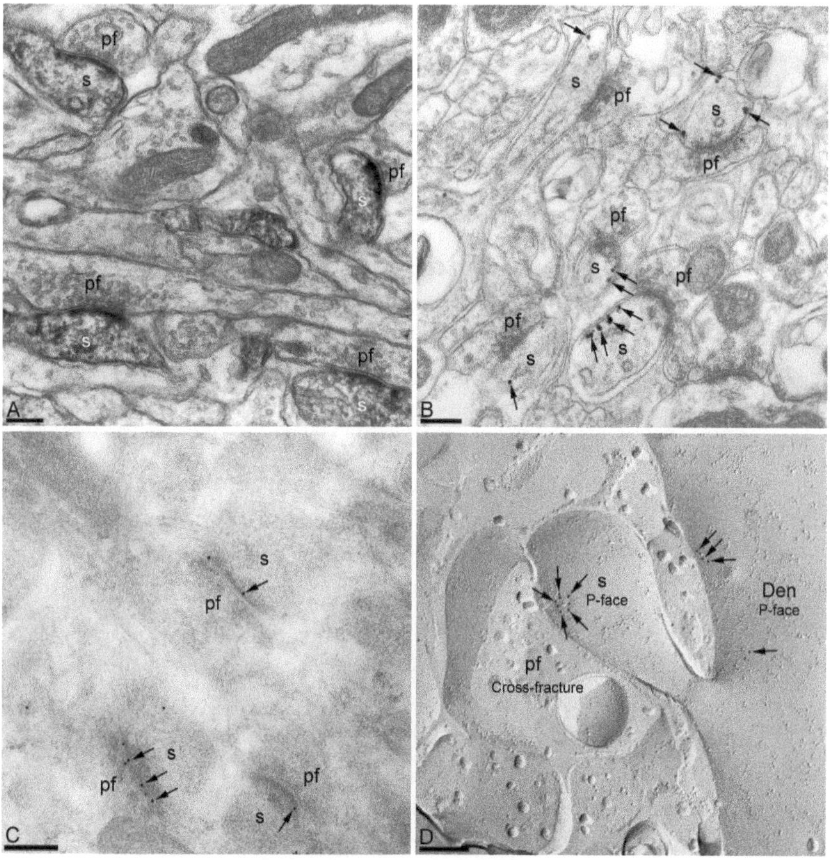

Figure 1 Immunoelectron microscopic techniques to study subcellular GIRK channel localization. Electron micrographs showing immunoreactivity for the GIRK3 subunit along the neuronal surface of Purkinje cells, as demonstrated using four different high-resolution immunohistochemical methods. (A) Preembedding immunoperoxidase method. Peroxidase reaction-end-product for GIRK3 fills dendritic spines (s) of Purkinje cells establishing synaptic contacts with parallel fiber axon terminals (pf). (B) Preembedding immunogold method. Immunoparticles for GIRK3 are distributed along the extrasynaptic plasma membrane (arrows) dendritic spines (s) of Purkinje cells establishing synaptic contacts with parallel fiber axon terminals (pf). (C) Postembedding immunogold method. Immunoparticles for GIRK3 are concentrated within postsynaptic densities (arrows) of dendritic spines (s) of Purkinje cells. (D) SDS-digested freeze-fracture replica labeling (SDS-FRL) technique. Freeze-fracture replicas prepared from mouse cerebellum were labeled with 10 nm (small black dots) immunoparticles to detect GIRK3. Clusters of immunoparticles for GIRK3 (arrows) are concentrated in dendritic spines (s) and shafts (Den) of Purkinje cells establishing synapses with parallel fiber terminals (pf). P-face, protoplasmatic face. Scale bars: A–D, 0.2 μm.

To localize any protein along the main body of the postsynaptic specialization, application of the *postembedding immunogold method* is required (Fig. 1C). Additionally, using different secondary antibodies conjugated with gold particles of distinct size (for instance, 5, 10, 15, or 20 nm) enables the simultaneous colocalization of diverse antigens in the same subcellular compartment (Fig. 1C). However, the main disadvantage of the post-embedding immunogold technique is low sensitivity compared to preembedding methods (Luján, 2010).

The SDS-FRL technique uses a combination of conventional freeze-fracture replica methodology and immunohistochemistry using colloidal gold. SDS-FRL permits unparalleled localization of membrane protein (Fig. 1D) beyond the limitations of the standard iEM techniques discussed above (Fujimoto, 1995). This method allows visualization of the two-dimensional distribution of membrane proteins retained on the surface of the replica after dissolving the attached biological tissue with sodium dodecyl sulfate (SDS), for which the technique is named (Masugi-Tokita & Shigemoto, 2007). Although the SDS detergent denatures surface proteins by unfolding their secondary, tertiary, and quaternary structures, antigenicity is often retained (Fujimoto, 1995) in a similar way to when immunoblot with SDS–polyacrylamide gel electrophoresis is employed. The main advantages of SDS-FRL are its high sensitivity (the highest among all iEM techniques) and allowance for quantitative studies of different markers/antigens along the whole neuronal surface (Allen et al., 2011; García-Negredo et al., 2014; Indriati et al., 2013; Kasugai et al., 2010; Fig. 1D). The main disadvantage is that fracture surface replicas are performed randomly, and due to the scarcity of morphological clues, it is difficult to identify cell types and subcellular compartments without the use of double labeling with specific markers (Allen et al., 2011; Indriati et al., 2013; Kasugai et al., 2010; Tarusawa et al., 2009).

In the last two decades, the *in situ* blotting technique, also known as the histoblot, has emerged as an attractive alternative for studying the regional distribution and expression levels of proteins, in general (Tonnes et al., 1999), and GIRK channel subunits in particular (Fernández-Alacid et al., 2011; Fig. 2). This method is a reliable and convenient way to compare the regional distribution of different proteins in brain samples without compromising the integrity of antibody-binding sites by tissue fixation, which is required for conventional immunohistochemistry (Tonnes et al., 1999). The histoblot technique requires similar experimental conditions to those of the immunoblot with SDS–polyacrylamide gel electrophoresis,

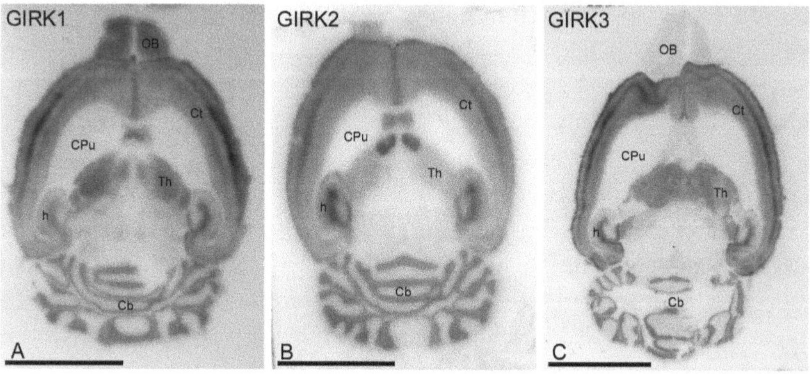

Figure 2 Regional distribution of GIRK1, GIRK2, and GIRK3 in the adult rodent brain. GIRK1 (A) and GIRK2 (B) protein distribution was visualized on histoblots of mouse brain horizontal sections, while GIRK3 protein distribution (C) was on histoblots of rat brain horizontal sections. The three GIRK channel subunits exhibited broad and overlapping distributions in the adult brain. In particular, strong immunoreactivity for GIRK1, GIRK2, and GIRK3 was detected in the cortex (Ct), cerebellum (Cb), hippocampus (h), olfactory bulb (OB), and thalamus (Th), with the lowest intensity in the caudate putamen (CPu). Scale bar: A–C, 0.5 cm.

except that for histoblots, protein transfer is achieved by mechanical pressure, and tissue samples are not homogenized. Therefore, the histoblot reflects the spatial pattern in which proteins are arranged within a brain section, rather than protein bands (Tonnes et al., 1999). The enhanced spatial resolution of the histoblot method compared to the immunoblot has allowed us to detect marked differences in the level of immunoreactivity and expression profiles in all brains regions of the transferred histological section. In addition, the histoblot affords quantitative analyses in different nuclei or in the different layers/subfields of a given brain region (Ciruela et al., 2010; Fernández-Alacid et al., 2011). The main disadvantage of the histoblot technique is that it does not have cellular or subcellular resolution.

4. REGIONAL DISTRIBUTION OF GIRK CHANNEL SUBUNITS IN THE CNS

In situ hybridization, histoblot, and immunohistochemical studies have revealed that GIRK1, GIRK2, and GIRK3 channel subunits are widely expressed throughout the CNS of rodents, showing distinct although partially overlapping patterns of mRNA and protein distribution (Aguado et al., 2008; Chen et al., 1997; Fernández-Alacid et al., 2009, 2011;

Inanobe et al., 1999; Karschin et al., 1996; Kobayashi et al., 1995; Koyrakh et al., 2005; Liao et al., 1996; Miyashita & Kubo, 1997; Ponce et al., 1996; Saenz del Burgo et al., 2008; Fig. 2). The GIRK4 subunit, on the other hand, is found primarily in cardiac tissue, with CNS distribution restricted to a few brain regions (Karschin et al., 1996; Perry et al., 2008; Wickman, Karschin, Karschin, Picciotto, & Clapham, 2000). It is important to mention that the anatomical distribution of immunoreactivity for the three neuronal GIRK channel subunits is highly comparable to that of their respective mRNAs, indicating that these proteins are mainly expressed in somatodendritic domains of neurons, near the site of protein synthesis (Karschin et al., 1996; Saenz del Burgo et al., 2008). Although rare, a few mismatches in location between the protein and the mRNA have been reported in some thalamic, midbrain, and brainstem nuclei (Karschin et al., 1996; Saenz del Burgo et al., 2008), and this suggests that the action site of GIRK channels is far from the site of synthesis, likely indicating their presence at the presynaptic level.

4.1 Distribution of GIRK1, GIRK2, and GIRK3 Subunits

GIRK1–3 mRNAs and proteins are found throughout the brain, with highest levels seen in the olfactory bulb, neocortex, hippocampus, and the granule cell layer of the cerebellum (Aguado et al., 2008; Fernández-Alacid et al., 2009, 2011; Karschin et al., 1996; Kobayashi et al., 1995; Koyrakh et al., 2005; Liao et al., 1996; Miyashita & Kubo, 1997; Ponce et al., 1996; Saenz del Burgo et al., 2008; Figs. 2 and 3A and B). Strong expression for the three subunits is also detected in the piriform cortex, the thalamus, and the hypothalamus. Only weak and diffuse expression is observed in the basal ganglia and globus pallidus.

While the distribution of neuronal GIRK mRNA transcripts and proteins in mouse (Kobayashi et al., 1995; Liao et al., 1996) is similar to that of in rat (DePaoli, Bell, & Stoffel, 1994; Karschin et al., 1996; Karschin & Karschin, 1997; Karschin et al., 1994; Liao et al., 1996), a few species differences have been described in the thalamus and cortex. For example, in the mouse thalamus, GIRK2 immunoreactivity is found in many thalamic nuclei, whereas in rat, GIRK2 is observed only in a few such nuclei (Karschin & Karschin, 1997; Liao et al., 1996). In addition, in the rat cortex, GIRK1, but not GIRK2, is detected in the barrels, whereas both GIRK1 and GIRK2 are found in the mouse cortex (Liao et al., 1996). Next, we will highlight GIRK subunit distributions in a few select brain

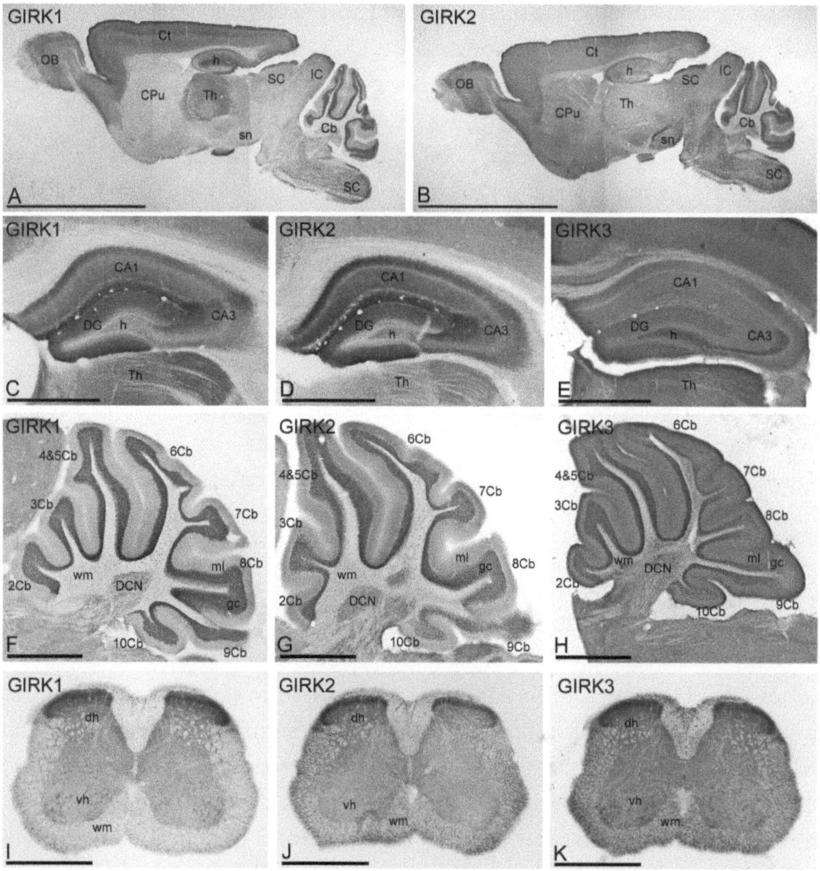

Figure 3 Distribution of immunoreactivity for GIRK channel subunits in adult mouse brain. (A and B) Parasagittal sections of the mouse brain showing immunoperoxidase labeling for GIRK1 and GIRK2. These two channel subunits exhibit broad and overlapping distributions in the brain. Note the strong immunolabeling of the olfactory bulb (OB), neocortex (Ct), hippocampus (Hp), and cerebellar cortex (Cb) as well as the moderate staining of the thalamus (Th), and the weak signal in the caudate putamen (CPu); GIRK2 is also strong in the substantia nigra (SN). (C–E) In the hippocampus, immunoreactivity for GIRK1 and GIRK2 is strong in all hippocampal subfield in the CA1 region, CA3 region, and dentate gyrus (DG), being weaker in the hilus (h). (F–H) In the cerebellum, immunoreactivity for GIRK1, GIRK2, and GIRK3 is very strong in the granule cell layer (gcl), moderate in the molecular layer (ml), and absent in the white matter (wm), throughout all cerebellar lobules. Cerebellar lobules: 2Cb, lobule 2; 3Cb, lobule 3; 4&5Cb, lobules 4 and 5; 6CB, lobule 6; 7Cb, lobule 7; 8Cb, lobule 8; 9Cb, lobule 9; 10Cb, lobule 10. (I–K) In the spinal cord, immunoreactivity for GIRK1, GIRK2, and GIRK3 is very strong in the dorsal horn (dh), weaker in the ventral horn (vh), and absent in the white matter (wm). Scale bars: A–B, 5 mm; C–K, 1 mm.

regions that illustrate key functional and morphological concepts related to GIRK assembly and distribution.

In the main olfactory bulb, GIRK1–3 transcripts are found in the granular and mitral cell layers, but only GIRK3 mRNA is abundant in glomerular cells (Karschin & Karschin, 1997). The anterior olfactory nucleus and the piriform (primary olfactory) cortex contain high levels of the three neuronal GIRK subunits (Karschin et al., 1996; Saenz del Burgo et al., 2008). In the neocortex, GIRK1–3 mRNAs and proteins are strongly expressed in all cortical areas and through all cortical layers, although density is highest for GIRK1 in layer IV, for GIRK2 in layer V, and for GIRK3 in layers II/III and IV (Fernández-Alacid et al., 2011; Karschin et al., 1996; Liao et al., 1996; Saenz del Burgo et al., 2008).

In the hippocampus, GIRK1, GIRK2, and GIRK3 are strongly expressed in CA1–CA3 pyramidal neurons and in dentate gyrus granule cells with overlapping distribution patterns, as assessed by *in situ* hybridization (Chen et al., 1997; Karschin et al., 1996; Karschin & Karschin, 1997; Kobayashi et al., 1995; Saenz del Burgo et al., 2008) and immunohistochemical studies (Fernández-Alacid et al., 2011; Kirizs et al., 2014; Koyrakh et al., 2005; Kulik et al., 2006; Saenz del Burgo et al., 2008; Fig. 3C–E). However, differences in expression among the distinct subunits have also been detected, as hilar neurons contain significantly higher levels of GIRK2 and GIRK3 than GIRK1 mRNAs (Saenz del Burgo et al., 2008). The overlapping distribution of GIRK subunits described at the regional level is also seen in the same layers or subfields. Thus, immunoreactivity for GIRK1 and GIRK2 shows similar patterns in dendritic layers of cornu ammonis (CA) areas and the dentate gyrus (Fig. 3C and D). In the CA1 region, immunolabeling for both proteins is intense in the *stratum lacunosum-moleculare* and in distal parts of the *stratum radiatum* and the *stratum oriens*. In the CA3 region, GIRK1 and GIRK2 immunoreactivity is strong in the dendritic layers, but moderate in the *stratum lucidum*. In the dentate gyrus, immunolabeling for GIRK1 and GIRK2 is strong in the hilus and moderate in the molecular layer (Fernández-Alacid et al., 2011; Koyrakh et al., 2005; Kulik et al., 2006; Fig. 3C and D). Conversely, GIRK3 immunoreactivity is moderate in the dendritic layers of CA1 and CA3, but strong in the *stratum lucidum* and the hilus (Fernández-Alacid et al., 2011; Grosse et al., 2003; Fig. 3E).

In the cerebellum, all three GIRK transcripts and proteins are abundantly expressed in the granule cell layer and to a lesser extent in the molecular layer, which also shows strong GIRK3 immunoreactivity (Aguado et al., 2008; Fernández-Alacid et al., 2011; Karschin et al., 1996; Saenz del Burgo et al., 2008; Fig. 3F–H). The large neurons in the deep cerebellar nuclei contain high levels of GIRK1 and GIRK3 mRNAs and very low levels of GIRK2 mRNA (Karschin et al., 1996). In the thalamus, GIRK1 and GIRK3 subunit mRNAs are abundant in all thalamic nuclei, and GIRK2 mRNA is present at high levels in the lateral and geniculate nuclei and at lower levels in the other thalamic nuclei (Karschin et al., 1996). GIRK1, GIRK2, and GIRK3 mRNAs are abundantly expressed in nuclei of the amygdala (Karschin et al., 1996).

The substantia nigra is composed of two structurally and functionally different subdivisions, the *pars reticulata* (SNr) and *pars compacta* (SNc). This area gives rise to the major output projections from the basal ganglia. In the substantia nigra *pars compacta* and the adjacent ventral tegmental area (VTA), GIRK2 mRNA and protein are present at high levels, while significantly lower levels of GIRK3 are detected here. GIRK3 mRNA is also present in the *pars reticulata* of the substantia nigra (Karschin et al., 1996; Saenz del Burgo et al., 2008). In the basal ganglia, the transcript for the GIRK2 subunit is absent, and only weak and diffuse expression is detected for GIRK1 and GIRK3 (Karschin et al., 1996; Saenz del Burgo et al., 2008). The inferior colliculus contains high levels of GIRK1 and GIRK3 mRNAs, but the transcript for GIRK2 is absent (Karschin et al., 1996). GIRK1, GIRK2, and GIRK3 transcripts and proteins are enriched in the superficial layers of the spinal cord dorsal horn (Karschin et al., 1996; Fig. 3I and J).

4.2 Distribution of GIRK4 Subunit

GIRK4 is expressed in the rodent brain, although at lower levels and in more restricted neuronal populations than the other three GIRK channel subunits (Chen et al., 1997; Iizuka, Tsunenari, Momota, Akiba, & Kono, 1997; Karschin et al., 1996; Karschin & Karschin, 1997; Karschin et al., 1994; Murer et al., 1997; Spauschus et al., 1996). To date, no specific antibodies have been developed that would allow for immunohistochemical techniques. Accordingly, GIRK4 brain distribution has been studied employing *in situ* hybridization with both radioactive and nonradioactive probes (Wickman et al., 2000) and also using transgenic mouse expressing enhanced green fluorescent protein (EGFP) under the control of the *Girk4* gene promoter (Aguado et al., 2008; Perry et al., 2008; Fig. 4). The most intense expression of GIRK4 is detected in the deep cortical pyramidal neurons, the endopiriform nucleus and claustrum of the insular cortex, the globus pallidus (Fig. 4A), the ventromedial hypothalamic nucleus, parafascicular

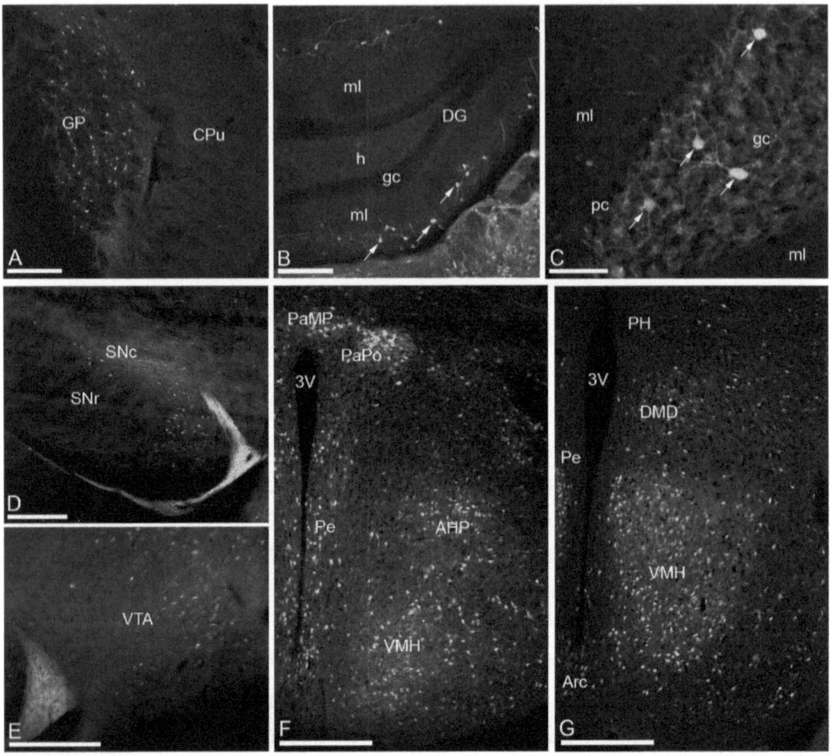

Figure 4 EGFP expression of GIRK4 in the brain of Tg(*Kcnj5*-EGFP)49Gsat mice. Images are representative of those taken from three different mice and confirm the restricted expression pattern of the GIRK4 subunit in the mouse brain. (A) In the globus pallidus (GP), strongly labeled neurons were detected, but very few in the caudate putamen (CPu). (B) In the hippocampus, a small subpopulation of neurons found in the hippocampal fissure and molecular layer of the dentate gyrus show clear EGFP labeling (arrows). (C) In the cerebellum, EGFP expression is found in small subsets of Golgi cells (arrows). (D and E) EGFP labeling in the substantia nigra and ventral tegmental area (VTA) is found in small subsets of cells. (F and G) In the hypothalamus, EGFP labeling was prominent and relatively uniform, showing the highest intensity in the ventromedial hypothalamic nucleus (VMN) and anterior hypothalamic area (AHP). Abbreviations: 3V, third ventricle; AHP, anterior hypothalamic area; Arc, arcuate nucleus; CPu, caudate putamen; DG, dentate gyrus; DMD, dorsomedial hypothalamic nucleus; gcl, granule cell layer; GP, globus pallidus; h, hilus; ml, molecular layer; PaMP, posterior magnocellular paraventricular hypothalamic nucleus; PaPo, posterior paraventricular hypothalamic nucleus; pc, Purkinje cell layer; Pe, periventricular hypothalamic nucleus; PH, posterior hypothalamic area; SNc, substantia nigra *pars compacta*; SNr, substantia nigra *pars reticulata*; VMN, ventromedial hypothalamic nucleus. These animals were provided by Dr. Kevin Wickman and used with his permission. Scale bars: A and B, 150 μm; C, 55 μm; D–G, 300 μm.

and paraventricular thalamic nuclei, and a few brainstem nuclei such as the inferior olive and vestibular nuclei. Low levels of GIRK4 mRNA are detected in the laterodorsal and lateral posterior thalamic nuclei, the nuclei of the diagonal band of the forebrain, the superior colliculus of the midbrain, and the medial vestibular and dorsal tegmental nuclei of the brainstem (Wickman et al., 2000).

Studies examining EGFP expression (under the control of the *Girk4* gene promoter) and Girk4 *in situ* hybridization indicated that the main olfactory bulb does not express GIRK4, with the exception of a few cells in the anterior olfactory nucleus displaying strong probe reactivity. In the neocortex, strongly labeled neurons are detected in layer VIb, also known as subplate neurons. In the hippocampus, GIRK4 mRNA and EGFP labeling are not detected in the principal cell layers, but are observed in several cells of the dentate gyrus molecular layer (Perry et al., 2008; Wickman et al., 2000; Fig. 4B). In the rat cerebellum, GIRK4 mRNA and protein are expressed in Purkinje cells (Iizuka et al., 1997; Murer et al., 1997), but are not detected in the molecular, granular, or Purkinje cell layers of the mouse cerebellum (Wickman et al., 2000). EGFP expression is found in small subsets of Golgi cells in the cerebellum (Aguado et al., 2008; Fig. 4C). Neurons in the lateral cerebellar nuclei, however, are labeled. EGFP labeling in the substantia nigra is also found in small subsets of cells (Fig. 4D) and is prominent and relatively uniform in the hypothalamus (Fig. 4F and G). EGFP intensity is highest in the ventromedial hypothalamic nucleus, although robust expression is also seen in the paraventricular nucleus and posterior aspect of the arcuate nucleus (Perry et al., 2008; Fig. 4F and G).

5. CELLULAR DISTRIBUTION OF GIRK CHANNEL SUBUNITS

Cell biological, biochemical, and electrophysiological approaches indicate that the dominant functional GIRK channel in the CNS contains GIRK1 and GIRK2 (Liao et al., 1996). This view is supported by data obtained using genetic studies, which show that ablation of GIRK2 results in loss of GIRK current in many brain regions, including the hippocampus (Koyrakh et al., 2005; Lüscher et al., 1997), cerebellum (Slesinger, Stoffel, Jan, & Jan, 1997), substantia nigra (Koyrakh et al., 2005), VTA (Arora et al., 2011; Labouebe et al., 2007), locus coeruleus (Cruz et al., 2008; Torrecilla et al., 2002), and spinal cord (Marker, Luján, Colón, & Wickman, 2006). The association of GIRK1 and GIRK2 *in vivo* is also supported by data showing reduced levels in the cerebellum and hippocampus of GIRK1 protein in GIRK2 KO mice and GIRK2 protein in GIRK1 KO mice (Aguado et al., 2008; Koyrakh et al., 2005; Signorini, Liao, Duncan, Jan, & Stoffel, 1997). However, many neuronal populations express GIRK3, yet still possess functional GIRK channels (Cruz et al., 2004; Jelacic et al., 2000). For example, the complete loss of opioid-induced currents in locus coeruleus neurons requires the ablation of both GIRK2 and GIRK3 (Torrecilla et al., 2002). Furthermore, GIRK3 ablation significantly reduces cocaine self-administration, implicating GIRK3 in reward behavior (Morgan, Carroll, Loth, Stoffel, & Wickman, 2003). These behavioral effects seen in GIRK3 KO animals support the idea that channels of varying subunit composition contribute to the physiological effects of GIRK channels (Luján et al., 2014; Lüscher & Slesinger, 2010). In addition, western blots and immunohistochemistry reveal that GIRK3 protein is significantly reduced in the cerebellum of both GIRK1 and GIRK2 KO mice (Aguado et al., 2008), which may reflect the association of the three subunits *in vivo*.

5.1 GIRK Channels in Neurochemically and Functionally Different Cell Populations

The localization of GIRK channel subunits in different cell types in the CNS has been addressed using double *in situ* hybridization (Saenz del Burgo et al., 2008), double or triple immunofluorescence (Aguado et al., 2008, 2013; Booker et al., 2013; Ciruela et al., 2010), and immunoelectron microscopic techniques (Aguado et al., 2008; Arora et al., 2011; Booker et al., 2013; Ciruela et al., 2010; Fernández-Alacid et al., 2009, 2011; Kirizs et al., 2014; Labouebe et al., 2007; Padgett et al., 2012). These approaches demonstrate that GIRK channel subunits are broadly distributed in neurons, although the GIRK3 subunit is also observed in glial cells in the cerebellum (Fernández-Alacid et al., 2009).

In the hippocampus, dentate gyrus, cerebral cortex, thalamus, and cerebellum, the mRNAs and proteins for GIRK1, GIRK2, and GIRK3 are expressed in the vast majority of glutamatergic neurons (Aguado et al., 2008; Ciruela et al., 2010; Fernández-Alacid et al., 2011; Karschin et al., 1996; Koyrakh et al., 2005; Saenz del Burgo et al., 2008), as well as in different subpopulations of GABAergic cells (Aguado et al., 2008; Booker et al., 2013; Karschin et al., 1996; Saenz del Burgo et al., 2008). One of the clearest examples of this can be found in the cerebellar cortex, which contains three types of neurochemically distinct interneurons: GABAergic, glycinergic, and GABAergic/glycinergic (Aguado et al., 2013). Using GlyT2-eGFP (marker of glycinergic and glycinergic/GABAergic interneurons) and GAD67-eGFP (marker of GABAergic and glycinergic/GABAergic interneurons) knock-in mice, in combination with other neurochemical markers, GIRK2 is expressed in glycinergic/GABAergic Golgi cells (Aguado et al., 2008). In the thalamus, the three primary neuronal GIRK channel subunits are mainly expressed in glutamatergic cells but occasionally in GABAergic cells, with the exception of the reticular thalamic nucleus where GIRK mRNAs always colocalize with GABAergic markers (Saenz del Burgo et al., 2008).

Using ChAT as a cholinergic marker, double *in situ* hybridization techniques show that almost all cholinergic cells express GIRK1 and GIRK2 mRNAs in the nucleus of the vertical limb of the diagonal band, whereas no GIRK3 mRNA is found in this region (Saenz del Burgo et al., 2008). Similarly to cholinergic cells, expression of GIRK channel subunits is described in serotonergic cells in the raphe nucleus using the serotonin transporter (ST) as a marker (Saenz del Burgo et al., 2008). Most serotonergic neurons, identified by the presence of ST, also coexpress transcripts for GIRK1, GIRK2, and GIRK3 (Saenz del Burgo et al., 2008). Using tyrosine hydroxylase as a marker, double-labeling immunohistochemistry shows that the GIRK2 subunit is expressed in dopaminergic neurons of the substantia nigra and VTA in the midbrain (Arora et al., 2011; Koyrakh et al., 2005; Labouebe et al., 2007; Padgett et al., 2012). Altogether, these data show that GIRK channels are found in neurons that control release of both excitatory and inhibitory neurotransmitters, and thus may play a role in the regulation of neuronal excitability of neurochemically distinct neurons throughout the CNS.

5.2 Cell Type-Dependent Localization of GIRK Channel Subunits

Parallel to their expression in different neurochemical neuron populations, a wide array of GIRK subunit combinations may be expressed in a highly cell-specific manner, even within the same neurochemical population, such as the GABAergic neurons. This potentially can result in a diversity of channel functions; the cerebellum exemplifies this molecular heterogeneity, with at least seven distinct GIRK subunit expression patterns in the various cerebellar cell types (Aguado et al., 2008). The cerebellar cortex contains GABAergic neurons (Purkinje cells, basket cells, stellate cells, Golgi cells, and Lugaro cells) and glutamatergic neurons (granule cells and unipolar brush cells), characterized by their very specific and uniform laminar arrangement and microcircuitries (Altman & Bayer, 1997). A combination of single-cell RT-PCR, immunohistochemistry at the light microscopic level, double-labeling immunofluorescence, and iEM indicated that Purkinje neurons express GIRK1/GIRK2/GIRK3 (although GIRK3 is the predominant subunit), basket cells express GIRK1/GIRK3, stellate cells express GIRK3, Golgi cells express GIRK2/GIRK4, Lugaro cells express GIRK1/GIRK2/GIRK3 (although at very low levels), granule cells express GIRK1/GIRK2/GIRK3 (at very high levels for all three), and unipolar brush cells express GIRK2/GIRK3 (Aguado et al., 2008; Fig. 5).

Midbrain dopamine neurons have a striking lack of GIRK1 expression. GIRK channels in these neurons are formed by heteromeric assembly of the alternatively spliced isoform GIRK2C with GIRK3, whereas GABAergic neurons in the same region express GIRK1, GIRK2C, and GIRK3 (Cruz et al., 2004; Labouebe et al., 2007). Other cell types show expression of only one subunit. For example, only GIRK2 is found in dopaminergic

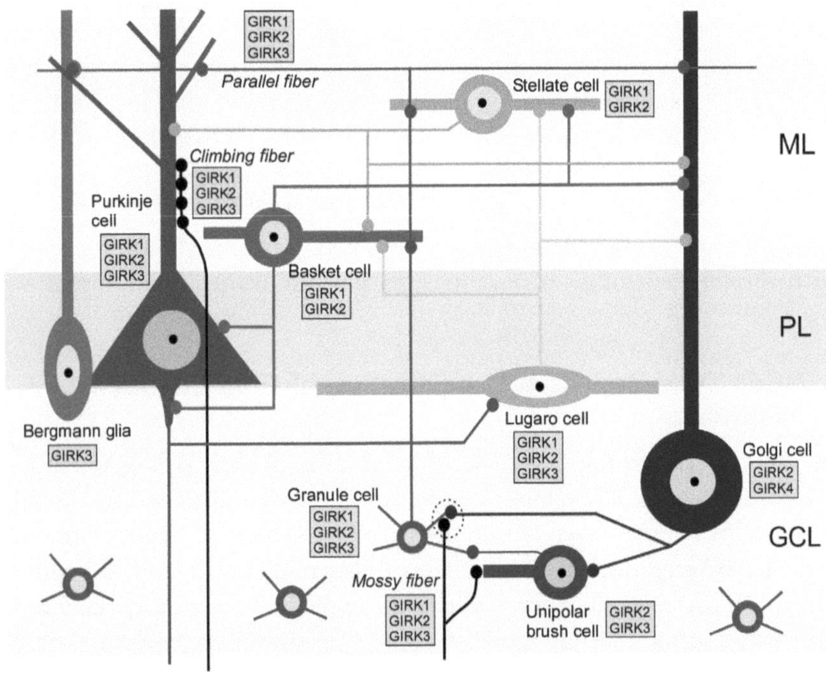

Figure 5 Cellular and subcellular localization of GIRK channel subunits in the cerebellum. Summary diagram illustrating the synaptic connectivity of cerebellar cell types and the subunit composition of GIRK subunits in the different cells and axon terminals. Each color (different shades of gray in the print version) denotes a different cell type.

cells of the substantia nigra *pars compacta* (SNc; Karschin et al., 1996); although strictly speaking, the channels in these cells are thought to be formed by combining GIRK2A and GIRK2C isoforms (Inanobe et al., 1999).

The two main neuron populations in the hippocampus, namely pyramidal cells (glutamatergic) and interneurons (GABAergic), express GIRK1, GIRK2, and GIRK3 (Booker et al., 2013; Fernández-Alacid et al., 2011; Koyrakh et al., 2005). However, only heteromeric complexes of GIRK1/GIRK2 are molecularly and functionally associated with $GABA_B$ receptors in pyramidal cells (Koyrakh et al., 2005), while GIRK3, although also present at the pyramidal cell plasma membrane, and thus might be also part of the channel complex, it is mostly localized at intracellular sites (Fernández-Alacid et al., 2011). In the subpopulation of parvalbumin-expressing interneurons, corresponding mainly to basket and chandelier cells, GIRK1, GIRK2, and GIRK3 share the same localization with very similar densities, suggesting that the $GABA_B$-evoked currents in interneurons are mediated by channels composed of all three subunits (Booker et al., 2013).

The functional consequences arising from heteromeric channel formation in different subcellular compartments of central neurons are mostly unknown. We know, however, that GIRK channel properties arise from the assembly of different subunits (Duprat et al., 1995; Kofuji, Davidson, & Lester, 1995; Krapivinsky et al., 1995). For example, expression of slow kinetics depends on the insertion of GIRK1 subunits into GIRK2-containing channels (Slesinger et al., 1996), and GIRK2/GIRK3 heteromultimers exhibit a decreased sensitivity to receptor activation (Cruz et al., 2004; Jelacic et al., 2000). Thus, the existing data on the different GIRK expression and distribution patterns, and the different subunit organizations of GIRK channels in given neurons, suggest a diversity of channel function.

6. SUBCELLULAR DISTRIBUTION OF GIRK CHANNEL SUBUNITS

The classical notion that the subcellular localization of ion channels in a particular cell type readily applies to all brain neurons is not well supported by available data. The emerging view is that each ion channel possesses its own unique subcellular distribution pattern in each cell type (Luján, 2010), suggesting that there are no simple rules concerning their subcellular organization. This is particularly true for GIRK channels. Indeed, available evidence suggests that GIRK channels can be located within any subcellular compartment at the neuronal surface, both at synaptic and extrasynaptic sites, in somata, dendritic shafts, dendritic spines, and at lower frequency in axons or axon terminals (Luján et al., 2014, 2009). However, the fact that they are not evenly distributed along the neuronal surface, but instead present with varying frequency in specific neuronal compartments, has many potential functional implications. In the above paragraphs, we have seen that GIRK channels are distributed in a cell-specific manner. In section 6.1, we highlight results of high-resolution immunolocalization studies demonstrating that the precise subcellular location of GIRK channels varies among neuronal compartments, inputs, and synapses, contributing to the ion channel molecular diversity in the brain. First, we describe GIRK channels in the somatodendritic domain of neurons, and then turn to GIRK channels that show axonal localization.

6.1 Compartment-Dependent Localization

One mechanism by which neurons control assembly of specific GIRK protein subunits may involve differential localization of these subunits in specific neuronal compartments (Luján et al., 2014, 2009). It is now widely accepted that the probability that a given GIRK channel is activated following GPCR stimulation depends not only on the intrinsic properties of the channel but also on its localization along the neuronal surface and the spatial relationship with GPCR and other associated proteins. Accordingly, a clear description of how GIRK channels are arranged along the neuronal surface and how they are targeted and anchored in their respective membrane domains is required to understand how these channels impact excitability and neuronal output.

Consistent with the expression and distribution of GIRK subunits described at the light microscopic level, iEM unequivocally reveals a high density of immunogold labeling for GIRK1 and GIRK2 along the extrasynaptic plasma membrane of dendritic shafts and spines, with lower density in axon terminals, of CA pyramidal cells in the hippocampus and Purkinje cells in the cerebellum (Aguado et al., 2008; Booker et al., 2013; Fernández-Alacid et al., 2009, 2011; Koyrakh et al., 2005; Kulik et al., 2006; Fig. 6). The question arises: is this pattern of localization true for all neuron populations, or are alternative distribution patterns possible? Although the extra/perisynaptic distribution for GIRK1 and GIRK2 has been seen in multiple neuron populations, the GIRK3 subunit is a potential influence and its appearance in the same neuron type increases molecular diversity.

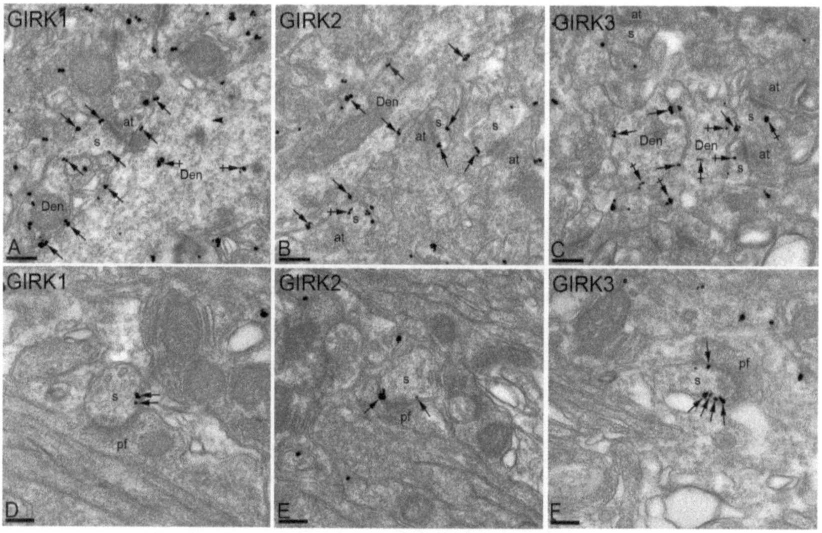

Figure 6 Subcellular localization of GIRK channel subunits in the hippocampus and cerebellum. (A–C) Electron micrographs show immunolabeling for GIRK1, GIRK2, and GIRK3 in pyramidal cells of the hippocampus using a preembedding immunogold method. (D–F) Electron micrographs show immunolabeling for GIRK1, GIRK2, and GIRK3 in Purkinje cells of the cerebellum using a preembedding immunogold method. Scale bars: A–F, 0.2 μm.

One way by which molecular diversity is increased can be described as a function of *presence* of GIRK subunits in the same subcellular compartment, but with varying density. This is the case of cerebellar Purkinje and granule cells which express GIRK1, GIRK2, and GIRK3 (Aguado et al., 2008; Ciruela et al., 2010; Fernández-Alacid et al., 2009). While dendrites of granule cells contain similar densities of GIRK1, GIRK2, and GIRK3, which are mainly found in close proximity to excitatory synapses and to $GABA_B$ receptors (Ciruela et al., 2010), Purkinje cell spines contain high-density GIRK3, and low levels of both GIRK1 and GIRK2 (Aguado et al., 2008; Fernández-Alacid et al., 2009). These data suggest that GIRK channels in granule cells are likely formed by GIRK1/GIRK2, GIRK2/GIRK3, and GIRK1/GIRK3, whereas most channels in Purkinje cells are likely formed by GIRK1/GIRK3 and GIRK2/GIRK3.

Other way for increasing molecular diversity is a function of *presence* or *absence* in a given compartment. For example, in addition to their main extrasynaptic localization, GIRK channels are also located at synaptic sites in excitatory synapses (Fernández-Alacid et al., 2009, 2011; Kulik et al.,

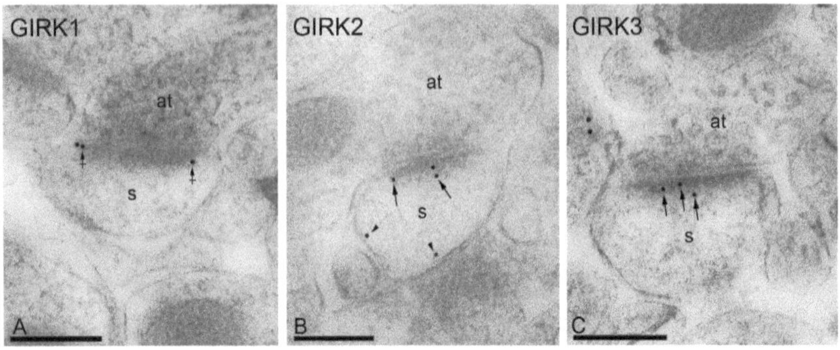

Figure 7 Subsynaptic localization of GIRK subunits in the hippocampus. Electron micrographs show immunolabeling for GIRK1, GIRK2, and GIRK3 in pyramidal cells using a postembedding immunogold method. (A) Immunoparticles for GIRK1 are not detected in postsynaptic density (PSDs) of dendritic spines (s), but rather at perisynaptic (crossed arrows) or extrasynaptic sites. (B and C) Immunoparticles for GIRK2 and GIRK3 are observed in the PSD of dendritic spines (s) (arrows) establishing asymmetrical synapses with axon terminals (at). Scale bars: A–C, 0.2 μm.

2006; Fig. 7A–C). In inhibitory synapses of the rodent hippocampus and cerebellum, however, immunogold particles for GIRK channel subunits have not been detected in the main body of postsynaptic specializations (Fernández-Alacid et al., 2009, 2011; Koyrakh et al., 2005; Kulik et al., 2006). In excitatory synapses, GIRK channel subunits show a variable distribution (Fig. 7A–C). For example, we used postembedding immunogold labeling of GIRK channel subunits in the cerebellar cortex and found that GIRK2 and GIRK3 are evenly distributed in excitatory synapses along the PSD of Purkinje cells (Fernández-Alacid et al., 2009). By contrast, the GIRK1 subunit is always absent from excitatory synapses and only observed at perisynaptic and extrasynaptic sites. Similarly, a differential subsynaptic distribution has been reported in the hippocampus (Fernández-Alacid et al., 2011; Koyrakh et al., 2005; Fig. 7A–C). This localization of GIRK2 and GIRK3 in the main body of the postsynaptic specialization is consistent with the presence of $GABA_B$ receptors in the same compartment, beautifully demonstrated using the highly sensitive SDS-FRL technique (Kulik et al., 2006), and also consistent with the synaptic localization of N-methyl-D-aspartate (NMDA) receptors (Somogyi, Tamás, Luján, & Buhl, 1998). From a functional point of view, stimulation protocols that evoke NMDA receptor-dependent long-term potentiation (LTP) of glutamatergic neurotransmission also strengthen synaptic $GABA_B$-GIRK signaling in hippocampal CA1 pyramidal neurons (Huang et al., 2005). In addition, activation of NMDA receptors in cultured hippocampal neurons increases surface expression of GIRK1 and GIRK2 subunits in the somatodendritic membrane (Chung, Qian, Ehlers, Jan, & Jan, 2009), which contributes to the molecular mechanisms underlying the depotentiation of excitatory LTP, as demonstrated using pharmacological or genetic ablation of GIRK signaling (Chung, Ge, et al., 2009). This suggests that NMDA receptor activation likely involves both extrasynaptic (GIRK1 and GIRK2) and synaptic (GIRK2) channels.

6.2 Localization of GIRK Channels Relative to Glutamate Release Sites

The preembedding immunogold method has further revealed the distribution patterns of GIRK channels relative to neurotransmitter release sites. Within the main extrasynaptic distribution of GIRK channels, they have a preferential localization toward the periphery of postsynaptic densities of asymmetrical synapses in several brain regions. For example, at the head of Purkinje cell spines in the mouse cerebellar cortex, approximately 10% of immunogold particles for GIRK1, GIRK2, and GIRK3 has been localized perisynaptically, i.e., within a 60-nm annulus surrounding the edge of synapses (Fig. 6D–F), whereas remaining particles are distributed extrasynaptically at more distant positions (Fernández-Alacid et al., 2009). At the head of CA1 pyramidal neuron spines, about 11–14% of the immunogold particles for GIRK1 and GIRK2 is also preferentially observed at the periphery of postsynaptic densities of asymmetrical synapses (Koyrakh et al., 2005; Fig. 6A–C). Interestingly, in both brain regions, around 75% of all immunoparticles resides within the first 300 nm of the excitatory synapse close to the glutamate release site (Fajardo-Serrano et al., 2013; Fernández-Alacid et al., 2009; Koyrakh et al., 2005). This association of GIRK channel subunits with asymmetrical synapses in CA1 pyramidal and Purkinje cells is consistent with the subcellular localization of $GABA_B$ receptors (Fajardo-Serrano et al., 2013; Fernández-Alacid et al., 2009; Koyrakh et al., 2005; Luján & Shigemoto, 2006). The activation of extrasynaptic GIRK channels through $GABA_B$ receptors requires the spillover of GABA from inhibitory synapses (Scanziani, 2000), and this is consistent with the $GABA_B$ receptor-mediated slow inhibitory postsynaptic potential (IPSPs) elicited by extracellular stimulation (Dutar & Nicoll, 1988), resulting in the control of glutamatergic transmission.

6.3 Input-Dependent Localization

Neurons receive glutamatergic inputs from several sources and express a large variety of neurotransmitter receptors and ion channels. It has been demonstrated that distinct glutamatergic inputs can act through different signaling proteins within a single neuron, contributing to input-specific functional requirements (reviewed by Luján, 2010). Such an input-dependent localization has been described for α-amino-3-hydroxy-5-methyl-4-isoxazolepropionic acid (AMPA) receptors and NMDA receptors (Nusser et al., 1998; Watanabe et al., 1998) and for small conductance Ca^{2+} activated K^{+} (SK) channels in the hippocampus (Ballesteros-Merino et al., 2012, 2014). The question then arises as to whether different GIRK channel subtypes can also be selectively associated with functionally distinct synaptic inputs.

The overlapping distribution of GIRK channels described at the regional level is also taking place in the same layers or subfields of a given region, suggesting the potential for considerable molecular diversity in a subcellular- and/or input-dependent manner. This is best illustrated in the hippocampus, where immunoreactivity for GIRK1 and GIRK2 shows extensive colocalization in the different dendritic layers (Fernández-Alacid et al., 2011; Kirizs et al., 2014; Koyrakh et al., 2005; Kulik et al., 2006). Electron microscopic immunogold localization experiments have revealed varying extrasynaptic densities of GIRK channel subunits among dendritic regions innervated by distinct synaptic inputs in CA1 pyramidal cells (Fernández-Alacid et al., 2011; Kirizs et al., 2014). The densities of GIRK1 and GIRK2 on pyramidal cell spines increase significantly from proximal (*stratum radiatum*) to distal dendrites (*stratum lacunosum-moleculare*), whereas GIRK3 is uniformly distributed along the surface of CA1 pyramidal cells. The three subunits are present at very low density in the cell body (*stratum pyramidale*) of pyramidal cells (Fernández-Alacid et al., 2011; Kirizs et al., 2014), which receive GABAergic synapses from basket cells. These results demonstrate that GIRK channels are mainly associated with glutamatergic inputs, especially those coming from the entorhinal cortex through the perforant pathway, which are distributed in the *stratum lacunosum-moleculare*.

6.4 Plasticity- and Drug-Dependent Changes in Distribution

The effects of several drugs of abuse, including opioids, psychostimulants (cocaine), and ethanol, are mediated or shaped by GIRK channel activity. For example, mice lacking GIRK channels exhibit dopamine-dependent hyperactivity and elevated responses to drugs that stimulate dopamine neurotransmission (Arora et al., 2011). Additionally, the subcellular localization of GIRK channels changes dynamically along the neuronal surface in response to drug exposure. For example, some drugs of abuse induce internalization of GIRK channels and associated GPCRs. This is the case for cocaine, which reduces $GABA_B$ receptor-GIRK signaling in dopaminergic neurons of the VTA, a phenomenon attributable to a reduction of GIRK2-containing channels at the plasma membrane (Arora et al., 2011). Similarly, cocaine suppresses $GABA_B$ receptor-dependent GIRK signaling in layer 5/6 pyramidal neurons of the medial prefrontal cortex and involves redistribution of $GABA_B$ receptors and GIRK channels from the plasma membrane to intracellular sites (Hearing et al., 2013). Administration of methamphetamine also reduces both synaptically evoked and baclofen-activated $GABA_B$ receptor-GIRK currents in GABAergic neurons of the VTA, which correlates with internalization of $GABA_{B1}$ and GIRK2 from the plasma membrane, as demonstrated using quantitative iEM (Padgett et al., 2012; Fig. 8A–D).

Other drugs, such as morphine, induce an increase of GIRK channels along the plasma membrane instead of their internalization. In cultured pyramidal cells of the hippocampus, morphine produced an elevation of GIRK2 protein and targeting to dendritic spines (Nassirpour et al., 2010). Thus, in morphine-treated neurons, the percentage of immunogold GIRK2 labeling increased on the spine plasma membrane, but decreases on the shaft and soma plasma membrane. The percentage of immunogold particles for GIRK2 was also raised in the somatic endoplasmic reticulum, suggesting an increase in transcription/translation of the GIRK channel subunit. This GIRK channel regulatory process is likely related to the GPCR signaling pathway, as functional analyses have revealed that morphine increases the size of serotonin-activated GIRK currents (Nassirpour et al., 2010).

6.5 Presynaptic Localization of GIRK Channel Subunits

Although GIRK channels are mainly concentrated at postsynaptic sites, many studies demonstrate the presence of GIRK1, GIRK2, and GIRK3 subunits in axon terminals in the brain (Aguado et al., 2008; Booker et al., 2013; Fajardo-Serrano et al., 2013; Fernández-Alacid et al., 2009, 2011; Grosse et al., 2003; Koyrakh et al., 2005; Kulik et al., 2006; Ladera et al., 2008; Marker et al., 2005; Morishige et al., 1996; Ponce et al., 1996; Fig. 9A–D). In presynaptic elements, GIRK channel subunits are similarly distributed relative to the neurotransmitter release site (Aguado et al.,

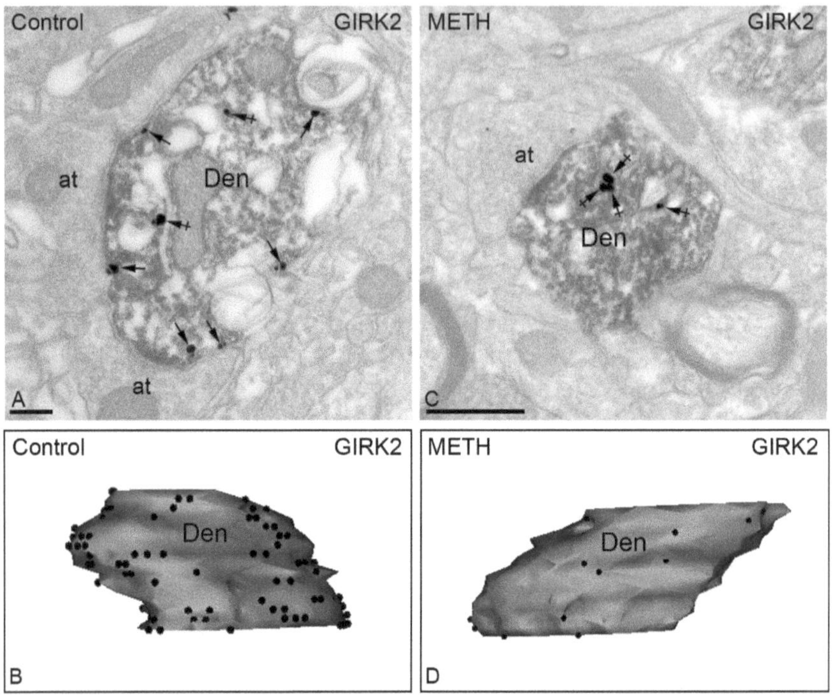

Figure 8 Reduced surface localization of GIRK2 in GABAergic neurons of methamphetamine-injected animals. Preembedding double-labeling method at EM level using GIRK2 and GAD65/67. (A and B) In control conditions, the peroxidase reaction product (GAD65/67 immunoreactivity) filled dendritic shafts (Den), whereas immunoparticles (GIRK2 immunoreactivity) were located along the extrasynaptic plasma membrane (arrows) and at intracellular sites (crossed arrows). Three-dimensional reconstruction of a labeled dendrite is shown in (B). Note that immunoparticles were abundantly distributed over the dendritic plasma membrane. (C and D) GIRK2 immunoparticles after injection of methamphetamine (METH) were significantly reduced over the dendritic plasma membrane and increased at intracellular sites (crossed arrows). Three-dimensional reconstruction of a labeled dendrite is shown in (D). Note that few immunoparticles are distributed along the dendritic plasma membrane. at, axon terminal. Scale bars: A, 0.2 μm; C, 0.5 μm.

2008; Fernández-Alacid et al., 2009, 2011; Koyrakh et al., 2005; Marker et al., 2005). GIRK1, GIRK2, and GIRK3 are often observed along the extrasynaptic membrane and also over the active zone of axon terminals (Fig. 9B–D; Fajardo-Serrano et al., 2013; Fernández-Alacid et al., 2009, 2011; Koyrakh et al., 2005; Marker et al., 2005), where they colocalize and likely form heteromeric channels (Fernández-Alacid et al., 2009).

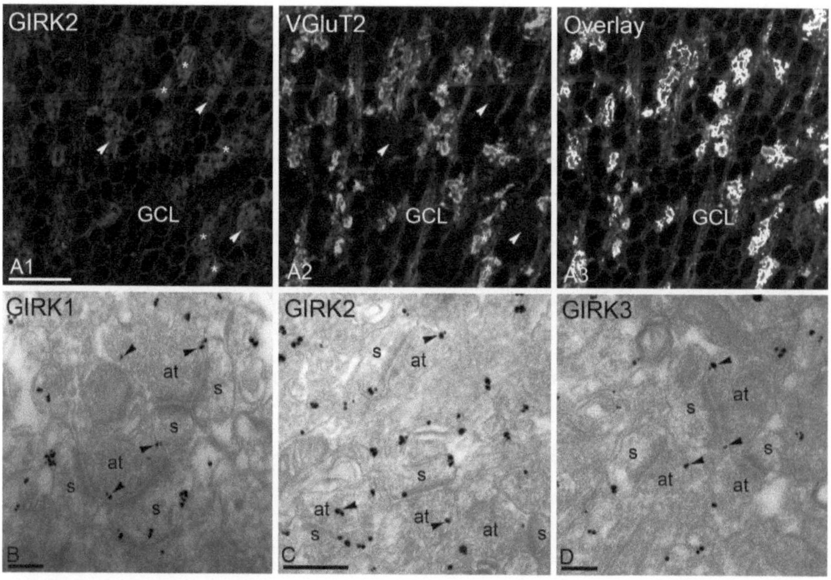

Figure 9 Presynaptic localization of GIRK channel subunits at the light and electron microscopic level. (A1–A3) Localization of GIRK2 at presynaptic sites in the granule cell layer (GCL) of the cerebellar cortex using double-labeling confocal microscopy. The vesicular glutamate transporter subtype VGluT2 is used as a marker of mossy fiber axon terminals, which establish asymmetrical synapses with granule cell dendrites at the cerebellar glomeruli. (B–D) Presynaptic localization of GIRK1, GIRK2, and GIRK3 in the *stratum radiatum* of the hippocampus using a preembedding immunogold technique. Immunoparticles for the GIRK subunits are localized to the extrasynaptic plasma membrane (arrowheads) and occasionally to the presynaptic membrane specialization of axon terminals (at) establishing putative excitatory synapses on spines (s). Scale bars: A, 20 μm; B–D, 0.2 μm.

The proximity of GIRK channel subunits to the presynaptic active zone suggests that they may also be involved in the presynaptic modulation of neuronal activity.

The mechanism of GIRK-dependent presynaptic modulation is not fully clear, but it might involve the inhibition of neurotransmitter release following activation of $GABA_B$ receptors (Fernández-Alacid et al., 2009; Ladera et al., 2008; Luján et al., 2014, 2009). Electrophysiological studies have shown that the binding of GABA to $GABA_B$ receptors activates GIRK channels at postsynaptic sites (Lüscher et al., 1997; Lüscher & Slesinger, 2010). These data do not support presynaptic GIRK channel activation as a primary mechanism by which $GABA_B$ receptors modulate neurotransmitter release (Lüscher et al., 1997). However, presynaptic GIRK channels are in very close proximity to $GABA_B$ receptors in the hippocampus and cerebellum (Fajardo-Serrano et al., 2013; Fernández-Alacid et al., 2009), and studies using synaptosomes prepared from cortex or cerebellum have demonstrated the functional association of GIRK and $GABA_B$ (Fernández-Alacid et al., 2009; Ladera et al., 2008). As a result of these functional and biochemical assays, two molecular mechanisms for $GABA_B$-mediated regulation of glutamate release have been proposed in different populations of axon terminals: (1) a mechanism insensitive to the GIRK channel blocker tertiapin-Q, whereby $GABA_B$ receptors act through G protein-mediated inhibition of N-type Ca^{2+} channels to reduce Ca^{2+} influx and inhibit glutamate release, and (2) a mechanism sensitive to tertiapin-Q, but insensitive to pertussis toxin, by which $GABA_B$ receptors activate GIRK channels which hyperpolarize the presynaptic plasma membrane, reducing excitability and glutamate release (Fernández-Alacid et al., 2009; Ladera et al., 2008). This functional coupling is supported by morphological data in the cerebellum, showing the upregulation of the GIRK3 subunit and $GABA_B$ receptors in parallel fiber terminals after genetic ablation of $GABA_{B1}$ and GIRK3, respectively (Fernández-Alacid et al., 2009).

6.6 Localization of GIRK-Associated Proteins

In the last few years, GIRK channels have been shown to interact physically with a variety of scaffolding and signaling proteins, and with integral membrane receptors. Increasing evidence is beginning to show the way in which these interacting proteins are responsible for the correct assembly and trafficking of GIRK channels, as well as their functional recycling and coupling to effector proteins (Hibino et al., 2010; Luján et al., 2014; Lüscher & Slesinger, 2010; Zylbergold, Ramakrishnan, & Hebert, 2010; Zylbergold et al., 2014).

Synapse-associated proteins (SAPs) are involved in targeting, clustering, and scaffolding of proteins at postsynaptic and presynaptic sites (Kim & Sheng, 2004). SAPs have a number of repeating protein–protein interaction modules known as post synaptic density protein, drosophila disc large tumor suppressor, and zonula occludens-1 protein (PDZ) domains, which bind via a T/SXV (where T is threonine, S is serine, X is any amino acid, V is valine) motif present in the C-terminus of the interacting protein. A PDZ-binding motif, located exclusively at the ends of C-termini of GIRK2C isoforms and GIRK3 subunits, seems to confer varying expression and/or function on GIRK channels. The PDZ-binding motif of GIRK3 is thought to negatively regulate the number of GIRK channels at the cell surface. This motif has been shown to confer interaction with sorting nexin 27 (SNX27), a protein that associates with the early endosome, and to underlie the higher rate of internalization of GIRK3-containing channels relative to those not containing GIRK3 (Lunn et al., 2007).

In addition to different regulatory enzymes like protein kinase A (PKA), protein phosphatase 1 (PP1) and PP2A, the GIRK signal complex contains GPCR, G, and regulators of G protein signaling (RGS) proteins. Among GPCRs, biochemical studies have demonstrated that $GABA_B$ receptors, β2-adrenergic receptors, and dopamine D2 and D4 receptors can be part of highly organized macromolecular complexes that contain GIRK channels (Zylbergold et al., 2010).

G protein signaling is controlled by a family of proteins called RGS, which increase the GTP hydrolase activity of Gα, thus accelerating the kinetics of ion channel modulation by GPCRs (reviewed by Luján et al., 2014; Lüscher & Slesinger, 2010). The cellular and subcellular localization of members of heterotrimeric G and RGS proteins is not as well characterized as that of GIRK channels and $GABA_B$ receptors. For most of these regulatory proteins, some information on expression in the brain is available, mainly based on *in situ* hybridization analyses of mRNA levels in brain sections (Gold, Ni, Dohlman, & Nestler, 1997). Immunohistochemical analyses of the subcellular localization of these subunit proteins have, for the most part, not been undertaken.

GIRK channels are synthesized in the rough endoplasmic reticulum (rER) and specifically transported to their functional site of action, mainly to the plasma membrane of dendritic spines and shafts. Once GIRK channels reach their final destination, the ion channel complex is retained by either interaction with other signaling proteins at the plasma membrane or scaffolding proteins. However, current data have suggested that the interaction between GIRK channels and other signaling molecules, such as $GABA_B$ receptors, G proteins, and RGS proteins, is formed shortly after biosynthesis, probably in the rER or rER–Golgi complex, thus taking place before cell surface targeting (Ciruela et al., 2010; Fajardo-Serrano et al., 2013). This might be critical and physiologically relevant, as the existence of these oligomeric complexes ensures rapid and specific $GABA_B$ receptor signaling. $GABA_B$ receptors and GIRK channel complexes have also been detected in native tissue, indicating that such interactions are not merely a consequence of the conditions of heterologous expression systems, but might have real physiological importance (Ciruela et al., 2010). There is also some evidence of protein–protein interactions between RGS and GIRK channels. For example, GIRK1 and GIRK4 are coimmunoprecipitated with RGS4

(Fujita, Inanobe, Chachin, Aizawa, & Kurachi, 2000). RGS7/Gβ5 can be coimmunoprecipitated with GIRK1, GIRK2, and GIRK3 (Xie et al., 2010), as well as with $GABA_B$ receptors and other proteins such as R7 family-binding protein (R7BP) (Ostrovskaya et al., 2014). Different Gβγ subunits show a similar ability to activate GIRK channels (Robitaille, Ramakrishnan, Baragli, & Hébert, 2009; Zylbergold et al., 2014). Supporting all these molecular and functional associations, iEM demonstrates that $GABA_B$ receptors, GIRK channels, and G and RGS proteins are localized in close proximity at the plasma membrane in different subcellular compartments (Ciruela et al., 2010; Fajardo-Serrano et al., 2013; Kulik et al., 2006).

7. DEVELOPMENTAL ASPECTS OF GIRK CHANNEL DISTRIBUTION

Brain development results from a temporospatial pattern of events, beginning with neuronal proliferation, followed by migration and differentiation and ending with synapse formation and circuit refinements. A growing body of evidence suggests that each step in the developmental sequence of the CNS involves the appropriate expression and function of neurotransmitters, their receptors, and ion channels (Luján, 2010; Luján, Shigemoto, & López-Bendito, 2005). For example, K_v channels are involved in the control of membrane potential by regulating Ca^{2+} entry into the cell to modulate growth cone guidance (Pollock et al., 2005), or in cell proliferation, by either preventing or stimulating cell division (Pardo, 2004). K2P channels (TASK-1 and TASK-3) contribute to the control of apoptosis (Trimarchi, Liu, Smith, & Keefe, 2002). Thus, understanding the developmental aspects of GIRK channel expression, transport, and function, may provide us with valuable information concerning maturation of neuronal networks. In this regard, the death of specific cell populations and neuronal migration abnormalities observed in *weaver* mice, a GIRK2 point mutant (Slesinger et al., 1996), suggest that GIRK2 channels are indeed critical to normal development and maturation.

7.1 Distribution of GIRK Channels During Pre- and Postnatal Development

Studying the evolving patterns of cellular and subcellular GIRK channel localization during development is fundamental to our understanding of how GIRK channel subunits contribute to developmental processes and function in the mature brain. A few recent studies have shed light on these important issues (Aguado et al., 2013; Fernández-Alacid et al., 2011). One indicator of the functional importance of GIRK channel diversity comes from examining the subunit mRNA or protein changes observed during development. However, to date, little information is available regarding onset of expression and precise localization of GIRK channels in embryonic and postnatal development. *In situ* hybridization studies have shown that GIRK1, GIRK2, and GIRK3 subunit mRNAs are strongly expressed in the embryonic rodent/rat brain, with mainly nonoverlapping expression patterns (Karschin & Karschin, 1997). Mitotic cells in the outer parts of the proliferative zones of the cerebellum and hippocampus do not express GIRK channel subunits, suggesting that they first appear after neuronal precursors have finished their last cell division (Aguado et al., 2013; Karschin & Karschin, 1997). By contrast, as brain development progresses during postnatal stages, expression of GIRK channel subunits begins to show overlapping mRNA patterns in the brain regions that also express the same subunits in adulthood.

Evidence exists for differential patterns of GIRK channel subunit expression during development (Aguado et al., 2013; Fernández-Alacid et al., 2011; Karschin & Karschin, 1997; Miyashita & Kubo, 1997); the exact changes in subunit expression vary with brain region. In most brain regions, GIRK1 and GIRK2 mRNAs are expressed at higher levels during development than in the adult animal. For instance, during postnatal days P2 and P10, both GIRK1 and GIRK2 are highly expressed in the cerebellum, pontine nucleus, and lateral reticular nucleus (Karschin & Karschin, 1997). In other regions, such as the hippocampus and septum, high levels of the two subunits remain until adulthood (Karschin & Karschin, 1997). GIRK3 mRNAs are the most widely expressed of all GIRK subunits, both in the entire developing brain and in adulthood (Karschin & Karschin, 1997). During embryonic development, GIRK3 is expressed throughout all structures, but from P2 to P10, its expression pattern is restricted to the cortex, olfactory bulb, hippocampus, septum, thalamus, ventromedial hypothalamic nucleus, inferior olive, and cerebellum. On the contrary, GIRK4 mRNAs are the most restricted of all GIRK subunits and, prenatally, are only detectable in the ventromedial hypothalamic nucleus (Karschin & Karschin, 1997). During postnatal development, GIRK4 is present at moderate levels in the ventromedial hypothalamic nucleus, paraventricular thalamic nuclei, and optic nerve layer of the superior colliculus. Between P5 and P21, transient GIRK4 mRNA is observed in the inferior olive and subthalamic nucleus (Karschin & Karschin, 1997).

Little information is available on the subcellular distribution of GIRK channels in the brain, although it is widely accepted that this information is crucial for elucidating their contribution to developmental processes and brain function. Nevertheless, recent findings have shed some light on this issue (Fernández-Alacid et al., 2011). These findings show a developmental shift in the subcellular localization of GIRK1, GIRK2, and GIRK3 in CA1 pyramidal cells. Early in development, the three GIRK subunits are predominantly localized to the rER in the pyramidal cell layer, but in the adult, GIRK1 and GIRK2 are almost exclusively located in the plasma membrane in dendrites and spines (Fernández-Alacid et al., 2011). The progressive trafficking of extrasynaptic GIRK1 and GIRK2 to dendritic compartments as a function of age is similar to the maturation of $GABA_B$ receptors (López-Bendito et al., 2004), in a process parallel to the establishment and maturation of excitatory synapses.

7.2 Acquisition of the GIRK2 Neurochemical Phenotype

The onset of GIRK channel appearance at the cell membrane in the developing brain is still unknown, and only some information has been provided for GIRK2 in two cerebellar neuron populations (granule cells and Golgi cells; Aguado et al., 2013). Granule cells are the most abundant glutamatergic neurons in the cerebellum. Their precursors leave the rhombic lip during early stages of embryonic development, migrate rostrally over the dorsal surface of the cerebellar anlage, and form the external granule cell layer. Then, granule cell precursors undergo proliferation, exit the cell cycle, and initiate inward radial migration across the molecular layer to form the internal granule cell layer (Altman & Bayer, 1997). Double-labeling immunofluorescence techniques with Pax6, a marker of granule cells derived from the rhombic lip, provided the first evidence that GIRK2 is highly expressed during all stages of granule cell development, from embryo to mature mouse (Aguado et al., 2013).

The GIRK2 subunit is also expressed in Golgi cells scattered throughout the granular layer (Aguado et al., 2008), which use GABA as a neurotransmitter. The main role of Golgi cells is to modulate mossy fiber afferents to the cerebellar cortex (Altman & Bayer, 1997). Golgi cells derive from progenitors that delaminate from the ventricular neuroepithelium. They undergo final mitosis both pre- and postnatally, after which they migrate through white matter, reaching the internal granule cell layer shortly after birth. There, they differentiate morphologically during the first postnatal week (Altman & Bayer, 1997). Double- and triple-labeling immunofluorescence techniques have provided evidence of GIRK2 expression in Golgi cells only during postmigratory stages, once they have arrived at the internal granular layer (Aguado et al., 2013). Therefore, the data demonstrate that GIRK2 expression in Golgi cells occurs ontogenetically later than in granule cells.

8. CONCLUSIONS AND FUTURE PERSPECTIVES

The electrical signaling of neurons depends largely on the abundance and function of a large diversity of neurotransmitter receptors and ion channels located at specific sites in neuronal somata, dendrites, and axons. The selective placement of GIRK channel subunits at precise locations in mammalian neurons and their dynamic regulation through specific signaling pathways allows for a wide variety of neuronal function in the brain. However, the cellular diversity of the brain and the morphological and functional complexity of neurons pose a challenge to understanding the specific role of GIRK channels in brain function and how their specific and discrete subcellular compartmentalization regulates signaling in mammalian neurons. Analysis of the cellular and subcellular localization of GIRK channels has become possible through the molecular cloning of these ion channels, the application of high-resolution immunohistochemical techniques, and the availability of specific antibodies.

The past decade has seen rapid advances in our understanding of GIRK channel distribution along the cell surface of different neurons. This has not only significantly advanced the GIRK channel field but also fundamentally changed our view of GIRK channel function and signaling through GPCRs, particularly through $GABA_B$ receptors. The use of neuroanatomical approaches, combined with biochemical and electrophysiological approaches and the development of genetically engineered mice that lack GIRK genes, has provided new insights into the physiological roles played by different types of GIRK channels in the mammalian brain. These techniques have revealed a molecular diversity of GIRK channels, and we now believe that their molecular variations are fundamentally important for fulfilling functional roles in neuronal signaling. Electron microscopy, in particular, has made an important contribution to the now widely accepted idea that GIRK channels exhibit distinct distributions at the neuronal surface, where they can be targeted to specific subcellular domains in a channel subunit-, region-, cell type- and compartment-dependent manner, reflecting their functional heterogeneity. This differential targeting to specific cell types and neuronal compartments reveals key associated proteins and how they are spatially and functionally coupled.

Considering the physiological importance of GIRK channels, we need to learn more about the molecular mechanisms that engender their unique subcellular distributions and their selective interactions with protein partners. Therefore, identification of the molecular components of neuronal compartments, as well as understanding the precise cellular and subcellular localization patterns of GIRK channels and their associated proteins, will offer new insights into the mechanisms that control GIRK channel signaling between neurons, both in normal and pathological brain activity. New knowledge gained may also validate GIRK channels as therapeutic drug targets. The mechanisms that lead to the specific distribution of GIRK channels in the neuronal membrane and how these mechanisms are regulated during development and in drug-dependent plasticity remain major topics of ongoing research.

ACKNOWLEDGMENTS

The authors would like to thank Alexandra Salewski, M.Sc., for the English revision of the manuscript and Mercedes Gil for her excellent technical assistance. We also thank the Spanish Ministry of Education and Science (BFU-2012-38348; Consolider-Ingenio CSD2008-00005), the European Union (HBP—Project Ref. 604102), and the Junta de Comunidades de Castilla-La Mancha (PPII-2014-005-P) for their generous support of our research. Additionally, our gratitude extends to all members of our laboratory for helpful discussions and critical comments. We apologize to all colleagues whose work, due to scope and space limitations, we may have unintentionally overlooked.

REFERENCES

Aguado, C., Colón, J., Ciruela, F., Schlaudraff, F., Cabañero, M. J., Perry, C., et al. (2008). Cell type-specific subunit composition of G protein-gated potassium channels in the cerebellum. *Journal of Neurochemistry*, *105*, 497–511.

Aguado, C., Fernández-Alacid, L., Cabañero, M. J., Yanagawa, Y., Schilling, K., Watanabe, M., et al. (2013). Differential maturation of GIRK2-expressing neurons in the mouse cerebellum. *Journal of Chemical Neuroanatomy*, *47*, 79–89.

Allen, D., Bond, C. T., Luján, R., Ballesteros-Merino, C., Lin, M. T., Wang, K., et al. (2011). The SK2-long isoform directs synaptic localization and function of SK2-containing channels. *Nature Neuroscience*, *14*, 744–749.

Altman, J., & Bayer, S. A. (1997). *Development of the cerebellar system in relation to its evolution, structure and functions*. New York: CRC.

Arora, D., Hearing, M., Haluk, D. M., Mirkovic, K., Fajardo-Serrano, A., Wessendorf, M. W., et al. (2011). Acute cocaine exposure weakens GABA(B) receptor-dependent G-protein-gated inwardly rectifying K^+ signaling in dopamine neurons of the ventral tegmental area. *Journal of Neuroscience*, *31*, 12251–12257.

Ballesteros-Merino, C., Lin, M., Wu, W. W., Ferrandiz-Huertas, C., Cabañero, M. J., Watanabe, M., et al. (2012). Developmental profile of SK2 channel expression and function in CA1 neurons. *Hippocampus*, *22*, 1467–1480.

Ballesteros-Merino, C., Watanabe, M., Shigemoto, R., Fukazawa, Y., Adelman, J. P., & Luján, R. (2014). Differential subcellular localization of SK3-containing channels in the hippocampus. *European Journal of Neuroscience*, *39*, 883–892.

Booker, S. A., Gross, A., Althof, D., Shigemoto, R., Bettler, B., Frotscher, M., et al. (2013). Differential GABAB-receptor-mediated effects in perisomatic- and dendrite-targeting parvalbumin interneurons. *Journal of Neuroscience*, *33*, 7961–7974.

Chen, S. C., Ehrhard, P., Goldowitz, D., & Smeyne, R. J. (1997). Developmental expression of the GIRK family of inward rectifying potassium channels: Implications for abnormalities in the weaver mutant mouse. *Brain Research*, *778*, 251–264.

Chung, H. J., Ge, W. P., Qian, X., Wiser, O., Jan, Y. N., & Jan, L. Y. (2009). G protein-activated inwardly rectifying potassium channels mediate depotentiation of long-term potentiation. *Proceedings of the National Academy of Sciences of the United States of America*, *106*, 635–640.

Chung, H. J., Qian, X., Ehlers, M., Jan, Y. N., & Jan, L. Y. (2009). Neuronal activity regulates phosphorylation-dependent surface delivery of G protein-activated inwardly rectifying potassium channels. *Proceedings of the National Academy of Sciences of the United States of America*, *106*, 629–634.

Ciruela, F., Fernández-Dueñas, V., Sahlholm, K., Fernández-Alacid, L., Nicolau, J. C., Watanabe, M., et al. (2010). Evidence for oligomerization between GABAB receptors and GIRK channels containing the GIRK1 and GIRK3 subunits. *European Journal of Neuroscience*, *32*, 1265–1277.

Cruz, H. G., Berton, F., Sollini, M., Blanchet, C., Pravetoni, M., Wickman, K., et al. (2008). Absence and rescue of morphine withdrawal in GIRK/Kir3 knock-out mice. *Journal of Neuroscience*, *28*, 4069–4077.

Cruz, H. G., Ivanova, T., Lunn, M. L., Stoffel, M., Slesinger, P. A., & Lüscher, C. (2004). Bi-directional effects of GABA(B) receptor agonists on the mesolimbic dopamine system. *Nature Neuroscience*, *7*, 153–159.

Curran, M. E. (1998). Potassium ion channels and human disease: Phenotypes to drug targets? *Current Opinion in Biotechnology*, *9*, 565–572.

Dascal, N. (1997). Signalling via the G protein-activated K^+ channels. *Cellular Signalling*, *9*, 551–573.

DePaoli, A. M., Bell, G. I., & Stoffel, M. (1994). G protein-activated inwardly rectifying potassium channel (GIRK1/KGA) mRNA in adult rat heart and brain by *in situ* hybridization histochemistry. *Molecular and Cellular Neuroscience*, *5*, 515–522.

Drake, C. T., Bausch, S. B., Milner, T. A., & Chavkin, C. (1997). GIRK1 immunoreactivity is present predominantly in dendrites, dendritic spines, and somata in the CA1 region of the hippocampus. *Proceedings of the National Academy of Sciences of the United States of America*, *94*, 1007–1012.

Duprat, F., Lesage, F., Guillemare, E., Fink, M., Hugnot, J. P., Bigay, J., et al. (1995). Heterologous multimeric assembly is essential for K+ channel activity of neuronal and cardiac G-protein-activated inward rectifiers. *Biochemical and Biophysical Research Communications*, *212*(2), 657–663. http://www.ncbi.nlm.nih.gov/pubmed/7626080.

Dutar, P., & Nicoll, R. A. (1988). A physiological role for GABAB receptors in the central nervous system. *Nature*, *332*, 156–158.

Fajardo-Serrano, A., Wydeven, N., Young, D., Watanabe, M., Shigemoto, R., Martemyanov, K. A., et al. (2013). Association of Rgs7/Gβ5 complexes with Girk channels and GABAB receptors in hippocampal CA1 pyramidal neurons. *Hippocampus*, *23*, 1231–1245.

Fernández-Alacid, L., Aguado, C., Ciruela, F., Martín, R., Colón, J., Cabañero, M. J., et al. (2009). Subcellular compartment-specific molecular diversity of pre- and post-synaptic GABA-activated GIRK channels in Purkinje cells. *Journal of Neurochemistry*, *110*, 1363–1376.

Fernández-Alacid, L., Watanabe, M., Molnár, E., Wickman, K., & Luján, R. (2011). Developmental regulation of G protein-gated inwardly-rectifying K^+ (GIRK/Kir3) channel subunits in the brain. *European Journal of Neuroscience*, *34*, 1724–1736.

Fujimoto, K. (1995). Freeze-fracture replica electron microscopy combined with SDS digestion for cytochemical labeling of integral membrane proteins. Application to the immunogold labeling of intercellular junctional complexes. *Journal of Cell Science*, *108*, 3443–3449.

Fujita, S., Inanobe, A., Chachin, M., Aizawa, Y., & Kurachi, Y. (2000). A regulator of G protein signalling (RGS) protein confers agonist-dependent relaxation gating to a G protein-gated K+ channel. *Journal of Physiology*, *526*, 341–347.

García-Negredo, G., Soto, D., Llorente, J., Morató, X., Galenkamp, K. M., Gómez-Soler, M., et al. (2014). Coassembly and coupling of SK2 channels and mGlu5 receptors. *Journal of Neuroscience*, *34*, 14793–14802.

Gold, S. J., Ni, Y. G., Dohlman, H. G., & Nestler, E. J. (1997). Regulators of G-protein signalling (RGS) proteins: Region-specific expression of nine subtypes in rat brain. *Journal of Neuroscience*, *17*, 8024–8037.

Grosse, G., Eulitz, D., Thiele, T., Pahner, I., Schröter, S., Takamori, S., et al. (2003). Axonal sorting of Kir3.3 defines a GABA-containing neuron in the CA3 region of rodent hippocampus. *Molecular and Cellular Neuroscience*, *24*, 709–724.

Gutman, G. A., Chandy, K. G., Grissmer, S., Lazdunski, M., McKinnon, D., Pardo, L. A., et al. (2005). International Union of Pharmacology. LIII. Nomenclature and molecular relationships of voltage-gated potassium channels. *Pharmacological Reviews*, *57*, 473–508.

Hearing, M., Kotecki, L., Marron Fernández de Velasco, E., Fajardo-Serrano, A., Chung, H. J., Luján, R., et al. (2013). Repeated cocaine weakens GABA(B)-Girk signaling in layer 5/6 pyramidal neurons in the prelimbic cortex. *Neuron*, *80*, 159–170.

Hibino, H., Inanobe, A., Furutani, K., Murakami, S., Findlay, I., & Kurachi, Y. (2010). Inwardly rectifying potassium channels: Their structure, function, and physiological roles. *Physiological Reviews*, *90*, 291–366.

Hille, B. (2001). *Ionic channels of excitable membranes*. Sunderland, MA: Sinauer.

Huang, C. S., Shi, S. H., Ule, J., Ruggiu, M., Barker, L. A., Darnell, R. B., et al. (2005). Common molecular pathways mediate long-term potentiation of synaptic excitation and slow synaptic inhibition. *Cell*, *123*, 105–118.

Iizuka, M., Tsunenari, I., Momota, Y., Akiba, I., & Kono, T. (1997). Localization of a G-protein-coupled inwardly rectifying K^+ channel, CIR, in the rat brain. *Neuroscience*, 77, 1–13.

Inanobe, A., Yoshimoto, Y., Horio, Y., Morishige, K. I., Hibino, H., Matsumoto, S., et al. (1999). Characterization of G-protein-gated K^+ channels composed of Kir3.2 subunits in dopaminergic neurons of the substantia nigra. *Journal of Neuroscience*, *19*, 1006–1017.

Indriati, D. W., Kamasawa, N., Matsui, K., Meredith, A. L., Watanabe, M., & Shigemoto, R. (2013). Quantitative localization of Cav2.1 (P/Q-type) voltage-dependent calcium channels in Purkinje cells: Somatodendritic gradient and distinct somatic coclustering with calcium-activated potassium channels. *Journal of Neuroscience*, *33*, 3668–3678.

Jaén, C., & Doupnik, C. A. (2005). Neuronal Kir3.1/Kir3.2a channels coupled to serotonin 1A and muscarinic m2 receptors are differentially modulated by the "short" RGS3 isoform. *Neuropharmacology*, *49*, 465–476.

Jelacic, T. M., Kennedy, M. E., Wickman, K., & Clapham, D. E. (2000). Functional and biochemical evidence for G-protein-gated inwardly rectifying K^+ (GIRK) channels composed of GIRK2 and GIRK3. *Journal of Biological Chemistry, 275*, 36211–36216.
Karschin, C., Dissmann, E., Stuhmer, W., & Karschin, A. (1996). IRK(1–3) and GIRK(1–4) inwardly rectifying K^+ channel mRNAs are differentially expressed in the adult rat brain. *Journal of Neuroscience, 16*, 3559–3570.
Karschin, C., & Karschin, A. (1997). Ontogeny of gene expression of Kir channel subunits in the rat. *Molecular and Cellular Neuroscience, 10*, 131–148.
Karschin, C., Schreibmayer, W., Dascal, N., Lester, H., Davidson, N., & Karschin, A. (1994). Distribution and localization of a G protein-coupled inwardly rectifying K^+ channel in the rat. *FEBS Letters, 348*, 139–144.
Kasugai, Y., Swinny, J. D., Roberts, J. D., Dalezios, Y., Fukazawa, Y., Sieghart, W., et al. (2010). Quantitative localisation of synaptic and extrasynaptic GABAA receptor subunits on hippocampal pyramidal cells by freeze-fracture replica immunolabelling. *European Journal of Neuroscience, 32*, 1868–1888.
Kim, E., & Sheng, M. (2004). PDZ domain proteins of synapses. *Nature Reviews. Neuroscience, 5*, 771–781.
Kirizs, T., Kerti-Szigeti, K., Lorincz, A., & Nusser, Z. (2014). Distinct axo-somato-dendritic distributions of three potassium channels in CA1 hippocampal pyramidal cells. *European Journal of Neuroscience, 39*, 1771–1783.
Kobayashi, T., Ikeda, K., Ichikawa, T., Abe, S., Togashi, S., & Kumanishi, T. (1995). Molecular cloning of a mouse G-protein-activated K^+ channel (mGIRK1) and distinct distributions of three GIRK (GIRK1, 2 and 3) mRNAs in mouse brain. *Biochemical and Biophysical Research Communications, 208*, 1166–1173.
Kofuji, P., Davidson, N., & Lester, H. A. (1995). Evidence that neuronal G-protein-gated inwardly rectifying K+ channels are activated by G beta gamma subunits and function as heteromultimers. *Proceedings of the National Academy of Sciences of the United States of America, 92*, 6542–6546.
Koyrakh, L., Luján, R., Colón, J., Karschin, C., Kurachi, Y., Karschin, A., et al. (2005). Molecular and cellular diversity of neuronal G-protein-gated potassium channels. *Journal of Neuroscience, 25*, 11468–11478.
Krapivinsky, G., Krapivinsky, L., Wickman, K., & Clapham, D. E. (1995). G beta gamma binds directly to the G protein-gated K^+ channel, IKACh. *Journal of Biological Chemistry, 270*, 29059–29062.
Kubo, Y., Adelman, J. P., Clapham, D. E., Jan, L. Y., Karschin, A., Kurachi, Y., et al. (2005). International Union of Pharmacology. LIV. Nomenclature and molecular relationships of inwardly rectifying potassium channels. *Pharmacological Reviews, 57*, 509–526.
Kulik, A., Vida, I., Fukazawa, Y., Guetg, N., Kasugai, Y., Marker, C. L., et al. (2006). Compartment-dependent colocalization of Kir3.2-containing K^+-channels and GABAB receptors in hippocampal pyramidal cells. *Journal of Neuroscience, 26*, 4289–4297.
Labouebe, G., Lomazzi, M., Cruz, H. G., Creton, C., Luján, R., Li, M., et al. (2007). RGS2 modulates coupling between GABAB receptors and GIRK channels in dopamine neurons of the ventral tegmental area. *Nature Neuroscience, 10*, 1559–1568.
Ladera, C., Godino, M. C., Cabañero, M. J., Torres, M., Watanabe, M., Luján, R., et al. (2008). Presynaptic GABAB receptors inhibit glutamate release through GIRK channels. *Journal of Neurochemistry, 107*, 1506–1517.
Liao, Y. J., Jan, Y. N., & Jan, L. Y. (1996). Heteromultimerization of G-protein-gated inwardly rectifying K^+ channel proteins GIRK1 and GIRK2 and their altered expression in weaver brain. *Journal of Neuroscience, 16*, 7137–7150.
Logothetis, D. E., Kurachi, Y., Galper, J., Neer, E. J., & Clapham, D. E. (1987). The beta gamma subunits of GTP-binding proteins activate the muscarinic K^+ channel in heart. *Nature, 325*, 321–326.

López-Bendito, G., Shigemoto, R., Kulik, A., Vida, I., Fairén, A., & Luján, R. (2004). Distribution of metabotropic GABA receptor subunits GABAB1a/b and GABAB2 in the rat hippocampus during prenatal and postnatal development. *Hippocampus*, *14*, 836–848.

Luján, R. (2010). Organisation of potassium channels on the neuronal surface. *Journal of Chemical Neuroanatomy*, *40*, 1–20.

Luján, R., Marron Fernandez de Velasco, E., Aguado, C., & Wickman, K. (2014). New insights into the therapeutic potential of Girk channels. *Trends in Neurosciences*, *37*, 20–29.

Luján, R., Maylie, J., & Adelman, J. P. (2009). New sites of action for GIRK and SK channels. *Nature Reviews. Neuroscience*, *10*, 475–480.

Luján, R., & Shigemoto, R. (2006). Localization of metabotropic GABA receptor subunits GABAB1 and GABAB2 relative to synaptic sites in the rat developing cerebellum. *European Journal of Neuroscience*, *23*, 1479–1490.

Luján, R., Shigemoto, R., & López-Bendito, G. (2005). Glutamate and GABA receptor signalling in the developing brain. *Neuroscience*, *130*, 567–580.

Lunn, M. L., Nassirpour, R., Arrabit, C., Tan, J., McLeod, I., Arias, C. M., et al. (2007). A unique sorting nexin regulates trafficking of potassium channels via a PDZ domain interaction. *Nature Neuroscience*, *101*, 1249–1259.

Lüscher, C., Jan, L. Y., Stoffel, M., Malenka, R. C., & Nicoll, R. A. (1997). G protein-coupled inwardly rectifying K^+ channels (GIRKs) mediate postsynaptic but not presynaptic transmitter actions in hippocampal neurons. *Neuron*, *19*, 687–695.

Lüscher, C., & Slesinger, P. A. (2010). Emerging roles for G protein-gated inwardly rectifying potassium (GIRK) channels in health and disease. *Nature Reviews. Neuroscience*, *11*, 301–315.

Marker, C., Luján, R., Colón, J., & Wickman, K. (2006). Distinct populations of spinal cord lamina II interneurons expressing G-protein gated potassium channels. *Journal of Neuroscience*, *26*, 12251–12259.

Marker, C., Luján, R., Loh, H., & Wickman, K. (2005). Spinal G protein-gated potassium channels contribute in a dose-dependent manner to the analgesic effect of mu and delta but not kappa opioids. *Journal of Neuroscience*, *25*, 3551–3559.

Masugi-Tokita, M., & Shigemoto, R. (2007). High-resolution quantitative visualization of glutamate and GABA receptors at central synapses. *Current Opinion in Neurobiology*, *17*, 387–393.

Miyashita, T., & Kubo, Y. (1997). Localization and developmental changes of the expression of two inward rectifying K+ channel proteins in the rat brain. *Brain Research*, *750*, 251–263.

Morgan, A. D., Carroll, M. E., Loth, A. K., Stoffel, M., & Wickman, K. (2003). Decreased cocaine self-administration in Kir3 potassium channel subunit knockout mice. *Neuropsychopharmacology*, *28*, 932–938.

Morishige, K. I., Inanobe, A., Takahashi, N., Yoshimoto, Y., Kurachi, H., Miyake, A., et al. (1996). G protein-gated K^+ channel (GIRK1) protein is expressed presynaptically in the paraventricular nucleus of the hypothalamus. *Biochemical and Biophysical Research Communications*, *220*, 300–305.

Murer, G., Adelbrecht, C., Lauritzen, I., Lesage, F., Lazdunski, M., Agid, Y., et al. (1997). An immunocytochemical study on the distribution of two G-protein-gated inward rectifier potassium channels (GIRK2 and GIRK4) in the adult rat brain. *Neuroscience*, *80*, 345–357.

Nassirpour, R., Bahima, L., Lalive, A. L., Lüscher, C., Luján, R., & Slesinger, P. A. (2010). Morphine- and CaMKII-dependent enhancement of GIRK channel signaling in hippocampal neurons. *Journal of Neuroscience*, *30*, 13419–13430.

North, A. (1989). Drug receptors and the inhibition of nerve cells. *British Journal of Pharmacology, 98*, 13–28.
Nusser, Z., Luján, R., Laube, G., Roberts, J. D., Molnar, E., & Somogyi, P. (1998). Cell type and pathway dependence of synaptic AMPA receptor number and variability in the hippocampus. *Neuron, 21*, 545–559.
Ostrovskaya, O., Xie, K., Masuho, I., Fajardo-Serrano, A., Luján, R., Wickman, K., et al. (2014). RGS7/Gβ5/R7BP complex regulates synaptic plasticity and memory by modulating hippocampal GABABR-GIRK signaling. *eLife, 3*, e02053.
Padgett, C. L., Lalive, A. L., Tan, K. R., Terunuma, M., Munoz, M. B., Pangalos, M. N., et al. (2012). Methamphetamine-evoked depression of GABA(B) receptor signaling in GABA neurons of the VTA. *Neuron, 73*, 978–989.
Pardo, L. A. (2004). Voltage-gated potassium channels in cell proliferation. *Physiology (Bethesda), 19*, 285–292.
Perry, C. A., Pravetoni, M., Teske, J. A., Aguado, C., Erickson, D. J., Medrano, J. F., et al. (2008). Predisposition to late-onset obesity in GIRK4 knockout mice. *Proceedings of the National Academy of Sciences of the United States of America, 105*, 8148–8153.
Pollock, N. S., Atkinson-Leadbeater, K., Johnston, J., Larouche, M., Wildering, W. C., & McFarlane, S. (2005). Voltage-gated potassium channels regulate the response of retinal growth cones to axon extension and guidance cues. *European Journal of Neuroscience, 22*, 569–578.
Ponce, A., Bueno, E., Kentros, C., Vega-Saenz de Miera, E., Chow, A., Hillman, D., et al. (1996). G-protein-gated inward rectifier K^+ channel proteins (GIRK1) are present in the soma and dendrites as well as in nerve terminals of specific neurons in the brain. *Journal of Neuroscience, 16*, 1990–2001.
Robitaille, M., Ramakrishnan, N., Baragli, A., & Hébert, T. E. (2009). Intracellular trafficking and assembly of specific Kir3 channel/G protein complexes. *Cellular Signalling, 21*, 488–501.
Saenz del Burgo, L., Cortes, R., Mengod, G., Zarate, J., Echevarria, E., & Salles, J. (2008). Distribution and neurochemical characterization of neurons expressing GIRK channels in the rat brain. *Journal of Comparative Neurology, 510*, 581–606.
Scanziani, M. (2000). GABA spillover activates postsynaptic GABAB receptors to control rhythmic hippocampal activity. *Neuron, 25*, 673–681.
Signorini, S., Liao, Y. J., Duncan, S. A., Jan, L. Y., & Stoffel, M. (1997). Normal cerebellar development but susceptibility to seizures in mice lacking G protein-coupled, inwardly rectifying K+ channel GIRK2. *Proceedings of the National Academy of Sciences of the United States of America, 94*, 923–927.
Slesinger, P. A., Patil, N., Liao, Y. J., Jan, Y. N., Jan, L. Y., & Cox, D. R. (1996). Functional effects of the mouse weaver mutation on G protein-gated inwardly rectifying K+ channels. *Neuron, 16*(2), 321–331.
Slesinger, P., Stoffel, M., Jan, Y., & Jan, L. (1997). Defective γ-aminobutyric acid type B receptor-activated inwardly rectifying K^+ currents in cerebellar granule cells isolated from weaver and Girk2 null mutant mice. *Proceedings of the National Academy of Sciences of the United States of America, 94*, 12210–12217.
Somogyi, P., Tamás, G., Luján, R., & Buhl, E. H. (1998). Salient features of synaptic organisation in the cerebral cortex. *Brain Research Reviews, 26*, 113–135.
Sosulina, L., Schwesig, G., Seifert, G., & Pape, H. C. (2008). Neuropeptide Y activates a G-protein-coupled inwardly rectifying potassium current and dampens excitability in the lateral amygdale. *Molecular and Cellular Neuroscience, 39*, 491–498.
Spauschus, A., Lentes, K. U., Wischmeyer, E., Dissmann, E., Karschin, C., & Karschin, A. (1996). A G-protein-activated inwardly rectifying K^+ channel (GIRK4) from human hippocampus associates with other GIRK channels. *Journal of Neuroscience, 16*, 930–938.

Tarusawa, E., Matsui, K., Budisantoso, T., Molnár, E., Watanabe, M., Matsui, M., et al. (2009). Input-specific intrasynaptic arrangements of ionotropic glutamate receptors and their impact on postsynaptic responses. *Journal of Neuroscience*, *29*, 12896–12908.

Tonnes, J., Stierli, B., Cerletti, C., Behrmann, J. T., Molnar, E., & Streit, P. (1999). Regional distribution and developmental changes of GluR1-flop protein revealed by monoclonal antibody in rat brain. *Journal of Neurochemistry*, *73*, 2195–2205.

Torrecilla, M., Marker, C. L., Cintora, S. C., Stoffel, M., Williams, J. T., & Wickman, K. (2002). G-protein-gated potassium channels containing Kir3.2 and Kir3.3 subunits mediate the acute inhibitory effects of opioids on locus ceruleus neurons. *Journal of Neuroscience*, *22*, 4328–4334.

Trimarchi, J. R., Liu, L., Smith, P. J., & Keefe, D. L. (2002). Apoptosis recruits two-pore domain potassium channels used for homeostatic volume regulation. *American Journal of Physiology. Cell Physiology*, *282*, C588–C594.

Watanabe, M., Fukaya, M., Sakimura, K., Manabe, T., Mishina, M., & Inoue, Y. (1998). Selective scarcity of NMDA receptor channel subunits in the stratum lucidum (mossy fibre-recipient layer) of the mouse hippocampal CA3 subfield. *European Journal of Neuroscience*, *10*, 478–487.

Wickman, K., & Clapham, D. E. (1995). Ion channel regulation by G proteins. *Physiological Reviews*, *75*, 865–885.

Wickman, K., Karschin, C., Karschin, A., Picciotto, M. R., & Clapham, D. E. (2000). Brain localization and behavioral impact of the G-protein-gated K^+ channel subunit GIRK4. *Journal of Neuroscience*, *20*, 5608–5615.

Witkowski, G., Szulczyk, B., Rola, R., & Szulczyk, P. (2008). D(1) dopaminergic control of G protein-dependent inward rectifier K(+) (GIRK)-like channel current in pyramidal neurons of the medial prefrontal cortex. *Neuroscience*, *155*, 53–63.

Xie, K., Allen, K. L., Kourrich, S., Colón-Saez, J., Thomas, M. J., Wickman, K., et al. (2010). Gbeta5 recruits R7 RGS proteins to GIRK channels to regulate the timing of neuronal inhibitory signaling. *Nature Neuroscience*, *13*, 661–663.

Yamada, M., Inanobe, A., & Kurachi, Y. (1998). G protein regulation of potassium ion channels. *Pharmacological Reviews*, *50*, 723–760.

Yang, J., Jan, Y. N., & Jan, L. Y. (1995). Determination of the subunit stoichiometry of an inwardly rectifying potassium channel. *Neuron*, *15*, 1441–1447.

Zylbergold, P., Ramakrishnan, N., & Hebert, T. (2010). The role of G proteins in assembly and function of Kir3 inwardly rectifying potassium channels. *Channels (Austin, Tex.)*, *4*, 411–421.

Zylbergold, P., Sleno, R., Khan, S. M., Jacobi, A. M., Belhke, M. A., & Hébert, T. E. (2014). Kir3 channel ontogeny—The role of Gβγ subunits in channel assembly and trafficking. *Frontiers in Cellular Neuroscience*, *8*, 108.

CHAPTER SIX

GIRK Channel Plasticity and Implications for Drug Addiction

Ezequiel Marron Fernandez de Velasco*, Nora McCall†, Kevin Wickman*,[1]

*Department of Pharmacology, University of Minnesota, Minneapolis, Minnesota, USA
†Department of Neuroscience, University of Minnesota, Minneapolis, Minnesota, USA
[1]Corresponding author: e-mail address: wickm002@umn.edu

Contents

1. Introduction 202
   1.1 Molecular and Functional Diversity of GIRK Channels 204
   1.2 GIRK Channels and Neuronal Excitability 206
2. GIRK Channels in the Mesocorticolimbic System 207
   2.1 VTA DA Neurons 209
   2.2 VTA GABA Neurons 210
   2.3 Layer 5/6 Pyramidal Neurons of the mPFC 211
3. Drug-Dependent Plasticity of GIRK Channel-Dependent Signaling 211
   3.1 Psychostimulant-Induced Suppression of GIRK Channel Activity in VTA DA Neurons 214
   3.2 Psychostimulant-Induced Suppression of GIRK Channel Activity in VTA GABA Neurons 214
   3.3 Cocaine-Induced Suppression of GIRK Channel Activity in mPFC Pyramidal Neurons 215
4. Mechanisms Underlying the Plasticity of GIRK Channel-Dependent Signaling 216
   4.1 Membrane Trafficking 216
   4.2 RGS Proteins 220
5. Concluding Thoughts and Future Directions 223
Acknowledgments 223
References 223

Abstract

Drugs of abuse can "hijack" synaptic plasticity, a physiological basis of learning and memory, establishing maladaptations that can promote drug addiction. A wealth of data supports the existence and importance of neuroadaptations in excitatory neurotransmission upon drug exposure. Recent discoveries, however, have shown that inhibitory neurotransmission mediated by G protein-gated inwardly rectifying potassium (K^+) (GIRK/Kir3) channels is also subject to adaptation triggered by exposure to drugs of abuse. GIRK channels are expressed in neuronal populations relevant to reward and

International Review of Neurobiology, Volume 123
ISSN 0074-7742
http://dx.doi.org/10.1016/bs.irn.2015.05.011

reward-related behaviors, where their activation by neurotransmitters such as GABA, dopamine, and adenosine reduces neuronal excitability. Studies in animal models have implicated GIRK channels in a number of behaviors including reward. Drugs of abuse also affect the inhibitory neurotransmission mediated by GIRK channels. These changes might be important for the development, maintenance, or relapse of addiction, making GIRK channels promising targets for novel addiction therapies.

1. INTRODUCTION

The correct function of the nervous system relies on a tight balance between excitatory and inhibitory neurotransmission (Aldenhoff, 1989; Catterall, Dib-Hajj, Meisler, & Pietrobon, 2008). Recent work has highlighted the importance of slow inhibitory neurotransmission mediated by G protein-gated inwardly rectifying potassium (K^+) (GIRK/Kir3) channels in maintaining this balance (Lujan, Marron Fernandez de Velasco, Aguado, & Wickman, 2014; Luscher & Slesinger, 2010). Studies of *Girk*$^{-/-}$ mice have provided some insight regarding the physiological relevance of these channels. For example, *Girk*$^{-/-}$ mice exhibited altered nociception (Marker, Stoffel, & Wickman, 2004; Mitrovic et al., 2003) and reduced anxiety-related behavior (Blednov, Stoffel, Chang, & Harris, 2001; Pravetoni & Wickman, 2008). Moreover, mice that overexpressed GIRK channels exhibited cognitive deficits (Cooper et al., 2012). *Girk*$^{-/-}$ mice also show altered reward-related behaviors such as decreased cocaine self-administration (Morgan, Carroll, Loth, Stoffel, & Wickman, 2003), increased cocaine-induced locomotor activity (Arora et al., 2010; Morgan et al., 2003), and decreased opioid withdrawal (Cruz et al., 2008), suggesting that GIRK channels might have a role in drug addiction (Table 1).

GIRK channels are altered by, and act as downstream mediators of, drugs of abuse such as cocaine (Arora et al., 2011; Hearing et al., 2013; Padgett et al., 2012), methamphetamine (METH) (Padgett et al., 2012; Sharpe, Varela, Bettinger, & Beckstead, 2014), morphine (Labouebe et al., 2007; Nassirpour et al., 2010), γ-hydroxybutyric acid (GHB) (Cruz et al., 2004; Labouebe et al., 2007), and ethanol (EtOH) (Blednov et al., 2001; Kobayashi et al., 1999). Opioids, cannabinoids, and GHB can affect GIRK channel activity by direct activation of G protein-coupled receptors (GPCRs) linked to inhibitory ($G_{i/o}$) G proteins, and indirectly via enhancing the levels of dopamine and GABA, which also stimulate inhibitory GPCRs in brain regions relevant to reward (Luscher & Slesinger, 2010). The goal of this chapter is to discuss GIRK channels, their presence and function in brain regions critical for reward, and how their activity is modified by exposure to drugs of abuse.

Table 1 GIRK Channels and Reward-Related Behavior in Mice

Manipulation	Behavior	Citation
GIRK1		
Girk1$^{-/-}$ mice	↑ Baseline and cocaine-induced locomotor activity	Arora et al. (2010)
Girk1$^{-/-}$ mice	↑ Morphine-induced locomotor activity	Kotecki et al. (2015)
Girk1$^{-/-}$ mice	↑ Intra-RMTg DAMGO locomotor activity	Kotecki et al. (2015)
GIRK2		
GIRK2 trisomy mice	↑ Sucrose preference	Cooper et al. (2012)
Girk2$^{-/-}$ mice	↓ In acute EtOH-elicited behaviors: stimulation of home cage motor activity anxiolytic behavior in elevated-plus maze handling-induced convulsions	Blednov et al. (2001)
Girk2$^{-/-}$ mice	↓ Cocaine self-administration	Morgan et al. (2003)
Girk2$^{-/-}$ mice	↑ Operant responding for food in food-restricted animals	Pravetoni and Wickman (2008)
Girk2$^{-/-}$ mice	↑ Baseline and cocaine-induced locomotor activity	Arora et al. (2010)
Girk2$^{-/-}$ mice	↑ Morphine-induced locomotor activity	Kotecki et al. (2015)
Girk2$^{-/-}$ mice (DA neuron-specific)	↑ Morphine-induced locomotor activity	Kotecki et al. (2015)
GIRK3		
Girk3$^{-/-}$ mice	↓ Cocaine self-administration	Morgan et al. (2003)
Girk3$^{-/-}$ mice	↓ Pentobarbitol- and zolpidem-induce handling-induced convulsions (withdrawal)	Kozell, Walter, Milner, Wickman, and Buck (2009)
Girk3$^{-/-}$ mice	↓ Morphine-induced locomotor activity	Kotecki et al. (2015)
Multiple subunits		
Girk2/Girk3$^{-/-}$ mice	↓ Morphine withdrawal behavior	Cruz et al. (2008)
GIRK1- and GIRK2- shRNA (mPFC) in mouse	↑ Cocaine-induced locomotor activity	Hearing et al. (2013)

This table summarizes the effects of genetic manipulation of GIRK channels on reward-related behaviors in mice.

1.1 Molecular and Functional Diversity of GIRK Channels

There are four mammalian GIRK channel subunits: GIRK1 (*Kir3.1/Kcnj3*), GIRK2 (*Kir3.2/Kcnj6*), GIRK3 (*Kir3.3/Kcnj9*), and GIRK4 (*Kir3.4/Kcnj5*) (Hibino et al., 2010; Lujan et al., 2014; Luscher & Slesinger, 2010). These subunits interact to form homo- or heterotetrameric GIRK channels (Hibino et al., 2010; Lujan et al., 2014; Luscher & Slesinger, 2010). GIRK1, GIRK2, and GIRK3 are broadly distributed throughout the nervous system, whereas GIRK4 expression is restricted to a few neuronal populations (Aguado et al., 2008; Fernandez-Alacid, Watanabe, Molnar, Wickman, & Lujan, 2011; Hearing et al., 2013; Karschin, Dissmann, Stuhmer, & Karschin, 1996; Perry et al., 2008).

The existence of splice variants of the GIRK1 and GIRK2 subunits contributes to the molecular diversity of GIRK channels (Wei et al., 1998; Zhu et al., 2001). Several splice variants of GIRK1 have been described; however, there is limited knowledge about the expression and functional roles of these isoforms (Nelson, Marino, & Allen, 1997; Steinecker, Rosker, & Schreibmayer, 2007; Wagner et al., 2010; Zhu et al., 2001). A number of splice variants of GIRK2 have been identified (Isomoto, Kondo, & Kurachi, 1997; Lesage et al., 1994, 1995; Wei et al., 1998), and three of these isoforms (GIRK2a, GIRK2b, and GIRK2c) are widely expressed in the brain (Isomoto et al., 1997; Wei et al., 1998). GIRK2 splice variants differ in the C-terminus. Notably, GIRK2c contains a class I PDZ-binding motif that is absent in GIRK2a and GIRK2b (Isomoto et al., 1997; Lesage et al., 1994, 1995; Wei et al., 1998). There is some overlap in the expression patterns of these splice variants, suggesting that more than one isoform could be present in the same cell in several regions of the brain (Wei et al., 1998).

Molecular differences between GIRK subunits and the specific subunit composition of the tetrameric complex have an influence on the trafficking of GIRK channels to the plasma membrane. GIRK2 and GIRK4 contain an endoplasmic reticulum export motif that allows them to form functional homotetrameric channels (Kofuji, Davidson, & Lester, 1995; Lesage et al., 1995; Ma et al., 2002; Slesinger et al., 1996). Conversely, GIRK1 and GIRK3 lack this export motif and thus require interaction with other subunits to form functional heterotetrameric channels (Jelacic, Kennedy, Wickman, & Clapham, 2000; Jelacic, Sims, & Clapham, 1999; Kennedy, Nemec, Corey, Wickman, & Clapham, 1999; Kofuji et al., 1995; Krapivinsky, Gordon, et al., 1995; Lesage et al., 1995; Ma et al., 2002). Interestingly, a single point mutation in the pore domain of GIRK1 allows for the expression of functional homomeric GIRK1 channels in heterologous expression systems (Chan, Sui, Vivaudou, & Logothetis, 1996), suggesting that other factors beyond the export motif can contribute to GIRK channel trafficking. As with GIRK2c, GIRK3 contains a class I PDZ-binding motif allowing them to interact with proteins containing PDZ domains (Lunn et al., 2007). GIRK3 also has a lysosomal targeting sequence, which promotes the delivery of GIRK3-containing channels to lysosomes, thereby reducing both channel surface expression and total protein levels (Ma et al., 2002). Accordingly, GIRK3 can negatively regulate trafficking of GIRK channels to the cell surface (Ma et al., 2002), potentially affecting cell excitability. GIRK channel subunit composition can also affect the biophysical and regulatory properties of the channel. For example, GIRK1-containing channels display longer open times and larger single-channel conductances than GIRK1-lacking channels (Jelacic et al., 1999; Kofuji et al., 1995; Labouebe et al., 2007; Lesage et al., 1995; Yamada, Inanobe, & Kurachi, 1998).

Considerable evidence supports the idea that the GIRK1/GIRK2 heterotetramer is a prominent neuronal GIRK channel subtype. Indeed, *in situ* hybridization, immunohistochemistry, and biochemical assays indicate that GIRK1 and GIRK2 subunits are coexpressed in a number of brain regions and that they interact with each other (Karschin et al., 1996; Koyrakh et al., 2005; Liao, Jan, & Jan, 1996). Furthermore, electrophysiological studies from *Girk*$^{-/-}$ mice have shown that GIRK1 and GIRK2 are the major contributors to GIRK channel function in several brain regions including the hippocampus, cerebellum, and ventral tegmental area (VTA) (Cruz et al., 2008; Hearing et al., 2013; Koyrakh et al., 2005; Labouebe et al., 2007; Luscher, Jan, Stoffel, Malenka, & Nicoll, 1997; Slesinger, Stoffel, Jan, & Jan, 1997). In addition, ablation of *Girk1* or *Girk2* produced a downregulation in the protein levels of GIRK2 and GIRK1, respectively, again suggesting an interaction between these subunits (Koyrakh et al., 2005; Signorini, Liao, Duncan, Jan, & Stoffel, 1997). Nevertheless, other subunit compositions have been confirmed in the brain, including GIRK2 homomers in the dopamine neurons of the substantia nigra (Inanobe et al., 1999; Koyrakh et al., 2005) and GIRK2/GIRK3 heteromers in the dopamine neurons of the VTA (Cruz et al., 2004; Labouebe et al., 2007). In brain regions, such as the hippocampus and the cerebellum, several subunit combinations are plausible as GIRK1, GIRK2, and GIRK3 expression has been detected (Aguado et al., 2008; Ciruela et al., 2010; Fernandez-Alacid et al., 2009, 2011; Karschin et al., 1996; Koyrakh et al., 2005; Saenz del Burgo et al., 2008). A comprehensive description of GIRK channel subunit expression can be found in chapter "Localization and Targeting of GIRK Channels in Mammalian Central Neurons" by Luján and Aguado.

1.2 GIRK Channels and Neuronal Excitability

GIRK channel gating is enhanced by direct binding of the Gβγ subunits released upon activation of GPCRs that couple to $G_{i/o}$ G proteins (Chan, Sui, Vivaudou, & Logothetis, 1997; He, Zhang, Mirshahi, & Logothetis, 1999; Hibino et al., 2010; Huang, Jan, & Jan, 1997; Huang, Slesinger, Casey, Jan, & Jan, 1995; Ivanina et al., 2003; Krapivinsky, Krapivinsky, Wickman, & Clapham, 1995; Logothetis, Kurachi, Galper, Neer, & Clapham, 1987; Reuveny et al., 1994; Slesinger, Reuveny, Jan, & Jan, 1995; Whorton & MacKinnon, 2013; Wickman et al., 1994). Interaction with PIP_2 is critical, as PIP_2 alone is able to activate the channel, whereas Gβγ binding alone cannot activate it in the absence of PIP_2 (Huang, Feng, & Hilgemann, 1998; Reuveny et al., 1994; Sui, Petit-Jacques, & Logothetis, 1998; Zhang, He, Yan, Mirshahi, & Logothetis, 1999). Recent structural data provide some insight in this process as the binding of PIP_2 positions the pore of the channel in a "pre-open" conformation state that enhances ionic flow upon Gβγ binding (Whorton & MacKinnon, 2011). The Gα subunit can also bind to GIRK channels and suppress basal activity, and this interaction might help explain the coupling specificity between GIRK channels and $G\alpha_{i/o}$ signaling pathways (Berlin et al., 2010, 2011; Clancy et al., 2005; Mase, Yokogawa, Osawa, & Shimada, 2012; Peleg, Varon, Ivanina, Dessauer, & Dascal, 2002; Rubinstein, Peleg, Berlin, Brass, & Dascal, 2007; Rubinstein et al., 2009). More detailed information on the process of GIRK channel regulation can be found in chapters "Unifying Mechanism of Controlling Kir3 Channel Activity by G Proteins and Phosphoinositides" by Logothetis et al. and "The Roles of Gβγ and Gα in Gating and Regulation of GIRK Channels" by Dascal and Kahanovitch.

Genetic ablation of GIRK channels has highlighted their contribution to intrinsic cell excitability, measured as decreased rheobase, increased action potential frequency, and increased resting membrane potential by an average of 10 mV (Hearing et al., 2013; Kim & Johnston, 2015; Koyrakh et al., 2005; Luscher et al., 1997; Torrecilla et al., 2002). This contribution may reflect a combination of basal/intrinsic activity of GIRK channels in the absence of GPCR stimulation and/or the activity evoked by tonic neurotransmitter release (Chen & Johnston, 2005; Kahanovitch et al., 2014; Rishal, Porozov, Yakubovich, Varon, & Dascal, 2005). GIRK channel activity can be enhanced following neurotransmitter-dependent GPCR activation and in this manner will further impact the excitability of the cell, observed as a hyperpolarization of the resting membrane potential and/or reduced frequency of neuronal firing (Johnson & North, 1992; Kim & Johnston, 2015; Luscher et al., 1997; Munoz & Slesinger, 2014; Pfaffinger, Martin, Hunter, Nathanson, & Hille, 1985; Seutin, Johnson, & North, 1994; Wickman, Nemec, Gendler, & Clapham, 1998).

GIRK channels are mainly localized postsynaptically (Kulik et al., 2006; Lujan, Maylie, & Adelman, 2009; Luscher et al., 1997), although presynaptic distribution has also been observed in some neurons (Jin & Lu, 1998; Ladera et al., 2008; Lujan et al., 2009) (For a detailed discussion of GIRK channel subcellular localization, see chapter "Localization and Targeting of GIRK Channels in Mammalian Central Neurons" by Luján and Aguado). Within the postsynaptic region, GIRK channels can be found in perisynaptic regions of the soma and dendrites and in the dendritic shaft and spines of particular cell types, including pyramidal neurons of the hippocampus (Aguado et al., 2008; Chen & Johnston, 2005; Fernandez-Alacid et al., 2009; Kulik et al., 2006). Activation of GIRK channels produces slow inhibitory postsynaptic currents that are responsible for the hyperpolarization of the neuron (Isaacson, Solis, & Nicoll, 1993; Luscher et al., 1997; Scanziani, 2000). The activation of GIRK channels can occur in three nonexclusive ways: (i) by dendro-dendritic release of neurotransmitter that affects the same neuron (Bacci, Huguenard, & Prince, 2004; Lacey, Mercuri, & North, 1987), (ii) by neurotransmitters released from neighboring neurons (Beckstead & Williams, 2007; Dutar, Vu, & Perkel, 1999; Newberry & Nicoll, 1985), and (iii) by dynamically regulated global extracellular levels of neurotransmitters (Pietersen, Lancaster, Patel, Hamilton, & Vreugdenhil, 2009; Sun, Huguenard, & Prince, 2002). Regardless of the origin of the neurotransmitter, GIRK channel activation reduces neuronal excitability (Lujan et al., 2014; Luscher & Slesinger, 2010).

2. GIRK CHANNELS IN THE MESOCORTICOLIMBIC SYSTEM

The network of interconnected brain regions responsible for encoding reward is referred to as the mesocorticolimbic system and includes the VTA, medial prefrontal cortex (mPFC), nucleus accumbens (NAc), hippocampus, amygdala, and other interconnected regions (Hearing, Zink, & Wickman, 2012; Luscher & Malenka, 2011). As there are numerous excellent reviews detailing the importance of connections within and between nuclei of the mesocorticolimbic system, especially pertaining to drugs of abuse (Chen, Hopf, & Bonci, 2010; Fields & Margolis, 2015;

Kauer & Malenka, 2007; Koob & Volkow, 2010; Luscher & Malenka, 2011; Luthi & Luscher, 2014; Nestler, 2001; Thomas, Kalivas, & Shaham, 2008), this chapter focuses on neuronal populations wherein GIRK channel-dependent signaling is subject to adaptation triggered by *in vivo* exposure to drugs of abuse. Notably, GIRK channel expression (Karschin et al., 1996; Liao et al., 1996; Marcott, Mamaligas, & Ford, 2014) and function (Marcott et al., 2014) within the NAc is minimal-to nonexistent; therefore, we will discuss neuron populations within the VTA and mPFC, regions wherein GIRK channels have been identified and that have been implicated in plasticity mediated by drugs of abuse (Fig. 1).

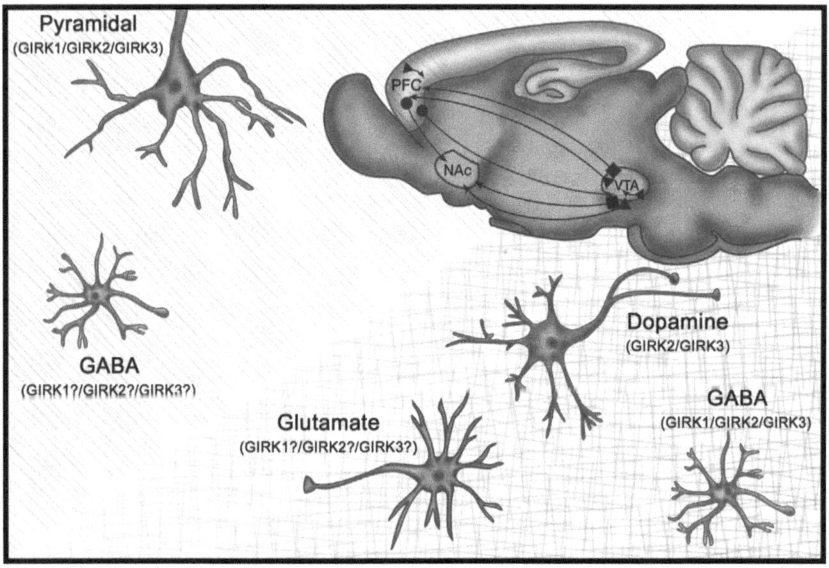

Figure 1 GIRK channels are expressed in brain areas relevant for reward. The schematic shows a sagittal view of a mouse brain highlighting some of the brain areas involved in reward and drug addiction: prefrontal cortex (PFC), nucleus accumbens (NAc), and ventral tegmental area (VTA). The schematic highlights connections within the mesocorticolimbic system that are particularly relevant to this discussion. The dopaminergic neurons (blue (black in the print version) square arrow) of the VTA project mainly to the NAc and PFC. VTA GABA interneurons (red (black in the print version) triangle arrow) control the activity of DA neurons and also project to NAc and PFC. Glutamatergic pyramidal neurons (green (black in the print version) circle arrows) from the PFC project to NAc and VTA, and their activity is also modulated by local GABA interneurons (red (black in the print version) triangle arrow). The schematic also depicts the major cell types found in these brain areas and the GIRK channel subunits expressed in each particular cell type.

2.1 VTA DA Neurons

Dopaminergic signaling from the VTA to the NAc and mPFC can be elicited by both natural and drug rewards (Schultz, 2002) and has been well studied in the context of drug abuse and addiction. Within this system, DA neurotransmission has been conceptualized as the "currency" of reward in the brain, as all major classes of drugs of abuse enhance DA neurotransmission within the mesocorticolimbic system, albeit through different mechanisms (Badiani, Belin, Epstein, Calu, & Shaham, 2011; Bayer & Glimcher, 2005; Nakahara, Itoh, Kawagoe, Takikawa, & Hikosaka, 2004; Schultz, Dayan, & Montague, 1997). The majority (>95%) of neurons in the VTA are DA or GABA neurons, while glutamatergic neurons comprise the smallest neuronal population (Fields, Hjelmstad, Margolis, & Nicola, 2007; Morales & Root, 2014). Intriguingly, recent data suggest that some VTA neurons can corelease DA and glutamate (Hnasko et al., 2010; Stuber, Hnasko, Britt, Edwards, & Bonci, 2010). While it is unknown at present whether GIRK channels are expressed in the glutamatergic neurons, both DA and GABA neurons in the VTA express GIRK channels (Cruz et al., 2004). As regulators of DA and GABA neuron excitability (Arora et al., 2010; Padgett et al., 2012), GIRK channels could play a key role in modulating the activity of the mesocorticolimbic system.

The excitability of VTA DA neurons can be regulated by GIRK channel-mediated postsynaptic inhibition, elicited through activation of $GABA_B$ receptors ($GABA_B$Rs) and D2 dopamine receptors (D2Rs). $GABA_B$Rs and D2Rs are located in the soma and dendrites of VTA DA neurons (Arora et al., 2011; Beckstead, Grandy, Wickman, & Williams, 2004; Beckstead & Williams, 2007; Cruz et al., 2004; Labouebe et al., 2007) and provide this neuronal population with paracrine (long-loop) and autocrine inhibitory feedback, respectively (Einhorn, Johansen, & White, 1988; Sugita, Johnson & North, 1992). $GABA_B$Rs on VTA DA neurons incorporate inhibitory feedback from the NAc, RMTg, and local interneurons, while D2Rs on DA neurons (i.e., auto-receptors) allow the local dendro-dendritic release of DA to temper DA neuron activity (Beaulieu & Gainetdinov, 2011; Fields et al., 2007; Ford, 2014; Neve, Seamans, & Trantham-Davidson, 2004; Sesack & Grace, 2010). Recent work suggests that VTA DA neurons are a heterogeneous population, with variability in spontaneous firing rates, hyperpolarization-activated cyclic nucleotide-gated currents (HCN; I_h), and D2R-GIRK currents (Lammel et al., 2008; Lammel, Ion, Roeper, & Malenka, 2011; Lammel, Lim, & Malenka, 2014; Lammel et al., 2012; Luthi & Luscher, 2014). These functional differences are associated with VTA DA neuron axonal projection targets; for example, mPFC-projecting DA neurons have higher firing rates and lack somatodendritic D2Rs (Lammel et al., 2008), while other populations, such as the basolateral amygdala- and NAc-projecting DA neurons, have lower spontaneous firing rates and possess somatodendritic D2Rs (Ford, Mark, & Williams, 2006). Interestingly, the GIRK channel subtype in VTA DA neurons is unique, consisting of GIRK2 and GIRK3; this channel has a lower sensitivity to Gβγ activation than GIRK1-containing channels (Cruz et al., 2004; Labouebe et al., 2007) and appears to modulate the sensitivity of the reward circuitry to drugs of abuse (Arora et al., 2010, 2011; Kotecki et al., 2015).

2.2 VTA GABA Neurons

VTA GABA neurons play an important role in regulating the activity of the mesocorticolimbic system (Carr & Sesack, 2000a; Dobi, Margolis, Wang, Harvey, & Morales, 2010; Johnson & North, 1992). While VTA GABA neurons have long been thought of as local regulators of VTA DA neurons activity, recent work combining optogenetics and behavioral analysis have demonstrated that these GABA neuron project to other reward-relevant brain regions, such as the NAc and PFC, and are involved in aversion (Carr & Sesack, 2000b; Fields et al., 2007; Tan et al., 2012; Van Bockstaele & Pickel, 1995; van Zessen, Phillips, Budygin, & Stuber, 2012; Walsh & Han, 2014).

GIRK channels in VTA GABA neurons are thought to consist of GIRK1, GIRK2, and GIRK3 subunits, based on gene expression analysis, immunolabeling, and the comparison of GPCR-GIRK signaling in VTA GABA neurons from wild-type and *Girk*$^{-/-}$ mice (Cruz et al., 2004; Labouebe et al., 2007). Along with other G protein-dependent effectors, GIRK channels mediate the inhibitory effect of $GABA_BR$ activation on VTA GABA neurons and, in this way, lead to decreased excitability (Cruz et al., 2004; Labouebe et al., 2007). Mu opioid receptor (MOR) agonists such as morphine and heroin are thought to increase DA release through a disinhibition mechanism, in which GABAergic input onto VTA DA neurons is decreased (Johnson & North, 1992; Lecca, Melis, Luchicchi, Muntoni, & Pistis, 2012; Steffensen et al., 2006). GIRK channels coupled to MOR in VTA GABA neurons have been proposed to mediate the opioid-induced disinhibition of VTA DA neurons and corresponding behaviors (Chieng, Azriel, Mohammadi, & Christie, 2011; Johnson &

North, 1992; Matsui, Jarvie, Robinson, Hentges, & Williams, 2014; Steffensen et al., 2006).

2.3 Layer 5/6 Pyramidal Neurons of the mPFC

The mPFC, like other cortical regions, has a layered organization comprised of glutamatergic pyramidal neurons and GABAergic interneurons (Conde, Maire-Lepoivre, Audinat, & Crepel, 1995; Gabbott & Bacon, 1997; Krettek & Price, 1977). Layer 5/6 pyramidal neurons in mice, analogous to layer 6 in primates, project to several subcortical regions, including the NAc and VTA (Sesack, Deutch, Roth, & Bunney, 1989; Sesack & Pickel, 1992; Taber, Das, & Fibiger, 1995; Taber & Fibiger, 1995). The mPFC can be divided into the more dorsal prelimbic cortex, which mainly projects to the core of the NAc, and the more ventral infralimbic cortex, which mainly projects to the NAc shell (Berendse, Galis-de Graaf, & Groenewegen, 1992). The mPFC is critical for executive function tasks such as decision-making, goal-directed, and reward-related behaviors (Goldstein & Volkow, 2011; Luscher & Malenka, 2011).

Alterations in the activation of glutamatergic projections from the mPFC to NAc are involved in the expression of drug-related behaviors, including reinstatement of drug-seeking (Baker et al., 2003; McFarland & Kalivas, 2001; McFarland, Lapish, & Kalivas, 2003; van Huijstee & Mansvelder, 2014) and locomotor sensitization (Luscher, 2013; Steketee & Kalivas, 2011), the progressive increase in drug-elicited locomotor activity following repeated drug administration. GIRK1, GIRK2, and GIRK3 subunits are expressed in the mPFC, and genetic ablation of GIRK channels affects the excitability of layer 5/6 pyramidal neurons of the prelimbic mPFC (PrLC) (Hearing et al., 2013). Interestingly, a sex difference in $GABA_BR$-GIRK signaling showing increased $GABA_BR$-GIRK currents in adolescent males has been recently described in this brain area (Marron Fernandez de Velasco et al., 2015).

3. DRUG-DEPENDENT PLASTICITY OF GIRK CHANNEL-DEPENDENT SIGNALING

Neuronal plasticity and the mechanisms that govern it, particularly those related to excitatory neurotransmission, have been widely investigated over the last few decades (Harris, 1999; Malinow, 2003; Nicoll, 2003; Sheng & Kim, 2002). Different forms of plasticity, such as short- and long-term potentiation and depression, are thought to be the basis for learning and memory. Addiction is sometimes conceptualized as an aberrant form of learning and memory, as exposure to drugs of abuse can elicit synaptic transmission or alter synaptic plasticity in different areas of the mesocorticolimbic system (Kauer & Malenka, 2007; Luscher & Malenka, 2011). *In vivo* exposure to addictive drugs, such as cocaine, morphine, nicotine, and benzodiazepines, increases the strength of excitatory synapses in midbrain DA neurons (Argilli, Sibley, Malenka, England, & Bonci, 2008; Saal, Dong, Bonci, & Malenka, 2003; Ungless, Whistler, Malenka, & Bonci, 2001).

Adaptations in response to drugs of abuse have also been described at inhibitory synapses. $GABA_AR$-mediated fast inhibitory neurotransmission at VTA DA neurons can undergo long-term potentiation (LTP_{GABA}) following high-frequency stimulation (Nugent, Penick, & Kauer, 2007). Drugs of abuse, including morphine, cocaine, and nicotine, can transiently prevent VTA DA neurons from expressing LTP_{GABA} (Niehaus, Murali, & Kauer, 2010; Nugent et al., 2007), suggesting that changes in fast inhibitory neurotransmission could contribute to early stages of addiction. Additionally, withdrawal from repeated cocaine reduces surface $GABA_AR$ expression and enhances LTP in mPFC layer 5 pyramidal neurons of the rat, suggesting that drug-induced adaptations in inhibitory signaling could contribute to later stages of addiction, such as drug craving (Lu, Cheng, Lim, Khoshnevisrad, & Poo, 2010).

Recently, plasticity of slow inhibitory neurotransmission mediated by GIRK channels has been described (Beckstead & Williams, 2007; Chung, Ge, et al., 2009; Chung, Qian, Ehlers, Jan, & Jan, 2009; Huang et al., 2005), revealing an additional mechanism for regulation of neuronal excitability and homeostasis. Interestingly, the plasticity of excitatory neurotransmission and the plasticity of GPCR-GIRK channel signaling share some common mechanisms, including NMDAR-, Ca^{2+}-, and CaMKII-dependence (Chung, Ge, et al., 2009; Chung, Qian, et al., 2009; Huang et al., 2005; Saal et al., 2003; Ungless et al., 2001). Furthermore, recent data have shown that the activity of VTA DA neurons can bidirectionally modulate GIRK channel activity (Lalive et al., 2014). In addition to being modified by synaptic and cellular activity, GIRK channel-dependent signaling can be altered by *in vivo* exposure to drugs of abuse (Table 2) such as cocaine (Arora et al., 2011; Hearing et al., 2013; Padgett et al., 2012) and METH (Padgett et al., 2012; Sharpe et al., 2014). In sections 3 and 4, we will discuss drug-induced plasticity of GIRK-dependent signaling in key neuronal populations within the mesocorticolimbic system.

Table 2 GIRK Channel Plasticity Elicited by Drugs of Abuse

Drug	Neuron Population	Effect on GIRK Channels	Citation
Cocaine (acute)	VTA (DA neuron, putative)	↓ $GABA_BR$-GIRK currents (transient)	Arora et al. (2011)
Cocaine (acute)	VTA (GABA neuron)	↓ $GABA_BR$-GIRK currents (transient)	Padgett et al. (2012)
Cocaine (repeated)	VTA (DA neuron, putative)	↓ $GABA_BR$-GIRK currents (transient)	Arora et al. (2011)
Cocaine (repeated)	Layer 5/6 prelimbic prefrontal cortex (pyramidal neuron)	↓ $GABA_BR$-GIRK currents (persistent)	Hearing et al. (2013)
Methamphetamine (acute)	VTA (DA neuron, defined)	↓ $GABA_BR$-GIRK currents (transient)	Padgett et al. (2012)
Methamphetamine (acute)	VTA (GABA neuron, defined)	↓ $GABA_BR$-GIRK currents (transient)	Padgett et al. (2012)
Methamphetamine (chronic)	Substantia nigra (DA neuron, putative)	↓ $GABA_BR$-GIRK currents ↓ DA-GIRK currents	Sharpe et al. (2014)
Methamphetamine (chronic)	VTA (DA neuron, putative)	↓ $GABA_BR$-GIRK currents ↓ DA-GIRK currents	Sharpe et al. (2014)
Morphine	Hippocampal pyramidal (culture)	↑ GIRK2 in dendritic spines ↑ 5-HT-GIRK currents ↑ Basal GIRK currents ↓ $GABA_BR$-GIRK currents	Nassirpour et al. (2010)
Morphine (repeated)	VTA (DA neuron, defined)	↑ In $GABA_BR$-GIRK coupling efficiency	Labouebe et al. (2007)
γ-Hydroxybutyric acid (GHB) (repeated)	VTA (DA neuron, defined)	↑ In $GABA_BR$-GIRK coupling efficiency	Labouebe et al. (2007)

This table summarizes the effect of drugs of abuse in GPCR-GIRK signaling in specific areas of the mesocorticolimbic system. For VTA studies, "putative" indicates that the dopamine neuron was identified by characteristic anatomical and electrophysiological properties, whereas "defined" indicates that the neuron populations were identified using a fluorescent reporter mouse line (i.e., *Pitx3-GFP* to identify midbrain dopamine neurons and *GAD67-GFP* to identify GABA neurons). With respect to drug exposure, "acute" indicates a single injection of drug, "repeated" indicates 5 days with one injection per day, and "chronic" indicates at least 2 weeks of stable self-administration. With respect to the effect of drugs on GIRK channel activity, "transient" indicates that the effect was absent at 7–10 days following the final drug administration, while "persistent" indicates that the effect is present 10 days or more following the final drug administration.

3.1 Psychostimulant-Induced Suppression of GIRK Channel Activity in VTA DA Neurons

Altering the excitability of VTA DA neurons can contribute to addictive behaviors (Brown, Korn, & Luscher, 2011; Marinelli, Cooper, Baker, & White, 2003), and both excitatory and inhibitory signaling in DA neurons of the VTA undergo adaptations in response to the psychostimulant cocaine (Hearing et al., 2012; Jones & Bonci, 2005; Luscher & Malenka, 2011; Thomas & Malenka, 2003). A single injection of cocaine produced a transient (4–7 days) increase in AMPAR-mediated current (Ungless et al., 2001) and decrease in $GABA_BR$-GIRK current in VTA DA neurons, a change that was accompanied by a reduction in GIRK channels at the plasma membrane (Arora et al., 2011). Repeated exposure (5 days) to cocaine produced a similar decrease in $GABA_BR$-GIRK signaling, indicating that a single cocaine dose is enough to saturate GIRK channel plasticity (Arora et al., 2011). Interestingly, this cocaine-induced decrease in $GABA_BR$-GIRK signaling was specific to the VTA, as it was not observed DA neurons of the neighboring substantia nigra pars compacta (Arora et al., 2011). The opposing effects of cocaine on excitatory (AMPA-mediated) and inhibitory ($GABA_BR$-GIRK) signaling suggests that these adaptations work together to increase the overall excitability of VTA DA neurons, which is predicted to increase DA neurotransmission at downstream targets like the NAc or mPFC. Because the timecourses of these adaptations are similar, it is tempting to speculate a shared mechanism of action. Interestingly, a single injection of a different psychostimulant (METH) also produced a transient (1 day) decrease in $GABA_BR$-GIRK signaling in VTA DA neurons (Padgett et al., 2012). In addition, prolonged (at least 2 weeks) METH self-administration decreased $GABA_BR$-GIRK signaling and D2R-GIRK signaling in midbrain DA neurons (Sharpe et al., 2014).

3.2 Psychostimulant-Induced Suppression of GIRK Channel Activity in VTA GABA Neurons

Psychostimulant-induced plasticity of GIRK channel activity is also observed in VTA GABA neurons. Acute administration of METH produced a more durable (present at 7 days) decrease in $GABA_BR$-GIRK signaling in VTA GABA neurons accompanied by a decrease in plasma membrane localization of $GABA_BR$ and GIRK channels (Padgett et al., 2012). The persistence of this effect was specific to GABA neurons, as acute treatment with METH caused a transient (1 day) decrease in

$GABA_BR$-GIRK signaling in VTA DA neurons (Padgett et al., 2012). Additionally, GABA neuron firing was no longer inhibited by $GABA_BR$ activation following acute METH exposure. As such, METH exposure weakened the control of VTA GABA neuron firing rate by $GABA_BR$ signaling, impairing this "brake" on GABA release and potentially leading to enhanced GABA-dependent inhibition in the VTA and downstream targets (Padgett et al., 2012). This overall increase in GABAergic transmission could represent an effort to reestablish basal levels of dopamine neurotransmission following psychostimulant exposure (Padgett et al., 2012).

3.3 Cocaine-Induced Suppression of GIRK Channel Activity in mPFC Pyramidal Neurons

Changes in glutamatergic signaling from the mPFC to the NAc are thought to mediate the long-lasting effects of drugs of abuse (Steketee & Kalivas, 2011), and GIRK channel-dependent signaling in the PrLC may contribute to this change (Hearing et al., 2013). While the effects of cocaine on GIRK channel activity in the VTA were readily apparent following a single drug injection (Arora et al., 2011; Padgett et al., 2012), adaptations in $GABA_BR$-GIRK signaling in layer 5/6 pyramidal neurons of the mPFC required repeated (5 days) cocaine administration (Hearing et al., 2013). Repeated cocaine persistently reduced $GABA_BR$-GIRK signaling and decreased the localization of $GABA_BR$ and GIRK channels at the plasma membrane of layer 5/6 mPFC pyramidal neurons (Hearing et al., 2013). These effects were specific to layer 5/6 pyramidal neurons in the PrLC, as no change in $GABA_BR$-GIRK signaling was observed in other cortical regions, including layer 5/6 pyramidal neurons of the adjacent infralimbic cortex (Hearing et al., 2013). This decrease in $GABA_BR$-GIRK signaling might diminish the ability of PrLC layer 5/6 pyramidal neurons to be inhibited by VTA GABA neurons and mPFC GABA interneurons, potentially enhancing their glutamatergic output to the NAc core. An increase in PrLC layer 5/6 pyramidal neuron excitability could have an impact on addiction-associated behaviors, as increases in glutamatergic signaling in the NAc core are required for reinstatement of cocaine-seeking behavior and output from the PrLC contributes to cocaine locomotor sensitization (Bowers, Chen, & Bonci, 2010; Steketee & Kalivas, 2011). Supporting this hypothesis, RNAi-mediated suppression of GIRK1/GIRK2 expression correlated with pre-sensitization to the locomotor-enhancing effects of cocaine (Hearing et al., 2013).

4. MECHANISMS UNDERLYING THE PLASTICITY OF GIRK CHANNEL-DEPENDENT SIGNALING

As discussed earlier, several recent studies have shown that drugs of abuse can modulate GIRK channel-dependent signaling in key neuronal populations within the mesocorticolimbic circuitry. These studies have also suggested intriguing potential neurochemical and molecular mechanisms that may work in conjunction with one another to alter GIRK channel-dependent signaling. In the following section, we will focus on mechanisms by which drugs of abuse may modulate GIRK channel-dependent signaling.

4.1 Membrane Trafficking

As previously discussed, adaptations in GIRK channel-dependent signaling after drug exposure have been described in VTA DA and GABA neurons and in layer 5/6 mPFC pyramidal neurons (Arora et al., 2011; Hearing et al., 2013; Padgett et al., 2012). In these different cell types, a common theme is a drug-induced redistribution of GIRK channels from the cell surface to internal sites, accompanied by a subsequent decrease in GPCR-GIRK signaling (Fig. 2). Here, we focus on how the interaction with other proteins and phosphorylation impact the membrane trafficking of GIRK channels.

4.1.1 Phosphorylation

Two recent studies have independently highlighted the importance of phosphorylation in the context of drug-induced plasticity of GIRK channel activity (Hearing et al., 2013; Padgett et al., 2012). In both studies, a psychostimulant-induced redistribution of $GABA_BRs$ and GIRK channels from the plasma membrane to internal sites was accompanied by a reduction in GIRK channel activity evoked by $GABA_BR$ activation. Interestingly, okadaic acid (OA), a potent protein phosphatase inhibitor, rapidly reversed the effects of psychostimulants on GIRK channel activity (Hearing et al., 2013; Padgett et al., 2012).

The phosphorylation event(s) responsible for the effects of psychostimulants on GIRK channel activity is/are unclear. Activity-dependent dephosphorylation of GIRK2 at Ser-9 produced an increase in GIRK channel-dependent current via an increase in forward trafficking of GIRK channels (Chung, Qian, et al., 2009). While phosphorylation of Ser-9 could be involved in the psychostimulant-dependent trafficking of

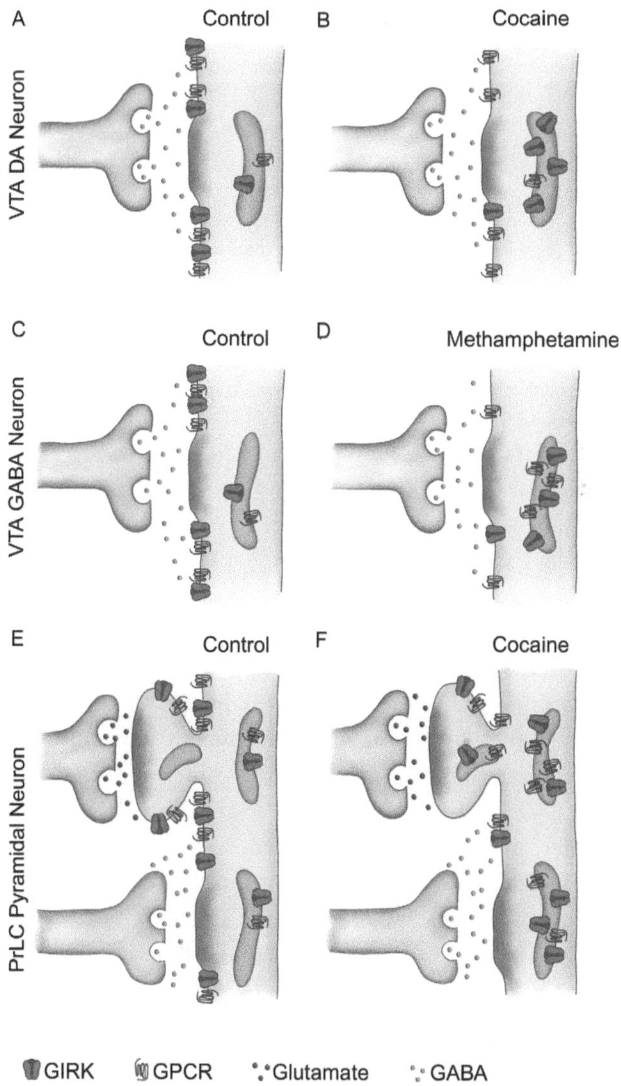

Figure 2 Drug-induced plasticity of GPCR-GIRK signaling. Psychostimulant exposure produced a suppression of $GABA_BR$-GIRK signaling in VTA dopamine (DA) neurons (A, B), VTA GABA neurons (C, D), and pyramidal neurons of the prelimbic cortex (PrLC). (A, B) A single exposure to cocaine produced a reduction in $GABA_BR$-GIRK currents in VTA DA neurons that correlated with a decrease in the surface trafficking of GIRK2-containing GIRK channels. (C, D) A single exposure to METH produced a reduction in $GABA_BR$-GIRK currents in VTA GABA neurons that correlated with a decrease in the surface trafficking of GIRK2-containing channels and $GABA_BRs$ through a phosphorylation-dependent mechanism. (E, F) Repeated exposure to cocaine produced a reduction in $GABA_BR$-GIRK currents in layer 5/6 PrLC pyramidal neurons that correlated with a decrease in the plasma membrane localization of GIRK2-containing channels and $GABA_BRs$ through a phosphorylation-dependent mechanism.

GIRK channels, this seems unlikely, as the expected effect of OA would be to drive GIRK2 channels to internal sites (Chung, Qian, et al., 2009). Even though other phosphorylation sites have been found within GIRK channel sequences (Clayton, Xu, & Chavkin, 2009; Müllner, Steinecker, Gorischek, & Schreibmayer, 2009; Rusinova et al., 2009), no information is available regarding a possible contribution to GIRK channel trafficking.

The membrane localization of GPCRs can also be regulated by phosphorylation (Marchese, Paing, Temple, & Trejo, 2008) and could be responsible for the changes in $GABA_BR$-GIRK signaling following psychostimulant administration. Indeed, $GABA_BRs$, as well as other GPCRs, can form macrocomplexes with GIRK channels, and this interaction could allow them to be trafficked together (Ciruela et al., 2010; Fajardo-Serrano et al., 2013; Koyrakh et al., 2005; Lavine et al., 2002; Lober, Pereira, & Lambert, 2006). Phosphorylation plays a key role in the trafficking of $GABA_BR$ (Terunuma, Pangalos, & Moss, 2010), and the trafficking of the $GABA_BR$ could have an effect on GIRK channel localization. Interestingly, $GABA_BRs$ function as obligate heterodimers formed by $GABA_BR1$ and $GABA_BR2$ subunits (Xu, Zhang, Rondard, Pin, & Liu, 2014), and they are both susceptible to phosphorylation. Higher levels of $GABA_BR1$ were found at internal sites following psychostimulant administration in both layer 5/6 pyramidal neurons of the PrLC and GABA neurons of the VTA (Hearing et al., 2013; Padgett et al., 2012). This change in localization was accompanied by a decrease in $GABA_BR2$ phosphorylation at residue Ser-783, which has been shown to be important for trafficking and can be regulated by protein phosphatase 2A (PP2A) activity (Terunuma et al., 2010). Thus, the effect of OA could represent an increase in the phosphorylation level of $GABA_BR2$ at residue Ser-783, which drives the receptor to the plasma membrane, reversing the effect of psychostimulant exposure. Phosphorylation of $GABA_BR1$ has also been shown to affect receptor trafficking. CaMKII-dependent phosphorylation of Ser-867 from $GABA_BR1$ promotes the internalization of the receptor (Guetg et al., 2010) and could contribute to the trafficking of GIRK channels and the OA effect. We cannot, however, rule out the possibility that other mechanisms are in play. Indeed, the cocaine-induced suppression of $GABA_BR$-GIRK signaling in VTA DA neurons was not accompanied by a redistribution of $GABA_BR1$ (Arora et al., 2011). Thus, drug-dependent regulation of $GABA_BR$-GIRK signaling could either be at the level of trafficking the receptor, the channel, or both.

4.1.2 Sorting Nexin 27

As previously discussed, GIRK2c and GIRK3, but not other GIRK subunits, contain a class I PDZ-binding motif in their C-terminus (-ESKV), which allows them to interact with SNX27 (Lunn et al., 2007), while interactions with other proteins bearing a PDZ domain, such as PSD-95 or SAP-97, do not occur (Inanobe et al., 1999; Kim & Sheng, 2004; Lunn et al., 2007; Nehring et al., 2000). The sorting nexin (SNX) protein family is large, consisting of approximately 33 members that are involved in several aspects of endocytosis and protein trafficking through different membrane compartments (Carlton, Bujny, Rutherford, & Cullen, 2005; Worby & Dixon, 2002). All members of the SNX family contain a characteristic lipid-binding phox (PX) domain, a small motif that selectively binds various phosphatidylinositol phosphates. The PX domain found in SNX27 preferentially binds PI3P and targets SNX27 to early endosomes (Gillooly et al., 2000; Lunn et al., 2007; Ponting, 1996). Other motifs are present in different members of the SNX family; however, only SNX27 contains an N-terminal PDZ domain (Lunn et al., 2007), which allows direct interactions with proteins containing PDZ-binding motifs. The specificity of the interaction between SNX27 and GIRK2c and GIRK3 relies on specific amino acid positions within and nearby of the PDZ-binding motif of GIRK2c and GIRK3 (Balana et al., 2011; Lunn et al., 2007).

SNX27 is expressed throughout the brain, including the cortex, hippocampus, and VTA (Balana et al., 2011; Lunn et al., 2007; Munoz & Slesinger, 2014). Two splice variants of SNX27 have been identified (SNX27a and SNX27b), which vary only in the last 15 amino acids of the C-terminus (Kajii et al., 2003). The SNX27b variant, also known as METH-responsive transcript 1, is upregulated in rats sensitized to METHs (Fujiyama, Kajii, Hiraoka, & Nishikawa, 2003; Kajii et al., 2003). Studies involving overexpression suggest an important role of SNX27b in the trafficking of GIRK3-containing GIRK channels; indeed, a SNX27b-dependent decrease in the plasma membrane localization of GIRK3-containing GIRK channels translates into a decrease in GIRK-dependent currents (Balana et al., 2011; Lunn et al., 2007; Nassirpour & Slesinger, 2007).

Recent data indicate that SNX27b plays a more complex role in trafficking of GIRK channels, suggesting a role in forward trafficking (Munoz & Slesinger, 2014) A similar role was described for trafficking of β-adrenergic receptors (Temkin et al., 2011). Ablation of *Snx27* specifically in DA neurons ($Snx27_{DA}{}^{-/-}$) decreased the plasma membrane localization of GIRK channels and $GABA_BR$-GIRK signaling (Munoz & Slesinger, 2014). Conversely, no effect was observed on D2R-GIRK currents, indicating that D2R and $GABA_BR$ signaling complexes might be differentially compartmentalized and/or regulated by SNX27b in VTA DA neurons (Munoz & Slesinger, 2014). The decrease in $GABA_BR$-GIRK currents was paralleled by a reduced ability of the $GABA_BR$ agonist baclofen to suppress the firing of VTA DA neurons. Re-expression of SNX27b or the expression of GIRK2a, which does not interact with SNX27, in VTA DA neurons of $Snx27_{DA}{}^{-/-}$ mice was sufficient to restore normal GIRK trafficking, $GABA_BR$-GIRK currents, and $GABA_BR$-dependent inhibition of action potential firing (Munoz & Slesinger, 2014).

The reduction of $GABA_BR$-GIRK currents in the $Snx27_{DA}{}^{-/-}$ mice provides a unique model to study the contribution of GIRK channel activity to reward-related behaviors. Psychostimulants such as cocaine induce locomotor sensitization, and intact VTA DA neuron signaling is needed for the expression of this behavior (Cornish & Kalivas, 2001; Ouachikh, Dieb, Durif, & Hafidi, 2013). Interestingly, $Snx27_{DA}{}^{-/-}$ mice showed an increase in locomotor activity following cocaine administration (Munoz & Slesinger, 2014). This phenotype was normalized following the selective re-expression of the SNX27-insensitive GIRK2 isoform (GIRK2a) in VTA DA neurons from $Snx27_{DA}{}^{-/-}$ mice. Given these results, it is tempting to speculate that decreased surface trafficking of GIRK channels in VTA DA neurons might presensitize the animal to the effects of psychostimulants. This idea is supported by the observation that psychostimulant treatment produces a decrease in the plasma membrane localization of GIRK2 in VTA DA neurons (Arora et al., 2011).

4.2 RGS Proteins

RGS proteins negatively regulate G protein-dependent signaling (Hollinger & Hepler, 2002; Traynor, 2010). While the RGS family is quite large (~37 members) and diverse (Kach, Sethakorn, & Dulin, 2012; Sethakorn, Yau, & Dulin, 2010), all RGS proteins share a common motif, the "RGS box" or RH homology domain. This motif interacts with Gα-GTP proteins, increasing the endogenous GTPase activity of the Gα subunit, thus accelerating its inactivation (Berman & Gilman, 1998; Neubig & Siderovski, 2002; Ross & Wilkie, 2000; Tesmer, Berman, Gilman, & Sprang, 1997). RGS proteins can regulate cellular excitability via their indirect modulation of GIRK channel and voltage-dependent calcium channel activity (Berman, Wilkie, & Gilman, 1996; Chuang, Yu,

Jan, & Jan, 1998; Doupnik, Davidson, Lester, & Kofuji, 1997; Hollinger & Hepler, 2002; Sondek & Siderovski, 2001; Zhou, Siderovski, & Miller, 2000). Some RGS proteins are expressed in brain areas important for reward, where they could modulate GIRK channel activity (Gold, Ni, Dohlman, & Nestler, 1997). Moreover, the expression of some RGS proteins, including RGS2, RGS4, RGS6, and RGS9, is regulated by drugs of abuse (Lomazzi, Slesinger, & Luscher, 2008; Traynor, 2010). More information about the role of RGS proteins in the regulation of GPCR-GIRK signaling can be found in chapter "RGS Redundancy and Implications in GPCR–GIRK Signaling" by Doupnik.

4.2.1 RGS2

RGS2 regulates the coupling efficiency (i.e., ability of GPCRs to activate GIRK channels) between $GABA_BRs$ and GIRK channels in VTA DA neurons (Labouebe et al., 2007). While GIRK channels are found in both DA (GIRK2/GIRK3) and GABA (GIRK1/GIRK2/GIRK3) neurons of the VTA (Cruz et al., 2004; Labouebe et al., 2007), coupling is less efficient in DA than in GABA neurons (Labouebe et al., 2007). Differences in expression of RGS2 have been found between VTA DA and GABA neurons (Labouebe et al., 2007). This differential expression, along with an RGS2-specific interaction with GIRK3, and the lower Gβγ affinity of the GIRK2/GIRK3 heterotetramer provide a possible explanation for the differential coupling efficiency between these two cell types (Labouebe et al., 2007). Indeed, genetic ablation of *Rgs2* or *Girk3* enhanced $GABA_BR$-GIRK coupling efficiency in VTA DA neurons toward the levels observed in VTA GABA neurons (Labouebe et al., 2007).

A selective change in coupling efficiency in VTA DA neurons could underlie the actions of some drugs of abuse, such as GHB (Brebner, Childress, & Roberts, 2002; Cousins, Roberts, & de Wit, 2002; Martellotta, Fattore, Cossu, & Fratta, 1997). Indeed, administration of GHB triggered a decrease in *Rgs2* expression in the VTA that was paralleled by a significant enhancement of $GABA_BR$-GIRK coupling efficiency in VTA DA neurons (Labouebe et al., 2007). The practical consequence of this change in coupling efficiency is best understood in the context of both DA and GABA neurons of the VTA. Since the coupling efficiency in VTA GABA neurons is higher than in VTA DA neurons, a low dose of a $GABA_BR$ agonist would preferentially affect GABA neurons, producing a disinhibition of VTA DA neurons. Conversely, a higher dose of the same agonist could directly silence DA neurons (Labouebe et al., 2007). Indeed, this aligns with a behavioral shift observed after chronic GHB administration, in which a dose of GHB that would normally be rewarding became aversive following a previous exposure to GHB, suggesting a possible mechanism of tolerance (Labouebe et al., 2007). Thus, RGS2 could represent a novel target for the treatment of drug addiction.

4.2.2 R7 RGS

The R7 family of RGS proteins includes RGS6, RGS7, RGS9, and RGS11 (Anderson, Posokhova, & Martemyanov, 2009). This family of proteins is unique in that they contain a domain homologous to the G protein Gγ subunit (Gγ-like domain, or GGL), which facilitates interaction with the atypical member of the Gβ protein family, Gβ5 (Anderson et al., 2009). The resultant complex thus resembles the Gβγ dimer (Cheever et al., 2008). Not surprisingly, R7 RGS/Gβ5 complexes can interact with GIRK channels and modulate their activity in CA1 pyramidal neurons of the hippocampus (RGS7/Gβ5) (Fajardo-Serrano et al., 2013; Ostrovskaya et al., 2014; Xie et al., 2010). In addition, R7 RGS/Gβ5 complexes can interact with adaptor proteins, including R7BP and R9BP, which can anchor them to the plasma membrane, potentially bringing the complex close to its site of action (Anderson et al., 2009).

The expression of several members of the R7 RGS family is regulated by exposure to drugs of abuse (Lomazzi et al., 2008; Traynor, 2010). One of the most studied examples, RGS9-2, is predominantly expressed in the striatum (Zachariou et al., 2003), where expression of GIRK channels is low (Karschin et al., 1996; Liao et al., 1996; Marcott et al., 2014; Olive & Newton, 2010). Though RGS9-2 can modulate GIRK channel activity in a heterologous system (Rahman et al., 2003), their functional interaction has not been established *in vivo*. RGS6 has been linked to EtOH addiction, as *Rgs6*$^{-/-}$ mice consumed significantly less EtOH and exhibited a reduction in EtOH preference (Stewart et al., 2015). RGS6 is expressed in VTA DA neurons, and its expression was dramatically increased after prolonged EtOH exposure (Stewart et al., 2015). RGS6/Gβ5 regulation of GIRK channel activity has been described in cerebellum (Maity et al., 2012). Finally, in CA1 pyramidal neurons of the hippocampus, ablation of RGS7 disrupted spatial learning and memory and produced significantly delayed deactivation kinetics of $GABA_BR$-induced GIRK currents, increased $GABA_BR$-GIRK coupling efficiency, and increased cell excitability (Ostrovskaya et al., 2014). Given the influence of RGS7 in hippocampal function and importance of the hippocampus in context-induced relapse (Pascoli, Turiault, & Luscher,

2012; Sesack & Grace, 2010), RGS7 regulation of GIRK channel activity in this region could contribute to this addiction-related behavior.

5. CONCLUDING THOUGHTS AND FUTURE DIRECTIONS

In this chapter, we have described drug-induced changes in GPCR-GIRK channel-mediated signaling and discussed key mechanisms that may underlie these adaptations. While it is clear that drugs of abuse can alter GIRK channel-dependent signaling in reward-relevant neuron populations, the link between these adaptations and the progression of addiction is unclear. If drug-mediated changes in GIRK channel-dependent signaling are necessary for the full expression of addictive behaviors, as studies with *Girk*$^{-/-}$ mice would suggest, targeting GIRK channels or mechanisms mediating the plasticity of GIRK-dependent might be an effective strategy for treating addiction. Such potential approaches could include pharmacologic manipulation of specific GIRK channel subtypes, RGS proteins, or SNX27. Future work testing the link between GIRK channel plasticity and drugs of abuse will yield exciting insights and inform the development of such novel therapeutics to attenuate or prevent the acquisition or development of additive behaviors.

ACKNOWLEDGMENTS

The authors would like to thank Dr. Patricia Alvarez-Muñoz for her assistance in the preparation of the figures and Dr. Megan Tipps for critical review of the manuscript. This work was supported by NIH Grants to N.M. (DA007234) and K.W. (MH061933 and DA034696).

REFERENCES

Aguado, C., Colon, J., Ciruela, F., Schlaudraff, F., Cabanero, M. J., Perry, C., et al. (2008). Cell type-specific subunit composition of G protein-gated potassium channels in the cerebellum. *Journal of Neurochemistry*, *105*(2), 497–511. http://dx.doi.org/10.1111/j.1471-4159.2007.05153.x.

Aldenhoff, J. B. (1989). Imbalance of neuronal excitability as a cause of psychic disorder. *Pharmacopsychiatry*, *22*(6), 227–240. http://dx.doi.org/10.1055/s-2007-1014606.

Anderson, G. R., Posokhova, E., & Martemyanov, K. A. (2009). The R7 RGS protein family: Multi-subunit regulators of neuronal G protein signaling. *Cell Biochemistry and Biophysics*, *54*(1-3), 33–46. http://dx.doi.org/10.1007/s12013-009-9052-9.

Argilli, E., Sibley, D. R., Malenka, R. C., England, P. M., & Bonci, A. (2008). Mechanism and time course of cocaine-induced long-term potentiation in the ventral tegmental area. *The Journal of Neuroscience*, *28*(37), 9092–9100. http://dx.doi.org/10.1523/jneurosci.1001-08.2008.

Arora, D., Haluk, D. M., Kourrich, S., Pravetoni, M., Fernandez-Alacid, L., Nicolau, J. C., et al. (2010). Altered neurotransmission in the mesolimbic reward system of *Girk* mice.

Journal of Neurochemistry, *114*(5), 1487–1497. http://dx.doi.org/10.1111/j.1471-4159.2010.06864.x.

Arora, D., Hearing, M., Haluk, D. M., Mirkovic, K., Fajardo-Serrano, A., Wessendorf, M. W., et al. (2011). Acute cocaine exposure weakens $GABA_B$ receptor-dependent G-protein-gated inwardly rectifying K^+ signaling in dopamine neurons of the ventral tegmental area. *The Journal of Neuroscience*, *31*(34), 12251–12257. http://dx.doi.org/10.1523/JNEUROSCI.0494-11.2011.

Bacci, A., Huguenard, J. R., & Prince, D. A. (2004). Long-lasting self-inhibition of neocortical interneurons mediated by endocannabinoids. *Nature*, *431*(7006), 312–316. http://dx.doi.org/10.1038/nature02913.

Badiani, A., Belin, D., Epstein, D., Calu, D., & Shaham, Y. (2011). Opiate versus psychostimulant addiction: The differences do matter. *Nature Reviews. Neuroscience*, *12*(11), 685–700. http://dx.doi.org/10.1038/nrn3104.

Baker, D. A., McFarland, K., Lake, R. W., Shen, H., Tang, X. C., Toda, S., et al. (2003). Neuroadaptations in cystine-glutamate exchange underlie cocaine relapse. *Nature Neuroscience*, *6*(7), 743–749. http://dx.doi.org/10.1038/nn1069.

Balana, B., Maslennikov, I., Kwiatkowski, W., Stern, K. M., Bahima, L., Choe, S., et al. (2011). Mechanism underlying selective regulation of G protein-gated inwardly rectifying potassium channels by the psychostimulant-sensitive sorting nexin 27. *Proceedings of the National Academy of Sciences of the United States of America*, *108*(14), 5831–5836. http://dx.doi.org/10.1073/pnas.1018645108.

Bayer, H. M., & Glimcher, P. W. (2005). Midbrain dopamine neurons encode a quantitative reward prediction error signal. *Neuron*, *47*(1), 129–141. http://dx.doi.org/10.1016/j.neuron.2005.05.020.

Beaulieu, J. M., & Gainetdinov, R. R. (2011). The physiology, signaling, and pharmacology of dopamine receptors. *Pharmacological Reviews*, *63*(1), 182–217. http://dx.doi.org/10.1124/pr.110.002642.

Beckstead, M. J., Grandy, D. K., Wickman, K., & Williams, J. T. (2004). Vesicular dopamine release elicits an inhibitory postsynaptic current in midbrain dopamine neurons. *Neuron*, *42*(6), 939–946. http://dx.doi.org/10.1016/j.neuron.2004.05.019.

Beckstead, M. J., & Williams, J. T. (2007). Long-term depression of a dopamine IPSC. *The Journal of Neuroscience*, *27*(8), 2074–2080. http://dx.doi.org/10.1523/jneurosci.3251-06.2007.

Berendse, H. W., Galis-de Graaf, Y., & Groenewegen, H. J. (1992). Topographical organization and relationship with ventral striatal compartments of prefrontal corticostriatal projections in the rat. *The Journal of Comparative Neurology*, *316*(3), 314–347. http://dx.doi.org/10.1002/cne.903160305.

Berlin, S., Keren-Raifman, T., Castel, R., Rubinstein, M., Dessauer, C. W., Ivanina, T., et al. (2010). G $alpha_i$ and G betagamma jointly regulate the conformations of a G betagamma effector, the neuronal G protein-activated K^+ channel (GIRK). *The Journal of Biological Chemistry*, *285*(9), 6179–6185. http://dx.doi.org/10.1074/jbc.M109.085944.

Berlin, S., Tsemakhovich, V. A., Castel, R., Ivanina, T., Dessauer, C. W., Keren-Raifman, T., et al. (2011). Two distinct aspects of coupling between $Galpha_i$ protein and G protein-activated K^+ channel (GIRK) revealed by fluorescently labeled $Galpha_{i3}$ protein subunits. *The Journal of Biological Chemistry*, *286*(38), 33223–33235. http://dx.doi.org/10.1074/jbc.M111.271056.

Berman, D. M., & Gilman, A. G. (1998). Mammalian RGS proteins: Barbarians at the gate. *The Journal of Biological Chemistry*, *273*(3), 1269–1272.

Berman, D. M., Wilkie, T. M., & Gilman, A. G. (1996). GAIP and RGS4 are GTPase-activating proteins for the G_i subfamily of G protein alpha subunits. *Cell*, *86*(3), 445–452.

Blednov, Y. A., Stoffel, M., Chang, S. R., & Harris, R. A. (2001). Potassium channels as targets for ethanol: Studies of G-protein-coupled inwardly rectifying potassium channel 2 (GIRK2) null mutant mice. *Journal of Pharmacology and Experimental Therapeutics, 298*(2), 521–530.

Bowers, M. S., Chen, B. T., & Bonci, A. (2010). AMPA receptor synaptic plasticity induced by psychostimulants: The past, present, and therapeutic future. *Neuron, 67*(1), 11–24. http://dx.doi.org/10.1016/j.neuron.2010.06.004.

Brebner, K., Childress, A. R., & Roberts, D. C. (2002). A potential role for $GABA_B$ agonists in the treatment of psychostimulant addiction. *Alcohol and Alcoholism, 37*(5), 478–484.

Brown, M. T., Korn, C., & Luscher, C. (2011). Mimicking synaptic effects of addictive drugs with selective dopamine neuron stimulation. *Channels (Austin, Tex.), 5*(6), 461–463. http://dx.doi.org/10.4161/chan.5.6.17120.

Carlton, J., Bujny, M., Rutherford, A., & Cullen, P. (2005). Sorting nexins–unifying trends and new perspectives. *Traffic, 6*(2), 75–82. http://dx.doi.org/10.1111/j.1600-0854.2005.00260.x.

Carr, D. B., & Sesack, S. R. (2000a). Projections from the rat prefrontal cortex to the ventral tegmental area: Target specificity in the synaptic associations with mesoaccumbens and mesocortical neurons. *The Journal of Neuroscience, 20*(10), 3864–3873.

Carr, D. B., & Sesack, S. R. (2000b). GABA-containing neurons in the rat ventral tegmental area project to the prefrontal cortex. *Synapse, 38*(2), 114–123. http://dx.doi.org/10.1002/1098-2396(200011)38:2<114::aid-syn2>3.0.co;2-r.

Catterall, W. A., Dib-Hajj, S., Meisler, M. H., & Pietrobon, D. (2008). Inherited neuronal ion channelopathies: New windows on complex neurological diseases. *The Journal of Neuroscience, 28*(46), 11768–11777. http://dx.doi.org/10.1523/jneurosci.3901-08.2008.

Chan, K. W., Sui, J. L., Vivaudou, M., & Logothetis, D. E. (1996). Control of channel activity through a unique amino acid residue of a G protein-gated inwardly rectifying K^+ channel subunit. *Proceedings of the National Academy of Sciences of the United States of America, 93*(24), 14193–14198.

Chan, K. W., Sui, J. L., Vivaudou, M., & Logothetis, D. E. (1997). Specific regions of heteromeric subunits involved in enhancement of G protein-gated K^+ channel activity. *The Journal of Biological Chemistry, 272*(10), 6548–6555.

Cheever, M. L., Snyder, J. T., Gershburg, S., Siderovski, D. P., Harden, T. K., & Sondek, J. (2008). Crystal structure of the multifunctional Gbeta5-RGS9 complex. *Nature Structural & Molecular Biology, 15*(2), 155–162. http://dx.doi.org/10.1038/nsmb.1377.

Chen, B. T., Hopf, F. W., & Bonci, A. (2010). Synaptic plasticity in the mesolimbic system: Therapeutic implications for substance abuse. *Annals of the New York Academy of Sciences, 1187*, 129–139. http://dx.doi.org/10.1111/j.1749-6632.2009.05154.x.

Chen, X., & Johnston, D. (2005). Constitutively active G-protein-gated inwardly rectifying K^+ channels in dendrites of hippocampal CA1 pyramidal neurons. *The Journal of Neuroscience, 25*(15), 3787–3792. http://dx.doi.org/10.1523/jneurosci.5312-04.2005.

Chieng, B., Azriel, Y., Mohammadi, S., & Christie, M. J. (2011). Distinct cellular properties of identified dopaminergic and GABAergic neurons in the mouse ventral tegmental area. *The Journal of Physiology, 589*(Pt. 15), 3775–3787. http://dx.doi.org/10.1113/jphysiol.2011.210807.

Chuang, H. H., Yu, M., Jan, Y. N., & Jan, L. Y. (1998). Evidence that the nucleotide exchange and hydrolysis cycle of G proteins causes acute desensitization of G-protein gated inward rectifier K^+ channels. *Proceedings of the National Academy of Sciences of the United States of America, 95*(20), 11727–11732.

Chung, H. J., Ge, W. P., Qian, X., Wiser, O., Jan, Y. N., & Jan, L. Y. (2009). G protein-activated inwardly rectifying potassium channels mediate depotentiation of long-term potentiation. *Proceedings of the National Academy of Sciences of the United States of America, 106*(2), 635–640. http://dx.doi.org/10.1073/pnas.0811685106.

Chung, H. J., Qian, X., Ehlers, M., Jan, Y. N., & Jan, L. Y. (2009). Neuronal activity regulates phosphorylation-dependent surface delivery of G protein-activated inwardly rectifying potassium channels. *Proceedings of the National Academy of Sciences of the United States of America, 106*(2), 629–634. http://dx.doi.org/10.1073/pnas.0811615106.

Ciruela, F., Fernandez-Duenas, V., Sahlholm, K., Fernandez-Alacid, L., Nicolau, J. C., Watanabe, M., et al. (2010). Evidence for oligomerization between $GABA_B$ receptors and GIRK channels containing the GIRK1 and GIRK3 subunits. *The European Journal of Neuroscience, 32*(8), 1265–1277. http://dx.doi.org/10.1111/j.1460-9568.2010.07356.x.

Clancy, S. M., Fowler, C. E., Finley, M., Suen, K. F., Arrabit, C., Berton, F., et al. (2005). Pertussis-toxin-sensitive Gα subunits selectively bind to C-terminal domain of neuronal GIRK channels: Evidence for a heterotrimeric G-protein-channel complex. *Molecular and Cellular Neuroscience, 28*(2), 375–389. http://dx.doi.org/10.1016/j.mcn.2004.10.009.

Clayton, C. C., Xu, M., & Chavkin, C. (2009). Tyrosine phosphorylation of Kir3 following kappa-opioid receptor activation of p38 MAPK causes heterologous desensitization. *The Journal of Biological Chemistry, 284*(46), 31872–31881. http://dx.doi.org/10.1074/jbc.M109.053793.

Conde, F., Maire-Lepoivre, E., Audinat, E., & Crepel, F. (1995). Afferent connections of the medial frontal cortex of the rat. II. Cortical and subcortical afferents. *The Journal of Comparative Neurology, 352*(4), 567–593. http://dx.doi.org/10.1002/cne.903520407.

Cooper, A., Grigoryan, G., Guy-David, L., Tsoory, M. M., Chen, A., & Reuveny, E. (2012). Trisomy of the G protein-coupled K^+ channel gene, *Kcnj6*, affects reward mechanisms, cognitive functions, and synaptic plasticity in mice. *Proceedings of the National Academy of Sciences of the United States of America, 109*(7), 2642–2647. http://dx.doi.org/10.1073/pnas.1109099109.

Cornish, J. L., & Kalivas, P. W. (2001). Repeated cocaine administration into the rat ventral tegmental area produces behavioral sensitization to a systemic cocaine challenge. *Behavioural Brain Research, 126*(1–2), 205–209.

Cousins, M. S., Roberts, D. C., & de Wit, H. (2002). $GABA_B$ receptor agonists for the treatment of drug addiction: A review of recent findings. *Drug and Alcohol Dependence, 65*(3), 209–220.

Cruz, H. G., Berton, F., Sollini, M., Blanchet, C., Pravetoni, M., Wickman, K., et al. (2008). Absence and rescue of morphine withdrawal in GIRK/*Kir3* knock-out mice. *The Journal of Neuroscience, 28*(15), 4069–4077. http://dx.doi.org/10.1523/jneurosci.0267-08.2008.

Cruz, H. G., Ivanova, T., Lunn, M.-L., Stoffel, M., Slesinger, P. A., & Lüscher, C. (2004). Bi-directional effects of $GABA_B$ receptor agonists on the mesolimbic dopamine system. *Nature Neuroscience,* 7(2), 153–159. http://dx.doi.org/10.1038/nn1181.

Dobi, A., Margolis, E. B., Wang, H. L., Harvey, B. K., & Morales, M. (2010). Glutamatergic and nonglutamatergic neurons of the ventral tegmental area establish local synaptic contacts with dopaminergic and nondopaminergic neurons. *The Journal of Neuroscience, 30*(1), 218–229. http://dx.doi.org/10.1523/jneurosci.3884-09.2010.

Doupnik, C. A., Davidson, N., Lester, H. A., & Kofuji, P. (1997). RGS proteins reconstitute the rapid gating kinetics of Gbetagamma-activated inwardly rectifying K^+ channels. *Proceedings of the National Academy of Sciences of the United States of America, 94*(19), 10461–10466.

Dutar, P., Vu, H. M., & Perkel, D. J. (1999). Pharmacological characterization of an unusual mGluR-evoked neuronal hyperpolarization mediated by activation of GIRK channels. *Neuropharmacology, 38*(4), 467–475.

Einhorn, L. C., Johansen, P. A., & White, F. J. (1988). Electrophysiological effects of cocaine in the mesoaccumbens dopamine system: Studies in the ventral tegmental area. *The Journal of Neuroscience, 8*(1), 100–112.

Fajardo-Serrano, A., Wydeven, N., Young, D., Watanabe, M., Shigemoto, R., Martemyanov, K. A., et al. (2013). Association of Rgs7/Gbeta5 complexes with Girk channels and $GABA_B$ receptors in hippocampal CA1 pyramidal neurons. *Hippocampus*, *23*(12), 1231–1245. http://dx.doi.org/10.1002/hipo.22161.

Fernandez-Alacid, L., Aguado, C., Ciruela, F., Martin, R., Colon, J., Cabanero, M. J., et al. (2009). Subcellular compartment-specific molecular diversity of pre- and post-synaptic $GABA_B$-activated GIRK channels in Purkinje cells: Pre- and post-synaptic $GABA_B$-activated GIRK channels in PCs. *Journal of Neurochemistry*, *110*(4), 1363–1376. http://dx.doi.org/10.1111/j.1471-4159.2009.06229.x.

Fernandez-Alacid, L., Watanabe, M., Molnar, E., Wickman, K., & Lujan, R. (2011). Developmental regulation of G protein-gated inwardly-rectifying K^+ (GIRK/Kir3) channel subunits in the brain. *The European Journal of Neuroscience*, *34*(11), 1724–1736. http://dx.doi.org/10.1111/j.1460-9568.2011.07886.x.

Fields, H. L., Hjelmstad, G. O., Margolis, E. B., & Nicola, S. M. (2007). Ventral tegmental area neurons in learned appetitive behavior and positive reinforcement. *Annual Review of Neuroscience*, *30*, 289–316. http://dx.doi.org/10.1146/annurev.neuro.30.051606.094341.

Fields, H. L., & Margolis, E. B. (2015). Understanding opioid reward. *Trends in Neurosciences*, *38*(4), 217–225. http://dx.doi.org/10.1016/j.tins.2015.01.002.

Ford, C. P. (2014). The role of D2-autoreceptors in regulating dopamine neuron activity and transmission. *Neuroscience*, *282C*, 13–22. http://dx.doi.org/10.1016/j.neuroscience.2014.01.025.

Ford, C. P., Mark, G. P., & Williams, J. T. (2006). Properties and opioid inhibition of mesolimbic dopamine neurons vary according to target location. *The Journal of Neuroscience*, *26*(10), 2788–2797. http://dx.doi.org/10.1523/jneurosci.4331-05.2006.

Fujiyama, K., Kajii, Y., Hiraoka, S., & Nishikawa, T. (2003). Differential regulation by stimulants of neocortical expression of mrt1, arc, and homer1a mRNA in the rats treated with repeated methamphetamine. *Synapse*, *49*(3), 143–149. http://dx.doi.org/10.1002/syn.10220.

Gabbott, P. L., & Bacon, S. J. (1997). Calcineurin immunoreactivity in prelimbic cortex (area 32) of the rat. *Brain Research*, *747*(2), 352–356.

Gillooly, D. J., Morrow, I. C., Lindsay, M., Gould, R., Bryant, N. J., Gaullier, J. M., et al. (2000). Localization of phosphatidylinositol 3-phosphate in yeast and mammalian cells. *EMBO Journal*, *19*(17), 4577–4588. http://dx.doi.org/10.1093/emboj/19.17.4577.

Gold, S. J., Ni, Y. G., Dohlman, H. G., & Nestler, E. J. (1997). Regulators of G-protein signaling (RGS) proteins: Region-specific expression of nine subtypes in rat brain. *The Journal of Neuroscience*, *17*(20), 8024–8037.

Goldstein, R. Z., & Volkow, N. D. (2011). Dysfunction of the prefrontal cortex in addiction: Neuroimaging findings and clinical implications. *Nature Reviews. Neuroscience*, *12*(11), 652–669. http://dx.doi.org/10.1038/nrn3119.

Guetg, N., Aziz, S. A., Holbro, N., Turecek, R., Rose, T., Seddik, R., et al. (2010). NMDA receptor-dependent $GABA_B$ receptor internalization via CaMKII phosphorylation of serine 867 in $GABA_B1$. *Proceedings of the National Academy of Sciences*, *107*(31), 13924–13929. http://dx.doi.org/10.1073/pnas.1000909107.

Harris, K. M. (1999). Structure, development, and plasticity of dendritic spines. *Current Opinion in Neurobiology*, *9*(3), 343–348.

He, C., Zhang, H., Mirshahi, T., & Logothetis, D. E. (1999). Identification of a potassium channel site that interacts with G protein betagamma subunits to mediate agonist-induced signaling. *The Journal of Biological Chemistry*, *274*(18), 12517–12524.

Hearing, M., Kotecki, L., Marron Fernandez de Velasco, E., Fajardo-Serrano, A., Chung, H. J., Lujan, R., et al. (2013). Repeated cocaine weakens $GABA_B$-Girk signaling in layer 5/6 pyramidal neurons in the prelimbic cortex. *Neuron, 80*(1), 159–170. http://dx.doi.org/10.1016/j.neuron.2013.07.019.

Hearing, M., Zink, A. N., & Wickman, K. (2012). Cocaine-induced adaptations in metabotropic inhibitory signaling in the mesocorticolimbic system. *Reviews in the Neurosciences, 23*(4), 325–351. http://dx.doi.org/10.1515/revneuro-2012-0045.

Hibino, H., Inanobe, A., Furutani, K., Murakami, S., Findlay, I., & Kurachi, Y. (2010). Inwardly rectifying potassium channels: Their structure, function, and physiological roles. *Physiological Reviews, 90*(1), 291–366. http://dx.doi.org/10.1152/physrev.00021.2009.

Hnasko, T. S., Chuhma, N., Zhang, H., Goh, G. Y., Sulzer, D., Palmiter, R. D., et al. (2010). Vesicular glutamate transport promotes dopamine storage and glutamate corelease *in vivo*. *Neuron, 65*(5), 643–656. http://dx.doi.org/10.1016/j.neuron.2010.02.012.

Hollinger, S., & Hepler, J. R. (2002). Cellular regulation of RGS proteins: Modulators and integrators of G protein signaling. *Pharmacological Reviews, 54*(3), 527–559.

Huang, C. L., Feng, S., & Hilgemann, D. W. (1998). Direct activation of inward rectifier potassium channels by PIP_2 and its stabilization by Gbetagamma. *Nature, 391*(6669), 803–806. http://dx.doi.org/10.1038/35882.

Huang, C. L., Jan, Y. N., & Jan, L. Y. (1997). Binding of the G protein betagamma subunit to multiple regions of G protein-gated inward-rectifying K^+ channels. *FEBS Letters, 405*(3), 291–298.

Huang, C. S., Shi, S. H., Ule, J., Ruggiu, M., Barker, L. A., Darnell, R. B., et al. (2005). Common molecular pathways mediate long-term potentiation of synaptic excitation and slow synaptic inhibition. *Cell, 123*(1), 105–118. http://dx.doi.org/10.1016/j.cell.2005.07.033.

Huang, C. L., Slesinger, P. A., Casey, P. J., Jan, Y. N., & Jan, L. Y. (1995). Evidence that direct binding of $G\beta\gamma$ to the GIRK1 G protein-gated inwardly rectifying K^+ channel is important for channel activation. *Neuron, 15*(5), 1133–1143.

Inanobe, A., Yoshimoto, Y., Horio, Y., Morishige, K. I., Hibino, H., Matsumoto, S., et al. (1999). Characterization of G-protein-gated K^+ channels composed of Kir3.2 subunits in dopaminergic neurons of the substantia nigra. *The Journal of Neuroscience, 19*(3), 1006–1017.

Isaacson, J. S., Solis, J. M., & Nicoll, R. A. (1993). Local and diffuse synaptic actions of GABA in the hippocampus. *Neuron, 10*(2), 165–175.

Isomoto, S., Kondo, C., & Kurachi, Y. (1997). Inwardly rectifying potassium channels: Their molecular heterogeneity and function. *The Japanese Journal of Physiology, 47*(1), 11–39.

Ivanina, T., Rishal, I., Varon, D., Mullner, C., Frohnwieser-Steinecke, B., Schreibmayer, W., et al. (2003). Mapping the Gbetagamma-binding sites in GIRK1 and GIRK2 subunits of the G protein-activated K^+ channel. *The Journal of Biological Chemistry, 278*(31), 29174–29183. http://dx.doi.org/10.1074/jbc.M304518200.

Jelacic, T. M., Kennedy, M. E., Wickman, K., & Clapham, D. E. (2000). Functional and biochemical evidence for G-protein-gated inwardly rectifying K+ (GIRK) channels composed of GIRK2 and GIRK3. *Journal of Biological Chemistry, 275*(46), 36211–36216. http://dx.doi.org/10.1074/jbc.M007087200.

Jelacic, T. M., Sims, S. M., & Clapham, D. E. (1999). Functional expression and characterization of G-protein-gated inwardly rectifying K^+ channels containing GIRK3. *The Journal of Membrane Biology, 169*(2), 123–129.

Jin, W., & Lu, Z. (1998). A novel high-affinity inhibitor for inward-rectifier K+ channels. *Biochemistry, 37*(38), 13291–13299.

Johnson, S. W., & North, R. A. (1992). Opioids excite dopamine neurons by hyperpolarization of local interneurons. *The Journal of Neuroscience, 12*(2), 483–488.

Jones, S., & Bonci, A. (2005). Synaptic plasticity and drug addiction. *Current Opinion in Pharmacology*, *5*(1), 20–25. http://dx.doi.org/10.1016/j.coph.2004.08.011.

Kach, J., Sethakorn, N., & Dulin, N. O. (2012). A finer tuning of G-protein signaling through regulated control of RGS proteins. *American Journal of Physiology. Heart and Circulatory Physiology*, *303*(1), H19–H35. http://dx.doi.org/10.1152/ajpheart.00764.2011.

Kahanovitch, U., Tsemakhovich, V., Berlin, S., Rubinstein, M., Styr, B., Castel, R., et al. (2014). Recruitment of Gbetagamma controls the basal activity of G-protein coupled inwardly rectifying potassium (GIRK) channels: Crucial role of distal C terminus of GIRK1. *The Journal of Physiology*, *592*(Pt 24), 5373–5390. http://dx.doi.org/10.1113/jphysiol.2014.283218.

Kajii, Y., Muraoka, S., Hiraoka, S., Fujiyama, K., Umino, A., & Nishikawa, T. (2003). A developmentally regulated and psychostimulant-inducible novel rat gene mrt1 encoding PDZ-PX proteins isolated in the neocortex. *Molecular Psychiatry*, *8*(4), 434–444. http://dx.doi.org/10.1038/sj.mp.4001258.

Karschin, C., Dissmann, E., Stuhmer, W., & Karschin, A. (1996). IRK(1-3) and GIRK(1-4) inwardly rectifying K^+ channel mRNAs are differentially expressed in the adult rat brain. *The Journal of Neuroscience*, *16*(11), 3559–3570.

Kauer, J. A., & Malenka, R. C. (2007). Synaptic plasticity and addiction. *Nature Reviews. Neuroscience*, *8*(11), 844–858. http://dx.doi.org/10.1038/nrn2234.

Kennedy, M. E., Nemec, J., Corey, S., Wickman, K., & Clapham, D. E. (1999). GIRK4 confers appropriate processing and cell surface localization to G-protein-gated potassium channels. *The Journal of Biological Chemistry*, *274*(4), 2571–2582. http://dx.doi.org/10.1074/jbc.274.4.2571.

Kim, C. S., & Johnston, D. (2015). A1 adenosine receptor-mediated GIRK channels contributes to the resting conductance of CA1 neurons in the dorsal hippocampus. *Journal of Neurophysiology*, *113*(7), 2511–2523. http://dx.doi.org/10.1152/jn.00951.2014.

Kim, E., & Sheng, M. (2004). PDZ domain proteins of synapses. *Nature Reviews. Neuroscience*, *5*(10), 771–781. http://dx.doi.org/10.1038/nrn1517.

Kobayashi, T., Ikeda, K., Kojima, H., Niki, H., Yano, R., Yoshioka, T., et al. (1999). Ethanol opens G-protein-activated inwardly rectifying K^+ channels. *Nature Neuroscience*, *2*(12), 1091–1097. http://dx.doi.org/10.1038/16019.

Kofuji, P., Davidson, N., & Lester, H. A. (1995). Evidence that neuronal G-protein-gated inwardly rectifying K^+ channels are activated by G beta gamma subunits and function as heteromultimers. *Proceedings of the National Academy of Sciences of the United States of America*, *92*(14), 6542–6546.

Koob, G. F., & Volkow, N. D. (2010). Neurocircuitry of addiction. *Neuropsychopharmacology*, *35*(1), 217–238. http://dx.doi.org/10.1038/npp.2009.110.

Kotecki, L., Hearing, M., McCall, N., Marron Fernandez de Velasco, E., Pravetoni, M., Arora, D., et al. (2015). GIRK channels modulate opioid-induced motor activity in a cell type- and subunit-dependent manner. *Journal of Neuroscience*, *35*(18), 7131–7142. http://dx.doi.org/10.1523/JNEUROSCI.5051-14.2015.

Koyrakh, L., Lujan, R., Colon, J., Karschin, C., Kurachi, Y., Karschin, A., et al. (2005). Molecular and cellular diversity of neuronal G-protein-gated potassium channels. *The Journal of Neuroscience*, *25*(49), 11468–11478. http://dx.doi.org/10.1523/JNEUROSCI.3484-05.2005.

Kozell, L. B., Walter, N. A., Milner, L. C., Wickman, K., & Buck, K. J. (2009). Mapping a barbiturate withdrawal lucus to a 0.44 Mb interval and analysis of a novel null mutant identify a role for *Kcnj9* (GIRK3) in withdrawal from pentobarbital, volpidem and ethanol. *The Journal of Neuroscience*, *29*(37), 11662–11673.

Krapivinsky, G., Gordon, E. A., Wickman, K., Velimirovic, B., Krapivinsky, L., & Clapham, D. E. (1995). The G-protein-gated atrial K^+ channel IK_{ACh} is a heteromultimer of two inwardly rectifying K^+-channel proteins. *Nature*, *374*(6518), 135–141. http://dx.doi.org/10.1038/374135a0.

Krapivinsky, G., Krapivinsky, L., Wickman, K., & Clapham, D. E. (1995). G beta gamma binds directly to the G protein-gated K^+ channel, IK_{ACh}. *The Journal of Biological Chemistry*, *270*(49), 29059–29062.

Krettek, J. E., & Price, J. L. (1977). The cortical projections of the mediodorsal nucleus and adjacent thalamic nuclei in the rat. *The Journal of Comparative Neurology*, *171*(2), 157–191. http://dx.doi.org/10.1002/cne.901710204.

Kulik, A., Vida, I., Fukazawa, Y., Guetg, N., Kasugai, Y., Marker, C. L., et al. (2006). Compartment-dependent colocalization of Kir3.2-containing K^+ channels and $GABA_B$ receptors in hippocampal pyramidal cells. *The Journal of Neuroscience*, *26*(16), 4289–4297. http://dx.doi.org/10.1523/JNEUROSCI.4178-05.2006.

Labouebe, G., Lomazzi, M., Cruz, H. G., Creton, C., Lujan, R., Li, M., et al. (2007). RGS2 modulates coupling between $GABA_B$ receptors and GIRK channels in dopamine neurons of the ventral tegmental area. *Nature Neuroscience*, *10*(12), 1559–1568. http://dx.doi.org/10.1038/nn2006.

Lacey, M. G., Mercuri, N. B., & North, R. A. (1987). Dopamine acts on D2 receptors to increase potassium conductance in neurones of the rat substantia nigra zona compacta. *The Journal of Physiology*, *392*, 397–416.

Ladera, C., del Carmen Godino, M., Jose Cabanero, M., Torres, M., Watanabe, M., Lujan, R., et al. (2008). Pre-synaptic GABA receptors inhibit glutamate release through GIRK channels in rat cerebral cortex. *Journal of Neurochemistry*, *107*(6), 1506–1517. http://dx.doi.org/10.1111/j.1471-4159.2008.05712.x.

Lalive, A. L., Munoz, M. B., Bellone, C., Slesinger, P. A., Luscher, C., & Tan, K. R. (2014). Firing modes of dopamine neurons drive bidirectional GIRK channel plasticity. *The Journal of Neuroscience*, *34*(15), 5107–5114. http://dx.doi.org/10.1523/JNEUROSCI.5203-13.2014.

Lammel, S., Hetzel, A., Hackel, O., Jones, I., Liss, B., & Roeper, J. (2008). Unique properties of mesoprefrontal neurons within a dual mesocorticolimbic dopamine system. *Neuron*, *57*(5), 760–773. http://dx.doi.org/10.1016/j.neuron.2008.01.022.

Lammel, S., Ion, D. I., Roeper, J., & Malenka, R. C. (2011). Projection-specific modulation of dopamine neuron synapses by aversive and rewarding stimuli. *Neuron*, *70*(5), 855–862. http://dx.doi.org/10.1016/j.neuron.2011.03.025.

Lammel, S., Lim, B. K., & Malenka, R. C. (2014). Reward and aversion in a heterogeneous midbrain dopamine system. *Neuropharmacology*, *76*, 351–359. http://dx.doi.org/10.1016/j.neuropharm.2013.03.019.

Lammel, S., Lim, B. K., Ran, C., Huang, K. W., Betley, M. J., Tye, K. M., et al. (2012). Input-specific control of reward and aversion in the ventral tegmental area. *Nature*, *491*(7423), 212–217. http://dx.doi.org/10.1038/nature11527.

Lavine, N., Ethier, N., Oak, J. N., Pei, L., Liu, F., Trieu, P., et al. (2002). G protein-coupled receptors form stable complexes with inwardly rectifying potassium channels and adenylyl cyclase. *The Journal of Biological Chemistry*, *277*(48), 46010–46019. http://dx.doi.org/10.1074/jbc.M205035200.

Lecca, S., Melis, M., Luchicchi, A., Muntoni, A. L., & Pistis, M. (2012). Inhibitory inputs from rostromedial tegmental neurons regulate spontaneous activity of midbrain dopamine cells and their responses to drugs of abuse. *Neuropsychopharmacology*, *37*(5), 1164–1176. http://dx.doi.org/10.1038/npp.2011.302.

Lesage, F., Duprat, F., Fink, M., Guillemare, E., Coppola, T., Lazdunski, M., et al. (1994). Cloning provides evidence for a family of inward rectifier and G-protein coupled K^+ channels in the brain. *FEBS Letters*, *353*(1), 37–42.

Lesage, F., Guillemare, E., Fink, M., Duprat, F., Heurteaux, C., Fosset, M., et al. (1995). Molecular properties of neuronal G-protein-activated inwardly rectifying K^+ channels. *The Journal of Biological Chemistry*, *270*(48), 28660–28667.

Liao, Y. J., Jan, Y. N., & Jan, L. Y. (1996). Heteromultimerization of G-protein-gated inwardly rectifying K^+ channel proteins GIRK1 and GIRK2 and their altered expression in weaver brain. *Journal of Neuroscience, 16*(22), 7137–7150.

Lober, R. M., Pereira, M. A., & Lambert, N. A. (2006). Rapid activation of inwardly rectifying potassium channels by immobile G-protein-coupled receptors. *The Journal of Neuroscience, 26*(48), 12602–12608. http://dx.doi.org/10.1523/jneurosci.4020-06.2006.

Logothetis, D. E., Kurachi, Y., Galper, J., Neer, E. J., & Clapham, D. E. (1987). The beta gamma subunits of GTP-binding proteins activate the muscarinic K^+ channel in heart. *Nature, 325*(6102), 321–326. http://dx.doi.org/10.1038/325321a0.

Lomazzi, M., Slesinger, P. A., & Luscher, C. (2008). Addictive drugs modulate GIRK-channel signaling by regulating RGS proteins. *Trends in Pharmacological Sciences, 29*(11), 544–549. http://dx.doi.org/10.1016/j.tips.2008.07.011.

Lu, H., Cheng, P. L., Lim, B. K., Khoshnevisrad, N., & Poo, M. M. (2010). Elevated BDNF after cocaine withdrawal facilitates LTP in medial prefrontal cortex by suppressing GABA inhibition. *Neuron, 67*(5), 821–833. http://dx.doi.org/10.1016/j.neuron.2010.08.012.

Lujan, R., Marron Fernandez de Velasco, E., Aguado, C., & Wickman, K. (2014). New insights into the therapeutic potential of Girk channels. *Trends in Neurosciences, 37*(1), 20–29. http://dx.doi.org/10.1016/j.tins.2013.10.006.

Lujan, R., Maylie, J., & Adelman, J. P. (2009). New sites of action for GIRK and SK channels. *Nature Reviews. Neuroscience, 10*(7), 475–480. http://dx.doi.org/10.1038/nrn2668.

Lunn, M. L., Nassirpour, R., Arrabit, C., Tan, J., McLeod, I., Arias, C. M., et al. (2007). A unique sorting nexin regulates trafficking of potassium channels via a PDZ domain interaction. *Nature Neuroscience, 10*(10), 1249–1259. http://dx.doi.org/10.1038/nn1953.

Luscher, C. (2013). Drug-evoked synaptic plasticity causing addictive behavior. *The Journal of Neuroscience, 33*(45), 17641–17646. http://dx.doi.org/10.1523/JNEUROSCI.3406-13.2013.

Luscher, C., Jan, L. Y., Stoffel, M., Malenka, R. C., & Nicoll, R. A. (1997). G protein-coupled inwardly rectifying K^+ channels (GIRKs) mediate postsynaptic but not presynaptic transmitter actions in hippocampal neurons. *Neuron, 19*(3), 687–695.

Luscher, C., & Malenka, R. C. (2011). Drug-evoked synaptic plasticity in addiction: From molecular changes to circuit remodeling. *Neuron, 69*(4), 650–663. http://dx.doi.org/10.1016/j.neuron.2011.01.017.

Luscher, C., & Slesinger, P. A. (2010). Emerging roles for G protein-gated inwardly rectifying potassium (GIRK) channels in health and disease. *Nature Reviews. Neuroscience, 11*(5), 301–315. http://dx.doi.org/10.1038/nrn2834.

Luthi, A., & Luscher, C. (2014). Pathological circuit function underlying addiction and anxiety disorders. *Nature Neuroscience, 17*(12), 1635–1643. http://dx.doi.org/10.1038/nn.3849.

Ma, D., Zerangue, N., Raab-Graham, K., Fried, S. R., Jan, Y. N., & Jan, L. Y. (2002). Diverse trafficking patterns due to multiple traffic motifs in G protein-activated inwardly rectifying potassium channels from brain and heart. *Neuron, 33*(5), 715–729.

Maity, B., Stewart, A., Yang, J., Loo, L., Sheff, D., Shepherd, A. J., et al. (2012). Regulator of G protein signaling 6 (RGS6) protein ensures coordination of motor movement by modulating $GABA_B$ receptor signaling. *The Journal of Biological Chemistry, 287*(7), 4972–4981. http://dx.doi.org/10.1074/jbc.M111.297218.

Malinow, R. (2003). AMPA receptor trafficking and long-term potentiation. *Philosophical Transactions of the Royal Society of London. Series B, Biological Sciences, 358*(1432), 707–714. http://dx.doi.org/10.1098/rstb.2002.1233.

Marchese, A., Paing, M. M., Temple, B. R., & Trejo, J. (2008). G protein-coupled receptor sorting to endosomes and lysosomes. *Annual Review of Pharmacology and Toxicology, 48*, 601–629. http://dx.doi.org/10.1146/annurev.pharmtox.48.113006.094646.

Marcott, P. F., Mamaligas, A. A., & Ford, C. P. (2014). Phasic dopamine release drives rapid activation of striatal D2-receptors. *Neuron, 84*(1), 164–176. http://dx.doi.org/10.1016/j.neuron.2014.08.058.

Marinelli, M., Cooper, D. C., Baker, L. K., & White, F. J. (2003). Impulse activity of midbrain dopamine neurons modulates drug-seeking behavior. *Psychopharmacology, 168*(1-2), 84–98. http://dx.doi.org/10.1007/s00213-003-1491-1.

Marker, C. L., Stoffel, M., & Wickman, K. (2004). Spinal G-protein-gated K^+ channels formed by GIRK1 and GIRK2 subunits modulate thermal nociception and contribute to morphine analgesia. *The Journal of Neuroscience, 24*(11), 2806–2812. http://dx.doi.org/10.1523/JNEUROSCI.5251-03.2004.

Marron Fernandez de Velasco, E., Hearing, M., Xia, Z., Victoria, N. C., Lujan, R., & Wickman, K. (2015). Sex differences in $GABA_BR$-GIRK signaling in layer 5/6 pyramidal neurons of the mouse prelimbic cortex. *Neuropharmacology, 95*, 353–360. http://dx.doi.org/10.1016/j.neuropharm.2015.03.029.

Martellotta, M. C., Fattore, L., Cossu, G., & Fratta, W. (1997). Rewarding properties of gamma-hydroxybutyric acid: An evaluation through place preference paradigm. *Psychopharmacology, 132*(1), 1–5.

Mase, Y., Yokogawa, M., Osawa, M., & Shimada, I. (2012). Structural basis for modulation of gating property of G protein-gated inwardly rectifying potassium ion channel (GIRK) by i/o-family G protein alpha subunit ($Galpha_{i/o}$). *The Journal of Biological Chemistry, 287*(23), 19537–19549. http://dx.doi.org/10.1074/jbc.M112.353888.

Matsui, A., Jarvie, B. C., Robinson, B. G., Hentges, S. T., & Williams, J. T. (2014). Separate GABA Afferents to Dopamine Neurons Mediate Acute Action of Opioids, Development of Tolerance, and Expression of Withdrawal. *Neuron, 82*(6), 1346–1356. http://dx.doi.org/10.1016/j.neuron.2014.04.030.

McFarland, K., & Kalivas, P. W. (2001). The circuitry mediating cocaine-induced reinstatement of drug-seeking behavior. *The Journal of Neuroscience, 21*(21), 8655–8663.

McFarland, K., Lapish, C. C., & Kalivas, P. W. (2003). Prefrontal glutamate release into the core of the nucleus accumbens mediates cocaine-induced reinstatement of drug-seeking behavior. *The Journal of Neuroscience, 23*(8), 3531–3537.

Mitrovic, I., Margeta-Mitrovic, M., Bader, S., Stoffel, M., Jan, L. Y., & Basbaum, A. I. (2003). Contribution of GIRK2-mediated postsynaptic signaling to opiate and alpha 2-adrenergic analgesia and analgesic sex differences. *Proceedings of the National Academy of Sciences of the United States of America, 100*(1), 271–276. http://dx.doi.org/10.1073/pnas.0136822100.

Morales, M., & Root, D. H. (2014). Glutamate neurons within the midbrain dopamine regions. *Neuroscience, 282c*, 60–68. http://dx.doi.org/10.1016/j.neuroscience.2014.05.032.

Morgan, A. D., Carroll, M. E., Loth, A. K., Stoffel, M., & Wickman, K. (2003). Decreased cocaine self-administration in Kir3 potassium channel subunit knockout mice. *Neuropsychopharmacology, 28*(5), 932–938. http://dx.doi.org/10.1038/sj.npp.1300100.

Müllner, C., Steinecker, B., Gorischek, A., & Schreibmayer, W. (2009). Identification of the structural determinant responsible for the phosphorylation of G-protein activated potassium channel 1 by cAMP-dependent protein kinase: PKA phosphorylation of GIRK1. *FEBS Journal, 276*(21), 6218–6226. http://dx.doi.org/10.1111/j.1742-4658.2009.07325.x.

Munoz, M. B., & Slesinger, P. A. (2014). Sorting nexin 27 regulation of G protein-gated inwardly rectifying K^+ channels attenuates *in vivo* cocaine response. *Neuron, 82*(3), 659–669. http://dx.doi.org/10.1016/j.neuron.2014.03.011.

Nakahara, H., Itoh, H., Kawagoe, R., Takikawa, Y., & Hikosaka, O. (2004). Dopamine neurons can represent context-dependent prediction error. *Neuron, 41*(2), 269–280.

Nassirpour, R., Bahima, L., Lalive, A. L., Luscher, C., Lujan, R., & Slesinger, P. A. (2010). Morphine- and CaMKII-dependent enhancement of GIRK channel signaling in hippocampal neurons. *Journal of Neuroscience*, *30*(40), 13419–13430. http://dx.doi.org/10.1523/JNEUROSCI.2966-10.2010.

Nassirpour, R., & Slesinger, P. A. (2007). Subunit-specific regulation of Kir3 channels by sorting nexin 27. *Channels (Austin, Tex.)*, *1*(5), 331–333.

Nehring, R. B., Wischmeyer, E., Doring, F., Veh, R. W., Sheng, M., & Karschin, A. (2000). Neuronal inwardly rectifying K^+ channels differentially couple to PDZ proteins of the PSD-95/SAP90 family. *The Journal of Neuroscience*, *20*(1), 156–162.

Nelson, C. S., Marino, J. L., & Allen, C. N. (1997). Cloning and characterization of Kir3.1 (GIRK1) C-terminal alternative splice variants. *Brain Research. Molecular Brain Research*, *46*(1-2), 185–196.

Nestler, E. J. (2001). Molecular basis of long-term plasticity underlying addiction. *Nature Reviews. Neuroscience*, *2*(2), 119–128. http://dx.doi.org/10.1038/35053570.

Neubig, R. R., & Siderovski, D. P. (2002). Regulators of G-protein signalling as new central nervous system drug targets. *Nature Reviews. Drug Discovery*, *1*(3), 187–197. http://dx.doi.org/10.1038/nrd747.

Neve, K. A., Seamans, J. K., & Trantham-Davidson, H. (2004). Dopamine receptor signaling. *Journal of Receptor and Signal Transduction Research*, *24*(3), 165–205.

Newberry, N. R., & Nicoll, R. A. (1985). Comparison of the action of baclofen with gamma-aminobutyric acid on rat hippocampal pyramidal cells in vitro. *The Journal of Physiology*, *360*, 161–185.

Nicoll, R. A. (2003). Expression mechanisms underlying long-term potentiation: A postsynaptic view. *Philosophical Transactions of the Royal Society of London. Series B, Biological Sciences*, *358*(1432), 721–726. http://dx.doi.org/10.1098/rstb.2002.1228.

Niehaus, J. L., Murali, M., & Kauer, J. A. (2010). Drugs of abuse and stress impair LTP at inhibitory synapses in the ventral tegmental area. *The European Journal of Neuroscience*, *32*(1), 108–117. http://dx.doi.org/10.1111/j.1460-9568.2010.07256.x.

Nugent, F. S., Penick, E. C., & Kauer, J. A. (2007). Opioids block long-term potentiation of inhibitory synapses. *Nature*, *446*(7139), 1086–1090. http://dx.doi.org/10.1038/nature05726.

Olive, M. F., & Newton, P. M. (2010). Protein kinase C isozymes as regulators of sensitivity to and self-administration of drugs of abuse-studies with genetically modified mice. *Behavioural Pharmacology*, *21*(5-6), 493–499. http://dx.doi.org/10.1097/FBP.0b013e32833d8bb7.

Ostrovskaya, O., Xie, K., Masuho, I., Fajardo-Serrano, A., Lujan, R., Wickman, K., et al. (2014). RGS7/Gbeta5/R7BP complex regulates synaptic plasticity and memory by modulating hippocampal $GABA_BR$-GIRK signaling. *eLife*, *3*. http://dx.doi.org/10.7554/eLife.02053, e02053.

Ouachikh, O., Dieb, W., Durif, F., & Hafidi, A. (2013). Differential behavioral reinforcement effects of dopamine receptor agonists in the rat with bilateral lesion of the posterior ventral tegmental area. *Behavioural Brain Research*, *252*, 24–31. http://dx.doi.org/10.1016/j.bbr.2013.05.042.

Padgett, C. L., Lalive, A. L., Tan, K. R., Terunuma, M., Munoz, M. B., Pangalos, M. N., et al. (2012). Methamphetamine-evoked depression of $GABA_B$ receptor signaling in GABA neurons of the VTA. *Neuron*, *73*(5), 978–989. http://dx.doi.org/10.1016/j.neuron.2011.12.031.

Pascoli, V., Turiault, M., & Luscher, C. (2012). Reversal of cocaine-evoked synaptic potentiation resets drug-induced adaptive behaviour. *Nature*, *481*(7379), 71–75. http://dx.doi.org/10.1038/nature10709.

Peleg, S., Varon, D., Ivanina, T., Dessauer, C. W., & Dascal, N. (2002). $Galpha_i$ controls the gating of the G protein-activated K^+ channel, GIRK. *Neuron*, *33*(1), 87–99.

Perry, C. A., Pravetoni, M., Teske, J. A., Aguado, C., Erickson, D. J., Medrano, J. F., et al. (2008). Predisposition to late-onset obesity in GIRK4 knockout mice. *Proceedings of the National Academy of Sciences of the United States of America*, *105*(23), 8148–8153. http://dx.doi.org/10.1073/pnas.0803261105.

Pfaffinger, P. J., Martin, J. M., Hunter, D. D., Nathanson, N. M., & Hille, B. (1985). GTP-binding proteins couple cardiac muscarinic receptors to a K channel. *Nature*, *317*(6037), 536–538.

Pietersen, A. N., Lancaster, D. M., Patel, N., Hamilton, J. B., & Vreugdenhil, M. (2009). Modulation of gamma oscillations by endogenous adenosine through A1 and A2A receptors in the mouse hippocampus. *Neuropharmacology*, *56*(2), 481–492. http://dx.doi.org/10.1016/j.neuropharm.2008.10.001.

Ponting, C. P. (1996). Novel domains in NADPH oxidase subunits, sorting nexins, and PtdIns 3-kinases: Binding partners of SH3 domains? *Protein Science*, *5*(11), 2353–2357. http://dx.doi.org/10.1002/pro.5560051122.

Pravetoni, M., & Wickman, K. (2008). Behavioral characterization of mice lacking GIRK/Kir3 channel subunits. *Genes, Brain, and Behavior*, 7(5), 523–531. http://dx.doi.org/10.1111/j.1601-183X.2008.00388.x.

Rahman, Z., Schwarz, J., Gold, S. J., Zachariou, V., Wein, M. N., Choi, K. H., et al. (2003). RGS9 modulates dopamine signaling in the basal ganglia. *Neuron*, *38*(6), 941–952.

Reuveny, E., Slesinger, P. A., Inglese, J., Morales, J. M., Iniguez-Lluhi, J. A., Lefkowitz, R. J., et al. (1994). Activation of the cloned muscarinic potassium channel by G protein beta gamma subunits. *Nature*, *370*(6485), 143–146. http://dx.doi.org/10.1038/370143a0.

Rishal, I., Porozov, Y., Yakubovich, D., Varon, D., & Dascal, N. (2005). Gbetagamma-dependent and Gbetagamma-independent basal activity of G protein-activated K^+ channels. *The Journal of Biological Chemistry*, *280*(17), 16685–16694. http://dx.doi.org/10.1074/jbc.M412196200.

Ross, E. M., & Wilkie, T. M. (2000). GTPase-activating proteins for heterotrimeric G proteins: Regulators of G protein signaling (RGS) and RGS-like proteins. *Annual Review of Biochemistry*, *69*, 795–827. http://dx.doi.org/10.1146/annurev.biochem.69.1.795.

Rubinstein, M., Peleg, S., Berlin, S., Brass, D., & Dascal, N. (2007). G_{i3} primes the G protein-activated K^+ channels for activation by coexpressed Gbeta in intact *Xenopus* oocytes. *The Journal of Physiology*, *581*(1), 17–32. http://dx.doi.org/10.1113/jphysiol.2006.125864.

Rubinstein, M., Peleg, S., Berlin, S., Brass, D., Keren-Raifman, T., Dessauer, C. W., et al. (2009). Divergent regulation of GIRK1 and GIRK2 subunits of the neuronal G protein gated K^+ channel by $Galpha_i$GDP and Gbetagamma. *The Journal of Physiology*, *587*(Pt 14), 3473–3491. http://dx.doi.org/10.1113/jphysiol.2009.173229.

Rusinova, R., Shen, Y. M., Dolios, G., Padovan, J., Yang, H., Kirchberger, M., et al. (2009). Mass spectrometric analysis reveals a functionally important PKA phosphorylation site in a Kir3 channel subunit. *Pflügers Archiv*, *458*(2), 303–314. http://dx.doi.org/10.1007/s00424-008-0628-9.

Saal, D., Dong, Y., Bonci, A., & Malenka, R. C. (2003). Drugs of abuse and stress trigger a common synaptic adaptation in dopamine neurons. *Neuron*, *37*(4), 577–582.

Saenz del Burgo, L., Cortes, R., Mengod, G., Zarate, J., Echevarria, E., & Salles, J. (2008). Distribution and neurochemical characterization of neurons expressing GIRK channels in the rat brain. *The Journal of Comparative Neurology*, *510*(6), 581–606. http://dx.doi.org/10.1002/cne.21810.

Scanziani, M. (2000). GABA spillover activates postsynaptic $GABA_B$ receptors to control rhythmic hippocampal activity. *Neuron*, *25*(3), 673–681.

Schultz, W. (2002). Getting formal with dopamine and reward. *Neuron*, *36*(2), 241–263.

Schultz, W., Dayan, P., & Montague, P. R. (1997). A neural substrate of prediction and reward. *Science*, *275*(5306), 1593–1599.

Sesack, S. R., Deutch, A. Y., Roth, R. H., & Bunney, B. S. (1989). Topographical organization of the efferent projections of the medial prefrontal cortex in the rat: An anterograde tract-tracing study with *Phaseolus vulgaris* leucoagglutinin. *The Journal of Comparative Neurology, 290*(2), 213–242. http://dx.doi.org/10.1002/cne.902900205.

Sesack, S. R., & Grace, A. A. (2010). Cortico-Basal Ganglia reward network: Microcircuitry. *Neuropsychopharmacology, 35*(1), 27–47. http://dx.doi.org/10.1038/npp.2009.93.

Sesack, S. R., & Pickel, V. M. (1992). Prefrontal cortical efferents in the rat synapse on unlabeled neuronal targets of catecholamine terminals in the nucleus accumbens septi and on dopamine neurons in the ventral tegmental area. *The Journal of Comparative Neurology, 320*(2), 145–160. http://dx.doi.org/10.1002/cne.903200202.

Sethakorn, N., Yau, D. M., & Dulin, N. O. (2010). Non-canonical functions of RGS proteins. *Cellular Signalling, 22*(9), 1274–1281. http://dx.doi.org/10.1016/j.cellsig.2010.03.016.

Seutin, V., Johnson, S. W., & North, R. A. (1994). Effect of dopamine and baclofen on N-methyl-D-aspartate-induced burst firing in rat ventral tegmental neurons. *Neuroscience, 58*(1), 201–206.

Sharpe, A. L., Varela, E., Bettinger, L., & Beckstead, M. J. (2014). Methamphetamine self-administration in mice decreases GIRK channel-mediated currents in midbrain dopamine neurons. *The International Journal of Neuropsychopharmacology*. http://dx.doi.org/10.1093/ijnp/pyu073.

Sheng, M., & Kim, M. J. (2002). Postsynaptic signaling and plasticity mechanisms. *Science, 298*(5594), 776–780. http://dx.doi.org/10.1126/science.1075333.

Signorini, S., Liao, Y. J., Duncan, S. A., Jan, L. Y., & Stoffel, M. (1997). Normal cerebellar development but susceptibility to seizures in mice lacking G protein-coupled, inwardly rectifying K^+ channel GIRK2. *Proceedings of the National Academy of Sciences of the United States of America, 94*(3), 923–927.

Slesinger, P. A., Patil, N., Liao, Y. J., Jan, Y. N., Jan, L. Y., & Cox, D. R. (1996). Functional effects of the mouse weaver mutation on G protein-gated inwardly rectifying K^+ channels. *Neuron, 16*(2), 321–331.

Slesinger, P. A., Reuveny, E., Jan, Y. N., & Jan, L. Y. (1995). Identification of structural elements involved in G protein gating of the GIRK1 potassium channel. *Neuron, 15*(5), 1145–1156.

Slesinger, P. A., Stoffel, M., Jan, Y. N., & Jan, L. Y. (1997). Defective gamma-aminobutyric acid type B receptor-activated inwardly rectifying K^+ currents in cerebellar granule cells isolated from weaver and Girk2 null mutant mice. *Proceedings of the National Academy of Sciences of the United States of America, 94*(22), 12210–12217.

Sondek, J., & Siderovski, D. P. (2001). Ggamma-like (GGL) domains: New frontiers in G-protein signaling and beta-propeller scaffolding. *Biochemical Pharmacology, 61*(11), 1329–1337.

Steffensen, S. C., Stobbs, S. H., Colago, E. E., Lee, R. S., Koob, G. F., Gallegos, R. A., et al. (2006). Contingent and non-contingent effects of heroin on mu-opioid receptor-containing ventral tegmental area GABA neurons. *Experimental Neurology, 202*(1), 139–151. http://dx.doi.org/10.1016/j.expneurol.2006.05.023.

Steinecker, B., Rosker, C., & Schreibmayer, W. (2007). The GIRK1 brain variant GIRK1d and its functional impact on heteromultimeric GIRK channels. *Journal of Receptor and Signal Transduction Research, 27*(5-6), 369–382. http://dx.doi.org/10.1080/10799890701713073.

Steketee, J. D., & Kalivas, P. W. (2011). Drug wanting: Behavioral sensitization and relapse to drug-seeking behavior. *Pharmacological Reviews, 63*(2), 348–365. http://dx.doi.org/10.1124/pr.109.001933.

Stewart, A., Maity, B., Anderegg, S. P., Allamargot, C., Yang, J., & Fisher, R. A. (2015). Regulator of G protein signaling 6 is a critical mediator of both reward-related behavioral and pathological responses to alcohol. *Proceedings of the National Academy of Sciences of the United States of America, 112*(7), E786–E795. http://dx.doi.org/10.1073/pnas.1418795112.

Stuber, G. D., Hnasko, T. S., Britt, J. P., Edwards, R. H., & Bonci, A. (2010). Dopaminergic terminals in the nucleus accumbens but not the dorsal striatum corelease glutamate. *The Journal of Neuroscience, 30*(24), 8229–8233. http://dx.doi.org/10.1523/jneurosci.1754-10.2010.

Sugita, S., Johnson, S. W., & North, R. A. (1992). Synaptic inputs to $GABA_A$ and $GABA_B$ receptors originate from discrete afferent neurons. *Neuroscience Letters, 134*(2), 207–211.

Sui, J. L., Petit-Jacques, J., & Logothetis, D. E. (1998). Activation of the atrial K_{ACh} channel by the betagamma subunits of G proteins or intracellular Na^+ ions depends on the presence of phosphatidylinositol phosphates. *Proceedings of the National Academy of Sciences of the United States of America, 95*(3), 1307–1312.

Sun, Q. Q., Huguenard, J. R., & Prince, D. A. (2002). Somatostatin inhibits thalamic network oscillations in vitro: actions on the GABAergic neurons of the reticular nucleus. *Journal of Neuroscience, 22*(13), 5374–5386.

Taber, M. T., Das, S., & Fibiger, H. C. (1995). Cortical regulation of subcortical dopamine release: Mediation via the ventral tegmental area. *Journal of Neurochemistry, 65*(3), 1407–1410.

Taber, M. T., & Fibiger, H. C. (1995). Electrical stimulation of the prefrontal cortex increases dopamine release in the nucleus accumbens of the rat: Modulation by metabotropic glutamate receptors. *The Journal of Neuroscience, 15*(5 Pt 2), 3896–3904.

Tan, K. R., Yvon, C., Turiault, M., Mirzabekov, J. J., Doehner, J., Labouebe, G., et al. (2012). GABA neurons of the VTA drive conditioned place aversion. *Neuron, 73*(6), 1173–1183. http://dx.doi.org/10.1016/j.neuron.2012.02.015.

Temkin, P., Lauffer, B., Jager, S., Cimermancic, P., Krogan, N. J., & von Zastrow, M. (2011). SNX27 mediates retromer tubule entry and endosome-to-plasma membrane trafficking of signalling receptors. *Nature Cell Biology, 13*(6), 715–721. http://dx.doi.org/10.1038/ncb2252.

Terunuma, M., Pangalos, M. N., & Moss, S. J. (2010). Functional modulation of $GABA_B$ receptors by protein kinases and receptor trafficking. *Advances in Pharmacology, 58*, 113–122, Elsevier.

Tesmer, J. J., Berman, D. M., Gilman, A. G., & Sprang, S. R. (1997). Structure of RGS4 bound to AlF4-activated Gialpha1: Stabilization of the transition state for GTP hydrolysis. *Cell, 89*(2), 251–261.

Thomas, M. J., Kalivas, P. W., & Shaham, Y. (2008). Neuroplasticity in the mesolimbic dopamine system and cocaine addiction. *British Journal of Pharmacology, 154*(2), 327–342. http://dx.doi.org/10.1038/bjp.2008.77.

Thomas, M. J., & Malenka, R. C. (2003). Synaptic plasticity in the mesolimbic dopamine system. *Philosophical Transactions of the Royal Society of London. Series B, Biological Sciences, 358*(1432), 815–819. http://dx.doi.org/10.1098/rstb.2002.1236.

Torrecilla, M., Marker, C. L., Cintora, S. C., Stoffel, M., Williams, J. T., & Wickman, K. (2002). G-protein-gated potassium channels containing Kir3.2 and Kir3.3 subunits mediate the acute inhibitory effects of opioids on locus ceruleus neurons. *Journal of Neuroscience, 22*(11), 4328–4334.

Traynor, J. (2010). Regulator of G protein-signaling proteins and addictive drugs. *Annals of the New York Academy of Sciences, 1187*, 341–352. http://dx.doi.org/10.1111/j.1749-6632.2009.05150.x.

Ungless, M. A., Whistler, J. L., Malenka, R. C., & Bonci, A. (2001). Single cocaine exposure *in vivo* induces long-term potentiation in dopamine neurons. *Nature, 411*(6837), 583–587. http://dx.doi.org/10.1038/35079077.

Van Bockstaele, E. J., & Pickel, V. M. (1995). GABA-containing neurons in the ventral tegmental area project to the nucleus accumbens in rat brain. *Brain Research*, *682*(1-2), 215–221.

van Huijstee, A. N., & Mansvelder, H. D. (2014). Glutamatergic synaptic plasticity in the mesocorticolimbic system in addiction. *Frontiers in Cellular Neuroscience*, *8*, 466. http://dx.doi.org/10.3389/fncel.2014.00466.

van Zessen, R., Phillips, J. L., Budygin, E. A., & Stuber, G. D. (2012). Activation of VTA GABA neurons disrupts reward consumption. *Neuron*, *73*(6), 1184–1194. http://dx.doi.org/10.1016/j.neuron.2012.02.016.

Wagner, V., Stadelmeyer, E., Riederer, M., Regitnig, P., Gorischek, A., Devaney, T., et al. (2010). Cloning and characterisation of GIRK1 variants resulting from alternative RNA editing of the *KCNJ3* gene transcript in a human breast cancer cell line. *Journal of Cellular Biochemistry*, *110*(3), 598–608. http://dx.doi.org/10.1002/jcb.22564.

Walsh, J. J., & Han, M. H. (2014). The heterogeneity of ventral tegmental area neurons: Projection functions in a mood-related context. *Neuroscience*, *282C*, 101–108. http://dx.doi.org/10.1016/j.neuroscience.2014.06.006.

Wei, J., Hodes, M. E., Piva, R., Feng, Y., Wang, Y., Ghetti, B., et al. (1998). Characterization of murine Girk2 transcript isoforms: Structure and differential expression. *Genomics*, *51*(3), 379–390. http://dx.doi.org/10.1006/geno.1998.5369.

Whorton, M. R., & MacKinnon, R. (2011). Crystal structure of the mammalian GIRK2 K^+ channel and gating regulation by G proteins, PIP_2, and sodium. *Cell*, *147*(1), 199–208. http://dx.doi.org/10.1016/j.cell.2011.07.046.

Whorton, M. R., & MacKinnon, R. (2013). X-ray structure of the mammalian GIRK2-betagamma G-protein complex. *Nature*, *498*(7453), 190–197. http://dx.doi.org/10.1038/nature12241.

Wickman, K., Iniguez-Lluhl, J. A., Davenport, P. A., Taussig, R., Krapivinsky, G. B., Linder, M. E., et al. (1994). Recombinant G-protein beta gamma-subunits activate the muscarinic-gated atrial potassium channel. *Nature*, *368*(6468), 255–257. http://dx.doi.org/10.1038/368255a0.

Wickman, K., Nemec, J., Gendler, S. J., & Clapham, D. E. (1998). Abnormal heart rate regulation in *GIRK4* knockout mice. *Neuron*, *20*(1), 103–114.

Worby, C. A., & Dixon, J. E. (2002). Sorting out the cellular functions of sorting nexins. *Nature Reviews. Molecular Cell Biology*, *3*(12), 919–931. http://dx.doi.org/10.1038/nrm974.

Xie, K., Allen, K. L., Kourrich, S., Colon-Saez, J., Thomas, M. J., Wickman, K., et al. (2010). Gbeta5 recruits R7 RGS proteins to GIRK channels to regulate the timing of neuronal inhibitory signaling. *Nature Neuroscience*, *13*(6), 661–663. http://dx.doi.org/10.1038/nn.2549.

Xu, C., Zhang, W., Rondard, P., Pin, J. P., & Liu, J. (2014). Complex $GABA_B$ receptor complexes: How to generate multiple functionally distinct units from a single receptor. *Frontiers in Pharmacology*, *5*, 12. http://dx.doi.org/10.3389/fphar.2014.00012.

Yamada, M., Inanobe, A., & Kurachi, Y. (1998). G protein regulation of potassium ion channels. *Pharmacological Reviews*, *50*(4), 723–760.

Zachariou, V., Georgescu, D., Sanchez, N., Rahman, Z., DiLeone, R., Berton, O., et al. (2003). Essential role for RGS9 in opiate action. *Proceedings of the National Academy of Sciences of the United States of America*, *100*(23), 13656–13661. http://dx.doi.org/10.1073/pnas.2232594100.

Zhang, H., He, C., Yan, X., Mirshahi, T., & Logothetis, D. E. (1999). Activation of inwardly rectifying K^+ channels by distinct PtdIns(4,5)P2 interactions. *Nature Cell Biology*, *1*(3), 183–188. http://dx.doi.org/10.1038/11103.

Zhou, J. Y., Siderovski, D. P., & Miller, R. J. (2000). Selective regulation of N-type Ca channels by different combinations of G-protein beta/gamma subunits and RGS proteins. *The Journal of Neuroscience*, *20*(19), 7143–7148.

Zhu, L., Wu, X., Wu, M. B., Chan, K. W., Logothetis, D. E., & Thornhill, W. B. (2001). Cloning and characterization of G protein-gated inward rectifier K^+ channel (GIRK1) isoforms from heart and brain. *Journal of Molecular Neuroscience*, *16*(1), 21–32. http://dx.doi.org/10.1385/jmn:16:1:21.

CHAPTER SEVEN

GIRK Channels: A Potential Link Between Learning and Addiction

Megan E. Tipps[*,†,1], **Kari J. Buck**[*,‡]
[*]Department of Behavioral Neuroscience, Oregon Health & Science University, Portland, Oregon, USA
[†]Department of Pharmacology, University of Minnesota, Minneapolis, Minnesota, USA
[‡]Portland Alcohol Research Center, Portland VA Medical Center, Portland, Oregon, USA
[1]Corresponding author: e-mail address: metipps@gmail.com

Contents

1. Introduction 240
2. GIRK Channels 241
   2.1 Neuronal GIRK Channels 241
   2.2 Modulators of GIRK Function 242
3. GIRK Channel Modulation by Drugs of Abuse 245
   3.1 Drugs That Activate $G_{i/o}$-Coupled GPCRs 247
   3.2 Drug Effects on Intracellular Modulators of GIRK Signaling 248
   3.3 Ethanol: A Direct Activator of GIRK Channels 249
4. Neuronal Plasticity 251
   4.1 GIRK Channels and Neuronal Plasticity 251
   4.2 GIRK Channels and Cognitive Dysfunction 256
   4.3 Drug-Induced Changes in GIRK Function 259
5. Summary and Future Directions 262
References 263

Abstract

The ability of drug-associated cues to reinitiate drug craving and seeking, even after long periods of abstinence, has led to the hypothesis that addiction represents a form of pathological learning, in which drugs of abuse hijack normal learning and memory processes to support long-term addictive behaviors. In this chapter, we review evidence suggesting that G protein-gated inwardly rectifying potassium (GIRK/Kir3) channels are one mechanism through which numerous drugs of abuse can modulate learning and memory processes. We will examine the role of GIRK channels in two forms of experience-dependent long-term changes in neuronal function: homeostatic plasticity and synaptic plasticity. We will also discuss how drug-induced changes in GIRK-mediated signaling can lead to changes that support the development and maintenance of addiction.

International Review of Neurobiology, Volume 123
ISSN 0074-7742
http://dx.doi.org/10.1016/bs.irn.2015.05.012

1. INTRODUCTION

Addiction is a chronic relapsing disorder that manifests as compulsive drug seeking and taking despite significant adverse consequences (Everitt, 2014; McLellan, Lewis, O'Brien, & Kleber, 2000). Drug cravings can persist even after long periods of abstinence, and one of the hallmarks of addiction is the cycle of drug use, attempted abstinence, and relapse. These key behavioral aspects are common across all drugs of abuse, despite the diverse mechanisms of action for different drug classes (Nestler, 2005). The ability of a wide range of drugs with a variety of molecular targets to generate such a similar behavioral profile suggests that a common set of neural processes altered by drugs of abuse support the state of addiction (Nestler, 2002, 2005).

Based on the chronic relapsing nature of addiction, a leading theory in the field hypothesizes that addiction represents a form of pathological learning, in which drugs of abuse hijack normal learning and memory processes to support long-term addictive behaviors (Hyman, 2005; Kelley, 2004; Nestler, 2013). In support of this hypothesis, a substantial body of evidence shows that key learning and memory-related mechanisms and brain regions are altered by drug exposure (Gould & Leach, 2014; Luo, Xue, Shen, & Lu, 2013; Spiga, Mulas, Piras, & Diana, 2014). In addition, a number of similarities exist between drug-induced changes in neural activity and the changes observed during normal learning and memory processes (Nestler, 2002, 2013). Perhaps most importantly, support for this theory can also be found in animal models of addiction, most of which rely on associative or instrumental learning (Arroyo, Markou, Robbins, & Everitt, 1998; Crombag, Bossert, Koya, & Shaham, 2008; Everitt, Dickinson, & Robbins, 2001; Everitt et al., 1999), and in the behavior of human addicts. Drug seeking in humans is markedly influenced by exposure to drug-associated cues and contexts that induce subjective craving states, reinitiate drug seeking, and trigger relapse after abstinence (Everitt, 2014; Li et al., 2014). This suggests that although abused substances have clear rewarding properties that drive initial consumption, the persistence of addictive behavior relies on the ability of the physiological effects of drugs to acquire motivational significance via associations with environmental cues (O'Brien, Childress, Ehrman, & Robbins, 1998). The drug-associated cues are then capable of predicting drug availability or inducing craving in abstinent users (Crombag & Shaham, 2002; Di Ciano & Everitt, 2003a), and may elicit and maintain drug seeking, making these cues powerful determinants of relapse (Kelley, 2004).

Based on this theory, there have been a number of investigations into mechanisms that could potentially serve as an interface between drugs of abuse and the systems that support learning and memory. In addition to expanding our understanding of how drugs of abuse alter neural processes, such mechanisms may also represent potential points of therapeutic intervention for addiction to a variety of drugs (Everitt, 2014; Kauer & Malenka, 2007; Nestler, 2005). In this chapter, we propose that G protein-gated inwardly rectifying potassium (GIRK) channels are one pathway through which numerous drugs of abuse can access learning and memory systems. The role of GIRK channels in neuronal excitability and neurotransmitter signaling, as well as the expression of GIRK channels in brain regions relevant to both learning processes and reward (Luscher & Slesinger, 2010), suggests that these channels may mediate the ability of abused drugs to influence learning and memory. In order for GIRK channels to be a potential route through which drugs of abuse alter learning and memory processes, however, a number of criteria should be considered: (i) GIRK channel-dependent signaling should be modulated by multiple drugs of abuse, making it common pathway for different drug classes; (ii) GIRK channels should play a role in the normal learning and memory processes thought to contribute to addiction development; and (iii) drug-induced alterations in GIRK channel-dependent signaling should lead to changes that would support the development and maintenance of addictive behaviors. In the following sections, we will provide a brief overview of GIRK channels in the brain and then discuss the evidence for GIRK channels as potential mediators of drug-induced changes in learning and memory processes using these criteria.

2. GIRK CHANNELS

2.1 Neuronal GIRK Channels

GIRK channels are members of a large family of inwardly rectifying potassium channels (Doupnik, Davidson, & Lester, 1995). They are primarily activated by the direct binding of $G\beta\gamma$ dimers associated with pertussis toxin-sensitive, $G_{i/o}$-linked G protein-coupled receptors (GPCRs) (Dascal, 1997; Mark & Herlitze, 2000; Sadja, Alagem, & Reuveny, 2003), including $GABA_B$, opioid, and dopamine 2 (D_2)-like receptors (Kim, Nakajima, & Nakajima, 1995; Padgett & Slesinger, 2010; Ulens, Daenens, & Tytgat, 1999). Although interactions with $G\beta\gamma$ represent the canonical pathway for GIRK channel activation (Mark & Herlitze, 2000), there is evidence that the $G\alpha$ subunit also modulates GIRK channel function

(Berlin et al., 2010) (for a detailed review of GIRK-G protein interactions, see chapter "The Roles of Gβγ and Gα in Gating and Regulation of GIRK Channels" by Dascal and Kahanovitch). Upon activation, GIRK channels conduct an outward potassium current that decreases neuronal excitability (Beckstead, Grandy, Wickman, & Williams, 2004; Dutar & Nicoll, 1988; Luscher, Jan, Stoffel, Malenka, & Nicoll, 1997).

In the mammalian nervous system, GIRK channels are tetramers composed of various combinations of four distinct subunits (GIRK1–GIRK4). GIRK1, GIRK2, and GIRK3 are the most prominent subunits in the brain (Karschin, Dissmann, Stuhmer, & Karschin, 1996; Koyrakh et al., 2005); however, there is evidence for the expression and functional relevance of GIRK4-containing channels in the nervous system as well (Kennedy, Nemec, Corey, Wickman, & Clapham, 1999; Spauschus et al., 1996; Wickman, Karschin, Karschin, Picciotto, & Clapham, 2000). The GIRK subunits form a mixture of heteromeric (e.g., GIRK1/2, GIRK1/3, GIRK2/3, GIRK1/4) and homomeric (GIRK2) channels that are expressed in distinct, but overlapping, patterns throughout the brain (Jelacic, Kennedy, Wickman, & Clapham, 2000; Karschin et al., 1996; Koyrakh et al., 2005), including key regions for both learning and memory (Chen & Johnston, 2005; Luscher et al., 1997; Sosulina, Schwesig, Seifert, & Pape, 2008) and reward (Cruz et al., 2004; Inanobe et al., 1999). The composition of GIRK channels can also vary across cell types, even within a single brain region (Aguado et al., 2008; Cruz et al., 2004; Fernandez-Alacid et al., 2009). Given that GIRK channels composed of different subunits exhibit distinct functional properties (Jelacic et al., 2000; Jelacic, Sims, & Clapham, 1999; Schoots et al., 1999; Wydeven, Young, Mirkovic, & Wickman, 2012), this diversity allows for varied GIRK-mediated signaling across brain regions and cell types (for a detailed discussion of GIRK channel diversity, see chapter "Localization and Targeting of GIRK Channels in Mammalian Central Neurons" by Luján and Aguado).

2.2 Modulators of GIRK Function

In addition to activation by G proteins, GIRK channel function is modulated by a number of intracellular mechanisms. For example, phosphatidylinositol 4,5-bisphosphate (PIP_2) is required for constitutive GIRK channel activity and plays a key role in agonist-dependent activation of these channels (Huang, Feng, & Hilgemann, 1998; Xiao, Zhen, & Yang, 2003; Zhang, He, Yan, Mirshahi, & Logothetis, 1999). GIRK channels are also sensitive to phosphorylation and a number of modulatory proteins. This section will provide a brief overview of the primary modulatory pathways that affect GIRK signaling. More detailed descriptions of these complex interactions can be found in other chapters ("Unifying Mechanism of Controlling Kir3 Channel Activity by G Proteins and Phosphoinositides" by Logothetis et al. and "RGS Redundancy and Implications in GPCR–GIRK Signaling" by Doupnik).

2.2.1 Phosphorylation of GIRK Channels

GIRK channels contain phosphorylation sites that make them sensitive to a number of intracellular kinases and phosphatases, with a variety of effects on GIRK channel function. For example, protein kinase A (PKA) facilitates GIRK channel activation (Rusinova et al., 2009), potentially through the phosphorylation of the GIRK1 subunit (Medina et al., 2000; Mullner, Yakubovich, Dessauer, Platzer, & Schreibmayer, 2003), while protein phosphatase 2A (PP2A) dephosphorylates GIRK channels and abolishes G protein-mediated activation (Medina et al., 2000; Mullner et al., 2003). Phospholipase C has the dual effect of stabilizing the open conformation of GIRK channels while simultaneously reducing the overall probability of the channel being open via activation of protein kinase C (PKC), which decreases GIRK currents by phosphorylating GIRK1 (Brown, Thomas, Dekker, Tinker, & Leaney, 2005; Leaney, Dekker, & Tinker, 2001; Lei, Talley, & Bayliss, 2001; Mao et al., 2004). These phosphorylation pathways make GIRK channels sensitive to a number of intracellular cascades and allow GIRK-mediated signaling to be indirectly modulated by the activation of non-$G_{i/o}$ GPCRs (Lei, Jones, Talley, Garrison, & Bayliss, 2003; Lei et al., 2001).

2.2.2 Regulators of G Protein Signaling

The kinetics and sensitivity of GPCR-GIRK signaling pathways are also modulated by a family of proteins known as Regulators of G protein Signaling (RGS). RGS proteins contain a GTPase-activating domain that accelerates the intrinsic GTP hydrolysis activity of Gα, thus promoting the formation of the inactive G protein heterotrimer, reducing the availability of active Gβγ (Doupnik, Jaen, & Zhang, 2004; Sjogren, 2011). RGS proteins can accelerate both the activation and deactivation rates of GIRK channels (Doupnik, Davidson, Lester, & Kofuji, 1997; Herlitze, Ruppersberg, & Mark, 1999; Mark & Herlitze, 2000; Saitoh et al., 1999) and influence GPCR-GIRK coupling (Doupnik et al., 2004; Labouebe et al., 2007). Further, RGS proteins increase the overall diversity of GIRK channel-dependent signaling by associating with GPCR-GIRK complexes in some cell types but not others (Labouebe et al., 2007; Xie & Palmer, 2007), showing preferential modulation of signaling via specific GPCRs (Wang, Liu, Mullah, Siderovski, & Neubig, 2002; Xu et al., 1999), and selectively modulating GIRK channels composed of specific subunits (Labouebe et al., 2007).

2.2.3 GIRK Channel Trafficking

The expression levels of GIRK channels at the cell surface can be modified by intracellular trafficking mechanisms. In addition to the endoplasmic reticulum export signal that allows it to form homomeric channels, the GIRK2 subunit also contains a Val-Leu (VL) internalization motif (Ma et al., 2002). A serine residue (Ser-9) next to this motif can be dephosphorylated by PP1 to alter the surface expression of GIRK2-containing channels (Chung, Qian, Ehlers, Jan, & Jan, 2009; Ma et al., 2002). In addition, the GIRK2 splice variant GIRK2c and the GIRK3 subunit both contain PDZ-binding domains that have been shown to associate with Sorting Nexin 27 (SNX27) (Lunn et al., 2007; Ma et al., 2002). SNX27 promotes channel trafficking from the cell surface to the endosome, thus reducing the availability of associated GIRK channels (Balana et al., 2011; Nassirpour & Slesinger, 2007), though recent findings suggest that the role of SNX27 in GIRK trafficking may be more complicated than initially thought (Munoz & Slesinger, 2014). Due to the subunit specificity of the PDZ-binding motif, this mechanism can be engaged to selectively modulate the surface expression of GIRK channels in specific cell types within the same brain region (Cruz et al., 2004; Lunn et al., 2007) or GIRK channels coupled to different GPCRs within the same cells (Munoz & Slesinger, 2014).

2.2.4 Modulation of GPCRs

In addition to mechanisms that target GIRK channels themselves, alterations in the function of GPCRs coupled to GIRK channels can also lead to changes in GIRK signaling. A subfamily of the potassium channel tetramerization domain-containing (KCTD) proteins constitutively associate with the C-terminal domain of the GABA(B2) subunit of $GABA_B$ receptors (Ivankova et al., 2013; Schwenk et al., 2010). These KCTD proteins regulate agonist potency as well as the kinetics and magnitude of the potassium currents activated by $GABA_B$ receptors in a subtype-specific manner (Ivankova et al., 2013; Schwenk et al., 2010). For example, the currents induced by receptors associated with KCTD16 and KCTD8 show very little desensitization over time, while those induced by $GABA_B$ receptors associated with KCTD12 and 12b have very pronounced desensitization (Schwenk et al., 2010). These differences are thought to be due to the presence of specific homology domains on the various KCTD subtypes (Seddik et al., 2012). KCTD proteins also bind to G proteins at the receptor, and this binding is responsible for the acceleration of the potassium currents activated by the $GABA_B$ receptor (Turecek et al., 2014). KCTD12, however, binds to activated $G\beta\gamma$ subunits and prevents their interaction with GIRK channels, thus representing a mechanism for fast, reversible desensitization (Adelfinger et al., 2014; Turecek et al., 2014).

Another family of proteins that interact with $GABA_B$ receptors is the 14-3-3 family, a group of abundant brain proteins involved in a variety of processes, including signal transduction and neuronal development (Couve et al., 2001; Fanger et al., 1998; Foote & Zhou, 2012; Fu, Subramanian, & Masters, 2000; Honda, Ohba, & Yasuda, 1997; Skoulakis & Davis, 1998; Zhang, Kobayashi, Graves, Piwnica-Worms, & Tonks, 1997). Two members of this family, 14-3-3η and 14-3-3ζ, interact directly with the $GABA_B1$ subunit of the $GABA_B$ receptor and disrupt its association with the $GABA_B2$ subunit, preventing the formation of the $GABA_B$ heterodimer (Couve et al., 2001). In addition, recent work has demonstrated that an increase in 14-3-3η protein in hippocampal cultures produces a functional decoupling between $GABA_B$ receptors and GIRK channels, thus suppressing $GABA_B$-GIRK signaling (Workman et al., 2015).

3. GIRK CHANNEL MODULATION BY DRUGS OF ABUSE

Despite the fact that drugs of abuse have distinct chemical structures and different primary targets, a number of prominent addiction-related neuroadaptations are observed following exposure to most, if not all, addictive drugs. This suggests that the common adaptations triggered by multiple drugs of abuse likely contribute to the shared features of addiction syndrome (Hyman, 2005; Nestler, 2013). Thus, in the search for mechanisms that might support addiction development or serve as points of therapeutic intervention, it is important to identify candidates that, through various mechanisms, can be modulated by a range of drugs. GIRK channels have been shown to play a role in both the cellular and behavioral effects of several abused drugs, making them a potential common effector across multiple drug classes. In this section, we will briefly discuss the various drug types associated with GIRK channels and the mechanisms through which abused drugs can alter GIRK channel-dependent signaling. A summary of

Table 1 Drugs That Modulate GIRK Channels

Drug Class	Drugs	Mechanism of Action	GIRK Modulation	References
Opioids	Morphine, heroin	μ-Opioid receptor agonist	Activation via opioid receptors	Nagi and Pineyro (2014)
Stimulants	Cocaine, methamphetamine	Enhance dopamine release by reducing/reversing reuptake	Activation via D_2-like receptors	Kim et al. (1995); Beckstead et al. (2004)
			Modulated by SNX27	Munoz and Slesinger (2014)
Cannabinoids	Marijuana (THC)	CB_1 receptor agonist	Activation via CB_1 receptor	Bacci, Huguenard, and Prince (2004); Guo and Ikeda (2004)
Sedatives/hypnotics	GHB	$GABA_B$ receptor agonist	Activation via $GABA_B$ receptor	Cruz et al. (2004)
			Modulated by RGS2	Labouebe et al. (2007)
	Ethanol	Multiple targets	Direct GIRK channel activation	Lewohl et al. (1999); Aryal, Dvir, Choe, and Slesinger (2009)
			Potential modulation by RGS6	Stewart et al. (2015)
Dissociatives	Ketamine	NMDA receptor antagonist	Reduces coupling to $GABA_B$	Workman et al. (2015)

List of drug classes known to affect GIRK channel signaling. Specific examples for each class are given, along with their primary mechanisms of action and the mechanisms through which they are thought to interact with GIRK channels.

these drug types and their mechanisms is shown in Table 1. A more detailed account of the molecular mechanisms that drive drug-induced GIRK channel plasticity and the role of GIRK channels in drug-induced behaviors can be found in other chapters ("GIRK Channel Plasticity and Implications for Drug Addiction" by de Velasco et al. and "Behavioral and Genetic Evidence for GIRK Channels in the CNS: Role in Physiology, Pathophysiology, and Drug Addiction" by Mayfield et al.).

3.1 Drugs That Activate $G_{i/o}$-Coupled GPCRs

Because they are downstream of $G_{i/o}$-coupled GPCRs, GIRK channels are in a position to be a common effector for a wide range of drug types. For example, GIRK channels are functionally coupled to μ-, κ-, and δ-opioid receptors in some neuron populations and thus mediate some of the inhibitory effects of morphine, heroin, and other opioid-based analgesics (Nagi & Pineyro, 2014). GIRK channels play an important role in μ-opioid receptor desensitization (Blanchet & Luscher, 2002; Blanchet, Sollini, & Luscher, 2003; Williams, 2014) and mediate the ability of opioid receptors to modulate neuronal activity in the dorsal raphe (Lemos et al., 2012), a critical mediator of opioid-dependent dysphoria (Land et al., 2008; Land et al., 2009), and the locus coeruleus (Torrecilla et al., 2002), which plays a key role in opioid withdrawal (Mazei-Robison & Nestler, 2012). Behavioral studies have implicated GIRK channels in opioid-induced analgesia (Blednov, Stoffel, Alva, & Harris, 2003; Marker, Lujan, Loh, & Wickman, 2005), seizure susceptibility during opioid withdrawal, opioid-induced locomotor activity, and opioid reward (Ikeda et al., 2002; Nagi & Pineyro, 2014). Further, μ-opioid receptor-mediated, GIRK-dependent signaling is altered by repeated morphine exposure, suggesting that drug-induced plasticity in GIRK signaling may contribute to the development of opioid tolerance (Ingram, Macey, Fossum, & Morgan, 2008).

Though less well characterized than the role of GPCR-GIRK signaling in opioid responses, there are several other GIRK-linked GPCRs that may potentially modulate drug effects. GIRK channels couple to cannabinoid 1 (CB_1) receptors, and endocannabinoids have been shown to activate a GIRK-mediated current in both neurons and heterologous systems (Bacci et al., 2004; Guo & Ikeda, 2004; McAllister, Griffin, Satin, & Abood, 1999). *Girk2*$^{-/-}$ mice show reduced sensitivity to the antinociceptive effect of CB_1 receptor agonists (Blednov et al., 2003), suggesting that GIRK channels may mediate other effects of cannabinoids. The activation of GIRK channels by D_2-like receptors (Kim et al., 1995) is an important mechanism for inhibitory dopamine signaling (Beckstead et al., 2004; Beckstead & Williams, 2007) and may contribute to the role of GIRK channels in the effects of stimulant drugs, such as cocaine and methamphetamine (Arora et al., 2011; Padgett et al., 2012). Serotonin 5-HT_{1A} receptors have also been implicated in the effects of stimulants (Muller, Carey, Huston, & De Souza Silva, 2007); however, a definitive role for 5-HT_{1A}-GIRK signaling in these effects has not yet been shown. Interestingly, changes in $GABA_B$-GIRK signaling are observed following exposure to several drugs of abuse, though for most drugs, this mechanism is not due to direct activation of $GABA_B$ receptors by drugs. Instead, changes in $GABA_B$-GIRK function represent a more generalized drug-induced plasticity in GIRK signaling, which will be discussed in later sections and is explored in detail in chapter "GIRK Channel Plasticity and Implications for Drug Addiction" by de Velasco et al. However, the club drug γ-hydroxybutyrate (GHB) activates $GABA_B$ receptors, and GIRK-mediated signaling is thought to drive the rewarding properties of this compound (Cruz et al., 2004; Labouebe et al., 2007).

3.2 Drug Effects on Intracellular Modulators of GIRK Signaling

In addition to direct GPCR activation, several of the intracellular mechanisms that modulate GIRK channel function also serve as potential routes through which drugs can alter GIRK channel-dependent signaling. RGS proteins have been implicated in the effects of a number of drugs of abuse (Bishop, Cullinan, Curran, & Gutstein, 2002; Hooks, Martemyanov, & Zachariou, 2008; Lomazzi, Slesinger, & Luscher, 2008; Traynor & Neubig, 2005). For example, the expression of RGS9-2, which regulates GPCR-GIRK signaling in heterologous systems, increases following acute morphine administration, but decreases following chronic exposure (Zachariou et al., 2003), and mice lacking RGS9-2 exhibit significantly altered morphine reward, analgesia, and withdrawal (Zachariou et al., 2003). *Rgs9-2*$^{-/-}$ mice also show enhanced cocaine reward and locomotor responses, while the overexpression of RGS9-2 reduces locomotor responses to cocaine (Rahman et al., 2003), indicating a role for RGS9-2 in the effects of multiple drugs of abuse.

RGS6, which plays an important role in shaping the kinetics of GIRK responses (Maity et al., 2012), is implicated in the effects of alcohol. The loss of RGS6 reduces alcohol consumption, conditioned alcohol preference, and the severity of alcohol withdrawal (Stewart et al., 2015). Additionally, RGS6 expression in the ventral tegmental area (VTA) is increased by alcohol exposure (Stewart et al., 2015). RGS2 expression is reduced by chronic exposure to both morphine and GHB, and the differential modulation of GIRK signaling in the dopaminergic and GABAergic neurons of the VTA by RGS2 is thought to contribute to the development of GHB tolerance (Labouebe et al., 2007). Further, the difference in $GABA_B$-GIRK coupling efficiency and RGS2 modulation in GABAergic and dopaminergic neurons of the VTA may explain why GHB is rewarding, and thus has potential abuse liability, while baclofen, another $GABA_B$ agonist, shows no rewarding properties (Cruz et al., 2004; Labouebe et al., 2007). Although not all of the described effects of RGS proteins have been definitively linked to GIRK channels, the available data support the hypothesis that changes in RGS expression and function are a potential route through which drugs can alter GIRK signaling.

The mechanisms that control GIRK channel trafficking may also play a role in drug-induced changes in GIRK-mediated signaling. GIRK channel trafficking via the PDZ/SNX27 mechanism is implicated in the psychostimulant-induced suppression of GIRK channel activity (Balana et al., 2011; Yao et al., 2004). Mice lacking SNX27 in VTA dopamine neurons show enhanced cocaine responses (Munoz & Slesinger, 2014), and SNX27b mRNA expression is upregulated following psychostimulant exposure (Kajii et al., 2003). This increase in SNX27, which would result in reduced GIRK channel expression at the cell surface, could contribute to the reduced $GABA_B$-GIRK signaling induced by exposure to these drugs (Arora et al., 2011; Padgett et al., 2012).

Finally, mechanisms that alter GPCR-GIRK coupling may also contribute to the effect of drugs on GIRK-mediated signaling. A recent report on the effects of the NMDA receptor antagonist ketamine found that the blockade of NMDA receptors resulted in a rapid decoupling of GIRK channels from $GABA_B$ receptors (Workman et al., 2015). This decoupling is mediated by the adaptor protein 14-3-3η and is critical for the antidepressant effects of ketamine. While the relationship between this mechanism and ketamine abuse is unclear, GPCR-GIRK coupling represents another potential pathway through which drugs that do not act on GPCRs might affect GIRK channel-dependent signaling.

3.3 Ethanol: A Direct Activator of GIRK Channels

Ethanol binds to GIRK channels and activates them in a $G\beta\gamma$-independent manner at concentrations relevant to human alcohol consumption (Aryal et al., 2009; Kobayashi et al., 1999; Lewohl et al., 1999). The binding site for ethanol is a hydrophobic pocket formed by the N-terminal domain and βD–βE loop of one GIRK channel subunit and the βL–βM loop of a neighboring subunit (Aryal et al., 2009). By interacting with multiple residues within this pocket, ethanol can stabilize the open state of GIRK channels, likely by increasing the affinity of the channel for PIP_2 (Bodhinathan & Slesinger, 2013, 2014). Interestingly, this ethanol binding site overlaps with the region that is thought to interact with Gβγ (Bodhinathan & Slesinger, 2013). For a detailed review of GIRK channel structure, including ethanol binding, see chapter "Structural Insights into GIRK Channel Function" by Glaaser and Slesinger. While a similar ethanol binding pocket has been reported for other types of channels (Howard et al., 2011; Mascia, Trudell, & Harris, 2000; Mihic et al., 1997), the ability to activate GIRK channels in the absence of a primary agonist is an especially interesting mechanism for ethanol, which is primarily thought to act as an allosteric modulator of several receptors and ion channels, rather than directly activating its molecular targets (Crews, Morrow, Criswell, & Breese, 1996; Trudell, Messing, Mayfield, & Harris, 2014).

The use of genetic mouse models lacking GIRK subunits has demonstrated a role for GIRK channels in numerous ethanol-mediated behaviors, including consumption (Blednov, Stoffel, Chang, & Harris, 2001), conditioned place preference (Hill, Alva, Blednov, & Cunningham, 2003), and withdrawal severity (Kozell, Walter, Milner, Wickman, & Buck, 2009). It is currently unclear, however, whether GIRK subunit ablation alters ethanol responses because GIRK channels represent a key binding site for ethanol or because GIRK channels are an important part of a larger ethanol-mediated circuit. Two particularly interesting aspects of GIRK channel function have been revealed by these ethanol studies. First, while the loss of GIRK2 decreases ethanol-conditioned place preference (Hill et al., 2003), the loss of GIRK3 has the opposite effect, with *Girk3*$^{-/-}$ mice showing increased preference for the ethanol-paired side compared to wild-type littermates (Megan E. Tipps and Kari J. Buck, unpublished data). This suggests that GIRK channels composed of different subunits may have distinct sensitivities to certain effects of drugs of abuse, similar to differences observed in the activation of GIRK channel subtypes by Gβγ (Cruz et al., 2004; Schoots et al., 1999). Second, the identification of *Girk3/Kcnj9* as a quantitative trait gene for sedative-hypnotic drug withdrawal (Kozell et al., 2009) suggests that differences in GIRK3 subunit expression levels may be one of the genetic factors that determine sensitivity to various drug effects and the potential development of addiction (Crabbe, Phillips, Harris, Arends, & Koob, 2006).

4. NEURONAL PLASTICITY

The longevity of addictive behavior suggests that long-term alterations in neural function are required for addiction (Luscher & Malenka, 2011). Neuronal plasticity, defined as long-lasting changes in signaling and excitability, is the primary cellular correlate for learning and memory processes and is associated with the development and maintenance of several long-term behaviors (Goosens & Maren, 2002; Maren & Baudry, 1995). Interestingly, the same general types of neuronal modifications in brain areas associated with behavioral memory (e.g., hippocampus and amygdala) also occur in reward-related brain regions, such as the VTA and nucleus accumbens (NAc) (Bonci & Malenka, 1999; Engblom et al., 2008; Kombian & Malenka, 1994), where they are important for mediating addiction development and maintenance (Luscher & Bellone, 2008; Nestler, 2013). There are two primary forms of neuronal plasticity that are thought to drive long-term behavioral changes based on external stimuli: homeostatic plasticity and synaptic plasticity. In this section, we will investigate the role of GIRK channels in both forms of plasticity and the modulation of these mechanisms by drugs of abuse.

4.1 GIRK Channels and Neuronal Plasticity

4.1.1 Homeostatic Plasticity

Homeostatic plasticity involves experience-dependent changes in the intrinsic excitability of an entire cell in a manner that is not synapse specific (Turrigiano, 1999). These changes maintain the activity of a cell within a finite range, without altering individual synapse strength. Further, homeostatic mechanisms allow neurons to use their own activity as a feedback signal to modify excitability, and thus play a key role in network stability (Kauer & Malenka, 2007; Turrigiano, 1999). While synaptic plasticity is often considered the primary cellular correlate of learning and memory, homeostatic plasticity also supports long-term changes, including enhanced and/or reduced sensitivity to drugs, such as the development of tolerance or sensitivity to repeated drug exposure (Kauer & Malenka, 2007; Nestler, 2013).

There are several examples throughout the brain of GIRK channels playing a critical role in modulating neuronal excitability, as well as data demonstrating that the regulation of excitability by GIRK-mediated signaling is important for neural functions, including learning and memory processes. For example, GIRK channels are expressed in the hippocampus

(Fajardo-Serrano et al., 2013; Karschin et al., 1996; Koyrakh et al., 2005; Luscher et al., 1997), a region that is critical for spatial navigation and contextual learning as well as a key contributor to context-induced relapse in rodent models of drug seeking (Bossert & Stern, 2014; Gould & Leach, 2014; Isokawa, 2012; Luo, Tahsili-Fahadan, Wise, Lupica, & Aston-Jones, 2011). In addition to mediating the postsynaptic inhibitory effects of neurotransmitters (Luscher et al., 1997), GIRK channels in CA1 hippocampal neurons have a resting potassium conductance (Kim & Johnston, 2015). Interestingly, this resting GIRK conductance is higher for neurons in the dorsal hippocampus compared to those in the ventral hippocampus, and this difference is thought to contribute to the more hyperpolarized resting membrane potential observed in dorsal CA1 neurons (Kim & Johnston, 2015). This suggests that GIRK-mediated currents contribute to the different functional properties of these two subregions and potentially the distinct roles they are thought to play in hippocampal-dependent behaviors (Fanselow & Dong, 2010).

A constitutive GIRK-mediated current is also observed in the pyramidal neurons of the medial prefrontal cortex (mPFC), a region that is crucial for cognitive functions such as decision making and inhibitory control (Matsumoto, Suzuki, & Tanaka, 2003; Miller & Cohen, 2001) and plays an important role in the valuation of rewards and establishing reward-related memories (Kelley, 2004). This current is inhibited by D_1 receptor agonists (Witkowski, Szulczyk, Rola, & Szulczyk, 2008) via the activation of PKA and cyclic adenosine monophosphate (Witkowski, Rola, & Szulczyk, 2012). The inhibition of GIRK channel activity results in a prolonged depolarization of mPFC pyramidal neurons, a phenomena known as the upstate, which is thought to be an electrophysiological substrate for working memory (Witkowski et al., 2008).

GIRK channels are also involved in the ability of neurons to regulate their own excitability in response to neuronal activity, a key aspect of homeostatic plasticity. The best example of this can be seen in the dopamine neurons of the VTA. Dopamine signaling in the VTA is one of the primary mechanisms connecting addiction with learning. Initially, dopamine was thought to function as a simple internal measure of a hedonic (pleasurable) state; however, work by Schultz and colleagues (Hollerman & Schultz, 1998; Schultz, 2002; Schultz, Dayan, & Montague, 1997) illustrates that the information encoded by dopamine signaling is more complex than a simple measure of the reward value of a stimulus. Although VTA dopamine neurons increase firing in response to an unexpected reward, Schultz et al.

also found that when a reward is paired with an external cue that predicts reward availability, the timing of dopamine firing changed. Specifically, the timing of the increase in dopamine neuron firing shifted such that the presentation of the cue, not the presentation of the reward, elicited dopamine increases (Schultz et al., 1997), suggesting that these neurons code for the anticipation of reward predicted by a specific cue. Further, if the reward-associated cue is presented but the expected reward is not delivered, dopamine signaling will decrease (Hollerman & Schultz, 1998), illustrating two distinct roles for dopamine: (1) reward prediction based on external cues and (2) encoding errors in previous predictions (Schultz, 2002).

These aspects of VTA dopamine signaling rely on the ability of dopamine neurons to switch between two primary firing modes, low tonic and high phasic bursting (Schultz, 2002). Interestingly, the occurrence of these two firing patterns alters GIRK channel activity within these cells. When dopamine neurons show high burst-firing activity, an increase in the GIRK-mediated slow inhibitory postsynaptic current (IPSC) amplitude is observed. Conversely, the amplitude of slow IPSCs is reduced during low tonic dopamine firing (Lalive et al., 2014), suggesting that changes in GIRK channel activity may be an intrinsic self-regulation mechanism that is engaged in response to dopamine neuron firing modes. These changes in GIRK-mediated IPSCs have been attributed to activity-dependent channel trafficking via SNX27 (Lunn et al., 2007), and are thought to represent an important mechanism through which VTA dopamine neurons can regulate their excitability (Lalive et al., 2014).

4.1.2 Synaptic Plasticity

A ubiquitous property of all synapses is the ability to undergo activity-dependent changes in synaptic strength, a phenomena referred to as synaptic plasticity. This process is used to refine the synapse-specific signals that correlate to experiences and associations regarding external stimuli (Malenka & Bear, 2004). The type of synaptic plasticity most commonly associated with behavioral learning involves changes in synaptic function in the form of either long-term potentiation (LTP) or long-term depression (LTD). These refer to the ability of synaptic stimulation to either increase or decrease, respectively, the strength of the postsynaptic response to subsequent stimulation. LTP and LTD are basic properties of most excitatory synapses in the central nervous system (Malenka & Bear, 2004), and the traditional forms of LTP and LTD both rely on glutamatergic signaling. For LTP, the activation of NMDA glutamate receptors coincident with neuronal depolarization results in an increase in intracellular calcium levels that activates internal mechanisms, such as calcium/calmodulin-dependent kinase II (CaMKII) (Nicoll & Malenka, 1999). These mechanisms trigger an increase in AMPA glutamate receptor expression at the cell surface, thus increasing the strength of the synaptic response to future glutamate signals. A similar mechanism is thought to underlie LTD; however, in this case, the activation of NMDA receptors or neuronal depolarization is weaker, resulting in a smaller calcium response and the activation of a different set of intracellular mechanisms (Malenka & Bear, 2004). The end result is the removal of AMPA receptors from the cell surface and a reduced response to future excitation by glutamate. LTP and LTD are hypothesized to play key roles in several forms of experience-dependent plasticity, including various forms of learning and memory (Malenka & Bear, 2004).

The molecular mechanisms driving these changes include the phosphorylation of proteins, alterations in the availability of receptors and ion channels, and the regulation of gene expression, all of which have also been tied to changes induced by drug exposure (Hyman, Malenka, & Nestler, 2006; Kelley, 2004; Nestler, 2013). The specificity of drug cues and their ability to induce a defined behavioral sequence (drug seeking) suggests that some of the changes associated with addiction must be associative and synapse-specific, rather than diffuse, generalized changes (Hyman, 2005). In agreement with this, drugs of abuse can induce both LTD- and LTP-like changes in synapses (Nestler, 2013; Saal, Dong, Bonci, & Malenka, 2003; Ungless, Whistler, Malenka, & Bonci, 2001), thus utilizing these mechanisms to facilitate long-lasting changes in neuronal processing (Kauer & Malenka, 2007). Both LTP and LTD are observed in the VTA and NAc, where they are thought to underlie the development of addiction (Bonci & Malenka, 1999; Kauer, 2004; Kombian & Malenka, 1994); however, the role of GIRK channels in the synaptic plasticity of these regions is unknown. As the current body of literature regarding GIRK channels and synaptic plasticity focuses on the hippocampus, the traditional region for LTP and LTD measurements, this region will be the focus of the current section.

GIRK channels in hippocampal neurons have a reciprocal relationship with NMDA receptor-dependent plasticity, in which NMDA receptor activation influences GIRK expression and GIRK-mediated signaling modulates the development of NMDA-based plasticity. Neuronal excitation via NMDA receptors regulates inhibitory signaling and GIRK channel expression. The NMDA receptor activation and subsequent CaMKII activity that induces excitatory LTP also potentiate $GABA_B$-GIRK-mediated slow

IPSCs (Huang et al., 2005), illustrating that the same signaling pathways that mediate the plasticity of excitatory synaptic potentials can also induce long-term changes in synaptic inhibition. Further, NMDA receptor activation in cultured hippocampal neurons increases the surface expression of GIRK1 and GIRK2 subunits via PP1-mediated dephosphorylation of GIRK2 (Chung, Qian, et al., 2009). Interestingly, NMDA receptor activation results in an increase in GIRK-mediated currents stimulated by A_1 receptors, but not by $GABA_B$ receptors. This implies that these two receptor types form distinct complexes with GIRK channels and that they may be differentially regulated by neuronal activity (Chung, Ge, et al., 2009). Together, these reports suggest that activity-induced changes in GIRK channel-dependent signaling enable the dynamic regulation of neuronal excitability. These activity-dependent changes are important for refining the coincidence detection that drives input-dependent plasticity by narrowing the window in which subsequent synaptic inputs would be able to affect responses (Chung, Qian, et al., 2009; Huang et al., 2005).

Additionally, GIRK channels modulate the formation of excitatory synaptic plasticity via a mechanism known as depotentiation. Depotentiation refers to the ability of a low-frequency stimulus to prevent LTP induction by a previous high-frequency stimulus (Huang, Liang, & Hsu, 1999; Staubli & Lynch, 1990). This mechanism prevents LTP saturation and the development of nonspecific LTP, thus increasing the storage capacity of neural circuits. Depotentiation is absent in mice lacking GIRK2 and can be prevented by Tertiapin-Q, a GIRK channel blocker (Chung, Ge, et al., 2009), suggesting that GIRK channels are critical for this phenomena. A potential role for GIRK channel modulation via RGS proteins in synaptic plasticity and depotentiation has also been reported. The loss of RGS7 significantly slows $GABA_B$-GIRK current deactivation, which should decrease the excitability of hippocampal pyramidal neurons (Ostrovskaya et al., 2014). *Rgs7*$^{-/-}$ mice also exhibit impaired LTD and depotentiation, though LTP was unchanged. These effects on plasticity and excitability were accompanied by impaired performance in contextual fear conditioning, a hippocampal-based associative learning and memory task (Ostrovskaya et al., 2014).

Finally, alternative forms of LTD exist that could imply a more direct role for GIRK channels in this mechanism. Some LTD is thought to be triggered by the activation of metabotropic glutamate receptors (mGluRs) (Huang, Rowan, & Anwyl, 1999; Ito, 1989), a subset of which couple directly to GIRK channels (Knoflach & Kemp, 1998). Another form of

LTD involves an increase in endocannabinoid synthesis, leading to enhanced activation of inhibitory CB_1 receptors (Chevaleyre, Takahashi, & Castillo, 2006) to depress further neurotransmitter release from the presynaptic cell. GIRK channels couple to CB_1 receptors in expression systems (Guo & Ikeda, 2004; McAllister et al., 1999), and the long-lasting self-inhibition mediated by endocannabinoids in neocortical interneurons is thought to rely on the activation of CB_1-GIRK signaling (Bacci et al., 2004). This suggests that GIRK-mediated signaling may contribute to CB_1-dependent LTD, though this hypothesis is currently untested. Thus, it seems likely that GIRK channels are involved in multiple forms of synaptic plasticity.

4.2 GIRK Channels and Cognitive Dysfunction

If GIRK channels play an important role in neuronal plasticity, changes in GIRK channel expression or function would be expected to result in deficits in the behavioral learning that relies on these processes. In agreement with this, alterations in GIRK-mediated signaling have been implicated in the cognitive impairments associated with a number of neurological disorders characterized by impaired neuronal plasticity, including Down syndrome, Alzheimer's disease, and Parkinson's disease.

4.2.1 Down Syndrome

Kcnj6, the gene that codes for the GIRK2 subunit, is found on human chromosome 21. The trisomy of chromosome 21 causes Down syndrome, a congenital disease characterized by mental retardation and reduced muscle tone (Antonarakis, Lyle, Dermitzakis, Reymond, & Deutsch, 2004). A number of animal models of Down syndrome have been generated by inducing a trisomy of segments of mouse chromosome 16, which is highly homologous with human chromosome 21. These animal models show a number of cognitive and motor deficits similar to those observed in Down syndrome and have been used to determine the roles if individual genes contained in this chromosomal region (Antonarakis et al., 2004; Reeves et al., 1995). The Ts65Dn mouse model, which is trisometric for a large portion of chromosome 16, exhibits overexpression of GIRK2 in the hippocampus, frontal cortex, substantia nigra (SN), and cerebellum (Harashima, Jacobowitz, Stoffel, et al., 2006; Harashima, Jacobowitz, Witta, et al., 2006). As might be expected, these mice show increased $GABA_B$-induced GIRK channel activation and neuronal inhibition (Best, Siarey, & Galdzicki, 2007). Further, Ts65Dn mice have significantly attenuated LTP (Siarey, Stoll,

Rapoport, & Galdzicki, 1997) and significantly enhanced LTD (Siarey et al., 1999). Similar effects were observed for the Ts1Cje mouse model, which is trisometric for a smaller segment of mouse chromosome 16 containing *Kcnj6* (Siarey, Villar, Epstein, & Galdzicki, 2005). While this model still shows cognitive deficits, they are less severe than those observed in Ts65Dn mice (Sago et al., 1998). Ts1Cje mice also show impaired LTP formation and enhanced LTD formation, and GIRK2 was suggested as the most likely mediator of these effects (Siarey et al., 2005).

While these data indicate a role for GIRK2-containing channels in Down syndrome, chromosome 16 contains a number of other genes that could also influence excitability and cognition. To address this, learning, plasticity, and GIRK signaling were investigated in a mouse model with a selective overexpression of GIRK2 (28E6/67; Smith, Zhu, Zhang, Cheng, & Rubin, 1995). 28E6/67 (GIRK2 trisomy) mice exhibit impaired hippocampal-dependent spatial memory (Chabert et al., 2004) and associative learning (Cooper et al., 2012). These deficits were accompanied by increased GIRK currents mediated by A_1 receptor activation, but not $GABA_B$ receptors, suggesting that the extra copy of *Kcnj6* translates into increased GIRK channel expression and activity for channels coupled to some, but not all, GPCRs (Cooper et al., 2012). Interestingly, the A_1-GIRK signaling pathway selectively affected by the trisomy of GIRK2 is also selectively upregulated by NMDA activity (Chung, Qian, et al., 2009). GIRK2 trisomy mice also showed reduced depotentiation, possibly because the increased basal GIRK expression limits the ability of NMDA activation to further increase GIRK activity, as well as an increased propensity to develop LTD (Cooper et al., 2012). However, LTP was unaffected in this model (Smith et al., 1997), suggesting that the LTP impairment observed in the Ts1Cje and Ts65Dn models may require co-overexpression of one or more of the other genes found on chromosome 16 (Siarey et al., 2005).

It is interesting to note that the loss of GIRK2 also results in altered synaptic plasticity, specifically enhanced LTP (Cramer, Best, Stoffel, Siarey, & Galdzicki, 2010). This implies that normal synaptic plasticity requires GIRK function to be tightly regulated, as both an increase and a decrease in GIRK activity will disrupt these processes. In agreement with this, GIRK channel trafficking has been implicated in Down syndrome. Another gene affected by the trisomy of chromosome 21, miR-155, downregulates a transcriptional factor for *SNX27*, leading to reduced SNX27 expression (Wang et al., 2013). Interestingly, upregulating SNX27, which traffics GIRK channels away from the cell surface (Lunn et al., 2007) while also trafficking

NMDA receptors toward the cell surface (Wang et al., 2013), rescues the synaptic and cognitive deficits in Ts65Dn mice (Wang et al., 2013), suggesting an important role for dynamic GIRK trafficking in this disorder.

4.2.2 Alzheimer's Disease

GIRK channels have also been implicated in the pathology of Alzheimer's disease, a disorder in which memory deficits and disorientation are primary symptoms (Swanberg, Tractenberg, Mohs, Thal, & Cummings, 2004). Synaptic function, including LTP, is drastically impaired in Alzheimer's disease, specifically in the septohippocampal network (Kumar, 2011). Amyloid-β peptide (Aβ), a key effector of Alzheimer's disease, induces aberrant patterns of neural activity, destabilizes neuronal networks, and impairs LTP formation (Huang & Mucke, 2012; Palop & Mucke, 2010; Shankar et al., 2008). Recent work has shown a link between Aβ peptides and GIRK channel function (Nava-Mesa, Jimenez-Diaz, Yajeya, & Navarro-Lopez, 2013). Specifically, the application of Aβ peptide in hippocampal slices significantly depolarized neurons and concurrently increased input resistance, suggesting that these peptides close constitutively active channels. Further, when GABAergic pathways were stimulated, Aβ selectively antagonized the slow inhibitory postsynaptic potential mediated by $GABA_B$ receptors. This effect was linked to the GIRK component of $GABA_B$-mediated signaling, and the authors suggested that one function of Aβ may be to decrease GIRK channel conductance (Nava-Mesa et al., 2013).

4.2.3 Parkinson's Disease

Although dopamine release from the VTA is the common focus of addiction research, GIRK channels are also expressed in the dopamine neurons of the SN (Koyrakh et al., 2005; Watts, Williams, & Henderson, 1996). While GIRK channel-dependent signaling here is not as well characterized as signaling in the VTA, data from the *weaver* mouse model suggest that GIRK channels play an important role in this region. The *weaver* mutation causes a single amino acid change in the GIRK2 subunit (Patil et al., 1995). This mutation results in GIRK channels that are nonselective for cations and have dramatically reduced sensitivity to Gβγ (Navarro et al., 1996). A prevalent phenotype in *weaver* mice is the loss of granule cell neurons in the cerebellum (Smeyne & Goldowitz, 1989) and dopamine neurons in the SN (Roffler-Tarlov, Martin, Graybiel, & Kauer, 1996; Schmidt et al., 1982), suggesting that GIRK channel function is required for the long-term viability of these cells. The *weaver* mutation results in severe locomotor abnormalities, including uneven gait, ataxia, and hyperactivity (Schmidt et al., 1982; Sidman, 1968), as well as an increased propensity for seizures (Eisenberg & Messer, 1989). Mice carrying the *weaver* mutation also show impairments in several assays for instrumental learning. Specifically, *weaver* mice earn fewer rewards and maintain a pattern of responding that is inappropriate to the contingencies of each task (Derenne, Arsenault, Austin, & Weatherly, 2007). Based on these phenotypes, the *weaver* mouse model has been used to investigate the role of SN dopamine neuron loss in Parkinson's disease, a disorder characterized by movement impairments and cognitive dysfunction, including impaired working memory and response inhibition (Matsumoto, 2015); however, it should be noted that the *weaver* GIRK mutation is not part of Parkinson's pathology in humans (Bandmann, Davis, Marsden, & Wood, 1996) and there are important differences in the overall neurological changes observed in the two models (Derenne et al., 2007). Alternatively, *weaver* mice have been proposed as a potential tool to differentiate the roles of distinct dopamine signaling pathways (Maharajan, Maharajan, Ravagnan, & Paino, 2001), as the mesolimbic dopamine projections are largely unaffected in this model.

4.3 Drug-Induced Changes in GIRK Function

Having established that GIRK-mediated signaling is sensitive to modulation by multiple drugs of abuse and that GIRK channels play an important role in the neuronal plasticity that underlies long-term behavioral changes, we now turn our attention to the third criteria supporting a role of GIRK channels in addiction development: the ability of drug-induced alterations in GIRK channel-dependent signaling to support the development and maintenance of addictive behaviors.

4.3.1 GPCR-GIRK Signaling

Drug-induced changes in GIRK function have been observed in several brain regions, and a detailed description of the mechanisms underlying these changes can be found in chapter "GIRK Channel Plasticity and Implications for Drug Addiction" by de Velasco et al. For example, $GABA_B$-dependent GIRK signaling in the dopamine neurons of the VTA is weakened by acute cocaine exposure (Arora et al., 2011). Acute methamphetamine and cocaine exposure also reduces $GABA_B$-mediated GIRK signaling in GABAergic VTA neurons, with the dampening effect of methamphetamine lasting for several days after exposure (Padgett et al., 2012). This reduction in $GABA_B$-GIRK function in GABA neurons would lead to reduced dopamine output from this region and may represent a mechanism through which

GABA neuron activity reduces dopamine signaling, and thus reward perception, in response to initial stimulant exposure (Luscher & Malenka, 2011; Padgett et al., 2012). As a result, GIRK channel-dependent signaling in this region contributes to the increase in dopamine output associated with reward perception and acts as an intrinsic inhibitory mechanism on that output, potentially serving as a "break" mechanism during early drug exposure. However, repeated methamphetamine administration decreases GIRK currents in VTA dopamine neurons (Sharpe, Varela, Bettinger, & Beckstead, 2014), resulting in reduced inhibition on dopamine release. Given that increases in VTA dopamine firing rates are observed following repeated stimulant exposure (White, 1996), this finding suggests that the ability of repeated stimulant exposure to overcome the inherent GIRK-mediated "break" mechanism may be an important part of addiction.

Repeated cocaine exposure also weakens GIRK channel activation by $GABA_B$ receptors in layer 5/6 mPFC pyramidal neurons (Hearing et al., 2013). Given the role of GIRK channels in tempering the excitability of these neurons, the cocaine-induced reduction in GIRK signaling would result in an increase in glutamatergic output from the mPFC. Interestingly, an increase in glutamatergic input to the VTA accompanied the reduced GIRK signaling in VTA dopamine neurons following acute cocaine exposure (Arora et al., 2011), suggesting that GIRK-dependent signaling may drive the drug-induced changes in both systems. Further, decreased D_2- and $GABA_B$-dependent modulation of glutamatergic output from the mPFC to the VTA and NAc contributes to the development of sensitization to the locomotor-stimulating effects of cocaine (Beyer & Steketee, 2002; Jayaram & Steketee, 2004). Supporting a role for GIRK channels in these behavioral changes, the persistent suppression of GIRK signaling in the mPFC via a viral-mediated reduction in GIRK channel expression resulted in an elevated response to cocaine-induced locomotor stimulation (Hearing et al., 2013). Thus, a persistent reduction in GIRK signaling, similar to the one observed following repeated cocaine exposure, may presensitize mice to the motor stimulatory effect of cocaine. These data suggest that GIRK channel-dependent signaling in the PFC may represent an inhibitory mechanism that can be disrupted by drugs of abuse to yield persistent adaptations in excitatory neurotransmission to mesolimbic areas that promote addiction-related behaviors associated with drug seeking and relapse.

4.3.2 Neuronal Plasticity

In addition to altering GPCR-mediated GIRK signaling in regions associated with learning and addiction, drug-induced changes in homeostatic excitability can also be directly linked to GIRK channel function. Chronic exposure to opioids increases the excitability of neurons in the locus coeruleus (Kogan, Nestler, & Aghajanian, 1992), a mechanism that is thought to drive some of the symptoms of opioid withdrawal. GIRK channels mediate the intrinsic excitability of neurons in the locus coeruleus, as well as the effects of acute opioids on these neurons (Torrecilla, Fernandez-Aedo, Arrue, Zumarraga, & Ugedo, 2013; Torrecilla et al., 2002). Additionally, the morphine withdrawal-induced increase in the spontaneous firing of locus coeruleus neurons is absent in *Girk2/3*$^{-/-}$ mice, and the behavioral withdrawal symptoms are attenuated (Cruz et al., 2008). This strongly suggests that the activation and regulation of GIRK channels via opioids is a key component of these drug-induced changes in the overall excitability of the locus coeruleus.

Synaptic plasticity mechanisms are also sensitive to drug exposure. Hippocampal LTP is disrupted by a number of drugs, including ethanol, stimulants, and opioids (Bao et al., 2007; McQuiston, 2007; Roberto, Nelson, Ur, & Gruol, 2002; Thompson, Gosnell, & Wagner, 2002). For example, LTP in the hippocampus is reduced by chronic exposure to morphine or heroin (Bao et al., 2007; Pu, Bao, Xu, Ma, & Pei, 2002), with a parallel impairment in hippocampal-based learning (Pu et al., 2002). Although the impact of drug-induced changes in GIRK signaling on LTP and LTD has not been directly assessed, GIRK channels do mediate at least some of the inhibition of hippocampal interneurons induced by opioids (Svoboda, Adams, & Lupica, 1999). Treatment with morphine induces a change in GIRK2 localization in cultured hippocampal neurons, specifically increasing the expression of GIRK channels in dendritic spines (Nassirpour et al., 2010). This results in an increase in both the basal and serotonin-induced GIRK currents, but decreases $GABA_B$-induced GIRK currents. Interestingly, these changes were dependent on CaMKII, implying that an increase in internal calcium is required (Nassirpour et al., 2010). This is similar to the NMDA receptor activity-dependent increase observed in GIRK expression (see above) and suggests that similar mechanisms may be responsible for GIRK channel modulation in response to both excitatory events and drug exposure.

5. SUMMARY AND FUTURE DIRECTIONS

In this review, we discussed the evidence supporting a role for GIRK channels as a common pathway through which drugs of abuse can alter learning and memory processes. While the studies reviewed here lay the groundwork for investigating this possibility, future work will be required to fully understand the role of GIRK channels in these processes. For example, although GIRK channels are implicated in drug seeking and voluntary consumptive behaviors (Blednov et al., 2001; Hill et al., 2003; Morgan, Carroll, Loth, Stoffel, & Wickman, 2003), the potential contribution of these channels to models of relapse triggered by reexposure to drug cues is unexplored. It will also be important to determine the role of GIRK channels in the drug-induced changes of normal learning and memory processes. Our lab recently demonstrated that acute ethanol withdrawal alters associative learning in a manner that is distinct from both acute intoxication and withdrawal from chronic ethanol exposure (Tipps, Raybuck, Buck, & Lattal, 2015). In a follow-up study, we found that the loss of GIRK3 prevented the acute withdrawal-induced changes without altering baseline learning (Megan E. Tipps and Kari J. Buck, unpublished data). In addition to illustrating that GIRK channels mediate at least some of the effects of alcohol on learning and memory processes, this work also highlights the potential utility of GIRK-targeted therapeutics for treating drug-induced behaviors without impairing general cognitive function.

As discussed in the introduction to this chapter, one of the primary reasons to identify common mechanisms of learning and drug abuse is to determine their utility as targets for treatment. Interestingly, GIRK channels are already a common factor among several currently used therapeutics for addiction, including $D_{2/3}$ receptor agonists for stimulant dependence (Ersche et al., 2011); mGluR agonists for cocaine and alcohol dependence (Hao, Martin-Fardon, & Weiss, 2010; Sidhpura, Weiss, & Martin-Fardon, 2010); and the $GABA_B$ receptor agonist baclofen for stimulant seeking, alcoholism, and cue-induced craving (Addolorato et al., 2011; Di Ciano & Everitt, 2003b; Young et al., 2014). Previous considerations of the utility of GIRK modulators as therapeutic tools have been hindered by the potential for undesirable side effects, particularly in cardiac function (Lujan, Marron Fernandez de Velasco, Aguado, & Wickman, 2014). However, the recent development of GIRK channel subtype-selective modulators that can be systemically administered have opened up new possibilities for GIRK channels as therapeutic targets (Kaufmann et al., 2013; Wen et al., 2013; Wydeven et al., 2014). In agreement with previous work, we propose that GIRK channels represent a novel and potentially highly effective route for the treatment of addiction disorders (Bodhinathan & Slesinger, 2014; Lujan, Marron Fernandez de Velasco, Aguado, & Wickman, 2014; Luscher & Slesinger, 2010) and may be particularly useful when combined with cognitive behavioral therapy to help reduce the ability of drug-associated cues to induce relapse (McCusker, 2001).

REFERENCES

Addolorato, G., Leggio, L., Ferrulli, A., Cardone, S., Bedogni, G., Caputo, F., et al. (2011). Dose-response effect of baclofen in reducing daily alcohol intake in alcohol dependence: Secondary analysis of a randomized, double-blind, placebo-controlled trial. *Alcohol and Alcoholism*, *46*(3), 312–317. http://dx.doi.org/10.1093/alcalc/agr017.

Adelfinger, L., Turecek, R., Ivankova, K., Jensen, A. A., Moss, S. J., Gassmann, M., et al. (2014). GABAB receptor phosphorylation regulates KCTD12-induced K(+) current desensitization. *Biochemical Pharmacology*, *91*(3), 369–379. http://dx.doi.org/10.1016/j.bcp.2014.07.013.

Aguado, C., Colon, J., Ciruela, F., Schlaudraff, F., Cabanero, M. J., Perry, C., et al. (2008). Cell type-specific subunit composition of G protein-gated potassium channels in the cerebellum. *Journal of Neurochemistry*, *105*(2), 497–511. http://dx.doi.org/10.1111/j.1471-4159.2007.05153.x.

Antonarakis, S. E., Lyle, R., Dermitzakis, E. T., Reymond, A., & Deutsch, S. (2004). Chromosome 21 and Down syndrome: From genomics to pathophysiology. *Nature Reviews. Genetics*, *5*(10), 725–738. http://dx.doi.org/10.1038/nrg1448.

Arora, D., Hearing, M., Haluk, D. M., Mirkovic, K., Fajardo-Serrano, A., Wessendorf, M. W., et al. (2011). Acute cocaine exposure weakens GABA(B) receptor-dependent G-protein-gated inwardly rectifying K+ signaling in dopamine neurons of the ventral tegmental area. *The Journal of Neuroscience: The Official Journal of the Society for Neuroscience*, *31*(34), 12251–12257. http://dx.doi.org/10.1523/JNEUROSCI.0494-11.2011.

Arroyo, M., Markou, A., Robbins, T. W., & Everitt, B. J. (1998). Acquisition, maintenance and reinstatement of intravenous cocaine self-administration under a second-order schedule of reinforcement in rats: Effects of conditioned cues and continuous access to cocaine. *Psychopharmacology*, *140*(3), 331–344.

Aryal, P., Dvir, H., Choe, S., & Slesinger, P. A. (2009). A discrete alcohol pocket involved in GIRK channel activation. *Nature Neuroscience*, *12*(8), 988–995. http://dx.doi.org/10.1038/nn.2358.

Bacci, A., Huguenard, J. R., & Prince, D. A. (2004). Long-lasting self-inhibition of neocortical interneurons mediated by endocannabinoids. *Nature*, *431*(7006), 312–316. http://dx.doi.org/10.1038/nature02913.

Balana, B., Maslennikov, I., Kwiatkowski, W., Stern, K. M., Bahima, L., Choe, S., et al. (2011). Mechanism underlying selective regulation of G protein-gated inwardly rectifying potassium channels by the psychostimulant-sensitive sorting nexin 27. *Proceedings of the National Academy of Sciences of the United States of America*, *108*(14), 5831–5836. http://dx.doi.org/10.1073/pnas.1018645108.

Bandmann, O., Davis, M. B., Marsden, C. D., & Wood, N. W. (1996). The human homologue of the weaver mouse gene in familial and sporadic Parkinson's disease. *Neuroscience*, *72*(4), 877–879.

Bao, G., Kang, L., Li, H., Li, Y., Pu, L., Xia, P., et al. (2007). Morphine and heroin differentially modulate in vivo hippocampal LTP in opiate-dependent rat. *Neuropsychopharmacology: Official Publication of the American College of Neuropsychopharmacology*, *32*(8), 1738–1749. http://dx.doi.org/10.1038/sj.npp.1301308.

Beckstead, M. J., Grandy, D. K., Wickman, K., & Williams, J. T. (2004). Vesicular dopamine release elicits an inhibitory postsynaptic current in midbrain dopamine neurons. *Neuron*, *42*(6), 939–946. http://dx.doi.org/10.1016/j.neuron.2004.05.019.

Beckstead, M. J., & Williams, J. T. (2007). Long-term depression of a dopamine IPSC. *The Journal of Neuroscience: The Official Journal of the Society for Neuroscience*, *27*(8), 2074–2080. http://dx.doi.org/10.1523/JNEUROSCI.3251-06.2007.

Berlin, S., Keren-Raifman, T., Castel, R., Rubinstein, M., Dessauer, C. W., Ivanina, T., et al. (2010). G alpha(i) and G betagamma jointly regulate the conformations of a G betagamma effector, the neuronal G protein-activated K+ channel (GIRK). *The Journal of Biological Chemistry*, *285*(9), 6179–6185. http://dx.doi.org/10.1074/jbc.M109.085944.

Best, T. K., Siarey, R. J., & Galdzicki, Z. (2007). Ts65Dn, a mouse model of Down syndrome, exhibits increased GABAB-induced potassium current. *Journal of Neurophysiology*, *97*(1), 892–900. http://dx.doi.org/10.1152/jn.00626.2006.

Beyer, C. E., & Steketee, J. D. (2002). Cocaine sensitization: Modulation by dopamine D2 receptors. *Cerebral Cortex*, *12*(5), 526–535.

Bishop, G. B., Cullinan, W. E., Curran, E., & Gutstein, H. B. (2002). Abused drugs modulate RGS4 mRNA levels in rat brain: Comparison between acute drug treatment and a drug challenge after chronic treatment. *Neurobiology of Disease*, *10*(3), 334–343.

Blanchet, C., & Luscher, C. (2002). Desensitization of mu-opioid receptor-evoked potassium currents: Initiation at the receptor, expression at the effector. *Proceedings of the National Academy of Sciences of the United States of America*, *99*(7), 4674–4679. http://dx.doi.org/10.1073/pnas.072075399.

Blanchet, C., Sollini, M., & Luscher, C. (2003). Two distinct forms of desensitization of G-protein coupled inwardly rectifying potassium currents evoked by alkaloid and peptide mu-opioid receptor agonists. *Molecular and Cellular Neurosciences*, *24*(2), 517–523.

Blednov, Y. A., Stoffel, M., Alva, H., & Harris, R. A. (2003). A pervasive mechanism for analgesia: Activation of GIRK2 channels. *Proceedings of the National Academy of Sciences of the United States of America*, *100*(1), 277–282. http://dx.doi.org/10.1073/pnas.012682399.

Blednov, Y. A., Stoffel, M., Chang, S. R., & Harris, R. A. (2001). Potassium channels as targets for ethanol: Studies of G-protein-coupled inwardly rectifying potassium channel 2 (GIRK2) null mutant mice. *The Journal of Pharmacology and Experimental Therapeutics*, *298*(2), 521–530.

Bodhinathan, K., & Slesinger, P. A. (2013). Molecular mechanism underlying ethanol activation of G-protein-gated inwardly rectifying potassium channels. *Proceedings of the National Academy of Sciences of the United States of America*, *110*(45), 18309–18314. http://dx.doi.org/10.1073/pnas.1311406110.

Bodhinathan, K., & Slesinger, P. A. (2014). Alcohol modulation of G-protein-gated inwardly rectifying potassium channels: From binding to therapeutics. *Frontiers in Physiology*, *5*, 76. http://dx.doi.org/10.3389/fphys.2014.00076.

Bonci, A., & Malenka, R. C. (1999). Properties and plasticity of excitatory synapses on dopaminergic and GABAergic cells in the ventral tegmental area. *The Journal of Neuroscience: The Official Journal of the Society for Neuroscience*, *19*(10), 3723–3730.

Bossert, J. M., & Stern, A. L. (2014). Role of ventral subiculum in context-induced reinstatement of heroin seeking in rats. *Addiction Biology*, *19*(3), 338–342. http://dx.doi.org/10.1111/adb.12015.

Brown, S. G., Thomas, A., Dekker, L. V., Tinker, A., & Leaney, J. L. (2005). PKC-delta sensitizes Kir3.1/3.2 channels to changes in membrane phospholipid levels after M3

receptor activation in HEK-293 cells. *American Journal of Physiology. Cell Physiology, 289*(3), C543–C556. http://dx.doi.org/10.1152/ajpcell.00025.2005.
Chabert, C., Jamon, M., Cherfouh, A., Duquenne, V., Smith, D. J., Rubin, E., et al. (2004). Functional analysis of genes implicated in Down syndrome: 1. Cognitive abilities in mice transpolygenic for Down Syndrome Chromosomal Region-1 (DCR-1). *Behavior Genetics, 34*(6), 559–569. http://dx.doi.org/10.1007/s10519-004-5584-3.
Chen, X., & Johnston, D. (2005). Constitutively active G-protein-gated inwardly rectifying K+ channels in dendrites of hippocampal CA1 pyramidal neurons. *The Journal of Neuroscience: The Official Journal of the Society for Neuroscience, 25*(15), 3787–3792. http://dx.doi.org/10.1523/JNEUROSCI.5312-04.2005.
Chevaleyre, V., Takahashi, K. A., & Castillo, P. E. (2006). Endocannabinoid-mediated synaptic plasticity in the CNS. *Annual Review of Neuroscience, 29*, 37–76. http://dx.doi.org/10.1146/annurev.neuro.29.051605.112834.
Chung, H. J., Ge, W. P., Qian, X., Wiser, O., Jan, Y. N., & Jan, L. Y. (2009). G protein-activated inwardly rectifying potassium channels mediate depotentiation of long-term potentiation. *Proceedings of the National Academy of Sciences of the United States of America, 106*(2), 635–640. http://dx.doi.org/10.1073/pnas.0811685106.
Chung, H. J., Qian, X., Ehlers, M., Jan, Y. N., & Jan, L. Y. (2009). Neuronal activity regulates phosphorylation-dependent surface delivery of G protein-activated inwardly rectifying potassium channels. *Proceedings of the National Academy of Sciences of the United States of America, 106*(2), 629–634. http://dx.doi.org/10.1073/pnas.0811615106.
Cooper, A., Grigoryan, G., Guy-David, L., Tsoory, M. M., Chen, A., & Reuveny, E. (2012). Trisomy of the G protein-coupled K+ channel gene, Kcnj6, affects reward mechanisms, cognitive functions, and synaptic plasticity in mice. *Proceedings of the National Academy of Sciences of the United States of America, 109*(7), 2642–2647. http://dx.doi.org/10.1073/pnas.1109099109.
Couve, A., Kittler, J. T., Uren, J. M., Calver, A. R., Pangalos, M. N., Walsh, F. S., et al. (2001). Association of GABA(B) receptors and members of the 14-3-3 family of signaling proteins. *Molecular and Cellular Neurosciences, 17*(2), 317–328. http://dx.doi.org/10.1006/mcne.2000.0938.
Crabbe, J. C., Phillips, T. J., Harris, R. A., Arends, M. A., & Koob, G. F. (2006). Alcohol-related genes: Contributions from studies with genetically engineered mice. *Addiction Biology, 11*(3-4), 195–269. http://dx.doi.org/10.1111/j.1369-1600.2006.00038.x.
Cramer, N. P., Best, T. K., Stoffel, M., Siarey, R. J., & Galdzicki, Z. (2010). GABAB-GIRK2-mediated signaling in Down syndrome. *Advances in Pharmacology, 58*, 397–426. http://dx.doi.org/10.1016/S1054-3589(10)58015-3.
Crews, F. T., Morrow, A. L., Criswell, H., & Breese, G. (1996). Effects of ethanol on ion channels. *International Review of Neurobiology, 39*, 283–367.
Crombag, H. S., Bossert, J. M., Koya, E., & Shaham, Y. (2008). Review. Context-induced relapse to drug seeking: A review. *Philosophical Transactions of the Royal Society of London. Series B, Biological Sciences, 363*(1507), 3233–3243. http://dx.doi.org/10.1098/rstb.2008.0090.
Crombag, H. S., & Shaham, Y. (2002). Renewal of drug seeking by contextual cues after prolonged extinction in rats. *Behavioral Neuroscience, 116*(1), 169–173.
Cruz, H. G., Berton, F., Sollini, M., Blanchet, C., Pravetoni, M., Wickman, K., et al. (2008). Absence and rescue of morphine withdrawal in GIRK/Kir3 knock-out mice. *The Journal of Neuroscience: The Official Journal of the Society for Neuroscience, 28*(15), 4069–4077. http://dx.doi.org/10.1523/JNEUROSCI.0267-08.2008.
Cruz, H. G., Ivanova, T., Lunn, M. L., Stoffel, M., Slesinger, P. A., & Luscher, C. (2004). Bi-directional effects of GABA(B) receptor agonists on the mesolimbic dopamine system. *Nature Neuroscience*, 7(2), 153–159. http://dx.doi.org/10.1038/nn1181.
Dascal, N. (1997). Signalling via the G protein-activated K+ channels. *Cellular Signalling, 9*(8), 551–573.

Derenne, A., Arsenault, M. L., Austin, D. P., & Weatherly, J. N. (2007). Weaver mutant mice exhibit long-term learning deficits under several measures of instrumental behavior. *Physiology & Behavior, 92*(5), 1002–1009. http://dx.doi.org/10.1016/j.physbeh.2007.07.015.

Di Ciano, P., & Everitt, B. J. (2003a). Differential control over drug-seeking behavior by drug-associated conditioned reinforcers and discriminative stimuli predictive of drug availability. *Behavioral Neuroscience, 117*(5), 952–960. http://dx.doi.org/10.1037/0735-7044.117.5.952.

Di Ciano, P., & Everitt, B. J. (2003b). The GABA(B) receptor agonist baclofen attenuates cocaine- and heroin-seeking behavior by rats. *Neuropsychopharmacology: Official Publication of the American College of Neuropsychopharmacology, 28*(3), 510–518. http://dx.doi.org/10.1038/sj.npp.1300088.

Doupnik, C. A., Davidson, N., & Lester, H. A. (1995). The inward rectifier potassium channel family. *Current Opinion in Neurobiology, 5*(3), 268–277.

Doupnik, C. A., Davidson, N., Lester, H. A., & Kofuji, P. (1997). RGS proteins reconstitute the rapid gating kinetics of gbetagamma-activated inwardly rectifying K+ channels. *Proceedings of the National Academy of Sciences of the United States of America, 94*(19), 10461–10466.

Doupnik, C. A., Jaen, C., & Zhang, Q. (2004). Measuring the modulatory effects of RGS proteins on GIRK channels. *Methods in Enzymology, 389*, 131–154. http://dx.doi.org/10.1016/S0076-6879(04)89009-8.

Dutar, P., & Nicoll, R. A. (1988). A physiological role for GABAB receptors in the central nervous system. *Nature, 332*(6160), 156–158. http://dx.doi.org/10.1038/332156a0.

Eisenberg, B., & Messer, A. (1989). Tonic/clonic seizures in a mouse mutant carrying the weaver gene. *Neuroscience Letters, 96*(2), 168–172.

Engblom, D., Bilbao, A., Sanchis-Segura, C., Dahan, L., Perreau-Lenz, S., Balland, B., et al. (2008). Glutamate receptors on dopamine neurons control the persistence of cocaine seeking. *Neuron, 59*(3), 497–508. http://dx.doi.org/10.1016/j.neuron.2008.07.010.

Ersche, K. D., Roiser, J. P., Abbott, S., Craig, K. J., Muller, U., Suckling, J., et al. (2011). Response perseveration in stimulant dependence is associated with striatal dysfunction and can be ameliorated by a D(2/3) receptor agonist. *Biological Psychiatry, 70*(8), 754–762. http://dx.doi.org/10.1016/j.biopsych.2011.06.033.

Everitt, B. J. (2014). Neural and psychological mechanisms underlying compulsive drug seeking habits and drug memories—Indications for novel treatments of addiction. *The European Journal of Neuroscience, 40*(1), 2163–2182. http://dx.doi.org/10.1111/ejn.12644.

Everitt, B. J., Dickinson, A., & Robbins, T. W. (2001). The neuropsychological basis of addictive behaviour. *Brain Research. Brain Research Reviews, 36*(2–3), 129–138.

Everitt, B. J., Parkinson, J. A., Olmstead, M. C., Arroyo, M., Robledo, P., & Robbins, T. W. (1999). Associative processes in addiction and reward. The role of amygdala-ventral striatal subsystems. *Annals of the New York Academy of Sciences, 877*, 412–438.

Fajardo-Serrano, A., Wydeven, N., Young, D., Watanabe, M., Shigemoto, R., Martemyanov, K. A., et al. (2013). Association of Rgs7/Gbeta5 complexes with Girk channels and GABAB receptors in hippocampal CA1 pyramidal neurons. *Hippocampus, 23*(12), 1231–1245. http://dx.doi.org/10.1002/hipo.22161.

Fanger, G. R., Widmann, C., Porter, A. C., Sather, S., Johnson, G. L., & Vaillancourt, R. R. (1998). 14-3-3 proteins interact with specific MEK kinases. *The Journal of Biological Chemistry, 273*(6), 3476–3483.

Fanselow, M. S., & Dong, H. W. (2010). Are the dorsal and ventral hippocampus functionally distinct structures? *Neuron, 65*(1), 7–19. http://dx.doi.org/10.1016/j.neuron.2009.11.031.

Fernandez-Alacid, L., Aguado, C., Ciruela, F., Martin, R., Colon, J., Cabanero, M. J., et al. (2009). Subcellular compartment-specific molecular diversity of pre- and post-synaptic GABA-activated GIRK channels in Purkinje cells. *Journal of Neurochemistry, 110*(4), 1363–1376. http://dx.doi.org/10.1111/j.1471-4159.2009.06229.x.

Foote, M., & Zhou, Y. (2012). 14-3-3 proteins in neurological disorders. *International Journal of Biochemistry and Molecular Biology, 3*(2), 152–164.

Fu, H., Subramanian, R. R., & Masters, S. C. (2000). 14-3-3 proteins: Structure, function, and regulation. *Annual Review of Pharmacology and Toxicology, 40*, 617–647. http://dx.doi.org/10.1146/annurev.pharmtox.40.1.617.

Goosens, K. A., & Maren, S. (2002). Long-term potentiation as a substrate for memory: Evidence from studies of amygdaloid plasticity and Pavlovian fear conditioning. *Hippocampus, 12*(5), 592–599. http://dx.doi.org/10.1002/hipo.10099.

Gould, T. J., & Leach, P. T. (2014). Cellular, molecular, and genetic substrates underlying the impact of nicotine on learning. *Neurobiology of Learning and Memory, 107*, 108–132. http://dx.doi.org/10.1016/j.nlm.2013.08.004.

Guo, J., & Ikeda, S. R. (2004). Endocannabinoids modulate N-type calcium channels and G-protein-coupled inwardly rectifying potassium channels via CB1 cannabinoid receptors heterologously expressed in mammalian neurons. *Molecular Pharmacology, 65*(3), 665–674. http://dx.doi.org/10.1124/mol.65.3.665.

Hao, Y., Martin-Fardon, R., & Weiss, F. (2010). Behavioral and functional evidence of metabotropic glutamate receptor 2/3 and metabotropic glutamate receptor 5 dysregulation in cocaine-escalated rats: Factor in the transition to dependence. *Biological Psychiatry, 68*(3), 240–248. http://dx.doi.org/10.1016/j.biopsych.2010.02.011.

Harashima, C., Jacobowitz, D. M., Stoffel, M., Chakrabarti, L., Haydar, T. F., Siarey, R. J., et al. (2006). Elevated expression of the G-protein-activated inwardly rectifying potassium channel 2 (GIRK2) in cerebellar unipolar brush cells of a Down syndrome mouse model. *Cellular and Molecular Neurobiology, 26*(4-6), 719–734. http://dx.doi.org/10.1007/s10571-006-9066-4.

Harashima, C., Jacobowitz, D. M., Witta, J., Borke, R. C., Best, T. K., Siarey, R. J., et al. (2006). Abnormal expression of the G-protein-activated inwardly rectifying potassium channel 2 (GIRK2) in hippocampus, frontal cortex, and substantia nigra of Ts65Dn mouse: A model of Down syndrome. *The Journal of Comparative Neurology, 494*(5), 815–833. http://dx.doi.org/10.1002/cne.20844.

Hearing, M., Kotecki, L., Marron Fernandez de Velasco, E., Fajardo-Serrano, A., Chung, H. J., et al. (2013). Repeated cocaine weakens GABA(B)-Girk signaling in layer 5/6 pyramidal neurons in the prelimbic cortex. *Neuron, 80*(1), 159–170. http://dx.doi.org/10.1016/j.neuron.2013.07.019.

Herlitze, S., Ruppersberg, J. P., & Mark, M. D. (1999). New roles for RGS2, 5 and 8 on the ratio-dependent modulation of recombinant GIRK channels expressed in Xenopus oocytes. *The Journal of Physiology, 517*(Pt. 2), 341–352.

Hill, K. G., Alva, H., Blednov, Y. A., & Cunningham, C. L. (2003). Reduced ethanol-induced conditioned taste aversion and conditioned place preference in GIRK2 null mutant mice. *Psychopharmacology, 169*(1), 108–114. http://dx.doi.org/10.1007/s00213-003-1472-4.

Hollerman, J. R., & Schultz, W. (1998). Dopamine neurons report an error in the temporal prediction of reward during learning. *Nature Neuroscience, 1*(4), 304–309. http://dx.doi.org/10.1038/1124.

Honda, R., Ohba, Y., & Yasuda, H. (1997). 14-3-3 zeta protein binds to the carboxyl half of mouse wee1 kinase. *Biochemical and Biophysical Research Communications, 230*(2), 262–265. http://dx.doi.org/10.1006/bbrc.1996.5933.

Hooks, S. B., Martemyanov, K., & Zachariou, V. (2008). A role of RGS proteins in drug addiction. *Biochemical Pharmacology, 75*(1), 76–84. http://dx.doi.org/10.1016/j.bcp.2007.07.045.

Howard, R. J., Murail, S., Ondricek, K. E., Corringer, P. J., Lindahl, E., Trudell, J. R., et al. (2011). Structural basis for alcohol modulation of a pentameric ligand-gated ion channel. *Proceedings of the National Academy of Sciences of the United States of America, 108*(29), 12149–12154. http://dx.doi.org/10.1073/pnas.1104480108.

Huang, C. L., Feng, S., & Hilgemann, D. W. (1998). Direct activation of inward rectifier potassium channels by PIP2 and its stabilization by Gbetagamma. *Nature, 391*(6669), 803–806. http://dx.doi.org/10.1038/35882.

Huang, C. C., Liang, Y. C., & Hsu, K. S. (1999). A role for extracellular adenosine in time-dependent reversal of long-term potentiation by low-frequency stimulation at hippocampal CA1 synapses. *The Journal of Neuroscience: The Official Journal of the Society for Neuroscience, 19*(22), 9728–9738.

Huang, Y., & Mucke, L. (2012). Alzheimer mechanisms and therapeutic strategies. *Cell, 148*(6), 1204–1222. http://dx.doi.org/10.1016/j.cell.2012.02.040.

Huang, L., Rowan, M. J., & Anwyl, R. (1999). Induction of long-lasting depression by (+)-alpha-methyl-4-carboxyphenylglycine and other group II mGlu receptor ligands in the dentate gyrus of the hippocampus in vitro. *European Journal of Pharmacology, 366*(2–3), 151–158.

Huang, C. S., Shi, S. H., Ule, J., Ruggiu, M., Barker, L. A., Darnell, R. B., et al. (2005). Common molecular pathways mediate long-term potentiation of synaptic excitation and slow synaptic inhibition. *Cell, 123*(1), 105–118. http://dx.doi.org/10.1016/j.cell.2005.07.033.

Hyman, S. E. (2005). Addiction: A disease of learning and memory. *The American Journal of Psychiatry, 162*(8), 1414–1422. http://dx.doi.org/10.1176/appi.ajp.162.8.1414.

Hyman, S. E., Malenka, R. C., & Nestler, E. J. (2006). Neural mechanisms of addiction: The role of reward-related learning and memory. *Annual Review of Neuroscience, 29*, 565–598. http://dx.doi.org/10.1146/annurev.neuro.29.051605.113009.

Ikeda, K., Kobayashi, T., Kumanishi, T., Yano, R., Sora, I., & Niki, H. (2002). Molecular mechanisms of analgesia induced by opioids and ethanol: Is the GIRK channel one of the keys? *Neuroscience Research, 44*(2), 121–131.

Inanobe, A., Yoshimoto, Y., Horio, Y., Morishige, K. I., Hibino, H., Matsumoto, S., et al. (1999). Characterization of G-protein-gated K+ channels composed of Kir3.2 subunits in dopaminergic neurons of the substantia nigra. *The Journal of Neuroscience: The Official Journal of the Society for Neuroscience, 19*(3), 1006–1017.

Ingram, S. L., Macey, T. A., Fossum, E. N., & Morgan, M. M. (2008). Tolerance to repeated morphine administration is associated with increased potency of opioid agonists. *Neuropsychopharmacology: Official Publication of the American College of Neuropsychopharmacology, 33*(10), 2494–2504. http://dx.doi.org/10.1038/sj.npp.1301634.

Isokawa, M. (2012). Cellular signal mechanisms of reward-related plasticity in the hippocampus. *Neural Plasticity, 2012*, 945373. http://dx.doi.org/10.1155/2012/945373.

Ito, M. (1989). Long-term depression. *Annual Review of Neuroscience, 12*, 85–102. http://dx.doi.org/10.1146/annurev.ne.12.030189.000505.

Ivankova, K., Turecek, R., Fritzius, T., Seddik, R., Prezeau, L., Comps-Agrar, L., et al. (2013). Up-regulation of GABA(B) receptor signaling by constitutive assembly with the K+ channel tetramerization domain-containing protein 12 (KCTD12). *The Journal of Biological Chemistry, 288*(34), 24848–24856. http://dx.doi.org/10.1074/jbc.M113.476770.

Jayaram, P., & Steketee, J. D. (2004). Effects of repeated cocaine on medial prefrontal cortical GABAB receptor modulation of neurotransmission in the mesocorticolimbic dopamine system. *Journal of Neurochemistry, 90*(4), 839–847. http://dx.doi.org/10.1111/j.1471-4159.2004.02525.x.

Jelacic, T. M., Kennedy, M. E., Wickman, K., & Clapham, D. E. (2000). Functional and biochemical evidence for G-protein-gated inwardly rectifying K+ (GIRK) channels composed of GIRK2 and GIRK3. *The Journal of Biological Chemistry, 275*(46), 36211–36216. http://dx.doi.org/10.1074/jbc.M007087200.

Jelacic, T. M., Sims, S. M., & Clapham, D. E. (1999). Functional expression and characterization of G-protein-gated inwardly rectifying K+ channels containing GIRK3. *The Journal of Membrane Biology, 169*(2), 123–129.

Kajii, Y., Muraoka, S., Hiraoka, S., Fujiyama, K., Umino, A., & Nishikawa, T. (2003). A developmentally regulated and psychostimulant-inducible novel rat gene mrt1 encoding PDZ-PX proteins isolated in the neocortex. *Molecular Psychiatry, 8*(4), 434–444. http://dx.doi.org/10.1038/sj.mp.4001258.

Karschin, C., Dissmann, E., Stuhmer, W., & Karschin, A. (1996). IRK(1-3) and GIRK(1-4) inwardly rectifying K+ channel mRNAs are differentially expressed in the adult rat brain. *The Journal of Neuroscience: The Official Journal of the Society for Neuroscience, 16*(11), 3559–3570.

Kauer, J. A. (2004). Learning mechanisms in addiction: Synaptic plasticity in the ventral tegmental area as a result of exposure to drugs of abuse. *Annual Review of Physiology, 66*, 447–475. http://dx.doi.org/10.1146/annurev.physiol.66.032102.112534.

Kauer, J. A., & Malenka, R. C. (2007). Synaptic plasticity and addiction. *Nature Reviews. Neuroscience, 8*(11), 844–858. http://dx.doi.org/10.1038/nrn2234.

Kaufmann, K., Romaine, I., Days, E., Pascual, C., Malik, A., Yang, L., et al. (2013). ML297 (VU0456810), the first potent and selective activator of the GIRK potassium channel, displays antiepileptic properties in mice. *ACS Chemical Neuroscience, 4*(9), 1278–1286. http://dx.doi.org/10.1021/cn400062a.

Kelley, A. E. (2004). Memory and addiction: Shared neural circuitry and molecular mechanisms. *Neuron, 44*(1), 161–179. http://dx.doi.org/10.1016/j.neuron.2004.09.016.

Kennedy, M. E., Nemec, J., Corey, S., Wickman, K., & Clapham, D. E. (1999). GIRK4 confers appropriate processing and cell surface localization to G-protein-gated potassium channels. *The Journal of Biological Chemistry, 274*(4), 2571–2582.

Kim, C. S., & Johnston, D. (2015). A1 adenosine receptor-mediated GIRK channels contribute to the resting conductance of CA1 neurons in the dorsal hippocampus. *Journal of Neurophysiology, 113*(7), 2511–2523. http://dx.doi.org/10.1152/jn.00951.2014.

Kim, K. M., Nakajima, Y., & Nakajima, S. (1995). G protein-coupled inward rectifier modulated by dopamine agonists in cultured substantia nigra neurons. *Neuroscience, 69*(4), 1145–1158.

Knoflach, F., & Kemp, J. A. (1998). Metabotropic glutamate group II receptors activate a G protein-coupled inwardly rectifying K+ current in neurones of the rat cerebellum. *The Journal of Physiology, 509*(Pt. 2), 347–354.

Kobayashi, T., Ikeda, K., Kojima, H., Niki, H., Yano, R., Yoshioka, T., et al. (1999). Ethanol opens G-protein-activated inwardly rectifying K+ channels. *Nature Neuroscience, 2*(12), 1091–1097. http://dx.doi.org/10.1038/16019.

Kogan, J. H., Nestler, E. J., & Aghajanian, G. K. (1992). Elevated basal firing rates and enhanced responses to 8-Br-cAMP in locus coeruleus neurons in brain slices from opiate-dependent rats. *European Journal of Pharmacology, 211*(1), 47–53.

Kombian, S. B., & Malenka, R. C. (1994). Simultaneous LTP of non-NMDA- and LTD of NMDA-receptor-mediated responses in the nucleus accumbens. *Nature, 368*(6468), 242–246. http://dx.doi.org/10.1038/368242a0.

Koyrakh, L., Lujan, R., Colon, J., Karschin, C., Kurachi, Y., Karschin, A., et al. (2005). Molecular and cellular diversity of neuronal G-protein-gated potassium channels. *The Journal of Neuroscience: The Official Journal of the Society for Neuroscience, 25*(49), 11468–11478. http://dx.doi.org/10.1523/JNEUROSCI.3484-05.2005.

Kozell, L. B., Walter, N. A., Milner, L. C., Wickman, K., & Buck, K. J. (2009). Mapping a barbiturate withdrawal locus to a 0.44 Mb interval and analysis of a novel null mutant identify a role for Kcnj9 (GIRK3) in withdrawal from pentobarbital, zolpidem, and ethanol. *The Journal of Neuroscience: The Official Journal of the Society for Neuroscience, 29*(37), 11662–11673. http://dx.doi.org/10.1523/JNEUROSCI.1413-09.2009.

Kumar, A. (2011). Long-term potentiation at CA3-CA1 hippocampal synapses with special emphasis on aging, disease, and stress. *Frontiers in Aging Neuroscience, 3*, 7. http://dx.doi.org/10.3389/fnagi.2011.00007.

Labouebe, G., Lomazzi, M., Cruz, H. G., Creton, C., Lujan, R., Li, M., et al. (2007). RGS2 modulates coupling between GABAB receptors and GIRK channels in dopamine neurons of the ventral tegmental area. *Nature Neuroscience, 10*(12), 1559–1568. http://dx.doi.org/10.1038/nn2006.

Lalive, A. L., Munoz, M. B., Bellone, C., Slesinger, P. A., Luscher, C., & Tan, K. R. (2014). Firing modes of dopamine neurons drive bidirectional GIRK channel plasticity. *The Journal of Neuroscience: The Official Journal of the Society for Neuroscience, 34*(15), 5107–5114. http://dx.doi.org/10.1523/JNEUROSCI.5203-13.2014.

Land, B. B., Bruchas, M. R., Lemos, J. C., Xu, M., Melief, E. J., & Chavkin, C. (2008). The dysphoric component of stress is encoded by activation of the dynorphin kappa-opioid system. *The Journal of Neuroscience: The Official Journal of the Society for Neuroscience, 28*(2), 407–414. http://dx.doi.org/10.1523/JNEUROSCI.4458-07.2008.

Land, B. B., Bruchas, M. R., Schattauer, S., Giardino, W. J., Aita, M., Messinger, D., et al. (2009). Activation of the kappa opioid receptor in the dorsal raphe nucleus mediates the aversive effects of stress and reinstates drug seeking. *Proceedings of the National Academy of Sciences of the United States of America, 106*(45), 19168–19173. http://dx.doi.org/10.1073/pnas.0910705106.

Leaney, J. L., Dekker, L. V., & Tinker, A. (2001). Regulation of a G protein-gated inwardly rectifying K+ channel by a Ca(2+)-independent protein kinase C. *The Journal of Physiology, 534*(Pt. 2), 367–379.

Lei, Q., Jones, M. B., Talley, E. M., Garrison, J. C., & Bayliss, D. A. (2003). Molecular mechanisms mediating inhibition of G protein-coupled inwardly-rectifying K+ channels. *Molecules and Cells, 15*(1), 1–9.

Lei, Q., Talley, E. M., & Bayliss, D. A. (2001). Receptor-mediated inhibition of G protein-coupled inwardly rectifying potassium channels involves G(alpha)q family subunits, phospholipase C, and a readily diffusible messenger. *The Journal of Biological Chemistry, 276*(20), 16720–16730. http://dx.doi.org/10.1074/jbc.M100207200.

Lemos, J. C., Roth, C. A., Messinger, D. I., Gill, H. K., Phillips, P. E., & Chavkin, C. (2012). Repeated stress dysregulates kappa-opioid receptor signaling in the dorsal raphe through a p38alpha MAPK-dependent mechanism. *The Journal of Neuroscience: The Official Journal of the Society for Neuroscience, 32*(36), 12325–12336. http://dx.doi.org/10.1523/JNEUROSCI.2053-12.2012.

Lewohl, J. M., Wilson, W. R., Mayfield, R. D., Brozowski, S. J., Morrisett, R. A., & Harris, R. A. (1999). G-protein-coupled inwardly rectifying potassium channels are targets of alcohol action. *Nature Neuroscience, 2*(12), 1084–1090. http://dx.doi.org/10.1038/16012.

Li, Q., Li, W., Wang, H., Wang, Y., Zhang, Y., Zhu, J., et al. (2014). Predicting subsequent relapse by drug-related cue-induced brain activation in heroin addiction: An event-related functional magnetic resonance imaging study. *Addiction Biology*. http://dx.doi.org/10.1111/adb.12182.

Lomazzi, M., Slesinger, P. A., & Luscher, C. (2008). Addictive drugs modulate GIRK-channel signaling by regulating RGS proteins. *Trends in Pharmacological Sciences, 29*(11), 544–549. http://dx.doi.org/10.1016/j.tips.2008.07.011.

Lujan, R., Marron Fernandez de Velasco, E., Aguado, C., & Wickman, K. (2014). New insights into the therapeutic potential of Girk channels. *Trends in Neurosciences*, *37*(1), 20–29. http://dx.doi.org/10.1016/j.tins.2013.10.006.

Lunn, M. L., Nassirpour, R., Arrabit, C., Tan, J., McLeod, I., Arias, C. M., et al. (2007). A unique sorting nexin regulates trafficking of potassium channels via a PDZ domain interaction. *Nature Neuroscience*, *10*(10), 1249–1259. http://dx.doi.org/10.1038/nn1953.

Luo, A. H., Tahsili-Fahadan, P., Wise, R. A., Lupica, C. R., & Aston-Jones, G. (2011). Linking context with reward: A functional circuit from hippocampal CA3 to ventral tegmental area. *Science*, *333*(6040), 353–357. http://dx.doi.org/10.1126/science.1204622.

Luo, Y. X., Xue, Y. X., Shen, H. W., & Lu, L. (2013). Role of amygdala in drug memory. *Neurobiology of Learning and Memory*, *105*, 159–173. http://dx.doi.org/10.1016/j.nlm.2013.06.017.

Luscher, C., & Bellone, C. (2008). Cocaine-evoked synaptic plasticity: A key to addiction? *Nature Neuroscience*, *11*(7), 737–738. http://dx.doi.org/10.1038/nn0708-737.

Luscher, C., Jan, L. Y., Stoffel, M., Malenka, R. C., & Nicoll, R. A. (1997). G protein-coupled inwardly rectifying K+ channels (GIRKs) mediate postsynaptic but not presynaptic transmitter actions in hippocampal neurons. *Neuron*, *19*(3), 687–695.

Luscher, C., & Malenka, R. C. (2011). Drug-evoked synaptic plasticity in addiction: From molecular changes to circuit remodeling. *Neuron*, *69*(4), 650–663. http://dx.doi.org/10.1016/j.neuron.2011.01.017.

Luscher, C., & Slesinger, P. A. (2010). Emerging roles for G protein-gated inwardly rectifying potassium (GIRK) channels in health and disease. *Nature Reviews. Neuroscience*, *11*(5), 301–315. http://dx.doi.org/10.1038/nrn2834.

Ma, D., Zerangue, N., Raab-Graham, K., Fried, S. R., Jan, Y. N., & Jan, L. Y. (2002). Diverse trafficking patterns due to multiple traffic motifs in G protein-activated inwardly rectifying potassium channels from brain and heart. *Neuron*, *33*(5), 715–729.

Maharajan, P., Maharajan, V., Ravagnan, G., & Paino, G. (2001). The weaver mutant mouse: A model to study the ontogeny of dopamine transmission systems and their role in drug addiction. *Progress in Neurobiology*, *64*(3), 269–276.

Maity, B., Stewart, A., Yang, J., Loo, L., Sheff, D., Shepherd, A. J., et al. (2012). Regulator of G protein signaling 6 (RGS6) protein ensures coordination of motor movement by modulating GABAB receptor signaling. *The Journal of Biological Chemistry*, *287*(7), 4972–4981. http://dx.doi.org/10.1074/jbc.M111.297218.

Malenka, R. C., & Bear, M. F. (2004). LTP and LTD: An embarrassment of riches. *Neuron*, *44*(1), 5–21. http://dx.doi.org/10.1016/j.neuron.2004.09.012.

Mao, J., Wang, X., Chen, F., Wang, R., Rojas, A., Shi, Y., et al. (2004). Molecular basis for the inhibition of G protein-coupled inward rectifier K(+) channels by protein kinase C. *Proceedings of the National Academy of Sciences of the United States of America*, *101*(4), 1087–1092. http://dx.doi.org/10.1073/pnas.0304827101.

Maren, S., & Baudry, M. (1995). Properties and mechanisms of long-term synaptic plasticity in the mammalian brain: Relationships to learning and memory. *Neurobiology of Learning and Memory*, *63*(1), 1–18. http://dx.doi.org/10.1006/nlme.1995.1001.

Mark, M. D., & Herlitze, S. (2000). G-protein mediated gating of inward-rectifier K+ channels. *European Journal of Biochemistry/FEBS*, *267*(19), 5830–5836.

Marker, C. L., Lujan, R., Loh, H. H., & Wickman, K. (2005). Spinal G-protein-gated potassium channels contribute in a dose-dependent manner to the analgesic effect of mu- and delta- but not kappa-opioids. *The Journal of Neuroscience: The Official Journal of the Society for Neuroscience*, *25*(14), 3551–3559. http://dx.doi.org/10.1523/JNEUROSCI.4899-04.2005.

Mascia, M. P., Trudell, J. R., & Harris, R. A. (2000). Specific binding sites for alcohols and anesthetics on ligand-gated ion channels. *Proceedings of the National Academy of Sciences of the United States of America*, *97*(16), 9305–9310. http://dx.doi.org/10.1073/pnas.160128797.

Matsumoto, M. (2015). Dopamine signals and physiological origin of cognitive dysfunction in Parkinson's disease. *Movement Disorders: Official Journal of the Movement Disorder Society*, *30*(4), 472–483. http://dx.doi.org/10.1002/mds.26177.

Matsumoto, K., Suzuki, W., & Tanaka, K. (2003). Neuronal correlates of goal-based motor selection in the prefrontal cortex. *Science*, *301*(5630), 229–232. http://dx.doi.org/10.1126/science.1084204.

Mazei-Robison, M. S., & Nestler, E. J. (2012). Opiate-induced molecular and cellular plasticity of ventral tegmental area and locus coeruleus catecholamine neurons. *Cold Spring Harbor Perspectives in Medicine*, *2*(7), a012070. http://dx.doi.org/10.1101/cshperspect.a012070.

McAllister, S. D., Griffin, G., Satin, L. S., & Abood, M. E. (1999). Cannabinoid receptors can activate and inhibit G protein-coupled inwardly rectifying potassium channels in a Xenopus oocyte expression system. *The Journal of Pharmacology and Experimental Therapeutics*, *291*(2), 618–626.

McCusker, C. G. (2001). Cognitive biases and addiction: An evolution in theory and method. *Addiction*, *96*(1), 47–56. http://dx.doi.org/10.1080/09652140020016950.

McLellan, A. T., Lewis, D. C., O'Brien, C. P., & Kleber, H. D. (2000). Drug dependence, a chronic medical illness: Implications for treatment, insurance, and outcomes evaluation. *JAMA: The Journal of the American Medical Association*, *284*(13), 1689–1695.

McQuiston, A. R. (2007). Effects of mu-opioid receptor modulation on GABAB receptor synaptic function in hippocampal CA1. *Journal of Neurophysiology*, *97*(3), 2301–2311. http://dx.doi.org/10.1152/jn.01179.2006.

Medina, I., Krapivinsky, G., Arnold, S., Kovoor, P., Krapivinsky, L., & Clapham, D. E. (2000). A switch mechanism for G beta gamma activation of I(KACh). *The Journal of Biological Chemistry*, *275*(38), 29709–29716. http://dx.doi.org/10.1074/jbc.M004989200.

Mihic, S. J., Ye, Q., Wick, M. J., Koltchine, V. V., Krasowski, M. D., Finn, S. E., et al. (1997). Sites of alcohol and volatile anaesthetic action on GABA(A) and glycine receptors. *Nature*, *389*(6649), 385–389. http://dx.doi.org/10.1038/38738.

Miller, E. K., & Cohen, J. D. (2001). An integrative theory of prefrontal cortex function. *Annual Review of Neuroscience*, *24*, 167–202. http://dx.doi.org/10.1146/annurev.neuro.24.1.167.

Morgan, A. D., Carroll, M. E., Loth, A. K., Stoffel, M., & Wickman, K. (2003). Decreased cocaine self-administration in Kir3 potassium channel subunit knockout mice. *Neuropsychopharmacology: Official Publication of the American College of Neuropsychopharmacology*, *28*(5), 932–938. http://dx.doi.org/10.1038/sj.npp.1300100.

Muller, C. P., Carey, R. J., Huston, J. P., & De Souza Silva, M. A. (2007). Serotonin and psychostimulant addiction: Focus on 5-HT1A-receptors. *Progress in Neurobiology*, *81*(3), 133–178. http://dx.doi.org/10.1016/j.pneurobio.2007.01.001.

Mullner, C., Yakubovich, D., Dessauer, C. W., Platzer, D., & Schreibmayer, W. (2003). Single channel analysis of the regulation of GIRK1/GIRK4 channels by protein phosphorylation. *Biophysical Journal*, *84*(2 Pt. 1), 1399–1409. http://dx.doi.org/10.1016/S0006-3495(03)74954-6.

Munoz, M. B., & Slesinger, P. A. (2014). Sorting nexin 27 regulation of G protein-gated inwardly rectifying K(+) channels attenuates in vivo cocaine response. *Neuron*, *82*(3), 659–669. http://dx.doi.org/10.1016/j.neuron.2014.03.011.

Nagi, K., & Pineyro, G. (2014). Kir3 channel signaling complexes: Focus on opioid receptor signaling. *Frontiers in Cellular Neuroscience*, *8*, 186. http://dx.doi.org/10.3389/fncel.2014.00186.

Nassirpour, R., Bahima, L., Lalive, A. L., Luscher, C., Lujan, R., & Slesinger, P. A. (2010). Morphine- and CaMKII-dependent enhancement of GIRK channel signaling in hippocampal neurons. *The Journal of Neuroscience: The Official Journal of the Society for Neuroscience*, *30*(40), 13419–13430. http://dx.doi.org/10.1523/JNEUROSCI.2966-10.2010.

Nassirpour, R., & Slesinger, P. A. (2007). Subunit-specific regulation of Kir3 channels by sorting nexin 27. *Channels*, *1*(5), 331–333.

Nava-Mesa, M. O., Jimenez-Diaz, L., Yajeya, J., & Navarro-Lopez, J. D. (2013). Amyloid-beta induces synaptic dysfunction through G protein-gated inwardly rectifying potassium channels in the fimbria-CA3 hippocampal synapse. *Frontiers in Cellular Neuroscience*, 7, 117. http://dx.doi.org/10.3389/fncel.2013.00117.

Navarro, B., Kennedy, M. E., Velimirovic, B., Bhat, D., Peterson, A. S., & Clapham, D. E. (1996). Nonselective and G betagamma-insensitive weaver K+ channels. *Science*, *272*(5270), 1950–1953.

Nestler, E. J. (2002). Common molecular and cellular substrates of addiction and memory. *Neurobiology of Learning and Memory*, *78*(3), 637–647.

Nestler, E. J. (2005). Is there a common molecular pathway for addiction? *Nature Neuroscience*, *8*(11), 1445–1449. http://dx.doi.org/10.1038/nn1578.

Nestler, E. J. (2013). Cellular basis of memory for addiction. *Dialogues in Clinical Neuroscience*, *15*(4), 431–443.

Nicoll, R. A., & Malenka, R. C. (1999). Expression mechanisms underlying NMDA receptor-dependent long-term potentiation. *Annals of the New York Academy of Sciences*, *868*, 515–525.

O'Brien, C. P., Childress, A. R., Ehrman, R., & Robbins, S. J. (1998). Conditioning factors in drug abuse: Can they explain compulsion? *Journal of Psychopharmacology*, *12*(1), 15–22.

Ostrovskaya, O., Xie, K., Masuho, I., Fajardo-Serrano, A., Lujan, R., Wickman, K., et al. (2014). RGS7/Gbeta5/R7BP complex regulates synaptic plasticity and memory by modulating hippocampal GABABR-GIRK signaling. *eLife*, *3*. http://dx.doi.org/10.7554/eLife.02053, e02053.

Padgett, C. L., Lalive, A. L., Tan, K. R., Terunuma, M., Munoz, M. B., Pangalos, M. N., et al. (2012). Methamphetamine-evoked depression of GABA(B) receptor signaling in GABA neurons of the VTA. *Neuron*, *73*(5), 978–989. http://dx.doi.org/10.1016/j.neuron.2011.12.031.

Padgett, C. L., & Slesinger, P. A. (2010). GABAB receptor coupling to G-proteins and ion channels. *Advances in Pharmacology*, *58*, 123–147. http://dx.doi.org/10.1016/S1054-3589(10)58006-2.

Palop, J. J., & Mucke, L. (2010). Amyloid-beta-induced neuronal dysfunction in Alzheimer's disease: From synapses toward neural networks. *Nature Neuroscience*, *13*(7), 812–818. http://dx.doi.org/10.1038/nn.2583.

Patil, N., Cox, D. R., Bhat, D., Faham, M., Myers, R. M., & Peterson, A. S. (1995). A potassium channel mutation in weaver mice implicates membrane excitability in granule cell differentiation. *Nature Genetics*, *11*(2), 126–129. http://dx.doi.org/10.1038/ng1095-126.

Pu, L., Bao, G. B., Xu, N. J., Ma, L., & Pei, G. (2002). Hippocampal long-term potentiation is reduced by chronic opiate treatment and can be restored by re-exposure to opiates. *The Journal of Neuroscience: The Official Journal of the Society for Neuroscience*, *22*(5), 1914–1921.

Rahman, Z., Schwarz, J., Gold, S. J., Zachariou, V., Wein, M. N., Choi, K. H., et al. (2003). RGS9 modulates dopamine signaling in the basal ganglia. *Neuron*, *38*(6), 941–952.

Reeves, R. H., Irving, N. G., Moran, T. H., Wohn, A., Kitt, C., Sisodia, S. S., et al. (1995). A mouse model for Down syndrome exhibits learning and behaviour deficits. *Nature Genetics*, *11*(2), 177–184. http://dx.doi.org/10.1038/ng1095-177.

Roberto, M., Nelson, T. E., Ur, C. L., & Gruol, D. L. (2002). Long-term potentiation in the rat hippocampus is reversibly depressed by chronic intermittent ethanol exposure. *Journal of Neurophysiology*, *87*(5), 2385–2397.

Roffler-Tarlov, S., Martin, B., Graybiel, A. M., & Kauer, J. S. (1996). Cell death in the midbrain of the murine mutation weaver. *The Journal of Neuroscience: The Official Journal of the Society for Neuroscience*, *16*(5), 1819–1826.

Rusinova, R., Shen, Y. M., Dolios, G., Padovan, J., Yang, H., Kirchberger, M., et al. (2009). Mass spectrometric analysis reveals a functionally important PKA phosphorylation site in a Kir3 channel subunit. *Pflugers Archiv: European Journal of Physiology*, *458*(2), 303–314. http://dx.doi.org/10.1007/s00424-008-0628-9.

Saal, D., Dong, Y., Bonci, A., & Malenka, R. C. (2003). Drugs of abuse and stress trigger a common synaptic adaptation in dopamine neurons. *Neuron*, *37*(4), 577–582.

Sadja, R., Alagem, N., & Reuveny, E. (2003). Gating of GIRK channels: Details of an intricate, membrane-delimited signaling complex. *Neuron*, *39*(1), 9–12.

Sago, H., Carlson, E. J., Smith, D. J., Kilbridge, J., Rubin, E. M., Mobley, W. C., et al. (1998). Ts1Cje, a partial trisomy 16 mouse model for Down syndrome, exhibits learning and behavioral abnormalities. *Proceedings of the National Academy of Sciences of the United States of America*, *95*(11), 6256–6261.

Saitoh, O., Kubo, Y., Odagiri, M., Ichikawa, M., Yamagata, K., & Sekine, T. (1999). RGS7 and RGS8 differentially accelerate G protein-mediated modulation of K+ currents. *The Journal of Biological Chemistry*, *274*(14), 9899–9904.

Schmidt, M. J., Sawyer, B. D., Perry, K. W., Fuller, R. W., Foreman, M. M., & Ghetti, B. (1982). Dopamine deficiency in the weaver mutant mouse. *The Journal of Neuroscience: The Official Journal of the Society for Neuroscience*, *2*(3), 376–380.

Schoots, O., Wilson, J. M., Ethier, N., Bigras, E., Hebert, T. E., & Van Tol, H. H. (1999). Co-expression of human Kir3 subunits can yield channels with different functional properties. *Cellular Signalling*, *11*(12), 871–883.

Schultz, W. (2002). Getting formal with dopamine and reward. *Neuron*, *36*(2), 241–263.

Schultz, W., Dayan, P., & Montague, P. R. (1997). A neural substrate of prediction and reward. *Science*, *275*(5306), 1593–1599.

Schwenk, J., Metz, M., Zolles, G., Turecek, R., Fritzius, T., Bildl, W., et al. (2010). Native GABA(B) receptors are heteromultimers with a family of auxiliary subunits. *Nature*, *465*(7295), 231–235. http://dx.doi.org/10.1038/nature08964.

Seddik, R., Jungblut, S. P., Silander, O. K., Rajalu, M., Fritzius, T., Besseyrias, V., et al. (2012). Opposite effects of KCTD subunit domains on GABA(B) receptor-mediated desensitization. *The Journal of Biological Chemistry*, *287*(47), 39869–39877. http://dx.doi.org/10.1074/jbc.M112.412767.

Shankar, G. M., Li, S., Mehta, T. H., Garcia-Munoz, A., Shepardson, N. E., Smith, I., et al. (2008). Amyloid-beta protein dimers isolated directly from Alzheimer's brains impair synaptic plasticity and memory. *Nature Medicine*, *14*(8), 837–842. http://dx.doi.org/10.1038/nm1782.

Sharpe, A. L., Varela, E., Bettinger, L., & Beckstead, M. J. (2014). Methamphetamine self-administration in mice decreases GIRK channel-mediated currents in midbrain dopamine neurons. *The International Journal of Neuropsychopharmacology/Official Scientific Journal of the Collegium Internationale Neuropsychopharmacologicum*. *18*(5). http://dx.doi.org/10.1093/ijnp/pyu073.

Siarey, R. J., Carlson, E. J., Epstein, C. J., Balbo, A., Rapoport, S. I., & Galdzicki, Z. (1999). Increased synaptic depression in the Ts65Dn mouse, a model for mental retardation in Down syndrome. *Neuropharmacology*, *38*(12), 1917–1920.

Siarey, R. J., Stoll, J., Rapoport, S. I., & Galdzicki, Z. (1997). Altered long-term potentiation in the young and old Ts65Dn mouse, a model for Down syndrome. *Neuropharmacology*, *36*(11–12), 1549–1554.

Siarey, R. J., Villar, A. J., Epstein, C. J., & Galdzicki, Z. (2005). Abnormal synaptic plasticity in the Ts1Cje segmental trisomy 16 mouse model of Down syndrome. *Neuropharmacology, 49*(1), 122–128. http://dx.doi.org/10.1016/j.neuropharm.2005.02.012.
Sidhpura, N., Weiss, F., & Martin-Fardon, R. (2010). Effects of the mGlu2/3 agonist LY379268 and the mGlu5 antagonist MTEP on ethanol seeking and reinforcement are differentially altered in rats with a history of ethanol dependence. *Biological Psychiatry, 67*(9), 804–811. http://dx.doi.org/10.1016/j.biopsych.2010.01.005.
Sidman, R. L. (1968). Development of interneuronal connections in brains of mutant mice. In F. D. Carlson (Ed.), *Physiological and biochemical aspects of nervous integration* (pp. 163–193). New Jersey: Prentice-Hall.
Sjogren, B. (2011). Regulator of G protein signaling proteins as drug targets: Current state and future possibilities. *Advances in Pharmacology, 62*, 315–347. http://dx.doi.org/10.1016/B978-0-12-385952-5.00002-6.
Skoulakis, E. M., & Davis, R. L. (1998). 14-3-3 proteins in neuronal development and function. *Molecular Neurobiology, 16*(3), 269–284. http://dx.doi.org/10.1007/BF02741386.
Smeyne, R. J., & Goldowitz, D. (1989). Development and death of external granular layer cells in the weaver mouse cerebellum: A quantitative study. *The Journal of Neuroscience: The Official Journal of the Society for Neuroscience, 9*(5), 1608–1620.
Smith, D. J., Stevens, M. E., Sudanagunta, S. P., Bronson, R. T., Makhinson, M., Watabe, A. M., et al. (1997). Functional screening of 2 Mb of human chromosome 21q22.2 in transgenic mice implicates minibrain in learning defects associated with Down syndrome. *Nature Genetics, 16*(1), 28–36. http://dx.doi.org/10.1038/ng0597-28.
Smith, D. J., Zhu, Y., Zhang, J., Cheng, J. F., & Rubin, E. M. (1995). Construction of a panel of transgenic mice containing a contiguous 2-Mb set of YAC/P1 clones from human chromosome 21q22.2. *Genomics, 27*(3), 425–434. http://dx.doi.org/10.1006/geno.1995.1073.
Sosulina, L., Schwesig, G., Seifert, G., & Pape, H. C. (2008). Neuropeptide Y activates a G-protein-coupled inwardly rectifying potassium current and dampens excitability in the lateral amygdala. *Molecular and Cellular Neurosciences, 39*(3), 491–498. http://dx.doi.org/10.1016/j.mcn.2008.08.002.
Spauschus, A., Lentes, K. U., Wischmeyer, E., Dissmann, E., Karschin, C., & Karschin, A. (1996). A G-protein-activated inwardly rectifying K+ channel (GIRK4) from human hippocampus associates with other GIRK channels. *The Journal of Neuroscience: The Official Journal of the Society for Neuroscience, 16*(3), 930–938.
Spiga, S., Mulas, G., Piras, F., & Diana, M. (2014). The "addicted" spine. *Frontiers in Neuroanatomy, 8*, 110. http://dx.doi.org/10.3389/fnana.2014.00110.
Staubli, U., & Lynch, G. (1990). Stable depression of potentiated synaptic responses in the hippocampus with 1-5 Hz stimulation. *Brain Research, 513*(1), 113–118.
Stewart, A., Maity, B., Anderegg, S. P., Allamargot, C., Yang, J., & Fisher, R. A. (2015). Regulator of G protein signaling 6 is a critical mediator of both reward-related behavioral and pathological responses to alcohol. *Proceedings of the National Academy of Sciences of the United States of America, 112*(7), E786–E795. http://dx.doi.org/10.1073/pnas.1418795112.
Svoboda, K. R., Adams, C. E., & Lupica, C. R. (1999). Opioid receptor subtype expression defines morphologically distinct classes of hippocampal interneurons. *The Journal of Neuroscience: The Official Journal of the Society for Neuroscience, 19*(1), 85–95.
Swanberg, M. M., Tractenberg, R. E., Mohs, R., Thal, L. J., & Cummings, J. L. (2004). Executive dysfunction in Alzheimer disease. *Archives of Neurology, 61*(4), 556–560. http://dx.doi.org/10.1001/archneur.61.4.556.
Thompson, A. M., Gosnell, B. A., & Wagner, J. J. (2002). Enhancement of long-term potentiation in the rat hippocampus following cocaine exposure. *Neuropharmacology, 42*(8), 1039–1042.

Tipps, M. E., Raybuck, J. D., Buck, K. J., & Lattal, K. M. (2015). Acute ethanol withdrawal impairs contextual learning and enhances cued learning. *Alcoholism, Clinical and Experimental Research*, *39*(2), 282–290. http://dx.doi.org/10.1111/acer.12614.

Torrecilla, M., Fernandez-Aedo, I., Arrue, A., Zumarraga, M., & Ugedo, L. (2013). Role of GIRK channels on the noradrenergic transmission in vivo: An electrophysiological and neurochemical study on GIRK2 mutant mice. *The International Journal of Neuropsychopharmacology/Official Scientific Journal of the Collegium Internationale Neuropsychopharmacologicum*, *16*(5), 1093–1104. http://dx.doi.org/10.1017/S1461145712000971.

Torrecilla, M., Marker, C. L., Cintora, S. C., Stoffel, M., Williams, J. T., & Wickman, K. (2002). G-protein-gated potassium channels containing Kir3.2 and Kir3.3 subunits mediate the acute inhibitory effects of opioids on locus ceruleus neurons. *The Journal of Neuroscience: The Official Journal of the Society for Neuroscience*, *22*(11), 4328–4334.

Traynor, J. R., & Neubig, R. R. (2005). Regulators of G protein signaling & drugs of abuse. *Molecular Interventions*, *5*(1), 30–41. http://dx.doi.org/10.1124/mi.5.1.7.

Trudell, J. R., Messing, R. O., Mayfield, J., & Harris, R. A. (2014). Alcohol dependence: Molecular and behavioral evidence. *Trends in Pharmacological Sciences*, *35*(7), 317–323. http://dx.doi.org/10.1016/j.tips.2014.04.009.

Turecek, R., Schwenk, J., Fritzius, T., Ivankova, K., Zolles, G., Adelfinger, L., et al. (2014). Auxiliary GABAB receptor subunits uncouple G protein betagamma subunits from effector channels to induce desensitization. *Neuron*, *82*(5), 1032–1044. http://dx.doi.org/10.1016/j.neuron.2014.04.015.

Turrigiano, G. G. (1999). Homeostatic plasticity in neuronal networks: The more things change, the more they stay the same. *Trends in Neurosciences*, *22*(5), 221–227.

Ulens, C., Daenens, P., & Tytgat, J. (1999). The dual modulation of GIRK1/GIRK2 channels by opioid receptor ligands. *European Journal of Pharmacology*, *385*(2–3), 239–245.

Ungless, M. A., Whistler, J. L., Malenka, R. C., & Bonci, A. (2001). Single cocaine exposure in vivo induces long-term potentiation in dopamine neurons. *Nature*, *411*(6837), 583–587. http://dx.doi.org/10.1038/35079077.

Wang, Q., Liu, M., Mullah, B., Siderovski, D. P., & Neubig, R. R. (2002). Receptor-selective effects of endogenous RGS3 and RGS5 to regulate mitogen-activated protein kinase activation in rat vascular smooth muscle cells. *The Journal of Biological Chemistry*, *277*(28), 24949–24958. http://dx.doi.org/10.1074/jbc.M203802200.

Wang, X., Zhao, Y., Zhang, X., Badie, H., Zhou, Y., Mu, Y., et al. (2013). Loss of sorting nexin 27 contributes to excitatory synaptic dysfunction by modulating glutamate receptor recycling in Down's syndrome. *Nature Medicine*, *19*(4), 473–480. http://dx.doi.org/10.1038/nm.3117.

Watts, A. E., Williams, J. T., & Henderson, G. (1996). Baclofen inhibition of the hyperpolarization-activated cation current, Ih, in rat substantia nigra zona compacta neurons may be secondary to potassium current activation. *Journal of Neurophysiology*, *76*(4), 2262–2270.

Wen, W., Wu, W., Romaine, I. M., Kaufmann, K., Du, Y., Sulikowski, G. A., et al. (2013). Discovery of 'molecular switches' within a GIRK activator scaffold that afford selective GIRK inhibitors. *Bioorganic & Medicinal Chemistry Letters*, *23*(16), 4562–4566. http://dx.doi.org/10.1016/j.bmcl.2013.06.023.

White, F. J. (1996). Synaptic regulation of mesocorticolimbic dopamine neurons. *Annual Review of Neuroscience*, *19*, 405–436. http://dx.doi.org/10.1146/annurev.ne.19.030196.002201.

Wickman, K., Karschin, C., Karschin, A., Picciotto, M. R., & Clapham, D. E. (2000). Brain localization and behavioral impact of the G-protein-gated K+ channel subunit GIRK4. *The Journal of Neuroscience: The Official Journal of the Society for Neuroscience*, *20*(15), 5608–5615.

Williams, J. T. (2014). Desensitization of functional micro-opioid receptors increases agonist off-rate. *Molecular Pharmacology, 86*(1), 52–61. http://dx.doi.org/10.1124/mol.114.092098.

Witkowski, G., Rola, R., & Szulczyk, P. (2012). Effect of cyclic adenosine monophosphate on the G protein-dependent inward rectifier K(+)-like channel current in medial prefrontal cortex pyramidal neurons. *Journal of Physiology and Pharmacology: An Official Journal of the Polish Physiological Society, 63*(5), 457–462.

Witkowski, G., Szulczyk, B., Rola, R., & Szulczyk, P. (2008). D(1) dopaminergic control of G protein-dependent inward rectifier K(+) (GIRK)-like channel current in pyramidal neurons of the medial prefrontal cortex. *Neuroscience, 155*(1), 53–63. http://dx.doi.org/10.1016/j.neuroscience.2008.05.021.

Workman, E. R., Haddick, P. C., Bush, K., Dilly, G. A., Niere, F., Zemelman, B. V., et al. (2015). Rapid antidepressants stimulate the decoupling of GABAB receptors from GIRK/Kir3 channels through increased protein stability of 14-3-3eta. *Molecular Psychiatry, 20*(3), 298–310. http://dx.doi.org/10.1038/mp.2014.165.

Wydeven, N., Marron Fernandez de Velasco, E., Du, Y., Benneyworth, M. A., Hearing, M. C., et al. (2014). Mechanisms underlying the activation of G-protein-gated inwardly rectifying K+ (GIRK) channels by the novel anxiolytic drug, ML297. *Proceedings of the National Academy of Sciences of the United States of America, 111*(29), 10755–10760. http://dx.doi.org/10.1073/pnas.1405190111.

Wydeven, N., Young, D., Mirkovic, K., & Wickman, K. (2012). Structural elements in the Girk1 subunit that potentiate G protein-gated potassium channel activity. *Proceedings of the National Academy of Sciences of the United States of America, 109*(52), 21492–21497. http://dx.doi.org/10.1073/pnas.1212019110.

Xiao, J., Zhen, X. G., & Yang, J. (2003). Localization of PIP2 activation gate in inward rectifier K+ channels. *Nature Neuroscience, 6*(8), 811–818. http://dx.doi.org/10.1038/nn1090.

Xie, G. X., & Palmer, P. P. (2007). How regulators of G protein signaling achieve selective regulation. *Journal of Molecular Biology, 366*(2), 349–365. http://dx.doi.org/10.1016/j.jmb.2006.11.045.

Xu, X., Zeng, W., Popov, S., Berman, D. M., Davignon, I., Yu, K., et al. (1999). RGS proteins determine signaling specificity of Gq-coupled receptors. *The Journal of Biological Chemistry, 274*(6), 3549–3556.

Yao, W. D., Gainetdinov, R. R., Arbuckle, M. I., Sotnikova, T. D., Cyr, M., Beaulieu, J. M., et al. (2004). Identification of PSD-95 as a regulator of dopamine-mediated synaptic and behavioral plasticity. *Neuron, 41*(4), 625–638.

Young, K. A., Franklin, T. R., Roberts, D. C., Jagannathan, K., Suh, J. J., Wetherill, R. R., et al. (2014). Nipping cue reactivity in the bud: Baclofen prevents limbic activation elicited by subliminal drug cues. *The Journal of Neuroscience: The Official Journal of the Society for Neuroscience, 34*(14), 5038–5043. http://dx.doi.org/10.1523/JNEUROSCI.4977-13.2014.

Zachariou, V., Georgescu, D., Sanchez, N., Rahman, Z., DiLeone, R., Berton, O., et al. (2003). Essential role for RGS9 in opiate action. *Proceedings of the National Academy of Sciences of the United States of America, 100*(23), 13656–13661. http://dx.doi.org/10.1073/pnas.2232594100.

Zhang, H., He, C., Yan, X., Mirshahi, T., & Logothetis, D. E. (1999). Activation of inwardly rectifying K+ channels by distinct PtdIns(4,5)P2 interactions. *Nature Cell Biology, 1*(3), 183–188. http://dx.doi.org/10.1038/11103.

Zhang, S. H., Kobayashi, R., Graves, P. R., Piwnica-Worms, H., & Tonks, N. K. (1997). Serine phosphorylation-dependent association of the band 4.1-related protein-tyrosine phosphatase PTPH1 with 14-3-3beta protein. *The Journal of Biological Chemistry, 272*(43), 27281–27287.

CHAPTER EIGHT

Behavioral and Genetic Evidence for GIRK Channels in the CNS: Role in Physiology, Pathophysiology, and Drug Addiction

Jody Mayfield[1], Yuri A. Blednov, R. Adron Harris
Waggoner Center for Alcohol and Addiction Research, The University of Texas at Austin, Austin, Texas, USA
[1]Corresponding author: e-mail address: jodymayfield@austin.utexas.edu

Contents

1. Introduction 280
2. GIRK Channels in CNS Disorders 281
   2.1 Cognitive Deficits 283
   2.2 Pain 285
   2.3 Motor Control 286
   2.4 Psychiatric Disorders 288
   2.5 Epilepsy 291
3. GIRK Channels in Addiction 291
   3.1 Ethanol 292
   3.2 Sedative/Hypnotics 297
   3.3 Psychostimulants 297
   3.4 Opioids 298
   3.5 Nicotine 299
   3.6 GIRK Modulators and Other Drugs of Abuse 299
4. Therapeutic Potential of GIRK Channel Modulators 300
5. Concluding Remarks and Future Directions 302
Acknowledgments 303
References 303

Abstract

G protein-coupled inwardly rectifying potassium (GIRK) channels are widely expressed throughout the brain and mediate the inhibitory effects of many neurotransmitters. As a result, these channels are important for normal CNS function and have also been implicated in Down syndrome, Parkinson's disease, psychiatric disorders, epilepsy, and drug addiction. Knockout mouse models have provided extensive insight into the significance of GIRK channels under these conditions. This review examines the behavioral and genetic evidence from animal models and genetic association studies in humans linking GIRK channels with CNS disorders. We further explore the possibility that

International Review of Neurobiology, Volume 123
ISSN 0074-7742
http://dx.doi.org/10.1016/bs.irn.2015.05.016

subunit-selective modulators and other advanced research tools will be instrumental in establishing the role of individual GIRK subunits in drug addiction and other relevant CNS diseases and in potentially advancing treatment options for these disorders.

1. INTRODUCTION

G protein-coupled inwardly rectifying K^+ (GIRK) channels are a family of ion channels that are activated via ligand-stimulated G protein-coupled receptors (GPCRs). Following ligand stimulation, activated G protein subunits are released that directly interact with and open GIRK channels so that they become permeable to K^+ ions. The outward K^+ current hyperpolarizes neuronal membranes and decreases neuronal excitability. GIRK channels are activated by a large family of GPCRs (reviewed in chapters "Unifying Mechanism of Controlling Kir3 Channel Activity by G Proteins and Phosphoinositides" by Logothetis et al. and "The Roles of Gβγ and Gα in Gating and Regulation of GIRK Channels" by Dascal and Kahanovitch including dopamine 2 (D_2), serotonin 1A (5-HT_{1A}), μ-, κ-, and δ-opioid, cannabinoid 1 (CB_1), and γ-aminobutyric acid type B ($GABA_B$) receptors.

There are four mammalian subunits (GIRK1–4) with overlapping but distinct expression patterns throughout the CNS that form heterotetrameric channels (Karschin, Dissmann, Stuhmer, & Karschin, 1996; Koyrakh et al., 2005). GIRK2 and GIRK4 subunits can also form functional homotetrameric channels (Koyrakh et al., 2005; Krapivinsky et al., 1995). GIRK1–3 are considered the predominant subunits in brain, while GIRK4 expression is more restricted (Perry et al., 2008; Wickman, Karschin, Karschin, Picciotto, & Clapham, 2000). GIRK2 appears to be an integral subunit of most neuronal GIRK channels (Cruz et al., 2004; Luscher, Jan, Stoffel, Malenka, & Nicoll, 1997; Slesinger, Stoffel, Jan, & Jan, 1997). The expression patterns of GIRK subunits vary in individual brain regions and even among subcellular compartments within individual neurons, ensuring discrete regional and cellular signaling (reviewed in chapter "Localization and Targeting of GIRK Channels in Mammalian Central Neurons" by Luján and Aguado). It is interesting that the unique subunit composition of GIRK channels in different neuronal populations may confer distinct functional properties (Jelacic, Kennedy, Wickman, & Clapham, 2000; Jelacic, Sims, & Clapham, 1999; Schoots et al., 1999) and drug sensitivities that mediate the rewarding effects of certain addictive drugs, such as γ-hydroxybutyrate (GHB) (Cruz et al., 2004; Labouebe et al., 2007).

GIRK channels have been implicated in both normal CNS functions and pathological states (Luscher & Slesinger, 2010). They control key neurological processes, such as neuronal plasticity and learning/memory, and are sensitive to different drugs of abuse (see chapters "GIRK Channel Plasticity and Implications for Drug Addiction" by de Velasco et al. and "GIRK Channels: A Potential Link Between Learning and Addiction" by Tipps and Buck), making them relevant targets to examine in behavioral studies of cognition and drug addiction. In this chapter, we examine behavioral evidence from mouse knockout models as well as genetic studies from animal models and humans that support a role for GIRK channels in different CNS processes. This review includes normal responses such as pain perception, motor control, and memory formation, as well as GIRK contributions to the pathophysiology of Parkinson's disease, Down syndrome, psychiatric diseases, and epilepsy. We also review the evidence for alcohol- and other drug-dependent behaviors that are mediated by GIRK-dependent signaling. Finally, we explore how recent progress in GIRK channel structural modeling (see chapter "Structural Insights into GIRK Channel Function" by Glaaser and Slesinger) and the development of subunit-selective channel modulators (Kaufmann et al., 2013; Ramos-Hunter et al., 2013; Wen et al., 2013) may advance understanding of channel function and, consequently, improve treatment options for many CNS diseases. These and other new research approaches may contribute to the design of better therapeutics for the CNS disorders that are associated with GIRK-dependent signaling.

2. GIRK CHANNELS IN CNS DISORDERS

Gene knockout mouse models have provided valuable insight into the role of GIRK channels in normal and pathological processes (Luscher & Slesinger, 2010). A primary concern of knockout lines is whether compensatory changes in the expression of other genes occur as a result of global deletion of individual genes. For example, GIRK1 protein levels are also decreased in mice lacking the *Girk2* gene (Signorini, Liao, Duncan, Jan, & Stoffel, 1997), and the lack of either GIRK1 or GIRK2 is correlated with lower expression of the other, suggesting a specific assembly between these subunits (Koyrakh et al., 2005; Marker, Stoffel, & Wickman, 2004). Although off-target effects of genetic deletion may be a confounding factor in knockout animals, these models have nevertheless been instrumental in generating information about the physiological relevance of GIRK channels. In sections 2 and 3, we outline the behavioral and genetic evidence in mice (Tables 1 and 2) and humans (Table 3), highlighting roles for GIRK channels in a variety of CNS disorders.

Table 1 Mouse *Girk* Genes Implicated in CNS Disorders

Behavioral Phenotypes	GIRK Genotypes	References
Learning/memory deficits	*Girk2* (*weaver*)	Derenne, Arsenault, Austin, and Weatherly (2007)
	Girk4 (−/−)	Wickman et al. (2000)
	Rgs7 (−/−)[a]	Ostrovskaya et al. (2014)
Down syndrome model	*Girk2* (triploid)	Cooper et al. (2012)
Reduced analgesia	*Girk1* (−/−)	Marker et al. (2004); Marker, Lujan, Loh, and Wickman (2005)
	Girk2 (*weaver*)	Ikeda, Kobayashi, Kumanishi, Niki, and Yano (2000); Kobayashi et al. (1999)
	Girk2 (−/−)	Blednov, Stoffel, Alva, and Harris (2003); Cruz et al. (2008); Marker, Cintora, Roman, Stoffel, and Wickman (2002); Marker et al. (2004, 2005); Mitrovic et al. (2003)
	Girk3 (−/−)	Marker et al. (2002); Smith et al. (2008)
	Girk2/3 (−/−)	Cruz et al. (2008); Marker et al. (2002)
Increased motor activity	*Girk1* (−/−)	Pravetoni and Wickman (2008)
	Girk2 (*weaver*)	Schmidt et al. (1982); Harkins and Fox (2002)
	Girk2 (−/−)	Blednov, Stoffel, Chang, and Harris (2001a); Blednov et al. (2002)
Parkinson's symptoms	*Girk2* (*weaver*)	Caviness and Rakic (1978); Coscia and Fentress (1993); Derenne et al. (2007); Schmidt et al. (1982)
Increased seizures	*Girk2* (*weaver*)	Eisenberg and Messer (1989)
	Girk2 (−/−)	Signorini et al. (1997)
	Girk2/3 (−/−)	Torrecilla et al. (2002)
Reduced anxiety-like behaviors	*Girk1* (−/−)	Pravetoni and Wickman (2008)
	Girk2 (−/−)	Blednov et al. (2001a); Pravetoni and Wickman (2008)
Increased depressive-resistant behaviors	*Girk2* (−/−)	Llamosas, Bruzos-Cidón, Rodríguez, Ugedo, and Torrecilla (2015)

Girk genes and associated phenotypes from *weaver*, knockout (−/−), and triploid mouse models are shown.

[a]*Rgs7* (regulator of G-protein signaling) is not a GIRK gene, but accelerates G-protein inactivation and negatively modulates GIRK responses (Lujan, Marron Fernandez de Velasco, Aguado, & Wickman, 2014). Parkinson's disease-like behavioral phenotypes include cognitive impairments, hyper-reactivity, ataxia, poor limb coordination, and tremors.

2.1 Cognitive Deficits

Several lines of evidence indicate that GIRK channels play a crucial role in cognitive function. GIRK channels are expressed in brain regions associated with learning and memory, including the hippocampus, amygdala, and prefrontal cortex (Ehrengruber et al., 1997; Hearing et al., 2013; Luscher et al., 1997) and modulate the development of synaptic plasticity (Chung et al., 2009). Furthermore, changes in GIRK signaling are associated with learning impairments in behavioral studies of mutant mice (Table 1). For example, *weaver* mice, which contain a single amino acid mutation in the pore of the GIRK2 subunit (Patil et al., 1995), exhibited long-term learning deficits in models of instrumental learning (Derenne et al., 2007). *Girk4*$^{-/-}$ mice showed impaired performance in the Morris water-maze test but did not differ from wild-type mice in the passive avoidance test, indicating some impairment in spatial learning and memory but not aversive learning (Wickman et al., 2000).

Modulators of GIRK channel function may also be important in synaptic plasticity and learning and memory. Regulators of G-protein signaling (RGS) proteins accelerate GTPase activity of the Gα subunit, thereby negatively modulating GPCR–GIRK activation (reviewed in chapter "RGS Redundancy and Implications in GPCR–GIRK Signaling" by Doupnik). The R7 family of RGS proteins has prominent roles in motor control, nociception, and reward-related behavior in mammals, similar to some of the known roles for GIRK channels (Anderson, Posokhova, & Martemyanov, 2009). RGS7 and its binding protein R7BP control $GABA_B$ receptor ($GABA_BR$)–GIRK signaling in hippocampal pyramidal neurons (Ostrovskaya et al., 2014). Deletion of the *Rgs7* gene or its binding protein in mice increased the sensitivity to baclofen activation and slowed GIRK current deactivation in these neurons. The enhanced $GABA_BR$–GIRK coupling sensitivity and slower deactivation kinetics decreased neuronal excitability and disrupted inhibitory forms of synaptic plasticity, likely contributing to the learning and memory deficits observed in *Rgs7*$^{-/-}$ mice (Ostrovskaya et al., 2014). These knockout mice showed deficits in context recognition after fear conditioning and impairments in different aspects of the Morris water-maze test. Both of these tests rely on hippocampal processing for memory formation, providing corroborating behavioral evidence that RGS proteins control $GABA_BR$–GIRK signaling and mediate hippocampal synaptic plasticity.

Gene duplication and overexpression of GIRK subunits can also cause cognitive impairments. GIRK channels are implicated in the pathology of

Down syndrome, a congenital disorder caused by an extra maternal copy (trisomy) of human chromosome 21 that is characterized by learning disabilities, craniofacial abnormalities, and hypotonia, although the extent of these phenotypes varies among individuals (Wiseman, Alford, Tybulewicz, & Fisher, 2009). The Down syndrome critical region (DSCR) of chromosome 21 contains several genes, including *KCNJ6*, which encodes the GIRK2 subunit (Toyoda et al., 2002). Transpolygenic mice carrying extra copies of chromosome 21 fragments covering the DSCR showed cognitive disabilities in most tests of the Morris water-maze and in the altered context stage in fear-conditioning tests (Chabert et al., 2004). The presence of *KCNJ6* in this region and the overexpression of GIRK2 and enhanced $GABA_B$R-dependent signaling may contribute to some of the mental disabilities in Down syndrome, as reviewed in Cramer, Best, Stoffel, Siarey, and Galdzicki (2010).

Different mouse models of Down syndrome have been generated that carry either a partial or full segment duplication of the analogous mouse chromosome 16, and these trisomy mice exhibit cognitive deficits (Liu et al., 2011). A partial trisomy 16 mouse model (Ts1Cje) containing the DSCR showed impaired performance levels in several tests in the Morris water-maze (Sago et al., 1998). Another partial trisomy model containing a larger region of chromosome 16 (Ts65Dn) showed greater learning impairment in Morris water-maze tests compared to Ts1Cje mice (Reeves et al., 1995). Ts65Dn mice were also more active in open-field tests and less responsive to environmental cues compared to controls (Coussons-Read & Crnic, 1996) and demonstrated deficits in a fear-conditioning test of associative learning (Costa, Scott-McKean, & Stasko, 2008). Ts1Rhr mice, which are trisomic for a small subset of the genes in Ts65Dn, and Ts1Cje mice showed cognitive deficits in novel object recognition and T-maze tests (Belichenko et al., 2009). Although all of these mice carry an extra copy of *Girk2*, other overexpressed genes on chromosome 16 may also contribute to the cognitive deficits observed in these models (Roubertoux & Carlier, 2010). The development of a mouse model with trisomy for *Girk2* alone provided more direct evidence for this gene in Down syndrome. These mice exhibited impaired hippocampal-dependent contextual-fear recall, altered responses to rewards, decreased excitatory synaptic plasticity, and increased long-term synaptic depression (Cooper et al., 2012). Recent evidence suggests that Keppen–Lubinsky syndrome, a rare disease characterized by physical abnormalities, lipodystrophy, developmental delays, intellectual disabilities, and microcephaly, is caused by mutations in *KCNJ6* (Masotti et al., 2015). Together, these studies highlight the importance of GIRK channels in cognition, in particular a role of GIRK2 (Tables 1 and 3).

2.2 Pain

Analgesics can target different types of GPCRs (e.g., opioid, cannabinoid, and $GABA_B$ receptors), and GIRK channels are a common effector of many analgesic medications (Pan et al., 2008). GIRK channels were first associated with pain perception based on the reduced analgesia observed in *weaver* mice after opioid or ethanol administration (Ikeda et al., 2000; Kobayashi et al., 1999). However, the *weaver* mutant is not an ideal model because the mutation causes neuronal degeneration rather than a selective loss of GIRK2 function (Patil et al., 1995). Viable mice lacking the *Girk2* gene were generated that lacked postsynaptic responses to neurotransmitters known to act through $G_{i/o}$-linked GPCRs, but retained normal presynaptic signaling (Luscher et al., 1997; Signorini et al., 1997; Slesinger et al., 1997). These knockout mice provided an improved model for studying the specific role of GIRK2-containing channels in neurophysiology and behavior.

In male *Girk2*$^{-/-}$ mice, there was a marked reduction or elimination of the antinociceptive effects of several compounds, including baclofen, ethanol, and a cannabinoid receptor agonist in the hot plate response test; however, analgesia induced by the NMDA receptor antagonist, ketamine, was not affected (Blednov et al., 2003). Although some of the drugs used in this study do not directly couple to GIRK channels, the loss of GIRK2 subunits may indirectly decrease neuronal sensitivity to these analgesics. Deletion of *Girk2* also blocked the endogenous opioid-dependent component of stress-induced analgesia, whereas nonopioid stress-induced analgesia was not altered. Other behavioral work corroborated these findings, showing that morphine- and clonidine-induced antinociception were reduced in male *Girk2*$^{-/-}$ mice (Mitrovic et al., 2003). Both studies provided evidence that GIRK2 contributes to sex differences in nociception and indicated that activation of GIRK2-containing channels underlies the analgesic effects of a diverse array of drugs (ethanol, opioids, nicotine, cannabinoids, as well as alpha adrenergic, muscarinic cholinergic, and $GABA_B$ receptor agonists suggesting a common mechanism for analgesic action.

Girk2$^{-/-}$ and *Girk3*$^{-/-}$ (but not *Girk4*$^{-/-}$) mice showed increased sensitivity to pain and blunted analgesic responses to morphine, with *Girk2*$^{-/-}$ mice being the least sensitive to morphine analgesia (Marker et al., 2002). *Girk2/3*$^{-/-}$ mice demonstrated greater hyperalgesia compared to the single mutants and were similar to *Girk2*$^{-/-}$ mice in blunted responses to morphine analgesia (Marker et al., 2002). Hyperalgesia and decreased analgesic responses following spinal administration of high doses of morphine were observed in *Girk1*$^{-/-}$ or *Girk2*$^{-/-}$ mice (Marker et al., 2004). Similar results were observed after administration of the GIRK channel blocker tertiapin in wild-type mice (Marker et al., 2004). GIRK1 and GIRK2 subunits are enriched in the superficial dorsal horn of the spinal cord (Marker et al., 2004), a key relay station for nociceptive processing, providing further evidence that spinal GIRK channels composed of GIRK1/2 subunits modulate thermal nociception and analgesia induced by high doses of morphine. In agreement with this, *Girk1*$^{-/-}$ or *Girk2*$^{-/-}$ mice showed blunted responses to μ- and ∂- but not κ-opioid receptor agonists, and tertiapin reduced the ability of the μ- and ∂- receptor agonists to increase the latency for tail withdrawal in the immersion tail-flick test (Marker et al., 2005).

Quantitative trait locus (QTL) mapping also identified *Girk3* as a candidate gene contributing to mouse strain-dependent differences in the analgesic effects of multiple drug classes, and *Girk3*$^{-/-}$ mice showed decreased analgesic responses to morphine and a cannabinoid receptor agonist (Smith et al., 2008). In humans, single nucleotide polymorphisms (SNPs) in *KCNJ6* (GIRK2) were associated with an increased requirement for opioid analgesics following abdominal surgery (Nishizawa et al., 2014, 2009). An SNP in *KCNJ6* was also associated with increased opioid requirement for analgesia and increased substitution therapy with methadone in former heroin addicts (Lotsch et al., 2010). A comparison of genetic variation in *KCNJ3* and *KCNJ6* identified SNPs in *KCNJ6* that were associated with a pain-related phenotype in patients following total knee arthroplasty with postoperative opioid analgesics (Bruehl et al., 2013). These studies suggest that genetic variation in GIRK-dependent signaling affects pain outcome in mice and humans (Tables 1 and 3). A review of GIRK signaling and therapeutic strategies in opioid-dependent analgesia is found in a recent review (Nagi & Pineyro, 2014).

2.3 Motor Control

Motor control is a dopamine-dependent behavior involving the nigrostriatal pathway. The dopamine neurons of the substantia nigra (SN), a key component of this pathway, express high levels of GIRK2-containing channels (Koyrakh et al., 2005; Reyes et al., 2012), suggesting that GIRK-mediated inhibition may contribute to motor activity. Increased motor activity was observed in both *weaver* (Schmidt et al., 1982) and *Girk2*$^{-/-}$ mice (Blednov et al., 2001a), supporting a role for GIRK2-containing channels in motor control (Table 1). *Girk2*$^{-/-}$ mice showed transient hyperactivity and slower habituation in an open-field test and increased spontaneous locomotor activity during the dark phase in their home cages (Blednov et al., 2002). Motor activity increased after habituation and was inhibited by the dopamine D_1 receptor antagonist SCH 23390 and increased by the D_1 partial agonist SKF 38393. SCH 23390 also inhibited basal activity levels in knockout and wild-type mice. These results suggested that D_1 receptor signaling is enhanced in *Girk2*$^{-/-}$ mice in a stressful environment, resulting in transient hyperactive behavior.

Other work showed that both *Girk1*$^{-/-}$ and *Girk2*$^{-/-}$ mice displayed increased motor activity, delayed habituation to an open field, and resistance to baclofen-induced ataxia in the rotarod test (Pravetoni & Wickman, 2008), but *Girk3*$^{-/-}$ and *Girk4*$^{-/-}$ mice did not differ from wild-type in locomotor activity (Pravetoni & Wickman, 2008; Wickman et al., 2000). In contrast, ML297 (a potent small molecule agonist of GIRK1-containing channels) suppressed motor activity in wild-type C57BL/6J mice in an open-field test at the highest dose tested (Wydeven et al., 2014).

Further evidence for a role of GIRK2 in motor control comes from *weaver* mice, which have been used as a model of Parkinson's disease (Table 1). Parkinson's disease produces progressive degeneration of dopamine neurons in the SN pars compacta, resulting in loss of motor coordination and some cognitive impairment (Rodriguez-Oroz et al., 2009). The GIRK2 mutation in *weaver* mice makes the channels nonselective for cations and less sensitive to $G_{\beta\gamma}$ (Kofuji et al., 1996; Navarro et al., 1996; Slesinger et al., 1996), producing constitutively active GIRK2 channels that may be responsible for the neuronal degeneration in dopaminergic neurons in the SN pars compacta and cerebellar granular neurons that are observed in these mice (Harkins & Fox, 2002; Schmidt et al., 1982; Smeyne & Goldowitz, 1989). The progressive dopaminergic degeneration causes gait instability, poor limb coordination, and tremors, similar to the phenotypes observed in Parkinson's disease and chronic drug addiction (Ebadi et al., 2005). However, the neurological changes in *weaver* mice do not completely mimic those of Parkinson's disease. Activation of GIRK2 channels and the resulting efflux of K^+ have also been linked to nerve growth factor-programmed cell death in dorsal root ganglion neurons, which is important in normal development of the nervous system (Coulson et al., 2008), indicating that GIRK2 subunits control both normal and pathological mechanisms of neuronal degeneration.

2.4 Psychiatric Disorders

2.4.1 Depression

Depression is the most common psychiatric disorder and a leading cause of disability worldwide (Saltiel & Silvershein, 2015). First-line treatment options for depression include selective serotonin reuptake inhibitors (SSRIs), serotonin–norepinephrine reuptake inhibitors, and norepinephrine–dopamine reuptake inhibitors (Saltiel & Silvershein, 2015). Although monoaminergic pathways are the primary targets of antidepressants, altered GIRK signaling may also be involved in their therapeutic action, considering that GIRK channels are the main inhibitory effectors of 5-HT_{1A} receptors. Deletion of GIRK2 subunits in mice was associated with depression-resistant behaviors combined with a reduced behavioral response to the antidepressant citalopram (an SSRI) and reduced electrophysiological responses to 5-HT_{1A} receptor agonists, suggesting that GIRK channels may be targets for treating depression by decreasing sensitivity of dorsal raphe neurons to serotonergic transmission (Llamosas et al., 2015) (Table 1). Furthermore, chronic administration of fluoxetine, an SSRI, restored anticipatory behavior in socially stressed rats in a rodent model of depression, and this response was partly mediated by the suppression of $GABA_B$R–GIRK signaling (Cornelisse et al., 2007). The effect of fluoxetine on GIRK currents was also present in control animals and was independent of the animal's depressed state. High concentrations of fluoxetine inhibited GIRK channels expressed in oocytes, whereas other SSRI antidepressants had little or no effect (Kobayashi, Washiyama, & Ikeda, 2004). The GIRK channel inhibitor tipepidine also acts as a novel antidepressive agent in the forced swimming test in rats, perhaps by enhancing dopaminergic or noradrenergic transmission (Kawaura et al., 2012). In humans, an epistatic interaction between *KCNJ6* (GIRK2) and CREB1 (cyclic adenosine 5′-phosphate (adenosine monophosphate)-response element binding protein) may influence rumination, a symptom of depression (Lazary et al., 2011).

A recent study showed that the rapid antidepressant effects of NMDA receptor antagonists cause $GABA_B$R–GIRK decoupling by increasing the stability of the adapter protein, 14-3-3η (Workman et al., 2015). Interestingly, 14-3-3η is implicated in other neurological diseases that are associated with GIRK channel function, including Parkinson's disease and schizophrenia (Foote & Zhou, 2012). Levels of $GABA_B$Rs and 14-3-3η decreased in the hippocampi of socially defeated rats in a model of depression (Workman et al., 2015). However, mice injected with NMDA antagonists (rapid antidepressants) showed increased $GABA_B$R and adaptor protein levels and decreased GIRK2 levels in hippocampal synaptoneurosomes. The elimination of $GABA_BR$–GIRK signaling via 14-3-3η was required for the rapid antidepressant efficacy of ketamine, suggesting that inhibition of GIRK signaling is a potential mechanism for treating depression, a condition associated with other psychiatric disorders and addictive behaviors.

2.4.2 Anxiety

There are many different types of anxiety disorders, and these disorders may increase the risk for comorbid mood and substance use disorders (Kessler, Ruscio, Shear, & Wittchen, 2010). Studies using knockout mice provide evidence for GIRK subunits in anxiety-related behaviors. For example, *Girk2*$^{-/-}$ mice demonstrated reduced anxiety with signs of hyperactivity in the elevated plus-maze, light/dark box, and "canopy" anxiety tests (Blednov et al., 2001a). In the elevated plus-maze, *Girk2*$^{-/-}$ mice spent a higher percentage of time in the open arms and had a greater number of total entries. A short period of social isolation decreased anxiety and increased total activity as shown by an increased number of open arm entries, whereas behavior in wild-type mice was not substantially altered by social isolation. In the light/dark box test, *Girk2*$^{-/-}$ mice demonstrated increased locomotion and a greater number of vertical explorations (rearings) in the light area. In the "canopy" test, increased locomotion in the exposed area and a trend to decrease the number of stretch attend postures in the most secure canopy area was observed in these mutants. Subsequent work showed decreased anxiety in the elevated plus-maze test in *Girk2*$^{-/-}$, and to a lesser extent in *Girk1*$^{-/-}$, but not *Girk3*$^{-/-}$ mice (Pravetoni & Wickman, 2008) (Table 1). As previously mentioned, deletion of GIRK2 subunits in mice resulted in decreased depressive-like behaviors (Llamosas et al., 2015), suggesting that GIRK2-containing channels may be relevant targets for treating both anxiety and depression.

Direct activation of GIRK1-containing channels by ML297 also reduced anxiety-related behavior in mice, without producing addictive or sedative effects (Wydeven et al., 2014). The anxiolytic effect of ML297 in the elevated plus-maze test was lost in *Girk1*$^{-/-}$ mice. Observing the same behavioral phenotype using null mutants or a GIRK agonist administered to wild-type mice could indicate that either too little or too much GIRK activity is anxiogenic. Studies of ML297, or other derivatives, may uncover a useful class of anxiolytic compounds with fewer side effects.

GIRK channels may also be involved in obsessive–compulsive disorder (OCD), an anxiety disorder characterized by obsessions or compulsions that cause distress or interfere with daily function (Decloedt & Stein, 2010). Although SSRIs are the first choice for treatment of OCD, approximately half of the patients with resistant OCD fail to respond to these drugs (Decloedt & Stein, 2010). In a mouse model of OCD that examined marble-burying behavior, tipepidine potently and dose-dependently reduced marble-burying behavior and was effective at doses that did not affect locomotor activity (Honda, Kawaura, Soeda, Shirasaki, & Takahama, 2011). Thus, inhibition of GIRK signaling, either using pharmacological channel inhibitors or genetic deletion, is associated with reduced anxiety-related behaviors and may represent a novel mechanism for the treatment of these disorders.

2.4.3 Schizophrenia

Schizophrenia is a mental disorder that can manifest as disorganized thoughts, hallucinations, or delusions, although symptoms vary dramatically between patients and even change over time in individual patients (Buckley, Miller, Lehrer, & Castle, 2009). The comorbidity of schizophrenia with other psychiatric disorders (e.g., anxiety, depression, and substance abuse) is well documented (Buckley et al., 2009). A genomewide association study (GWAS) of schizophrenia in a Japanese population identified an SNP in *KCNJ3* (GIRK1) (Yamada et al., 2011), and a subsequent GWAS of *KCNJ3* in a Chinese population identified nine SNPs that were associated with schizophrenia (Yamada et al., 2012). The initial SNP marker in the Japanese population also showed significant association in the Chinese population. Furthermore, analysis of transcript levels in the dorsolateral prefrontal cortex from postmortem brains of patients with schizophrenia or bipolar disorder revealed lower expression of *KCNJ3* compared to control subjects (Yamada et al., 2012). Thus, *KCNJ3* may represent a susceptibility gene for schizophrenia in Asian populations (Table 3).

2.4.4 Attention Deficit Hyperactivity Disorder

Attention deficit hyperactivity disorder (ADHD) is a prevalent neurodevelopmental psychiatric disorder characterized by impulsivity, inattention, and/or hyperactivity (Wilens & Spencer, 2010). ADHD affects cognitive development and can have long-lasting effects on academic success and social relationships. The neurobiological causes of its symptoms are unclear, but there is evidence for overlap between ADHD and schizophrenia, mood disorders, and substance abuse (Hamshere et al., 2013; Larsson et al., 2013; Wilens & Spencer, 2010), all of which appear to involve GIRK signaling. Tipepidine, which inhibits GIRK channel activity, is used clinically as a nonnarcotic antitussive and was recently evaluated for its ability to improve ADHD symptoms in pediatric patients. A preliminary pilot study indicated that tipepidine was well tolerated and improved ADHD rating scale scores (Sasaki et al., 2014). This suggests that inhibition of GIRK-dependent signaling might offer treatment options for ADHD that are safer than the current use of psychostimulants, which have undesirable side effects. Collectively, studies using tipepidine provide preliminary evidence for its role in different psychiatric disorders, including depression, anxiety, and ADHD.

2.5 Epilepsy

Epilepsy is a neurological disorder characterized by recurring epileptic seizures of varying duration and severity, which have no immediate underlying cause (Chang & Lowenstein, 2003). Mouse models have provided behavioral evidence that GIRK channels may be relevant targets for the treatment of seizure disorders (Table 1). The mutant GIRK2 channels in *weaver* mice are not selective for K^+ and upon activation, depolarize rather than hyperpolarize neurons, thus increasing neuronal excitability, likely contributing to the development of sporadic seizures observed in these mice (Eisenberg & Messer, 1989). *Girk2*$^{-/-}$ mice also developed spontaneous seizures and showed increased sensitivity to pentylenetetrazole-induced convulsions (Signorini et al., 1997). *Girk3*$^{-/-}$ mice did not develop seizures, but *Girk2/3*$^{-/-}$ mice experienced spontaneous and lethal seizures (Torrecilla et al., 2002). Seizures in these knockout mice may be explained by the overall loss of GIRK function and the decreased neuronal inhibition, which in turn, increases neuronal excitability. In addition, spinal administration of tertiapin, which blocks GIRK channels, had proconvulsant effects (Mazarati et al., 2006). Conversely, ML297, a selective activator of GIRK1-containing channels, had antiseizure efficacy in rat models of epilepsy (Kaufmann et al., 2013). Specifically, ML297 delayed seizure onset in an electroshock model and prevented convulsions and death in a chemical model of epilepsy (Kaufmann et al., 2013).

3. GIRK CHANNELS IN ADDICTION

The brain circuitry underlying addiction and the rewarding properties of drugs of abuse involves the mesocorticolimbic dopamine system, which consists of the ventral tegmental area (VTA), medial prefrontal cortex (mPFC), and nucleus accumbens (NAc) (van Huijstee & Mansvelder, 2014). GIRK channels are expressed in these regions, and GIRK signaling is altered by exposure to different types of addictive drugs, indicating that these channels are a common effector of drugs of abuse and likely mediate the neuroadaptations believed to be important in the development and progression of addiction. The location, cellular specificity, and long-lasting GIRK-mediated neuroadaptations triggered by different drugs of abuse suggest that GIRK channels contribute to expression of drug-addictive behaviors, such as drug seeking, craving, and relapse. In the following sections, we review the behavioral and genetic evidence from mouse models (Table 2) and genetic studies in humans (Table 3) that link GIRK channels with different drugs of abuse.

3.1 Ethanol

Alcohol initially produces intoxication, anxiolysis, and a sense of reward, presumably through direct action on specific targets such as ion channels or signaling cascades (Howard, Trudell, & Harris, 2014; Trudell, Messing, Mayfield, & Harris, 2014). After prolonged and repeated exposure, alcohol-induced changes in gene expression and synaptic function are thought to contribute to the development of altered behaviors such as tolerance, sensitization, and compulsive consumption, the hallmark of addiction (Gilpin & Koob, 2008).

GIRK channels are implicated in ethanol action as discussed in a recent review (Bodhinathan & Slesinger, 2014). $GABA_BR$–GIRK transmission in dopamine neurons of the VTA (Federici, Nistico, Giustizieri, Bernardi, & Mercuri, 2009) and $GABA_BR$–GIRK currents in cultured cerebellar granule cells were enhanced by ethanol (Lewohl et al., 1999). Ethanol also directly activated exogenously expressed GIRK channels at concentrations that are intoxicating in humans (Aryal, Dvir, Choe, & Slesinger, 2009; Kobayashi et al., 1999; Lewohl et al., 1999). GIRK2 is a prominent GIRK subunit in brain (and the VTA), and GIRK2-containing channels expressed in oocytes were more sensitive to ethanol than GIRK1/4 or GIRK4 channels (Lewohl et al., 1999).

The direct alcohol sensitivity of GIRK channels and their location in brain regions that are implicated in drug and natural reward circuitry indicate that they are relevant targets for alcohol action *in vivo*, and mouse models have been fundamental for deciphering their role in ethanol responses

Table 2 Role of Mouse *Girk* Genes in Mediating the Behavioral Effects of Drugs of Abuse

Drugs of Abuse	GIRK Genotypes	Behavioral Phenotypes	References
Ethanol	*Girk2* (*weaver*), *Girk2* (−/−)	↓ Analgesia	Kobayashi et al. (1999); Blednov et al. (2003)
	Girk2 (−/−)	↓ CTA/CPP	Hill, Alva, Blednov, and Cunningham (2003)
	Girk2 (−/−), *Girk3* (−/−)	↓ Acute withdrawal severity	Blednov, Stoffel, Chang, and Harris (2001b); Kozell, Walter, Milner, Wickman, and Buck (2009); Herman et al. (2015)
	Girk2 (−/−)	↑ Intake[a]	Blednov et al. (2001b)
	Girk3 (−/−)	↑ Binge-like drinking	Herman et al. (2015)
Pentobarbital Zolpidem	*Girk3* (−/−)	↓ Withdrawal	Kozell et al. (2009)
Cocaine	*Girk2* (−/−)	↑ Motor activity	Arora et al. (2010)
	Girk2 (−/−), *Girk3* (−/−)	↓ Self-administration	Morgan, Carroll, Loth, Stoffel, and Wickman (2003)
Opioids	*Girk2* (*weaver*)	↓ Morphine analgesia	Ikeda et al. (2000)
	Girk1 (−/−), *Girk2* (−/−)	↓ Opioid analgesia	Blednov et al. (2003); Marker et al. (2004,2005); Mitrovic et al. (2003)
	Girk2/3 (−/−), *Girk3* (−/−)	↓ Opioid analgesia	Cruz et al. (2008); Marker et al. (2002); Smith et al. (2008)
	Girk2 (−/−)	↑ Morphine-induced motor activity	Kotecki et al. (2015)
	Girk3 (−/−)	↓ Morphine-induced motor activity	Kotecki et al. (2015)
	Girk2/3 (−/−)	↓ Morphine withdrawal	Cruz et al. (2008)

The behavioral effects of drugs of abuse and the related *Girk* genes from *weaver* and knockout (−/−) mice are shown.
CTA, conditioned taste aversion; CPP, conditioned place preference.
[a]Ethanol intake only increased when the ethanol bottles were available in the preferred location, but not when the positions were alternated daily to control for side preferences. Kotecki et al. (2015) also showed that selective ablation of *Girk2* in VTA dopamine neurons increased morphine-induced motor activity, and the diminished activity in *Girk3* (−/−) mice was rescued by restoring GIRK3 expression in the VTA.

(Table 2). The first behavioral studies examining the role of GIRK channels in ethanol responses used *weaver* mice, which lacked acute ethanol-induced analgesia, but not the sedative, hypothermic, and locomotor-activating effects of ethanol (Kobayashi et al., 1999). Reduced ethanol analgesia was also observed in *Girk2*$^{-/-}$ mice (Blednov et al., 2003), and *Girk2*$^{-/-}$ mice displayed greater ethanol-stimulated activity in an open-field test (Blednov et al., 2001b). Although the withdrawal severity after acute administration of ethanol was reduced in these mice, there was no difference in withdrawal severity following a chronic ethanol diet when the amount of ethanol given to knockout mice was matched with the amount consumed by wild-type. There were also no genotype differences in ethanol-induced sleep time or acute functional tolerance.

The role of GIRK channels in ethanol reward has been investigated using two common models of addictive behavior: ethanol consumption and conditioned place preference. Animal models of voluntary ethanol administration are valuable for profiling behavioral and genetic determinants in human alcoholics, who exhibit excessive consumption as a hallmark of the disease (Hyman, Malenka, & Nestler, 2006). Animal models of conditioned place preference and conditioned taste aversion provide insight into the rewarding and aversive effects of ethanol, while models of withdrawal profile dependence and the symptoms that contribute to susceptibility for continued drug abuse (Chester & Cunningham, 2002; Green & Grahame, 2008; Metten et al., 1998). Ethanol consumption and preference did not differ in wild-type and *Girk2*$^{-/-}$ mice in the standard two-bottle choice test where the bottle positions were alternated daily to control for position preferences. However, when the ethanol bottles were always available in the preferred location, *Girk2*$^{-/-}$ mice consumed more ethanol compared to wild-type (Blednov et al., 2001b). In addition, *Girk2*$^{-/-}$ mice showed reduced conditioned taste aversion for 2.0 and 2.5 g/kg ethanol (Hill et al., 2003). Unlike wild-type mice, *Girk2*$^{-/-}$ mice failed to develop a conditioned place preference for ethanol (Hill et al., 2003). Preliminary evidence reported by Tipps and Buck in chapter "GIRK Channels: A Potential Link Between Learning and Addiction" indicates that *Girk3*$^{-/-}$ mice have increased preference for the ethanol-paired side compared to wild-type littermates. Thus, loss of GIRK2 or GIRK3 may produce opposite effects on ethanol-induced conditioned place preference, suggesting that the sensitivity to ethanol may depend upon the subunit composition of GIRK channels. Further evidence for subunit selectivity and ethanol sensitivity can be found in a recent study showing that deletion of GIRK3 increased limited access but not continuous access voluntary drinking and decreased acute withdrawal severity, but did not affect the metabolic, sedative, hypothermic, or ataxic effects of ethanol (Herman et al., 2015). Overexpression of GIRK3 in the VTA reversed the binge-drinking phenotype and reduced drinking in wild-type mice. Deletion of GIRK3 also decreased ethanol-induced excitation of VTA dopamine neurons and dopamine release in the NAc, providing additional evidence that GIRK3 is required for activation of the mesolimbic dopaminergic pathway by ethanol (Herman et al., 2015). These results point to a role for GIRK3 in the rewarding properties of ethanol and as a potential target for regulating binge-like drinking.

GIRK3 may also be required for other ethanol responses. For example, QTL mapping identified the *Girk3* gene in a region of mouse chromosome 1 that was associated with withdrawal from ethanol and other sedative–hypnotics (Kozell et al., 2009). DBA/2J and chromosome 1 congenic mice have a small QTL interval containing *Girk3* from the DBA/2J strain in a genetic C57BL/6J background and exhibit more severe withdrawal from ethanol and other drugs of abuse than C57BL/6J mice. *Girk3* expression in the brain is greater in these mice compared with C57BL/6J mice (Kozell et al., 2009). Furthermore, *Girk3*$^{-/-}$ mice demonstrate less severe withdrawal from ethanol than their wild-type littermates (Herman et al., 2015; Kozell et al., 2009). Interestingly, the region of chromosome 1 containing *Girk3* also contains QTLs for ethanol drinking (Tarantino, McClearn, Rodriguez, & Plomin, 1998), ethanol-conditioned aversion (Risinger & Cunningham, 1998), and acute sensitivity to ethanol (Crabbe, Belknap, Mitchell, & Crawshaw, 1994; Demarest, McCaughran, Mahjubi, Cipp, & Hitzemann, 1999), suggesting that this and other nearby genes may be involved in several ethanol-related behaviors.

A multivariate analysis of alcohol phenotypes in 37 different mouse mutant lines and their wild-type controls revealed that *Girk2* is part of a gene cluster associated with taste, and this cluster is driven by decreased ethanol and saccharin consumption (Blednov, Mayfield, Belknap, & Harris, 2012). The *Girk2* mutation was tested on two different backgrounds (B6x129, B6N6), which resulted in its placement in different gene clusters, suggesting that the genetic background also plays an important role in the function of this gene in alcohol-related phenotypes.

Human genetic studies have provided additional support for GIRK channels in alcohol-dependent phenotypes (Table 3). The Collaborative

Table 3 Human *KCNJ* Genes Implicated in CNS Disorders

Behavioral Phenotypes	Genotypes	References
Down syndrome	*KCNJ6*	Toyoda et al. (2002)
Keppen–Lubinsky syndrome	*KCNJ6*	Masotti et al. (2015)
Schizophrenia	*KCNJ3* SNPs	Yamada et al. (2011, 2012)
Increased opioids required for analgesia	*KCNJ6* SNPs	Bruehl et al. (2013); Lotsch, Pruss, Veh, and Doehring (2010); Nishizawa et al. (2009, 2014)
Reduced opioid withdrawal	*KCNJ6* SNPs	Lotsch et al. (2010)
Nicotine dependence	*KCNJ6* SNPs	Nishizawa et al. (2014); Saccone et al. (2007)
Alcohol dependence (adults) Hazardous drinking (adolescents)	*KCNJ6* SNPs	Clarke et al. (2011)
EROs in alcohol dependence	*KCNJ6* SNPs	Kang et al. (2012)

The human *KCNJ* genes associated with CNS disorders and drugs of abuse are shown. *KCNJ3* and *KCNJ6* correspond to the mouse *Girk1* and *Girk2* genes, which encode the GIRK1 and GIRK2 subunits, respectively.
EROs, event-related oscillations; SNPs, single nucleotide polymorphisms.

Study on the Genetics of Alcoholism (COGA) examined event-related oscillations (EROs) in electroencephalogram recordings, which signify cognitive processes during normal and pathological brain function (Basar, Basar-Eroglu, Karakas, & Schurmann, 2001). These brain oscillations are stable, highly heritable (van Beijsterveldt, Molenaar, de Geus, & Boomsma, 1996) and are shared between alcohol dependence and related disorders (Porjesz et al., 2005). A COGA family-based GWAS was performed for a specific ERO phenotype using SNPs genotyped in families affected by alcohol use disorder. The GWAS identified several SNPs in *KCNJ6* (GIRK2) that may account for the ERO phenotype (Kang et al., 2012). SNPs in the promoter region of *KCNJ6* were also associated with alcohol dependence in adults and hazardous drinking behavior in adolescents who were exposed to early life stress (Clarke et al., 2011). Overall, genetic variations, subunit composition, and cell specificity can all be critical determinants of ethanol action on GIRK channel function and behavioral responses.

3.2 Sedative/Hypnotics

Sedative–hypnotic drugs depress CNS function and are used to reduce anxiety or induce sleep. Although alcohol, opioids, and GHB fall into this general category, barbiturates and benzodiazepines are considered the two major classes of sedative–hypnotics. A QTL associated with pentobarbital withdrawal was mapped to a region containing 15 genes of mouse chromosome 1, and *Girk3* was identified as a particularly promising candidate (Kozell et al., 2009). Less severe pentobarbital withdrawal was associated with lower *Girk3* mRNA expression, suggesting that a *Girk3* null mutation would decrease pentobarbital withdrawal compared to that of wild-type mice. As observed for ethanol, *Girk3*$^{-/-}$ mice experienced less severe withdrawal from pentobarbital and zolpidem (Table 2), providing evidence for GIRK3-containing channels in acute withdrawal from barbiturates and benzodiazepines (Kozell et al., 2009). *Girk3*$^{-/-}$ and wild-type mice did not differ in pentobarbital-induced sedation and hypothermia, suggesting that *Girk3* mediates a subset of sedative–hypnotic effects.

3.3 Psychostimulants

Psychostimulants, such as cocaine and amphetamine, can produce a continuum of behavioral and cognitive effects, with low doses producing beneficial cognitive effects and high doses producing addiction and psychosis (Wood, Sage, Shuman, & Anagnostaras, 2014). Behavioral studies in mice indicate that GIRK signaling may be involved in the early stages of addiction to cocaine (Table 2). *Girk2*$^{-/-}$ mice demonstrated enhanced locomotor responses to cocaine (Arora et al., 2010). In addition, mice given lentiviral RNAi infusions to suppress GIRK1 and GIRK2 expression in the mPFC showed elevated motor activity in response to an initial injection of cocaine (Hearing et al., 2013), suggesting that persistent suppression of GIRK signaling can presensitize mice to the motor-stimulatory effect of cocaine. Effects on drug seeking for psychostimulants were also observed in *Girk2*$^{-/-}$ and *Girk3*$^{-/-}$ mice. These mice exhibited reduced intravenous self-administration of cocaine compared to wild-type counterparts (Morgan et al., 2003). Interestingly, *Girk2/3*$^{-/-}$ mice self-administered more cocaine than *Girk2*$^{-/-}$ or *Girk3*$^{-/-}$ mice, perhaps due to differing compensatory mechanisms in the single versus double subunit knockouts.

Sorting Nexin 27 (SNX27) regulates GIRK channel trafficking and its expression is upregulated by cocaine and methamphetamine (Kajii et al.,

2003). Mice lacking SNX27 showed a specific reduction in $GABA_BR$–GIRK currents in VTA dopamine neurons and were hypersensitive to the locomotor-stimulating effects of cocaine (Munoz & Slesinger, 2014). These effects were reversed by expression of GIRK2a, an SNX27-insensitive splice variant. Inhibition of $GABA_BR$–GIRK signaling, and the resulting increased dopamine neuron excitability in the VTA, may be a cellular mechanism promoting addiction to psychostimulants.

3.4 Opioids

GIRK channels are coupled to μ-, κ-, and δ-opioid receptors (Nagi & Pineyro, 2014), and as discussed earlier, animal and human studies provide corroborating evidence that GIRK channels mediate opioid analgesia (Tables 2 and 3). In addition, GIRK signaling is important for the motor-stimulating effects of morphine. GIRK2/3 channels in VTA dopamine neurons were required for the motor-stimulatory effect of systemic morphine, whereas GIRK1/2 channels in VTA GABA neurons were not involved (Kotecki et al., 2015). Thus, GIRK channels appear to regulate opioid-induced motor activity in a cell- and subunit-dependent manner. This work illustrates the utility of selective ablation of GIRK subunits in individual neurons to decipher cellular- and subunit-specific GIRK signaling. GIRK2/3 channels in VTA dopamine neurons are also important in mediating the motor-stimulatory effects of cocaine (Munoz & Slesinger, 2014) and may be a common target for other drugs of abuse (Cruz et al., 2004; Herman et al., 2015; Labouebe et al., 2007).

GIRK channels are also implicated in dependence after chronic exposure to morphine (Tables 2 and 3). For example, morphine withdrawal symptoms were greatly reduced in mice lacking *Girk2/3*$^{-/-}$ (Cruz et al., 2008). Electrophysiological responses in brain slices from these mice lacked the increased spontaneous firing that is associated with morphine withdrawal, and postsynaptic GIRK currents were abolished. In humans, an SNP in *KCNJ6* (GIRK2) was associated with increased opioid requirements for analgesia and a lack of opioid withdrawal symptoms (Lotsch et al., 2010).

Interestingly, tolerance to repeated morphine administration may be associated with increased potency and sensitization of opioid receptors. Opioids have a biphasic effect on GIRK currents from periaqueductal gray neurons (which contribute to opioid antinociception and tolerance) in morphine-tolerant rats (Ingram, Macey, Fossum, & Morgan, 2008). GIRK currents were initially potentiated by met-enkephalin and inhibited by a μ-opioid antagonist in brain slices from morphine-pretreated rats, suggesting that repeated morphine exposure *in vivo* enhances agonist stimulation of μ-opioid receptors; however, peak GIRK currents in slices from morphine-tolerant rats exhibited greater desensitization. The altered μ-opioid–GIRK signaling may contribute to the development of opioid tolerance.

3.5 Nicotine

The first candidate gene study examining genetic risk variants of nicotine dependence identified a marker in *KCNJ6* (GIRK2) as one of the top signals (Saccone et al., 2007). We previously discussed the association between polymorphisms in *KCNJ6* and individual postoperative sensitivity to opioid analgesia in humans. One unique SNP (rs2835859) was also associated with susceptibility to nicotine dependence in a Japanese population (Nishizawa et al., 2014). Carriers of the C allele of this SNP were less sensitive to pain, required less opioid analgesics postoperatively, had higher susceptibility to nicotine dependence, and required a greater number of trials in order to stop smoking (Table 3). Although there is no known connection between the nicotinic acetylcholine receptors activated by nicotine and GIRK channels, it is possible that these studies reflect a general shift in reward sensitivity mediated by differences in GIRK-dependent signaling.

3.6 GIRK Modulators and Other Drugs of Abuse

GIRK channels are implicated in ethanol and opiate withdrawal and cocaine seeking. It is interesting that RGS proteins are also associated with mediating effects of alcohol (Stewart et al., 2015), cocaine (Rahman et al., 2003), and morphine (Zachariou et al., 2003), although it is unknown whether these effects are related specifically to the modulatory influence of RGS proteins on GIRK-dependent signaling. In addition, GIRK channels may mediate some of the rewarding effects of tetrahydrocannabiol (the main component in marijuana), given that they are activated by endocannabinoids (Guo & Ikeda, 2004) and are implicated in cannabinoid-induced nociception and analgesia in knockout mice as previously discussed. GHB may prove to be another example in the list of addictive drugs that activate GIRK signaling. The different subunit composition of GIRK channels and $GABA_BR$–GIRK coupling efficiency in different types of VTA neurons may account for the cellular and behavioral effects of the $GABA_BR$ agonists, GHB (abused drug) and baclofen (anticraving drug) (Cruz et al., 2004; Labouebe et al., 2007). Unique neuronal populations and functional sensitivities may thus confer distinct regional and cellular control of GIRK channel function in brain. Based on their overall sensitivity (cellular, behavioral, genetic) to different types of drugs of abuse, GIRK channels likely constitute a common target in the addictive process with the potential to affect treatment outcome.

4. THERAPEUTIC POTENTIAL OF GIRK CHANNEL MODULATORS

The suggested roles for GIRK channels in CNS diseases are heavily based on studies using mouse knockout models. Although animal models have greatly advanced understanding of GIRK channel function in normal and disease states, extending studies of the genotypes and related phenotypes to humans are a necessary bridge. Behavioral evaluation of drug targets in animals, combined with analysis of genetic variants in humans, may provide an effective strategy for advancing therapeutics for drug dependence and other polygenic diseases. For example, a human genetic link between certain peroxisome proliferator-activated receptors (PPARs) and alcohol-related phenotypes corroborated studies in mice showing that specific PPAR agonists reduced ethanol consumption (Blednov et al., 2015). Combining both animal and human data to systematically evaluate and nominate specific GIRK subunits may be a beneficial approach for determining the significance of these channels in disease and their potential to affect treatment outcome.

Currently, effective pharmacotherapies for drug addiction disorders are lacking. Disulfiram, naltrexone (opioid antagonist), and acamprosate are FDA-approved for treating alcohol addiction, but have limited efficacy and are not routinely prescribed as therapeutics (Zindel & Kranzler, 2014). As discussed in this chapter, the $GABA_BR$–GIRK signaling pathway is involved in alcohol, cocaine, and GHB responses in animal studies. Interestingly, the $GABA_B$ agonist baclofen has been approved to treat alcohol addiction in France and is under clinical trials in the United States (Addolorato et al., 2011). A review of potential therapeutics for drug use disorders highlights the beneficial uses of baclofen and other $GABA_BR$ modulators (Addolorato, Leggio, Hopf, Diana, & Bonci, 2012). Baclofen shows promise in managing alcohol-withdrawal symptoms, reducing alcohol craving, and promoting alcohol abstinence in preclinical animal models and human alcoholics. $GABA_BR$ agonists and positive allosteric modulators may also be effective for combating addiction to other drugs of abuse, including cocaine, methamphetamine, nicotine, and opioids (Phillips & Reed, 2014). In humans, baclofen reduced subliminal cue-induced mesolimbic activation in cocaine-dependent individuals, suggesting that it may be promising in preventing relapse (Young et al., 2014). Baclofen was also noted in a single case report as a potential treatment for GHB withdrawal (LeTourneau, Hagg, & Smith, 2008).

In addition to a role in mediating the effects of drugs of abuse, we discussed how $GABA_BR$–GIRK signaling is involved in mood, memory, and nociception. This suggests that there are other therapeutic applications for novel $GABA_BR$ modulators, as well as modulators of other GPCRs with GIRK channel effectors. For example, the development of novel opioid analgesics has the potential to improve chronic pain management while reducing side effects. An ideal opioid ligand might be one with selectivity for δ-opioid receptors that mediates its analgesic effects primarily via GIRK channels without inducing cellular tolerance or other unwanted effects involving other pathways (Nagi & Pineyro, 2014).

Preliminary results indicate that the GIRK channel inhibitor tipepidine may be effective in treating ADHD in children (Sasaki, Hashimoto, Tachibana, Kurata, Okawada, et al., 2014), and preclinical and preliminary clinical studies suggest that it may have uses in other psychiatric disorders, such as anxiety and depression (Honda et al., 2011; Kawaura et al., 2012; Sasaki et al., 2014). In addition, new classes of potent, subunit-selective GIRK channel compounds have been identified that could enable pharmacological targeting of particular brain regions and behaviors. For example, ML297 activates recombinant neuronal GIRK channels containing the GIRK1 subunit and decreases anxiety-related behavior without sedative or overt addictive (rewarding) effects (Wydeven et al., 2014). ML297, or another subunit-selective derivative, may provide better therapeutics for seizure and/or anxiety disorders and also help decipher the contribution of GIRK channels in other diseases. Ideally, new classes of potent compounds designed for therapeutic applications in CNS disease will offer both subtype selectivity and an enhanced ability to penetrate the blood brain barrier. These criteria would advance the possibility of finding specific modulators that can alter access to alcohol and other drug sites on GIRK subunits in brain. Inhibiting access to binding pockets in channel proteins that are targeted by drugs of abuse, without altering channel gating, offers a selective mechanism for treating drug abuse with decreased side effects. Identification of small selective molecules, along with advances in X-ray crystallography of channel structure (discussed below), are promising tools for drug design and manipulation of discrete sites of channel action. Targeted molecules, acting on GIRK or other channels that are modulated by drugs of abuse, make the goal of therapeutics with decreased side effects an exciting prospect for addiction research.

Furthermore, other technologies (knock-in animal models, conditional and cell-specific knockouts, light activation of GPCRs, and chemogenetic applications for GPCRs, such as Designer Receptors Exclusively Activated by Designer Drugs or DREADDs) (Urban & Roth, 2015) provide discriminating tools for the future and allow more targeted approaches compared to studies of global null mutants. Human genetic association studies are also needed to determine candidate genes and more accurately assess the role of GIRK channels in CNS disorders. X-ray crystal structures have shown ethanol bound to ion channels and provide evidence for alcohol-binding cavities in GIRK (Aryal et al., 2009) and other channel proteins (Howard et al., 2014). The rapid progress being made in crystal structures will be key for modeling the interactions of drugs with GIRK subunits and diagramming the rules of engagement (see chapter "Structural Insights into GIRK Channel Function" by Glaaser and Slesinger). The physical, chemical, and modulatory properties of drug-binding pockets may reveal mechanisms or smart molecules that can displace and inhibit drug action on GIRK channels (Bodhinathan & Slesinger, 2014). Applying structural data with the aforementioned approaches can help connect molecular models to function and behavior and enhance translational research. However, these goals and ideals are tempered by the fact that full activation or inhibition of GIRK function, even in discrete areas, could profoundly alter the balance of excitatory/inhibitory signaling and produce unwanted consequences.

5. CONCLUDING REMARKS AND FUTURE DIRECTIONS

Mouse knockout models have been invaluable for determining the roles of GIRK channels in many different CNS processes, but continued progress likely requires more combinatorial approaches to bridge animal and human studies, as well as the implementation of new tools. Collectively, the resources and approaches described above (i.e., subunit-selective channel modulators, conditional knockouts, crystallography, etc.) will be crucial for deciphering the role of GIRK channels in disease and in drug design for individual pathological conditions. Advances in these different areas of research will be quite significant in determining if GIRK channels are indeed potential targets for treating CNS disorders. Although GIRK channel function is altered in many CNS diseases, other protein targets and signaling mechanisms are also affected, and the specific role of GIRK channels and their modulators must be convincing before they can be considered for therapeutic benefit.

Biological systems, including the pathological processes operating in CNS disorders, function within a framework of inter-connected pathways. Ultimately, successful treatment of complex-trait disorders will depend on a systems-level approach to disease. While the study of individual genes is informative, using network-centric and systems biology approaches to identify inter-related gene networks and pathways that are more likely representative of the array of processes operating in CNS diseases are also warranted. Rapid advances in genomic and proteomic techniques can transform our ability to analyze complex disease processes and decode large data sets into more meaningful biological processes (Gorini, Harris, & Mayfield, 2014). Because drug addiction and other CNS diseases represent multifactorial processes with genetic and environmental determinants and neuroadaptations related to disease progression (Renthal & Nestler, 2008), moving beyond the significance of individual candidate genes to include the relevant gene and protein networks may better ascertain the role of GIRK channels and their associated biological systems in different stages of disease.

ACKNOWLEDGMENTS

The authors thank Dr. Megan Tipps for helpful comments and suggestions. R.A.H. acknowledges funding from NIAAA Grant AA006399.

REFERENCES

Addolorato, G., Leggio, L., Ferrulli, A., Cardone, S., Bedogni, G., Caputo, F., et al. (2011). Dose–response effect of baclofen in reducing daily alcohol intake in alcohol dependence: Secondary analysis of a randomized, double-blind, placebo-controlled trial. *Alcohol and Alcoholism*, *46*(3), 312–317. http://dx.doi.org/10.1093/alcalc/agr017.

Addolorato, G., Leggio, L., Hopf, F. W., Diana, M., & Bonci, A. (2012). Novel therapeutic strategies for alcohol and drug addiction: Focus on GABA, ion channels and transcranial magnetic stimulation. *Neuropsychopharmacology: Official Publication of the American College of Neuropsychopharmacology*, *37*(1), 163–177. http://dx.doi.org/10.1038/npp.2011.216.

Anderson, G. R., Posokhova, E., & Martemyanov, K. A. (2009). The R7 RGS protein family: Multi-subunit regulators of neuronal G protein signaling. *Cell Biochemistry and Biophysics*, *54*(1–3), 33–46. http://dx.doi.org/10.1007/s12013-009-9052-9.

Arora, D., Haluk, D. M., Kourrich, S., Pravetoni, M., Fernandez-Alacid, L., Nicolau, J. C., et al. (2010). Altered neurotransmission in the mesolimbic reward system of Girk mice. *Journal of Neurochemistry*, *114*(5), 1487–1497. http://dx.doi.org/10.1111/j.1471-4159.2010.06864.x.

Aryal, P., Dvir, H., Choe, S., & Slesinger, P. A. (2009). A discrete alcohol pocket involved in GIRK channel activation. *Nature Neuroscience*, *12*(8), 988–995. http://dx.doi.org/10.1038/nn.2358.

Basar, E., Basar-Eroglu, C., Karakas, S., & Schurmann, M. (2001). Gamma, alpha, delta, and theta oscillations govern cognitive processes. *International Journal of Psychophysiology: Official Journal of the International Organization of Psychophysiology*, *39*(2–3), 241–248.

Belichenko, N. P., Belichenko, P. V., Kleschevnikov, A. M., Salehi, A., Reeves, R. H., & Mobley, W. C. (2009). The "Down syndrome critical region" is sufficient in the mouse model to confer behavioral, neurophysiological, and synaptic phenotypes characteristic of Down syndrome. *The Journal of Neuroscience: The Official Journal of the Society for Neuroscience*, *29*(18), 5938–5948. http://dx.doi.org/10.1523/JNEUROSCI.1547-09.2009.

Blednov, Y. A., Benavidez, J. M., Black, M., Ferguson, L. B., Schoenhard, G. L., Goate, A. M., et al. (2015). Peroxisome proliferator-activated receptors alpha and gamma are linked with alcohol consumption in mice and withdrawal and dependence in humans. *Alcoholism, Clinical and Experimental Research*, *39*(1), 136–145. http://dx.doi.org/10.1111/acer.12610.

Blednov, Y. A., Mayfield, R. D., Belknap, J., & Harris, R. A. (2012). Behavioral actions of alcohol: Phenotypic relations from multivariate analysis of mutant mouse data. *Genes, Brain, and Behavior*, *11*(4), 424–435. http://dx.doi.org/10.1111/j.1601-183X.2012.00780.x.

Blednov, Y. A., Stoffel, M., Alva, H., & Harris, R. A. (2003). A pervasive mechanism for analgesia: Activation of GIRK2 channels. *Proceedings of the National Academy of Sciences of the United States of America*, *100*(1), 277–282. http://dx.doi.org/10.1073/pnas.012682399.

Blednov, Y. A., Stoffel, M., Chang, S. R., & Harris, R. A. (2001a). GIRK2 deficient mice. Evidence for hyperactivity and reduced anxiety. *Physiology & Behavior*, *74*(1–2), 109–117.

Blednov, Y. A., Stoffel, M., Chang, S. R., & Harris, R. A. (2001b). Potassium channels as targets for ethanol: Studies of G-protein-coupled inwardly rectifying potassium channel 2 (GIRK2) null mutant mice. *The Journal of Pharmacology and Experimental Therapeutics*, *298*(2), 521–530.

Blednov, Y. A., Stoffel, M., Cooper, R., Wallace, D., Mane, N., & Harris, R. A. (2002). Hyperactivity and dopamine D1 receptor activation in mice lacking girk2 channels. *Psychopharmacology*, *159*(4), 370–378. http://dx.doi.org/10.1007/s00213-001-0937-6.

Bodhinathan, K., & Slesinger, P. A. (2014). Alcohol modulation of G-protein-gated inwardly rectifying potassium channels: From binding to therapeutics. *Frontiers in Physiology*, *5*, 76. http://dx.doi.org/10.3389/fphys.2014.00076.

Bruehl, S., Denton, J. S., Lonergan, D., Koran, M. E., Chont, M., Sobey, C., et al. (2013). Associations between KCNJ6 (GIRK2) gene polymorphisms and pain-related phenotypes. *Pain*, *154*(12), 2853–2859. http://dx.doi.org/10.1016/j.pain.2013.08.026.

Buckley, P. F., Miller, B. J., Lehrer, D. S., & Castle, D. J. (2009). Psychiatric comorbidities and schizophrenia. *Schizophrenia Bulletin*, *35*(2), 383–402. http://dx.doi.org/10.1093/schbul/sbn135.

Caviness, V. S., Jr., & Rakic, P. (1978). Mechanisms of cortical development: A view from mutations in mice. *Annual Review of Neuroscience*, *1*, 297–326. http://dx.doi.org/10.1146/annurev.ne.01.030178.001501.

Chabert, C., Jamon, M., Cherfouh, A., Duquenne, V., Smith, D. J., Rubin, E., et al. (2004). Functional analysis of genes implicated in Down syndrome: 1. Cognitive abilities in mice transpolygenic for Down Syndrome Chromosomal Region-1 (DCR-1). *Behavior Genetics*, *34*(6), 559–569. http://dx.doi.org/10.1007/s10519-004-5584-3.

Chang, B. S., & Lowenstein, D. H. (2003). Epilepsy. *The New England Journal of Medicine, 349*(13), 1257–1266. http://dx.doi.org/10.1056/NEJMra022308.

Chester, J. A., & Cunningham, C. L. (2002). GABA(A) receptor modulation of the rewarding and aversive effects of ethanol. *Alcohol, 26*(3), 131–143.

Chung, H. J., Ge, W. P., Qian, X., Wiser, O., Jan, Y. N., & Jan, L. Y. (2009). G protein-activated inwardly rectifying potassium channels mediate depotentiation of long-term potentiation. *Proceedings of the National Academy of Sciences of the United States of America, 106*(2), 635–640. http://dx.doi.org/10.1073/pnas.0811685106.

Clarke, T. K., Laucht, M., Ridinger, M., Wodarz, N., Rietschel, M., Maier, W., et al. (2011). KCNJ6 is associated with adult alcohol dependence and involved in gene x early life stress interactions in adolescent alcohol drinking. *Neuropsychopharmacology: Official Publication of the American College of Neuropsychopharmacology, 36*(6), 1142–1148. http://dx.doi.org/10.1038/npp.2010.247.

Cooper, A., Grigoryan, G., Guy-David, L., Tsoory, M. M., Chen, A., & Reuveny, E. (2012). Trisomy of the G protein-coupled K+ channel gene, Kcnj6, affects reward mechanisms, cognitive functions, and synaptic plasticity in mice. *Proceedings of the National Academy of Sciences of the United States of America, 109*(7), 2642–2647. http://dx.doi.org/10.1073/pnas.1109099109.

Cornelisse, L. N., Van der Harst, J. E., Lodder, J. C., Baarendse, P. J., Timmerman, A. J., Mansvelder, H. D., et al. (2007). Reduced 5-HT1A- and GABAB receptor function in dorsal raphe neurons upon chronic fluoxetine treatment of socially stressed rats. *Journal of Neurophysiology, 98*(1), 196–204. http://dx.doi.org/10.1152/jn.00109.2007.

Coscia, E. M., & Fentress, J. C. (1993). Neurological dysfunction expressed in the grooming behavior of developing weaver mutant mice. *Behavior Genetics, 23*(6), 533–541.

Costa, A. C., Scott-McKean, J. J., & Stasko, M. R. (2008). Acute injections of the NMDA receptor antagonist memantine rescue performance deficits of the Ts65Dn mouse model of Down syndrome on a fear conditioning test. *Neuropsychopharmacology: Official Publication of the American College of Neuropsychopharmacology, 33*(7), 1624–1632. http://dx.doi.org/10.1038/sj.npp.1301535.

Coulson, E. J., May, L. M., Osborne, S. L., Reid, K., Underwood, C. K., Meunier, F. A., et al. (2008). p75 neurotrophin receptor mediates neuronal cell death by activating GIRK channels through phosphatidylinositol 4,5-bisphosphate. *The Journal of Neuroscience: The Official Journal of the Society for Neuroscience, 28*(1), 315–324. http://dx.doi.org/10.1523/JNEUROSCI.2699-07.2008.

Coussons-Read, M. E., & Crnic, L. S. (1996). Behavioral assessment of the Ts65Dn mouse, a model for Down syndrome: Altered behavior in the elevated plus maze and open field. *Behavior Genetics, 26*(1), 7–13.

Crabbe, J. C., Belknap, J. K., Mitchell, S. R., & Crawshaw, L. I. (1994). Quantitative trait loci mapping of genes that influence the sensitivity and tolerance to ethanol-induced hypothermia in BXD recombinant inbred mice. *The Journal of Pharmacology and Experimental Therapeutics, 269*(1), 184–192.

Cramer, N. P., Best, T. K., Stoffel, M., Siarey, R. J., & Galdzicki, Z. (2010). GABAB-GIRK2-mediated signaling in Down syndrome. *Advances in Pharmacology, 58*, 397–426. http://dx.doi.org/10.1016/S1054-3589(10)58015-3.

Cruz, H. G., Berton, F., Sollini, M., Blanchet, C., Pravetoni, M., Wickman, K., et al. (2008). Absence and rescue of morphine withdrawal in GIRK/Kir3 knock-out mice. *The Journal of Neuroscience: The Official Journal of the Society for Neuroscience, 28*(15), 4069–4077. http://dx.doi.org/10.1523/JNEUROSCI.0267-08.2008.

Cruz, H. G., Ivanova, T., Lunn, M. L., Stoffel, M., Slesinger, P. A., & Luscher, C. (2004). Bi-directional effects of GABA(B) receptor agonists on the mesolimbic dopamine system. *Nature Neuroscience*, 7(2), 153–159. http://dx.doi.org/10.1038/nn1181.

Decloedt, E. H., & Stein, D. J. (2010). Current trends in drug treatment of obsessive-compulsive disorder. *Neuropsychiatric Disease and Treatment, 6*, 233–242.

Demarest, K., McCaughran, J., Jr., Mahjubi, E., Cipp, L., & Hitzemann, R. (1999). Identification of an acute ethanol response quantitative trait locus on mouse chromosome 2. *The Journal of Neuroscience: The Official Journal of the Society for Neuroscience, 19*(2), 549–561.

Derenne, A., Arsenault, M. L., Austin, D. P., & Weatherly, J. N. (2007). Weaver mutant mice exhibit long-term learning deficits under several measures of instrumental behavior. *Physiology & Behavior, 92*(5), 1002–1009. http://dx.doi.org/10.1016/j.physbeh.2007.07.015.

Ebadi, M., Brown-Borg, H., El Refaey, H., Singh, B. B., Garrett, S., Shavali, S., et al. (2005). Metallothionein-mediated neuroprotection in genetically engineered mouse models of Parkinson's disease. *Brain Research. Molecular Brain Research, 134*(1), 67–75. http://dx.doi.org/10.1016/j.molbrainres.2004.09.011.

Ehrengruber, M. U., Doupnik, C. A., Xu, Y., Garvey, J., Jasek, M. C., Lester, H. A., et al. (1997). Activation of heteromeric G protein-gated inward rectifier K+ channels overexpressed by adenovirus gene transfer inhibits the excitability of hippocampal neurons. *Proceedings of the National Academy of Sciences of the United States of America, 94*(13), 7070–7075.

Eisenberg, B., & Messer, A. (1989). Tonic/clonic seizures in a mouse mutant carrying the weaver gene. *Neuroscience Letters, 96*(2), 168–172.

Federici, M., Nistico, R., Giustizieri, M., Bernardi, G., & Mercuri, N. B. (2009). Ethanol enhances GABAB-mediated inhibitory postsynaptic transmission on rat midbrain dopaminergic neurons by facilitating GIRK currents. *European Journal of Neuroscience, 29*(7), 1369–1377. http://dx.doi.org/10.1111/j.1460-9568.2009.06700.x.

Foote, M., & Zhou, Y. (2012). 14-3-3 proteins in neurological disorders. *International Journal of Biochemistry and Molecular Biology, 3*(2), 152–164.

Gilpin, N. W., & Koob, G. F. (2008). Neurobiology of alcohol dependence: Focus on motivational mechanisms. *Alcohol Research & Health: The Journal of the National Institute on Alcohol Abuse and Alcoholism, 31*(3), 185–195.

Gorini, G., Harris, R. A., & Mayfield, R. D. (2014). Proteomic approaches and identification of novel therapeutic targets for alcoholism. *Neuropsychopharmacology: Official Publication of the American College of Neuropsychopharmacology, 39*(1), 104–130. http://dx.doi.org/10.1038/npp.2013.182.

Green, A. S., & Grahame, N. J. (2008). Ethanol drinking in rodents: Is free-choice drinking related to the reinforcing effects of ethanol? *Alcohol, 42*(1), 1–11. http://dx.doi.org/10.1016/j.alcohol.2007.10.005.

Guo, J., & Ikeda, S. R. (2004). Endocannabinoids modulate N-type calcium channels and G-protein-coupled inwardly rectifying potassium channels via CB1 cannabinoid receptors heterologously expressed in mammalian neurons. *Molecular Pharmacology, 65*(3), 665–674. http://dx.doi.org/10.1124/mol.65.3.665.

Hamshere, M. L., Stergiakouli, E., Langley, K., Martin, J., Holmans, P., Kent, L., et al. (2013). Shared polygenic contribution between childhood attention-deficit hyperactivity disorder and adult schizophrenia. *The British Journal of Psychiatry: The Journal of Mental Science, 203*(2), 107–111. http://dx.doi.org/10.1192/bjp.bp.112.117432.

Harkins, A. B., & Fox, A. P. (2002). Cell death in weaver mouse cerebellum. *Cerebellum, 1*(3), 201–206. http://dx.doi.org/10.1080/14734220260418420.

Hearing, M., Kotecki, L., Marron Fernandez de Velasco, E., Fajardo-Serrano, A., Chung, H. J., et al. (2013). Repeated cocaine weakens GABA(B)-Girk signaling in layer 5/6 pyramidal neurons in the prelimbic cortex. *Neuron, 80*(1), 159–170. http://dx.doi.org/10.1016/j.neuron.2013.07.019.

Herman, M. A., Sidhu, H., Stouffer, D. G., Kreifeldt, M., Le, D., Cates-Gatto, C., et al. (2015). GIRK3 gates activation of the mesolimbic dopaminergic pathway by ethanol. *Proceedings of the National Academy of Sciences of the United States of America*, in press.

Hill, K. G., Alva, H., Blednov, Y. A., & Cunningham, C. L. (2003). Reduced ethanol-induced conditioned taste aversion and conditioned place preference in GIRK2 null mutant mice. *Psychopharmacology*, *169*(1), 108–114. http://dx.doi.org/10.1007/s00213-003-1472-4.

Honda, S., Kawaura, K., Soeda, F., Shirasaki, T., & Takahama, K. (2011). The potent inhibitory effect of tipepidine on marble-burying behavior in mice. *Behavioural Brain Research*, *216*(1), 308–312. http://dx.doi.org/10.1016/j.bbr.2010.08.010.

Howard, R. J., Trudell, J. R., & Harris, R. A. (2014). Seeking structural specificity: Direct modulation of pentameric ligand-gated ion channels by alcohols and general anesthetics. *Pharmacological Reviews*, *66*(2), 396–412. http://dx.doi.org/10.1124/pr.113.007468.

Hyman, S. E., Malenka, R. C., & Nestler, E. J. (2006). Neural mechanisms of addiction: The role of reward-related learning and memory. *Annual Review of Neuroscience*, *29*, 565–598. http://dx.doi.org/10.1146/annurev.neuro.29.051605.113009.

Ikeda, K., Kobayashi, T., Kumanishi, T., Niki, H., & Yano, R. (2000). Involvement of G-protein-activated inwardly rectifying K (GIRK) channels in opioid-induced analgesia. *Neuroscience Research*, *38*(1), 113–116.

Ingram, S. L., Macey, T. A., Fossum, E. N., & Morgan, M. M. (2008). Tolerance to repeated morphine administration is associated with increased potency of opioid agonists. *Neuropsychopharmacology: Official Publication of the American College of Neuropsychopharmacology*, *33*(10), 2494–2504. http://dx.doi.org/10.1038/sj.npp.1301634.

Jelacic, T. M., Kennedy, M. E., Wickman, K., & Clapham, D. E. (2000). Functional and biochemical evidence for G-protein-gated inwardly rectifying K+ (GIRK) channels composed of GIRK2 and GIRK3. *The Journal of Biological Chemistry*, *275*(46), 36211–36216. http://dx.doi.org/10.1074/jbc.M007087200.

Jelacic, T. M., Sims, S. M., & Clapham, D. E. (1999). Functional expression and characterization of G-protein-gated inwardly rectifying K+ channels containing GIRK3. *The Journal of Membrane Biology*, *169*(2), 123–129.

Kajii, Y., Muraoka, S., Hiraoka, S., Fujiyama, K., Umino, A., & Nishikawa, T. (2003). A developmentally regulated and psychostimulant-inducible novel rat gene mrt1 encoding PDZ-PX proteins isolated in the neocortex. *Molecular Psychiatry*, *8*(4), 434–444. http://dx.doi.org/10.1038/sj.mp.4001258.

Kang, S. J., Rangaswamy, M., Manz, N., Wang, J. C., Wetherill, L., Hinrichs, T., et al. (2012). Family-based genome-wide association study of frontal theta oscillations identifies potassium channel gene KCNJ6. *Genes, Brain, and Behavior*, *11*(6), 712–719. http://dx.doi.org/10.1111/j.1601-183X.2012.00803.x.

Karschin, C., Dissmann, E., Stuhmer, W., & Karschin, A. (1996). IRK(1–3) and GIRK(1–4) inwardly rectifying K+ channel mRNAs are differentially expressed in the adult rat brain. *The Journal of Neuroscience: The Official Journal of the Society for Neuroscience*, *16*(11), 3559–3570.

Kaufmann, K., Romaine, I., Days, E., Pascual, C., Malik, A., Yang, L., et al. (2013). ML297 (VU0456810), the first potent and selective activator of the GIRK potassium channel, displays antiepileptic properties in mice. *ACS Chemical Neuroscience*, *4*(9), 1278–1286. http://dx.doi.org/10.1021/cn400062a.

Kawaura, K., Miki, R., Urashima, Y., Kawahara, R., Soeda, F., Shirasaki, T., et al. (2012). Pharmacological mechanisms of antidepressant-like effect of tipepidine in the forced swimming test. *Behavioural Brain Research*, *226*(2), 381–385. http://dx.doi.org/10.1016/j.bbr.2011.09.031.

Kessler, R. C., Ruscio, A. M., Shear, K., & Wittchen, H. U. (2010). Epidemiology of anxiety disorders. *Current Topics in Behavioral Neuroscience*, *2*, 21–35.

Kobayashi, T., Ikeda, K., Kojima, H., Niki, H., Yano, R., Yoshioka, T., et al. (1999). Ethanol opens G-protein-activated inwardly rectifying K+ channels. *Nature Neuroscience*, *2*(12), 1091–1097. http://dx.doi.org/10.1038/16019.

Kobayashi, T., Washiyama, K., & Ikeda, K. (2004). Modulators of G protein-activated inwardly rectifying K+ channels: Potentially therapeutic agents for addictive drug users. *Annals of the New York Academy of Sciences*, *1025*, 590–594. http://dx.doi.org/10.1196/annals.1316.073.

Kofuji, P., Hofer, M., Millen, K. J., Millonig, J. H., Davidson, N., Lester, H. A., et al. (1996). Functional analysis of the weaver mutant GIRK2 K+ channel and rescue of weaver granule cells. *Neuron*, *16*(5), 941–952.

Kotecki, L., Hearing, M., McCall, N. M., Marron Fernandez de Velasco, E., Pravetoni, M., et al. (2015). GIRK channels modulate opioid-induced motor activity in a cell type- and subunit-dependent manner. *The Journal of Neuroscience: The Official Journal of the Society for Neuroscience*, *35*(18), 7131–7142. http://dx.doi.org/10.1523/JNEUROSCI.5051-14.2015.

Koyrakh, L., Lujan, R., Colon, J., Karschin, C., Kurachi, Y., Karschin, A., et al. (2005). Molecular and cellular diversity of neuronal G-protein-gated potassium channels. *The Journal of Neuroscience: The Official Journal of the Society for Neuroscience*, *25*(49), 11468–11478. http://dx.doi.org/10.1523/JNEUROSCI.3484-05.2005.

Kozell, L. B., Walter, N. A., Milner, L. C., Wickman, K., & Buck, K. J. (2009). Mapping a barbiturate withdrawal locus to a 0.44 Mb interval and analysis of a novel null mutant identify a role for Kcnj9 (GIRK3) in withdrawal from pentobarbital, zolpidem, and ethanol. *The Journal of Neuroscience: The Official Journal of the Society for Neuroscience*, *29*(37), 11662–11673. http://dx.doi.org/10.1523/JNEUROSCI.1413-09.2009.

Krapivinsky, G., Gordon, E. A., Wickman, K., Velimirovic, B., Krapivinsky, L., & Clapham, D. E. (1995). The G-protein-gated atrial K+ channel IKACh is a heteromultimer of two inwardly rectifying K(+)-channel proteins. *Nature*, *374*(6518), 135–141. http://dx.doi.org/10.1038/374135a0.

Labouebe, G., Lomazzi, M., Cruz, H. G., Creton, C., Lujan, R., Li, M., et al. (2007). RGS2 modulates coupling between GABAB receptors and GIRK channels in dopamine neurons of the ventral tegmental area. *Nature Neuroscience*, *10*(12), 1559–1568. http://dx.doi.org/10.1038/nn2006.

Larsson, H., Ryden, E., Boman, M., Langstrom, N., Lichtenstein, P., & Landen, M. (2013). Risk of bipolar disorder and schizophrenia in relatives of people with attention-deficit hyperactivity disorder. *The British Journal of Psychiatry: The Journal of Mental Science*, *203*(2), 103–106. http://dx.doi.org/10.1192/bjp.bp.112.120808.

Lazary, J., Juhasz, G., Anderson, I. M., Jacob, C. P., Nguyen, T. T., Lesch, K. P., et al. (2011). Epistatic interaction of CREB1 and KCNJ6 on rumination and negative emotionality. *European Neuropsychopharmacology*, *21*(1), 63–70. http://dx.doi.org/10.1016/j.euroneuro.2010.09.009.

LeTourneau, J. L., Hagg, D. S., & Smith, S. M. (2008). Baclofen and gamma-hydroxybutyrate withdrawal. *Neurocritical Care*, *8*(3), 430–433. http://dx.doi.org/10.1007/s12028-008-9062-2.

Lewohl, J. M., Wilson, W. R., Mayfield, R. D., Brozowski, S. J., Morrisett, R. A., & Harris, R. A. (1999). G-protein-coupled inwardly rectifying potassium channels are targets of alcohol action. *Nature Neuroscience*, *2*(12), 1084–1090. http://dx.doi.org/10.1038/16012.

Liu, C., Belichenko, P. V., Zhang, L., Fu, D., Kleschevnikov, A. M., Baldini, A., et al. (2011). Mouse models for Down syndrome-associated developmental cognitive disabilities. *Developmental Neuroscience*, *33*(5), 404–413. http://dx.doi.org/10.1159/000329422.

Llamosas, N., Bruzos-Cidón, C., Rodríguez, J. J., Ugedo, L., & Torrecilla, M. (2015). Deletion of GIRK2 subunit of GIRK channels alters the 5-HT1A receptor-mediated signaling and results in a depression-resistant behavior. *The International Journal of Neuropsychopharmacology*. http://dx.doi.org/10.1093/ijnp/pyv051. [Epub ahead of print].

Lotsch, J., Pruss, H., Veh, R. W., & Doehring, A. (2010). A KCNJ6 (Kir3.2, GIRK2) gene polymorphism modulates opioid effects on analgesia and addiction but not on pupil size. *Pharmacogenetics and Genomics*, *20*(5), 291–297. http://dx.doi.org/10.1097/FPC.0b013e3283386bda.

Lujan, R., Marron Fernandez de Velasco, E., Aguado, C., & Wickman, K. (2014). New insights into the therapeutic potential of Girk channels. *Trends in Neurosciences*, *37*(1), 20–29. http://dx.doi.org/10.1016/j.tins.2013.10.006.

Luscher, C., Jan, L. Y., Stoffel, M., Malenka, R. C., & Nicoll, R. A. (1997). G protein-coupled inwardly rectifying K+ channels (GIRKs) mediate postsynaptic but not presynaptic transmitter actions in hippocampal neurons. *Neuron*, *19*(3), 687–695.

Luscher, C., & Slesinger, P. A. (2010). Emerging roles for G protein-gated inwardly rectifying potassium (GIRK) channels in health and disease. *Nature Reviews. Neuroscience*, *11*(5), 301–315. http://dx.doi.org/10.1038/nrn2834.

Marker, C. L., Cintora, S. C., Roman, M. I., Stoffel, M., & Wickman, K. (2002). Hyperalgesia and blunted morphine analgesia in G protein-gated potassium channel subunit knockout mice. *Neuroreport*, *13*(18), 2509–2513. http://dx.doi.org/10.1097/01.wnr.0000048541.12213.bb.

Marker, C. L., Lujan, R., Loh, H. H., & Wickman, K. (2005). Spinal G-protein-gated potassium channels contribute in a dose-dependent manner to the analgesic effect of mu- and delta- but not kappa-opioids. *The Journal of Neuroscience: The Official Journal of the Society for Neuroscience*, *25*(14), 3551–3559. http://dx.doi.org/10.1523/JNEUROSCI.4899-04.2005.

Marker, C. L., Stoffel, M., & Wickman, K. (2004). Spinal G-protein-gated K+ channels formed by GIRK1 and GIRK2 subunits modulate thermal nociception and contribute to morphine analgesia. *The Journal of Neuroscience: The Official Journal of the Society for Neuroscience*, *24*(11), 2806–2812. http://dx.doi.org/10.1523/JNEUROSCI.5251-03.2004.

Masotti, A., Uva, P., Davis-Keppen, L., Basel-Vanagaite, L., Cohen, L., Pisaneschi, E., et al. (2015). Keppen-Lubinsky syndrome is caused by mutations in the inwardly rectifying K+ channel encoded by KCNJ6. *American Journal of Human Genetics*, *96*(2), 295–300. http://dx.doi.org/10.1016/j.ajhg.2014.12.011.

Mazarati, A., Lundstrom, L., Sollenberg, U., Shin, D., Langel, U., & Sankar, R. (2006). Regulation of kindling epileptogenesis by hippocampal galanin type 1 and type 2 receptors: The effects of subtype-selective agonists and the role of G-protein-mediated signaling. *The Journal of Pharmacology and Experimental Therapeutics*, *318*(2), 700–708. http://dx.doi.org/10.1124/jpet.106.104703.

Metten, P., Phillips, T. J., Crabbe, J. C., Tarantino, L. M., McClearn, G. E., Plomin, R., et al. (1998). High genetic susceptibility to ethanol withdrawal predicts low ethanol consumption. *Mammalian Genome: Official Journal of the International Mammalian Genome Society*, *9*(12), 983–990.

Mitrovic, I., Margeta-Mitrovic, M., Bader, S., Stoffel, M., Jan, L. Y., & Basbaum, A. I. (2003). Contribution of GIRK2-mediated postsynaptic signaling to opiate and alpha 2-adrenergic analgesia and analgesic sex differences. *Proceedings of the National Academy of Sciences of the United States of America*, *100*(1), 271–276. http://dx.doi.org/10.1073/pnas.0136822100.

Morgan, A. D., Carroll, M. E., Loth, A. K., Stoffel, M., & Wickman, K. (2003). Decreased cocaine self-administration in Kir3 potassium channel subunit knockout mice.

Neuropsychopharmacology: Official Publication of the American College of Neuropsychopharmacology, 28(5), 932–938. http://dx.doi.org/10.1038/sj.npp.1300100.

Munoz, M. B., & Slesinger, P. A. (2014). Sorting nexin 27 regulation of G protein-gated inwardly rectifying K(+) channels attenuates in vivo cocaine response. *Neuron, 82*(3), 659–669. http://dx.doi.org/10.1016/j.neuron.2014.03.011.

Nagi, K., & Pineyro, G. (2014). Kir3 channel signaling complexes: Focus on opioid receptor signaling. *Frontiers in Cellular Neuroscience, 8,* 186. http://dx.doi.org/10.3389/fncel.2014.00186.

Navarro, B., Kennedy, M. E., Velimirovic, B., Bhat, D., Peterson, A. S., & Clapham, D. E. (1996). Nonselective and G betagamma-insensitive weaver K+ channels. *Science, 272*(5270), 1950–1953.

Nishizawa, D., Fukuda, K., Kasai, S., Ogai, Y., Hasegawa, J., Sato, N., et al. (2014). Association between KCNJ6 (GIRK2) gene polymorphism rs2835859 and post-operative analgesia, pain sensitivity, and nicotine dependence. *Journal of Pharmacological Sciences, 126*(3), 253–263.

Nishizawa, D., Nagashima, M., Katoh, R., Satoh, Y., Tagami, M., Kasai, S., et al. (2009). Association between KCNJ6 (GIRK2) gene polymorphisms and postoperative analgesic requirements after major abdominal surgery. *PLoS One, 4*(9), e7060. http://dx.doi.org/10.1371/journal.pone.0007060.

Ostrovskaya, O., Xie, K., Masuho, I., Fajardo-Serrano, A., Lujan, R., Wickman, K., et al. (2014). RGS7/Gbeta5/R7BP complex regulates synaptic plasticity and memory by modulating hippocampal GABABR-GIRK signaling. *eLife, 3,* e02053. http://dx.doi.org/10.7554/eLife.02053.

Pan, H. L., Wu, Z. Z., Zhou, H. Y., Chen, S. R., Zhang, H. M., & Li, D. P. (2008). Modulation of pain transmission by G-protein-coupled receptors. *Pharmacology and Therapeutics, 117*(1), 141–161. http://dx.doi.org/10.1016/j.pharmthera.2007.09.003.

Patil, N., Cox, D. R., Bhat, D., Faham, M., Myers, R. M., & Peterson, A. S. (1995). A potassium channel mutation in weaver mice implicates membrane excitability in granule cell differentiation. *Nature Genetics, 11*(2), 126–129. http://dx.doi.org/10.1038/ng1095-126.

Perry, C. A., Pravetoni, M., Teske, J. A., Aguado, C., Erickson, D. J., Medrano, J. F., et al. (2008). Predisposition to late-onset obesity in GIRK4 knockout mice. *Proceedings of the National Academy of Sciences of the United States of America, 105*(23), 8148–8153. http://dx.doi.org/10.1073/pnas.0803261105.

Phillips, T. J., & Reed, C. (2014). Targeting GABAB receptors for anti-abuse drug discovery. *Expert Opinion on Drug Discovery, 9*(11), 1307–1317. http://dx.doi.org/10.1517/17460441.2014.956076.

Porjesz, B., Rangaswamy, M., Kamarajan, C., Jones, K. A., Padmanabhapillai, A., & Begleiter, H. (2005). The utility of neurophysiological markers in the study of alcoholism. *Clinical Neurophysiology: Official Journal of the International Federation of Clinical Neurophysiology, 116*(5), 993–1018. http://dx.doi.org/10.1016/j.clinph. 2004.12.016.

Pravetoni, M., & Wickman, K. (2008). Behavioral characterization of mice lacking GIRK/Kir3 channel subunits. *Genes, Brain, and Behavior,* 7(5), 523–531. http://dx.doi.org/10.1111/j.1601-183X.2008.00388.x.

Rahman, Z., Schwarz, J., Gold, S. J., Zachariou, V., Wein, M. N., Choi, K. H., et al. (2003). RGS9 modulates dopamine signaling in the basal ganglia. *Neuron, 38*(6), 941–952.

Ramos-Hunter, S. J., Engers, D. W., Kaufmann, K., Du, Y., Lindsley, C. W., Weaver, C. D., et al. (2013). Discovery and SAR of a novel series of GIRK1/2 and GIRK1/4 activators. *Bioorganic and Medicinal Chemistry Letters, 23*(18), 5195–5198. http://dx.doi.org/10.1016/j.bmcl.2013.07.002.

Reeves, R. H., Irving, N. G., Moran, T. H., Wohn, A., Kitt, C., Sisodia, S. S., et al. (1995). A mouse model for Down syndrome exhibits learning and behaviour deficits. *Nature Genetics, 11*(2), 177–184. http://dx.doi.org/10.1038/ng1095-177.

Renthal, W., & Nestler, E. J. (2008). Epigenetic mechanisms in drug addiction. *Trends in Molecular Medicine, 14*(8), 341–350. http://dx.doi.org/10.1016/j.molmed.2008.06.004.

Reyes, S., Fu, Y., Double, K., Thompson, L., Kirik, D., Paxinos, G., et al. (2012). GIRK2 expression in dopamine neurons of the substantia nigra and ventral tegmental area. *The Journal of Comparative Neurology, 520*(12), 2591–2607. http://dx.doi.org/10.1002/cne.23051.

Risinger, F. O., & Cunningham, C. L. (1998). Ethanol-induced conditioned taste aversion in BXD recombinant inbred mice. *Alcoholism, Clinical and Experimental Research, 22*(6), 1234–1244.

Rodriguez-Oroz, M. C., Jahanshahi, M., Krack, P., Litvan, I., Macias, R., Bezard, E., et al. (2009). Initial clinical manifestations of Parkinson's disease: Features and pathophysiological mechanisms. *The Lancet. Neurology, 8*(12), 1128–1139. http://dx.doi.org/10.1016/S1474-4422(09)70293-5.

Roubertoux, P. L., & Carlier, M. (2010). Mouse models of cognitive disabilities in trisomy 21 (Down syndrome). *American Journal of Medical Genetics. Part C, Seminars in Medical Genetics, 154C*(4), 400–416. http://dx.doi.org/10.1002/ajmg.c.30280.

Saccone, S. F., Hinrichs, A. L., Saccone, N. L., Chase, G. A., Konvicka, K., Madden, P. A., et al. (2007). Cholinergic nicotinic receptor genes implicated in a nicotine dependence association study targeting 348 candidate genes with 3713 SNPs. *Human Molecular Genetics, 16*(1), 36–49. http://dx.doi.org/10.1093/hmg/ddl438.

Sago, H., Carlson, E. J., Smith, D. J., Kilbridge, J., Rubin, E. M., Mobley, W. C., et al. (1998). Ts1Cje, a partial trisomy 16 mouse model for Down syndrome, exhibits learning and behavioral abnormalities. *Proceedings of the National Academy of Sciences of the United States of America, 95*(11), 6256–6261.

Saltiel, P. F., & Silvershein, D. I. (2015). Major depressive disorder: Mechanism-based prescribing for personalized medicine. *Neuropsychiatric Disease and Treatment, 11*, 875–888. http://dx.doi.org/10.2147/NDT.S73261.

Sasaki, T., Hashimoto, K., Tachibana, M., Kurata, T., Kimura, H., Komatsu, H., et al. (2014). Tipepidine in adolescent patients with depression: A 4-week, open-label, preliminary study. *Neuropsychiatric Disease and Treatment, 10*, 719–722. http://dx.doi.org/10.2147/NDT.S63075.

Sasaki, T., Hashimoto, K., Tachibana, M., Kurata, T., Okawada, K., Ishikawa, M., et al. (2014). Tipepidine in children with attention deficit/hyperactivity disorder: A 4-week, open-label, preliminary study. *Neuropsychiatric Disease and Treatment, 10*, 147–151. http://dx.doi.org/10.2147/NDT.S58480.

Schmidt, M. J., Sawyer, B. D., Perry, K. W., Fuller, R. W., Foreman, M. M., & Ghetti, B. (1982). Dopamine deficiency in the weaver mutant mouse. *The Journal of Neuroscience: The Official Journal of the Society for Neuroscience, 2*(3), 376–380.

Schoots, O., Wilson, J. M., Ethier, N., Bigras, E., Hebert, T. E., & Van Tol, H. H. (1999). Co-expression of human Kir3 subunits can yield channels with different functional properties. *Cellular Signalling, 11*(12), 871–883.

Signorini, S., Liao, Y. J., Duncan, S. A., Jan, L. Y., & Stoffel, M. (1997). Normal cerebellar development but susceptibility to seizures in mice lacking G protein-coupled, inwardly rectifying K+ channel GIRK2. *Proceedings of the National Academy of Sciences of the United States of America, 94*(3), 923–927.

Slesinger, P. A., Patil, N., Liao, Y. J., Jan, Y. N., Jan, L. Y., & Cox, D. R. (1996). Functional effects of the mouse weaver mutation on G protein-gated inwardly rectifiying K+ channels. *Neuron, 16*(2), 321–331.

Slesinger, P. A., Stoffel, M., Jan, Y. N., & Jan, L. Y. (1997). Defective gamma-aminobutyric acid type B receptor-activated inwardly rectifying K+ currents in cerebellar granule cells isolated from weaver and Girk2 null mutant mice. *Proceedings of the National Academy of Sciences of the United States of America, 94*(22), 12210–12217.

Smeyne, R. J., & Goldowitz, D. (1989). Development and death of external granular layer cells in the weaver mouse cerebellum: A quantitative study. *The Journal of Neuroscience: The Official Journal of the Society for Neuroscience*, *9*(5), 1608–1620.

Smith, S. B., Marker, C. L., Perry, C., Liao, G., Sotocinal, S. G., Austin, J. S., et al. (2008). Quantitative trait locus and computational mapping identifies Kcnj9 (GIRK3) as a candidate gene affecting analgesia from multiple drug classes. *Pharmacogenetics and Genomics*, *18*(3), 231–241. http://dx.doi.org/10.1097/FPC.0b013e3282f55ab2.

Stewart, A., Maity, B., Anderegg, S. P., Allamargot, C., Yang, J., & Fisher, R. A. (2015). Regulator of G protein signaling 6 is a critical mediator of both reward-related behavioral and pathological responses to alcohol. *Proceedings of the National Academy of Sciences of the United States of America*, *112*(7), E786–E795. http://dx.doi.org/10.1073/pnas.1418795112.

Tarantino, L. M., McClearn, G. E., Rodriguez, L. A., & Plomin, R. (1998). Confirmation of quantitative trait loci for alcohol preference in mice. *Alcoholism, Clinical and Experimental Research*, *22*(5), 1099–1105.

Torrecilla, M., Marker, C. L., Cintora, S. C., Stoffel, M., Williams, J. T., & Wickman, K. (2002). G-protein-gated potassium channels containing Kir3.2 and Kir3.3 subunits mediate the acute inhibitory effects of opioids on locus ceruleus neurons. *The Journal of Neuroscience: The Official Journal of the Society for Neuroscience*, *22*(11), 4328–4334.

Toyoda, A., Noguchi, H., Taylor, T. D., Ito, T., Pletcher, M. T., Sakaki, Y., et al. (2002). Comparative genomic sequence analysis of the human chromosome 21 Down syndrome critical region. *Genome Research*, *12*(9), 1323–1332. http://dx.doi.org/10.1101/gr.153702.

Trudell, J. R., Messing, R. O., Mayfield, J., & Harris, R. A. (2014). Alcohol dependence: Molecular and behavioral evidence. *Trends in Pharmacological Sciences*, *35*(7), 317–323. http://dx.doi.org/10.1016/j.tips.2014.04.009.

Urban, D. J., & Roth, B. L. (2015). DREADDs (designer receptors exclusively activated by designer drugs): Chemogenetic tools with therapeutic utility. *Annual Review of Pharmacology and Toxicology*, *55*, 399–417. http://dx.doi.org/10.1146/annurev-pharmtox-010814-124803.

van Beijsterveldt, C. E., Molenaar, P. C., de Geus, E. J., & Boomsma, D. I. (1996). Heritability of human brain functioning as assessed by electroencephalography. *American Journal of Human Genetics*, *58*(3), 562–573.

van Huijstee, A. N., & Mansvelder, H. D. (2014). Glutamatergic synaptic plasticity in the mesocorticolimbic system in addiction. *Frontiers in Cellular Neuroscience*, *8*, 466. http://dx.doi.org/10.3389/fncel.2014.00466.

Wen, W., Wu, W., Romaine, I. M., Kaufmann, K., Du, Y., Sulikowski, G. A., et al. (2013). Discovery of 'molecular switches' within a GIRK activator scaffold that afford selective GIRK inhibitors. *Bioorganic & Medicinal Chemistry Letters*, *23*(16), 4562–4566. http://dx.doi.org/10.1016/j.bmcl.2013.06.023.

Wickman, K., Karschin, C., Karschin, A., Picciotto, M. R., & Clapham, D. E. (2000). Brain localization and behavioral impact of the G-protein-gated K+ channel subunit GIRK4. *The Journal of Neuroscience: The Official Journal of the Society for Neuroscience*, *20*(15), 5608–5615.

Wilens, T. E., & Spencer, T. J. (2010). Understanding attention-deficit/hyperactivity disorder from childhood to adulthood. *Postgraduate Medicine*, *122*(5), 97–109. http://dx.doi.org/10.3810/pgm.2010.09.2206.

Wiseman, F. K., Alford, K. A., Tybulewicz, V. L., & Fisher, E. M. (2009). Down syndrome–recent progress and future prospects. *Human Molecular Genetics*, *18*(R1), R75–R83. http://dx.doi.org/10.1093/hmg/ddp010.

Wood, S., Sage, J. R., Shuman, T., & Anagnostaras, S. G. (2014). Psychostimulants and cognition: A continuum of behavioral and cognitive activation. *Pharmacological Reviews*, *66*(1), 193–221. http://dx.doi.org/10.1124/pr.112.007054.

Workman, E. R., Haddick, P. C., Bush, K., Dilly, G. A., Niere, F., Zemelman, B. V., et al. (2015). Rapid antidepressants stimulate the decoupling of GABAB receptors from GIRK/Kir3 channels through increased protein stability of 14-3-3eta. *Molecular Psychiatry*, *20*(3), 298–310. http://dx.doi.org/10.1038/mp.2014.165.

Wydeven, N., Marron Fernandez de Velasco, E., Du, Y., Benneyworth, M. A., Hearing, M. C., et al. (2014). Mechanisms underlying the activation of G-protein-gated inwardly rectifying K+ (GIRK) channels by the novel anxiolytic drug, ML297. *Proceedings of the National Academy of Sciences of the United States of America*, *111*(29), 10755–10760. http://dx.doi.org/10.1073/pnas.1405190111.

Yamada, K., Iwayama, Y., Hattori, E., Iwamoto, K., Toyota, T., Ohnishi, T., et al. (2011). Genome-wide association study of schizophrenia in Japanese population. *PLoS One*, *6*(6), e20468. http://dx.doi.org/10.1371/journal.pone.0020468.

Yamada, K., Iwayama, Y., Toyota, T., Ohnishi, T., Ohba, H., Maekawa, M., et al. (2012). Association study of the KCNJ3 gene as a susceptibility candidate for schizophrenia in the Chinese population. *Human Genetics*, *131*(3), 443–451. http://dx.doi.org/10.1007/s00439-011-1089-3.

Young, K. A., Franklin, T. R., Roberts, D. C., Jagannathan, K., Suh, J. J., Wetherill, R. R., et al. (2014). Nipping cue reactivity in the bud: Baclofen prevents limbic activation elicited by subliminal drug cues. *The Journal of Neuroscience: The Official Journal of the Society for Neuroscience*, *34*(14), 5038–5043. http://dx.doi.org/10.1523/JNEUROSCI.4977-13.2014.

Zachariou, V., Georgescu, D., Sanchez, N., Rahman, Z., DiLeone, R., Berton, O., et al. (2003). Essential role for RGS9 in opiate action. *Proceedings of the National Academy of Sciences of the United States of America*, *100*(23), 13656–13661. http://dx.doi.org/10.1073/pnas.2232594100.

Zindel, L. R., & Kranzler, H. R. (2014). Pharmacotherapy of alcohol use disorders: Seventy-five years of progress. *Journal of Studies on Alcohol and Drugs. Supplement*, *75*(Suppl. 17), 79–88.

INDEX

Note: Page numbers followed by "*f*" indicate figures and "*t*" indicate tables.

A

Activation kinetics, 43–45
Addiction
  drugs of abuse, 299–300
  ethanol, 292–296
  GIRK modulators, 299–300
  nicotine, 299
  opioids, 298–299
  psychostimulants, 297–298
  sedative/hypnotics, 297
Adrenal gland, 54–55
Adrenocortical NCI-H295R cells, 54–55
Agonist-evoked current (*I*evoked)
  activation, 42
  activation kinetics, 43–45
  deactivation, 42
  desensitization, 42–43
  $G\alpha_{i/o}$GTP, 47–50
  magnitude, 45–47
  PTX-insensitive Gα, 50–51
A-kinase anchoring proteins (AKAPs), 5–7
Alanine (R201A), 137–138
Allosteric Monod–Wyman–Changeux-type model, 46–47
Alzheimer's disease, 258
Andersen–Tawil syndrome, 131–132
Anxiety, 289–290
Arginine (R201), 137–138
Atrial myocytes, 43–44
Attention deficit hyperactivity disorder (ADHD), 290–291

B

β-adrenergic-PKA cascade, 50
Bradyopsia, 105

C

Ca^{2+}-dependent protein kinases C (cPKC), 50–51
Calcium/calmodulin-dependent kinase II (CaMKII), 253–254
Cardiac GPCR-GIRK channel signaling, 100–101
Cardiac potassium (Kir3) channel activity
  Gβγ action, 18–21
  G proteins, 3–4
  membrane-delimited activation, 2, 3*f*
  membrane-delimited signaling mechanisms, 9–12
  molecular constituents, 4–5
  molecular mechanism of PIP_2, 12–17
  PIPs, 5–9
  posttranslational modification, 17–18
Central nervous system (CNS), 51–52
Cerebellar granule neurons (CGNs), 97
ChAT, 177
Chimeric GIRK1-prokaryotic Kirbac channel, 132
Chimeric KirBac/GIRK1 chimera, 49
CNS disorders
  ADHD, 290–291
  anxiety, 289–290
  cognitive deficits, 283–285
  depression, 288–289
  epilepsy, 291
  motor control, 286–287
  mouse Girk genes, 281, 282*t*
  pain, 285–286
  schizophrenia, 290
Cocaine, 141, 184–185
Cocaine-induced suppression, 215
Cognitive deficits, 283–285
Compartment-dependent localization, GIRK channels, 180–183
Cytoplasmic domains (CTDs), 125–126

D

D2 dopamine receptors (D2Rs), 209–210
Depotentiation, 255
Depression, 288–289
Diacylglycerol (DAG), 5–7
Direct "membrane-delimited" GPCR-to-ion channel signaling, 89–90

Double-labeling immunofluorescence techniques, 192
Down syndrome, 256–258
Down syndrome critical region (DSCR), 283–284
Drug-dependent plasticity, GIRK channels
  addiction, 211–212
  drugs of abuse, 212, 213*t*
  $GABA_AR$-mediated fast inhibitory neurotransmission, 212
  mPFC pyramidal neurons, 215
  short- and long-term potentiation and depression, 211–212
  VTA DA neurons, 214
  VTA GABA neurons, 214–215
Drugs of abuse, 299–300
  cellular and behavioral effects, 245–247
  ethanol, 249–250
  $G_{i/o}$-coupled GPCRs, 247–248
  GIRK modulation, 245–247, 246*t*
  intracellular modulators, 248–249

E

Electroretinography (ERG), 95
Epilepsy, 105, 291
Ethanol, 249–250, 292–296
Exponential activation time constant, 43–44

F

F3 (1-chloro-1,2,2-trifluorocyclobutane), 141–142
Full-length eukaryotic GIRK2 homomeric channel, 128

G

$GABA_B$ receptor, 184–185, 187–188
GABAergic neurons, 177–178
Gα interaction sites
  mapping, 35–37
  NMR spectroscopy, 37–39
  NT and distal CT, 39–40
Gating
  activation by alcohol, 138–140
  bundle crossing, 130–131
  G loop, 131–132
  G protein activation, 132–138
  structural determinants, 140–143
  trafficking motifs, 143–144
Gβγ action, 18–21
$G_{i/o}$-coupled GPCRs, 247–248
GIRK channels. *See* G protein-gated inwardly rectifying K^+ (GIRK) channels
$GIRK2_{C321V}$ mutant, 37
GIRK1/GIRK4 heterotetramer, 133–134
GIRK modulators, 299–300
GIRK2-R201A, 130–131
G-loop gate, 13, 14*f*
Glutamate E242, 37–38
Glycine, 130
GPCR–GIRK channel gating kinetics, 90–94
GPCRs. *See* G protein-coupled receptors (GPCRs)
G protein activation, 132–138
G protein-coupled receptors (GPCRs), 162, 202
G protein-gated inwardly rectifying K^+ (GIRK) channels
  acetylcholine, 118–119
  addiction, 240, 291–300
  agonist-evoked current, 42–51
  atrial myocytes, 120
  binding/mutagenesis *vs.* NMR data, 40–41
  cDNA, 119–120
  cell type-dependent localization, 177–179
  CNS disorders, 281–291
  compartment-dependent localization, 180–183
  distal C-terminus, 34
  drug addiction, 202, 203*t*
  drug-dependent plasticity, 211–215
  drugs of abuse, 245–250
  $G\alpha^{GDP}$ and Gβγ, 56–60
  Gα interaction sites, 35–40
  gating, 130–144
  Gβγ, 32–33
  GIRK4 distribution, 173–175
  GIRK1, GIRK2, and GIRK3 distribution, 170–173
  GIRK2 neurochemical phenotype acquisition, 192–193
  glutamate release sites, 183
  GPCRs, 202
  G protein subunits, 63–65

heterologous expression studies, 55–56
heterologous expression system, 119–120
I_{basal} physiology, 51–55
initial segment of NT, 35
input-dependent localization, 184
inward rectification, 118
Kir structures, 120–121, 122*t*
ligand stimulation, 280
mammalian subunits, 280
membrane-attached proteins, 28–30
membrane trafficking, 216–220
mesocorticolimbic system, 207–211
modulators, 242–245
molecular and functional diversity, 204–206
molecular cloning, 162–163
molecular organization and heterogeneity, 163–164
multiprotein signaling complexes, 30–31
muscarinic-type cholinergic receptors, 118–119
neuroanatomical approaches, 164–169
neurochemically and functionally different cell populations, 176–177
neuronal, 241–242
neuronal excitability, 206–207
neuronal plasticity, 251–261
neurotransmitter-evoked GIRK response, 31–32
plasticity- and drug-dependent changes, 184–185
pre- and postnatal development, 190–192
preformed signaling complex, 60–63
presynaptic localization, 185–188
protein kinase A, 30–31
proteins localization, 188–190
rectification, 128–130
reward-relevant neuron populations, 223
RGS proteins, 220–223
selectivity and permeation, 121–128
slow inhibitory neurotransmission, 202
specificity, 65–67
structural and phylogenetic relationship, 162
structural homology, 120
structure–function experiments, 120–121
structure of, 28, 29*f*
therapeutic potential, 300–302
G protein subunits, 63–65
GTPase-activating proteins, 35, 88

H

Halothane-insensitive Kir2.1, 141–142
Heart, 52–54
HEK293 cells, 57
High-resolution immunohistochemical techniques, 166
Hill coefficient, 46–47
Homeostatic plasticity, 251–253
Human genetic association studies, 302
Hypertension, 105

I

I_{basal} physiology
adrenal gland, 54–55
central nervous system, 51–52
heart, 52–54
Ifenprodil, 142
Immunoelectron microscopy (iEM), 165
Immunohistochemistry, 168
Inositol trisphosphate (IP_3), 5–7
Input-dependent localization, GIRK channels, 184
In situ blotting technique, 168–169
Isoflurane, 141–142
Isolated pleckstrin homology (PH), 9–10

K

Kcnj6 gene, 256–257

L

Layer 5/6 pyramidal neurons, 211
Local anesthetics, 140–141
Long-term potentiation (LTP)/long-term depression (LTD), 253–254

M

Membrane-delimited signaling mechanisms, 9–12, 90
Membrane trafficking
phosphorylation, 216–218
sorting Nexin 27, 219–220
Mesocorticolimbic system, GIRK channels
brain areas, 207–208, 208*f*
layer 5/6 pyramidal neurons, 211

Mesocorticolimbic system, GIRK channels (*Continued*)
  VTA DA neurons, 209–210
  VTA GABA neurons, 210–211
Metabotropic glutamate receptors (mGluRs), 255–256
Methane-thiol-sulfonate hydroxyethyl (MTS-HE) treatment, 140
2-methyl-2,4-pentanediol (MPD), 139–140
Midbrain dopamine neurons, 178–179
ML297, 143
Modulators of GIRK function
  modulation of GPCRs, 244–245
  phosphorylation, 243
  regulators of G protein signaling, 243–244
  trafficking, 244
Morphine, 185
Motor control, 286–287
Mouse Girk genes, 281, 282*t*
mPFC pyramidal neurons, 215
Mu opioid receptor (MOR) agonists, 210–211

N

Naringin, 143
Neuroanatomical approaches
  high-resolution immunohistochemical techniques, 166
  immunoelectron microscopy, 165
  postembedding immunogold method, 167*f*, 168
  preembedding immunogold method, 166, 167*f*
  preembedding immunoperoxidase method, 166, 167*f*
  SDS-FRL technique, 168
  in situ blotting technique, 168–169
  subunit-specific molecular tools, 164–165
Neuronal GIRK channels, 241–242
Neuronal GPCR–GIRK channel signaling, 96–99
Neuronal plasticity
  Alzheimer's disease, 258
  Down syndrome, 256–258
  drug-induced changes, 259–261
  homeostatic plasticity, 251–253
  Parkinson's disease, 258–259
  synaptic plasticity, 253–256
Nicotine, 299
NIP-142, 143
N-methyl-D-aspartate (NMDA) receptors, 181–183
N-terminal histidine, 132–133

O

On-bipolar neurons (OBNs), 90
Opioids, 298–299

P

Pain, 285–286
Parkinson's disease, 258–259
Pertussis toxin (PTX), 44–45
Phosphatidyl inositol (PI), 5–7
Phosphatidylinositol 4,5-bisphosphate (PIP_2), 101–102, 242–243
Phosphoinositides (PIPs), 5–9
Phosphorylation, 216–218, 243
Plasticity- and drug-dependent changes, GIRK channels, 184–185
Polyamines, 128
Postembedding immunogold method, 167*f*, 168
Potassium channel tetramerization domain-containing (KCTD) proteins, 244–245
Preembedding immunogold method, 166, 167*f*
Preembedding immunoperoxidase method, 166, 167*f*
Presynaptic localization , GIRK channels, 185–188
Protein kinase A (PKA), 30–31
Protein kinase C (PKC), 5–7, 30–31
Psychostimulant-induced suppression
  VTA DA neurons, 214
  VTA GABA neurons, 214–215
Psychostimulants, 297–298
PTX-insensitive Gα, 50–51
PTX-sensitive G proteins, 118–119

Q

Quantitative trait locus (QTL) mapping, 286

R

Rate-limiting factor, 44–45
R7-binding protein (R7BP), 97–98
Regulators of G protein signaling (RGS proteins), 243–244
  bradyopsia, 105

cardiac GPCR-GIRK channel signaling, 100–101
direct "membrane-delimited" GPCR-to-ion channel signaling, 89–90
encoded full-length proteins, 88–89
genetically engineered mice, 102–105, 103*t*
GPCR–GIRK channel gating kinetics, 90–94
G_q-mediated changes, 101–102
GTPase-activating proteins, 88
hypertension, 105
mammalian nervous system, 88
neuronal GPCR–GIRK channel signaling, 96–99
retinal neurons, 94–96
R7 RGS, 222–223
Regulators of G protein signaling (RGS proteins)RGS2, 221–222
Retinal neurons, 94–96
R7 family-binding protein (R7BP), 189–190
RGS6, 248–249
RGS proteins. *See* Regulators of G protein signaling (RGS proteins)

S

Schizophrenia, 290
SDS-FRL technique, 168
Sedative/hypnotics, 297
Single nucleotide polymorphisms (SNPs), 286
SNX27 regulation, 143–144
Sorting Nexin 27 (SNX27), 219–220
Subcellular localization, 166, 167*f*
Synapse-associated proteins (SAPs), 188–189
Synaptic plasticity, 253–256

T

Tertiapin, 142–143
Tipepidine, 142

V

Volatile anesthetics, 141–142
VTA DA neurons
drug-dependent plasticity, 214
mesocorticolimbic system, 209–210
VTA GABA neurons
drug-dependent plasticity, 214–215
mesocorticolimbic system, 210–211

X

Xenopus oocyte studies, 47–48

CONTENTS OF RECENT VOLUMES

Volume 37

Section I: Selectionist Ideas and Neurobiology

Selectionist and Instructionist Ideas in Neuroscience
Olaf Sporns

Population Thinking and Neuronal Selection: Metaphors or Concepts?
Ernst Mayr

Selection and the Origin of Information
Manfred Eigen

Section II: Development and Neuronal Populations

Morphoregulatory Molecules and Selectional Dynamics during Development
Kathryn L. Crossin

Exploration and Selection in the Early Acquisition of Skill
Esther Thelen and Daniela Corbetta

Population Activity in the Control of Movement
Apostolos P. Georgopoulos

Section III: Functional Segregation and Integration in the Brain

Reentry and the Problem of Cortical Integration
Giulio Tononi

Coherence as an Organizing Principle of Cortical Functions
Wolf Singerl

Temporal Mechanisms in Perception
Ernst Pöppel

Section IV: Memory and Models

Selection versus Instruction: Use of Computer Models to Compare Brain Theories
George N. Reeke, Jr.

Memory and Forgetting: Long-Term and Gradual Changes in Memory Storage
Larry R. Squire

Implicit Knowledge: New Perspectives on Unconscious Processes
Daniel L. Schacter

Section V: Psychophysics, Psychoanalysis, and Neuropsychology

Phantom Limbs, Neglect Syndromes, Repressed Memories, and Freudian Psychology
V. S. Ramachandran

Neural Darwinism and a Conceptual Crisis in Psychoanalysis
Arnold H. Modell

A New Vision of the Mind
Oliver Sacks

INDEX

Volume 38

Regulation of $GABA_A$ Receptor Function and Gene Expression in the Central Nervous System
A. Leslie Morrow

Genetics and the Organization of the Basal Ganglia
Robert Hitzemann, Yeang Olan, Stephen Kanes, Katherine Dains, and Barbara Hitzemann

Structure and Pharmacology of Vertebrate $GABA_A$ Receptor Subtypes
Paul J. Whiting, Ruth M. McKernan, and Keith A. Wafford

Neurotransmitter Transporters: Molecular Biology, Function, and Regulation
Beth Borowsky and Beth J. Hoffman

Presynaptic Excitability
Meyer B. Jackson

Monoamine Neurotransmitters in Invertebrates and Vertebrates: An Examination of the Diverse Enzymatic Pathways Utilized to Synthesize and Inactivate Biogenic Amines
B. D. Sloley and A. V. Juorio

Neurotransmitter Systems in Schizophrenia
Gavin P. Reynolds

Physiology of Bergmann Glial Cells
Thomas Müller and Helmut Kettenmann

INDEX

Volume 39

Modulation of Amino Acid-Gated Ion Channels by Protein Phosphorylation
Stephen J. Moss and Trevor G. Smart

Use-Dependent Regulation of $GABA_A$ Receptors
Eugene M. Barnes, Jr.

Synaptic Transmission and Modulation in the Neostriatum
David M. Lovinger and Elizabeth Tyler

The Cytoskeleton and Neurotransmitter Receptors
Valerie J. Whatley and R. Adron Harris

Endogenous Opioid Regulation of Hippocampal Function
Michele L. Simmons and Charles Chavkin

Molecular Neurobiology of the Cannabinoid Receptor
Mary E. Abood and Billy R. Martin

Genetic Models in the Study of Anesthetic Drug Action
Victoria J. Simpson and Thomas E. Johnson

Neurochemical Bases of Locomotion and Ethanol Stimulant Effects
Tamara J. Phillips and Elaine H. Shen

Effects of Ethanol on Ion Channels
Fulton T. Crews, A. Leslie Morrow, Hugh Criswell, and George Breese

INDEX

Volume 40

Mechanisms of Nerve Cell Death: Apoptosis or Necrosis after Cerebral Ischemia
R. M. E. Chalmers-Redman, A. D. Fraser, W. Y. H. Ju, J. Wadia, N. A. Tatton, and W. G. Tatton

Changes in Ionic Fluxes during Cerebral Ischemia
Tibor Kristian and Bo K. Siesjo

Techniques for Examining Neuroprotective Drugs *in Vitro*
A. Richard Green and Alan J. Cross

Techniques for Examining Neuroprotective Drugs *in Vivo*
Mark P. Goldberg, Uta Strasser, and Laura L. Dugan

Calcium Antagonists: Their Role in Neuroprotection
A. Jacqueline Hunter

Sodium and Potassium Channel Modulators: Their Role in Neuroprotection
Tihomir P. Obrenovich

NMDA Antagonists: Their Role in Neuroprotection
Danial L. Small

Development of the NMDA Ion-Channel Blocker, Aptiganel Hydrochloride, as a Neuroprotective Agent for Acute CNS Injury
Robert N. McBurney

The Pharmacology of AMPA Antagonists and Their Role in Neuroprotection
Rammy Gill and David Lodge

GABA and Neuroprotection
Patrick D. Lyden

Adenosine and Neuroprotection
Bertil B. Fredholm

Interleukins and Cerebral Ischemia
Nancy J. Rothwell, Sarah A. Loddick, and Paul Stroemer

Nitrone-Based Free Radical Traps as Neuroprotective Agents in Cerebral Ischemia and Other Pathologies
Kenneth Hensley, John M. Carney, Charles A. Stewart, Tahera Tabatabaie, Quentin Pye, and Robert A. Floyd

Neurotoxic and Neuroprotective Roles of Nitric Oxide in Cerebral Ischemia
Turgay Dalkara and Michael A. Moskowitz

A Review of Earlier Clinical Studies on Neuroprotective Agents and Current Approaches
Nils-Gunnar Wahlgren

INDEX

Volume 41

Section I: Historical Overview

Rediscovery of an Early Concept
Jeremy D. Schmahmann

Section II: Anatomic Substrates

The Cerebrocerebellar System
Jeremy D. Schmahmann and Deepak N. Pandya

Cerebellar Output Channels
Frank A. Middleton and Peter L. Strick

Cerebellar-Hypothalamic Axis: Basic Circuits and Clinical Observations
Duane E. Haines, Espen Dietrichs, Gregory A. Mihailoff, and E. Frank McDonald

Section III. Physiological Observations

Amelioration of Aggression: Response to Selective Cerebellar Lesions in the Rhesus Monkey
Aaron J. Berman

Autonomic and Vasomotor Regulation
Donald J. Reis and Eugene V. Golanov

Associative Learning
Richard F. Thompson, Shaowen Bao, Lu Chen, Benjamin D. Cipriano, Jeffrey S. Grethe, Jeansok J. Kim, Judith K. Thompson, Jo Anne Tracy, Martha S. Weninger, and David J. Krupa

Visuospatial Abilities
Robert Lalonde

Spatial Event Processing
Marco Molinari, Laura Petrosini, and Liliana G. Grammaldo

Section IV: Functional Neuroimaging Studies

Linguistic Processing
Julie A. Fiez and Marcus E. Raichle

Sensory and Cognitive Functions
Lawrence M. Parsons and Peter T. Fox

Skill Learning
Julien Doyon

Section V: Clinical and Neuropsychological Observations

Executive Function and Motor Skill Learning
Mark Hallett and Jordon Grafman

Verbal Fluency and Agrammatism
Marco Molinari, Maria G. Leggio, and Maria C. Silveri

Classical Conditioning
Diana S. Woodruff-Pak

Early Infantile Autism
Margaret L. Bauman, Pauline A. Filipek, and Thomas L. Kemper

Olivopontocerebellar Atrophy and Friedreich's Ataxia: Neuropsychological Consequences of Bilateral versus Unilateral Cerebellar Lesions
Thérèse Botez-Marquard and Mihai I. Botez

Posterior Fossa Syndrome
Ian F. Pollack

Cerebellar Cognitive Affective Syndrome
Jeremy D. Schmahmann and Janet C. Sherman

Inherited Cerebellar Diseases
Claus W. Wallesch and Claudius Bartels

Neuropsychological Abnormalities in Cerebellar Syndromes—Fact or Fiction?
Irene Daum and Hermann Ackermann

Section VI: Theoretical Considerations

Cerebellar Microcomplexes
Masao Ito

Control of Sensory Data Acquisition
James M. Bower

Neural Representations of Moving Systems
Michael Paulin

How Fibers Subserve Computing Capabilities: Similarities between Brains and Machines
Henrietta C. Leiner and Alan L. Leiner

Cerebellar Timing Systems
Richard Ivry

Attention Coordination and Anticipatory Control
Natacha A. Akshoomoff, Eric Courchesne, and Jeanne Townsend

Context-Response Linkage
W. Thomas Thach

Duality of Cerebellar Motor and Cognitive Functions
James R. Bloedel and Vlastislav Bracha

Section VII: Future Directions

Therapeutic and Research Implications
Jeremy D. Schmahmann

Volume 42

Alzheimer Disease
Mark A. Smith

Neurobiology of Stroke
W. Dalton Dietrich

Free Radicals, Calcium, and the Synaptic Plasticity-Cell Death Continuum: Emerging Roles of the Trascription Factor NFκB
Mark P. Mattson

AP-I Transcription Factors: Short- and Long-Term Modulators of Gene Expression in the Brain
Keith Pennypacker

Ion Channels in Epilepsy
Istvan Mody

Posttranslational Regulation of Ionotropic Glutamate Receptors and Synaptic Plasticity
Xiaoning Bi, Steve Standley, and Michel Baudry

Heritable Mutations in the Glycine, $GABA_A$, and Nicotinic Acetylcholine Receptors Provide New Insights into the Ligand-Gated Ion Channel Receptor Superfamily
Behnaz Vafa and Peter R. Schofield

INDEX

Volume 43

Early Development of the *Drosophila* Neuromuscular Junction: A Model for Studying Neuronal Networks in Development
Akira Chiba

Development of Larval Body Wall Muscles
Michael Bate, Matthias Landgraf, and Mar Ruiz Gómez Bate

Development of Electrical Properties and Synaptic Transmission at the Embryonic Neuromuscular Junction
Kendal S. Broadie

Ultrastructural Correlates of Neuromuscular Junction Development
Mary B. Rheuben, Motojiro Yoshihara, and Yoshiaki Kidokoro

Assembly and Maturation of the *Drosophila* Larval Neuromuscular Junction
L. Sian Gramates and Vivian Budnik

Second Messenger Systems Underlying Plasticity at the Neuromuscular Junction
Frances Hannan and Yi Zhong

Mechanisms of Neurotransmitter Release
J. Troy Littleton, Leo Pallanck, and Barry Ganetzky

Vesicle Recycling at the *Drosophila* Neuromuscular Junction
Daniel T. Stimson and Mani Ramaswami

Ionic Currents in Larval Muscles of *Drosophila*
Satpal Singh and Chun-Fang Wu

Development of the Adult Neuromuscular System
Joyce J. Fernandes and Haig Keshishian

Controlling the Motor Neuron
James R. Trimarchi, Ping Jin, and Rodney K. Murphey

Volume 44

Human Ego-Motion Perception
A. V. van den Berg

Optic Flow and Eye Movements
M. Lappe and K.-P. Hoffman

The Role of MST Neurons during Ocular Tracking in 3D Space
K. Kawano, U. Inoue, A. Takemura, Y. Kodaka, and F. A. Miles

Visual Navigation in Flying Insects
M. V. Srinivasan and S.-W. Zhang

Neuronal Matched Filters for Optic Flow Processing in Flying Insects
H. G. Krapp

A Common Frame of Reference for the Analysis of Optic Flow and Vestibular Information
B. J. Frost and D. R. W. Wylie

Optic Flow and the Visual Guidance of Locomotion in the Cat
H. Sherk and G. A. Fowler

Stages of Self-Motion Processing in Primate Posterior Parietal Cortex
F. Bremmer, J.-R. Duhamel, S. B. Hamed, and W. Graf

Optic Flow Analysis for Self-Movement Perception
C. J. Duffy

Neural Mechanisms for Self-Motion Perception in Area MST
R. A. Andersen, K. V. Shenoy, J. A. Crowell, and D. C. Bradley

Computational Mechanisms for Optic Flow Analysis in Primate Cortex
M. Lappe

Human Cortical Areas Underlying the Perception of Optic Flow: Brain Imaging Studies
M. W. Greenlee

What Neurological Patients Tell Us about the Use of Optic Flow
L. M. Vaina and S. K. Rushton

INDEX

Volume 45

Mechanisms of Brain Plasticity: From Normal Brain Function to Pathology
Philip. A. Schwartzkroin

Brain Development and Generation of Brain Pathologies
Gregory L. Holmes and Bridget McCabe

Maturation of Channels and Receptors: Consequences for Excitability
David F. Owens and Arnold R. Kriegstein

Neuronal Activity and the Establishment of Normal and Epileptic Circuits during Brain Development
John W. Swann, Karen L. Smith, and Chong L. Lee

The Effects of Seizures of the Hippocampus of the Immature Brain
Ellen F. Sperber and Solomon L. Moshe

Abnormal Development and Catastrophic Epilepsies: The Clinical Picture and Relation to Neuroimaging
Harry T. Chugani and Diane C. Chugani

Cortical Reorganization and Seizure Generation in Dysplastic Cortex
G. Avanzini, R. Preafico, S. Franceschetti, G. Sancini, G. Battaglia, and V. Scaioli

Rasmussen's Syndrome with Particular Reference to Cerebral Plasticity: A Tribute to Frank Morrell
Fredrick Andermann and Yuonne Hart

Structural Reorganization of Hippocampal Networks Caused by Seizure Activity
Daniel H. Lowenstein

Epilepsy-Associated Plasticity in gamma-Amniobutyric Acid Receptor Expression, Function and Inhibitory Synaptic Properties
Douglas A. Coulter

Synaptic Plasticity and Secondary Epileptogenesis
Timothy J. Teyler, Steven L. Morgan, Rebecca N. Russell, and Brian L. Woodside

Synaptic Plasticity in Epileptogenesis: Cellular Mechanisms Underlying Long-Lasting Synaptic Modifications that Require New Gene Expression
Oswald Steward, Christopher S. Wallace, and Paul F. Worley

Cellular Correlates of Behavior
Emma R. Wood, Paul A. Dudchenko, and Howard Eichenbaum

Mechanisms of Neuronal Conditioning
David A. T. King, David J. Krupa, Michael R. Foy, and Richard F. Thompson

Plasticity in the Aging Central Nervous System
C. A. Barnes

Secondary Epileptogenesis, Kindling, and Intractable Epilepsy: A Reappraisal from the Perspective of Neuronal Plasticity
Thomas P. Sutula

Kindling and the Mirror Focus
Dan C. McIntyre and Michael O. Poulter

Partial Kindling and Behavioral Pathologies
Robert E. Adamec

The Mirror Focus and Secondary Epileptogenesis
B. J. Wilder

Hippocampal Lesions in Epilepsy: A Historical Review
Robert Naquet

Clinical Evidence for Secondary Epileptogensis
Hans O. Luders

Epilepsy as a Progressive (or Nonprogressive "Benign") Disorder
John A. Wada

Pathophysiological Aspects of Landau-Kleffner Syndrome: From the Active Epileptic Phase to Recovery
Marie-Noelle Metz-Lutz, Pierre Maquet, Annd De Saint Martin, Gabrielle Rudolf, Norma Wioland, Edouard Hirsch, and Chriatian Marescaux

Local Pathways of Seizure Propagation in Neocortex
Barry W. Connors, David J. Pinto, and Albert E. Telefeian

Multiple Subpial Transection: A Clinical Assessment
C. E. Polkey

The Legacy of Frank Morrell
Jerome Engel, Jr.

Volume 46

Neurosteroids: Beginning of the Story
Etienne E. Baulieu, P. Robel, and M. Schumacher

Biosynthesis of Neurosteroids and Regulation of Their Synthesis
Synthia H. Mellon and Hubert Vaudry

Neurosteroid 7-Hydroxylation Products in the Brain
Robert Morfin and Luboslav Stárka

Neurosteroid Analysis
Ahmed A. Alomary, Robert L. Fitzgerald, and Robert H. Purdy

Role of the Peripheral-Type Benzodiazepine Receptor in Adrenal and Brain Steroidogenesis
Rachel C. Brown and Vassilios Papadopoulos

Formation and Effects of Neuroactive Steroids in the Central and Peripheral Nervous System
Roberto Cosimo Melcangi, Valerio Magnaghi, Mariarita Galbiati, and Luciano Martini

Neurosteroid Modulation of Recombinant and Synaptic $GABA_A$ Receptors
Jeremy J. Lambert, Sarah C. Harney, Delia Belelli, and John A. Peters

$GABA_A$-Receptor Plasticity during Long-Term Exposure to and Withdrawal from Progesterone
Giovanni Biggio, Paolo Follesa, Enrico Sanna, Robert H. Purdy, and Alessandra Concas

Stress and Neuroactive Steroids
Maria Luisa Barbaccia, Mariangela Serra, Robert H. Purdy, and Giovanni Biggio

Neurosteroids in Learning and Memory Processes
Monique Vallée, Willy Mayo, George F. Koob, and Michel Le Moal

Neurosteroids and Behavior
Sharon R. Engel and Kathleen A. Grant

Ethanol and Neurosteroid Interactions in the Brain
A. Leslie Morrow, Margaret J. VanDoren, Rebekah Fleming, and Shannon Penland

Preclinical Development of Neurosteroids as Neuroprotective Agents for the Treatment of Neurodegenerative Diseases
Paul A. Lapchak and Dalia M. Araujo

Clinical Implications of Circulating Neurosteroids
Andrea R. Genazzani, Patrizia Monteleone, Massimo Stomati, Francesca Bernardi, Luigi Cobellis, Elena Casarosa, Michele Luisi, Stefano Luisi, and Felice Petraglia

Neuroactive Steroids and Central Nervous System Disorders
Mingde Wang, Torbjörn Bäckström, Inger Sundström, Göran Wahlström, Tommy Olsson, Di Zhu, Inga-Maj Johansson, Inger Björn, and Marie Bixo

Neuroactive Steroids in Neuropsychopharmacology
Rainer Rupprecht and Florian Holsboer

Current Perspectives on the Role of Neurosteroids in PMS and Depression
Lisa D. Griffin, Susan C. Conrad, and Synthia H. Mellon

Index

Volume 47

Introduction: Studying Gene Expression in Neural Tissues by *in Situ* Hybridization
W. Wisden and B. J. Morris

Part I: *In Situ* Hybridization with Radiolabelled Oligonucleotides
In Situ Hybridization with Oligonucleotide Probes
Wl. Wisden and B. J. Morris

Cryostat Sectioning of Brains
Victoria Revilla and Alison Jones

Processing Rodent Embryonic and Early Postnatal Tissue for *in Situ* Hybridization with Radiolabelled Oligonucleotides
David J. Laurie, Petra C. U. Schrotz, Hannah Monyer, and Ulla Amtmann

Processing of Retinal Tissue for *in Situ* Hybridization
Frank Müller

Processing the Spinal Cord for *in Situ* Hybridization with Radiolabelled Oligonucleotides
A. Berthele and T. R. Tölle

Processing Human Brain Tissue for *in Situ* Hybridization with Radiolabelled Oligonucleotides
Louise F. B. Nicholson

In Situ Hybridization of Astrocytes and Neurons Cultured *in Vitro*
L. A. Arizza-McNaughton, C. De Felipe, and S. P. Hunt

In Situ Hybridization on Organotypic Slice Cultures
A. Gerfin-Moser and H. Monyer

Quantitative Analysis of *in Situ* Hybridization Histochemistry
Andrew L. Gundlach and Ross D. O'Shea

Part II: Nonradioactive *in Situ* hybridization

Nonradioactive *in Situ* Hybridization Using Alkaline Phosphatase-Labelled Oligonucleotides
S. J. Augood, E. M. McGowan, B. R. Finsen, B. Heppelmann, and P. C. Emson

Combining Nonradioactive *in Situ* Hybridization with Immunohistological and Anatomical Techniques
Petra Wahle

Nonradioactive *in Situ* Hybridization: Simplified Procedures for Use in Whole Mounts of Mouse and Chick Embryos
Linda Ariza-McNaughton and Robb Krumlauf

INDEX

Volume 48

Assembly and Intracellular Trafficking of $GABA_A$ Receptors Eugene
Barnes

Subcellular Localization and Regulation of $GABA_A$ Receptors and Associated Proteins Bernhard Lüscher and Jean-Marc Fritschy D_1 Dopamine Receptors
Richard Mailman

Molecular Modeling of Ligand-Gated Ion Channels: Progress and Challenges
Ed Bertaccini and James R. Trudel

Alzheimer's Disease: Its Diagnosis and Pathogenesis
Jillian J. Kril and Glenda M. Halliday

DNA Arrays and Functional Genomics in Neurobiology
Christelle Thibault, Long Wang, Li Zhang, and Michael F. Miles

INDEX

Volume 49

What Is West Syndrome?
Olivier Dulac, Christine Soufflet, Catherine Chiron, and Anna Kaminski

The Relationship between encephalopathy and Abnormal Neuronal Activity in the Developing Brain
Frances E. Jensen

Hypotheses from Functional Neuroimaging Studies
Csaba Juhász, Harry T. Chugani, Ouo Muzik, and Diane C. Chugani

Infantile Spasms: Unique Sydrome or General Age-Dependent Manifestation of a Diffuse Encephalopathy?
M. A. Koehn and M. Duchowny

Histopathology of Brain Tissue from Patients with Infantile Spasms
Harry V. Vinters

Generators of Ictal and Interictal Electroencephalograms Associated with Infantile Spasms: Intracellular Studies of Cortical and Thalamic Neurons
M. Steriade and I. Timofeev

Cortical and Subcortical Generators of Normal and Abnormal Rhythmicity
David A. McCormick

Role of Subcortical Structures in the Pathogenesis of Infantile Spasms: What Are Possible Subcortical Mediators?
F. A. Lado and S. L. Moshé

What Must We Know to Develop Better Therapies?
Jean Aicardi

The Treatment of Infantile Spasms: An Evidence-Based Approach
Mark Mackay, Shelly Weiss, and O. Carter Snead III

ACTH Treatment of Infantile Spasms: Mechanisms of Its Effects in Modulation of Neuronal Excitability
K. L. Brunson, S. Avishai-Eliner, and T. Z. Baram

Neurosteroids and Infantile Spasms: The Deoxycorticosterone Hypothesis
Michael A. Rogawski and Doodipala S. Reddy

Are there Specific Anatomical and/or Transmitter Systems (Cortical or Subcortical) That Should Be Targeted?
Phillip C. Jobe

Medical versus Surgical Treatment: Which Treatment When
W. Donald Shields

Developmental Outcome with and without Successful Intervention
Rochelle Caplan, Prabha Siddarth, Gary Mathern, Harry Vinters, Susan Curtiss, Jennifer Levitt, Robert Asarnow, and W. Donald Shields

Infantile Spasms versus Myoclonus: Is There a Connection?
Michael R. Pranzatelli

Tuberous Sclerosis as an Underlying Basis for Infantile Spasm
Raymond S. Yeung

Brain Malformation, Epilepsy, and Infantile Spasms
M. Elizabeth Ross

Brain Maturational Aspects Relevant to Pathophysiology of Infantile Spasms
G. Auanzini, F. Panzica, and S. Franceschetti

Gene Expression Analysis as a Strategy to Understand the Molecular Pathogenesis of Infantile Spasms
Peter B. Crino

Infantile Spasms: Criteria for an Animal Model
Carl E. Stafstrom and Gregory L. Holmes

INDEX

Volume 50

Part I: Primary Mechanisms

How Does Glucose Generate Oxidative Stress In Peripheral Nerve?
Irina G. Obrosova

Glycation in Diabetic Neuropathy: Characteristics, Consequences, Causes, and Therapeutic Options
Paul J. Thornalley

Part II: Secondary Changes

Protein Kinase C Changes in Diabetes: Is the Concept Relevant to Neuropathy?
Joseph Eichberg

Are Mitogen-Activated Protein Kinases Glucose Transducers for Diabetic Neuropathies?
Tertia D. Purves and David R. Tomlinson

Neurofilaments in Diabetic Neuropathy
Paul Fernyhough and Robert E. Schmidt

Apoptosis in Diabetic Neuropathy
Aviva Tolkovsky

Nerve and Ganglion Blood Flow in Diabetes: An Appraisal
Douglas W. Zochodne

Part III: Manifestations

Potential Mechanisms of Neuropathic Pain in Diabetes
Nigel A. Calcutt

Electrophysiologic Measures of Diabetic Neuropathy: Mechanism and Meaning
Joseph C. Arezzo and Elena Zotova

Neuropathology and Pathogenesis of Diabetic Autonomic Neuropathy
Robert E. Schmidt

Role of the Schwann Cell in Diabetic Neuropathy
Luke Eckersley

Part IV: Potential Treatment

Polyol Pathway and Diabetic Peripheral Neuropathy
Peter J. Oates

Nerve Growth Factor for the Treatment of Diabetic Neuropathy: What Went Wrong, What Went Right, and What Does the Future Hold?
Stuart C. Apfel

Angiotensin-Converting Enzyme Inhibitors: Are there Credible Mechanisms for Beneficial Effects in Diabetic Neuropathy?
Rayaz A. Malik and David R. Tomlinson

Clinical Trials for Drugs Against Diabetic Neuropathy: Can We Combine Scientific Needs With Clinical Practicalities?
Dan Ziegler and Dieter Luft

INDEX

Volume 51

Energy Metabolism in the Brain
Leif Hertz and Gerald A. Dienel

The Cerebral Glucose-Fatty Acid Cycle: Evolutionary Roots, Regulation, and (Patho) physiological Importance
Kurt Heininger

Expression, Regulation, and Functional Role of Glucose Transporters (GLUTs) in Brain
Donard S. Dwyer, Susan J. Vannucci, and Ian A. Simpson

Insulin-Like Growth Factor-1 Promotes Neuronal Glucose Utilization During Brain Development and Repair Processes
Carolyn A. Bondy and Clara M. Cheng

CNS Sensing and Regulation of Peripheral Glucose Levels
Barry E. Levin, Ambrose A. Dunn-Meynell, and Vanessa H. Routh

Glucose Transporter Protein Syndromes
Darryl C. De Vivo, Dong Wang, Juan M. Pascual, and Yuan Yuan Ho

Glucose, Stress, and Hippocampal Neuronal Vulnerability
Lawrence P. Reagan

Glucose/Mitochondria in Neurological Conditions
John P. Blass

Energy Utilization in the Ischemic/Reperfused Brain
John W. Phillis and Michael H. O'Regan

Diabetes Mellitus and the Central Nervous System
Anthony L. McCall

Diabetes, the Brain, and Behavior: Is There a Biological Mechanism Underlying the Association between Diabetes and Depression?
A. M. Jacobson, J. A. Samson, K. Weinger, and C. M. Ryan

Schizophrenia and Diabetes
David C. Henderson and Elissa R. Ettinger

Psychoactive Drugs Affect Glucose Transport and the Regulation of Glucose Metabolism
Donard S. Dwyer, Timothy D. Ardizzone, and Ronald J. Bradley

INDEX

Volume 52

Neuroimmune Relationships in Perspective
Frank Hucklebridge and Angela Clow

Sympathetic Nervous System Interaction with the Immune System
Virginia M. Sanders and Adam P. Kohm

Mechanisms by Which Cytokines Signal the Brain
Adrian J. Dunn

Neuropeptides: Modulators of Immune Responses in Health and Disease
David S. Jessop

Brain–Immune Interactions in Sleep
Lisa Marshall and Jan Born

Neuroendocrinology of Autoimmunity
Michael Harbuz

Systemic Stress-Induced Th2 Shift and Its Clinical Implications
Ibia J. Elenkov

Neural Control of Salivary S-IgA Secretion
Gordon B. Proctor and Guy H. Carpenter

Stress and Secretory Immunity
Jos A. Bosch, Christopher Ring, Eco J. C. de Geus, Enno C. I. Veerman, and Arie V. Nieuw Amerongen

Cytokines and Depression
Angela Clow

Immunity and Schizophrenia: Autoimmunity, Cytokines, and Immune Responses
Fiona Gaughran

Cerebral Lateralization and the Immune System
Pierre J. Neveu

Behavioral Conditioning of the Immune System
Frank Hucklebridge

Psychological and Neuroendocrine Correlates of Disease Progression
Julie M. Turner-Cobb

The Role of Psychological Intervention in Modulating Aspects of Immune Function in Relation to Health and Well-Being
J. H. Gruzelier

INDEX

Volume 53

Section I: Mitochondrial Structure and Function

Mitochondrial DNA Structure and Function
Carlos T. Moraes, Sarika Srivastava, Ilias Kirkinezos, Jose Oca-Cossio, Corina van Waveren, Markus Woischnick, and Francisca Diaz

Oxidative Phosphorylation: Structure, Function, and Intermediary Metabolism
Simon J. R. Heales, Matthew E. Gegg, and John B. Clark

Import of Mitochondrial Proteins
Matthias F. Bauer, Sabine Hofmann, and Walter Neupert

Section II: Primary Respiratory Chain Disorders

Mitochondrial Disorders of the Nervous System: Clinical, Biochemical, and Molecular Genetic Features
Dominic Thyagarajan and Edward Byrne

Section III: Secondary Respiratory Chain Disorders

Friedreich's Ataxia
J. M. Cooper and J. L. Bradley

Wilson Disease
C. A. Davie and A. H. V. Schapira

Hereditary Spastic Paraplegia
Christopher J. McDermott and Pamela J. Shaw

Cytochrome c Oxidase Deficiency
Giacomo P. Comi, Sandra Strazzer, Sara Galbiati, and Nereo Bresolin

Section IV: Toxin Induced Mitochondrial Dysfunction

Toxin-Induced Mitochondrial Dysfunction
Susan E. Browne and M. Flint Beal

Section V: Neurodegenerative Disorders

Parkinson's Disease
L. V. P. Korlipara and A. H. V. Schapira

Huntington's Disease: The Mystery Unfolds?
Åsa Petersén and Patrik Brundin

Mitochondria in Alzheimer's Disease
Russell H. Swerdlow and Stephen J. Kish

Contributions of Mitochondrial Alterations, Resulting from Bad Genes and a Hostile Environment, to the Pathogenesis of Alzheimer's Disease
Mark P. Mattson

Mitochondria and Amyotrophic Lateral Sclerosis
Richard W. Orrell and Anthony H. V. Schapira

Section VI: Models of Mitochondrial Disease

Models of Mitochondrial Disease
Danae Liolitsa and Michael G. Hanna

Section VII: Defects of β Oxidation Including Carnitine Deficiency

Defects of β Oxidation Including Carnitine Deficiency
K. Bartlett and M. Pourfarzam

Section VIII: Mitochondrial Involvement in Aging

The Mitochondrial Theory of Aging: Involvement of Mitochondrial DNA Damage and Repair
Nadja C. de Souza-Pinto and Vilhelm A. Bohr

INDEX

Volume 54

Unique General Anesthetic Binding Sites Within Distinct Conformational States of the Nicotinic Acetylcholine Receptor
Hugo R. Ariaas, William, R. Kem, James R. Truddell, and Michael P. Blanton

Signaling Molecules and Receptor Transduction Cascades That Regulate NMDA Receptor-Mediated Synaptic Transmission
Suhas. A. Kotecha and John F. MacDonald

Behavioral Measures of Alcohol Self-Administration and Intake Control: Rodent Models
Herman H. Samson and Cristine L. Czachowski

Dopaminergic Mouse Mutants: Investigating the Roles of the Different Dopamine Receptor Subtypes and the Dopamine Transporter
Shirlee Tan, Bettina Hermann, and Emiliana Borrelli

Drosophila melanogaster, A Genetic Model System for Alcohol Research
Douglas J. Guarnieri and Ulrike Heberlein

INDEX

Volume 55

Section I: Virsu Vectors For Use in the Nervous System

Non-Neurotropic Adenovirus: a Vector for Gene Transfer to the Brain and Gene Therapy of Neurological Disorders
P. R. Lowenstein, D. Suwelack, J. Hu, X. Yuan, M. Jimenez-Dalmaroni, S. Goverdhama, and M.G. Castro

Adeno-Associated Virus Vectors
E. Lehtonen and L. Tenenbaum

Problems in the Use of Herpes Simplex Virus as a Vector
L. T. Feldman

Lentiviral Vectors
J. Jakobsson, C. Ericson, N. Rosenquist, and C. Lundberg

Retroviral Vectors for Gene Delivery to Neural Precursor Cells
K. Kageyama, H. Hirata, and J. Hatakeyama

Section II: Gene Therapy with Virus Vectors for Specific Disease of the Nervous System

The Principles of Molecular Therapies for Glioblastoma
G. Karpati and J. Nalbatonglu

Oncolytic Herpes Simplex Virus
J. C. C. Hu and R. S. Coffin

Recombinant Retrovirus Vectors for Treatment of Brain Tumors
N. G. Rainov and C. M. Kramm

Adeno-Associated Viral Vectors for Parkinson's Disease
I. Muramatsu, L. Wang, K. Ikeguchi, K-i Fujimoto, T. Okada, H. Mizukami, Y. Hanazono, A. Kume, I. Nakano, and K. Ozawa

HSV Vectors for Parkinson's Disease
D. S. Latchman

Gene Therapy for Stroke
K. Abe and W. R. Zhang

Gene Therapy for Mucopolysaccharidosis
A. Bosch and J. M. Heard

INDEX

Volume 56

Behavioral Mechanisms and the Neurobiology of Conditioned Sexual Responding
Mark Krause

NMDA Receptors in Alcoholism
Paula L. Hoffman

Processing and Representation of Species-Specific Communication Calls in the Auditory System of Bats
George D. Pollak, Achim Klug, and Eric E. Bauer

Central Nervous System Control of Micturition
Gert Holstege and Leonora J. Mouton

The Structure and Physiology of the Rat Auditory System: An Overview
Manuel Malmierca

Neurobiology of Cat and Human Sexual Behavior
Gert Holstege and J. R. Georgiadis

INDEX

Volume 57

Cumulative Subject Index of Volumes 1–25

Volume 58

Cumulative Subject Index of Volumes 26–50

Volume 59

Loss of Spines and Neuropil
Liesl B. Jones

Schizophrenia as a Disorder of Neuroplasticity
Robert E. McCullumsmith, Sarah M. Clinton, and James H. Meador-Woodruff

The Synaptic Pathology of Schizophrenia: Is Aberrant Neurodevelopment and Plasticity to Blame?
Sharon L. Eastwood

Neurochemical Basis for an Epigenetic Vision of Synaptic Organization
E. Costa, D. R. Grayson, M. Veldic, and A. Guidotti

Muscarinic Receptors in Schizophrenia: Is There a Role for Synaptic Plasticity?
Thomas J. Raedler

Serotonin and Brain Development
Monsheel S. K. Sodhi and Elaine Sanders-Bush

Presynaptic Proteins and Schizophrenia
William G. Honer and Clint E. Young

Mitogen-Activated Protein Kinase Signaling
Svetlana V. Kyosseva

Postsynaptic Density Scaffolding Proteins at Excitatory Synapse and Disorders of Synaptic Plasticity: Implications for Human Behavior Pathologies
Andrea de Bartolomeis and Germano Fiore

Prostaglandin-Mediated Signaling in Schizophrenia
S. Smesny

Mitochondria, Synaptic Plasticity, and Schizophrenia
Dorit Ben-Shachar and Daphna Laifenfeld

Membrane Phospholipids and Cytokine Interaction in Schizophrenia
Jeffrey K. Yao and Daniel P. van Kammen

Neurotensin, Schizophrenia, and Antipsychotic Drug Action
Becky Kinkead and Charles B. Nemeroff

Schizophrenia, Vitamin D, and Brain Development
Alan Mackay-Sim, François FÉron, Darryl Eyles, Thomas Burne, and John McGrath

Possible Contributions of Myelin and Oligodendrocyte Dysfunction to Schizophrenia
Daniel G. Stewart and Kenneth L. Davis

Brain-Derived Neurotrophic Factor and the Plasticity of the Mesolimbic Dopamine Pathway
Oliver Guillin, Nathalie Griffon, Jorge Diaz, Bernard Le Foll, Erwan Bezard, Christian Gross, Chris Lammers, Holger Stark, Patrick Carroll, Jean-Charles Schwartz, and Pierre Sokoloff

S100B in Schizophrenic Psychosis
Matthias Rothermundt, Gerald Ponath, and Volker Arolt

Oct-6 Transcription Factor
Maria Ilia

NMDA Receptor Function, Neuroplasticity, and the Pathophysiology of Schizophrenia
Joseph T. Coyle and Guochuan Tsai

INDEX

Volume 60

Microarray Platforms: Introduction and Application to Neurobiology
Stanislav L. Karsten, Lili C. Kudo, and Daniel H. Geschwind

Experimental Design and Low-Level Analysis of Microarray Data
B. M. Bolstad, F. Collin, K. M. Simpson, R. A. Irizarry, and T. P. Speed

Brain Gene Expression: Genomics and Genetics
Elissa J. Chesler and Robert W. Williams

DNA Microarrays and Animal Models of Learning and Memory
Sebastiano Cavallaro

Microarray Analysis of Human Nervous System Gene Expression in Neurological Disease
Steven A. Greenberg

DNA Microarray Analysis of Postmortem Brain Tissue
Károly Mirnics, Pat Levitt, and David A. Lewis

INDEX

Volume 61

Section I: High-Throughput Technologies

Biomarker Discovery Using Molecular Profiling Approaches
Stephen J. Walker and Arron Xu

Proteomic Analysis of Mitochondrial Proteins
Mary F. Lopez, Simon Melov, Felicity Johnson, Nicole Nagulko, Eva Golenko, Scott Kuzdzal, Suzanne Ackloo, and Alvydas Mikulskis

Section II: Proteomic Applications

NMDA Receptors, Neural Pathways, and Protein Interaction Databases
Holger Husi

Dopamine Transporter Network and Pathways
Rajani Maiya and R. Dayne Mayfield

Proteomic Approaches in Drug Discovery and Development
Holly D. Soares, Stephen A. Williams, Peter J. Snyder, Feng Gao, Tom Stiger, Christian Rohlff, Athula Herath, Trey Sunderland, Karen Putnam, and W. Frost White

Section III: Informatics

Proteomic Informatics
Steven Russell, William Old, Katheryn Resing, and Lawrence Hunter

Section IV: Changes in the Proteome by Disease

Proteomics Analysis in Alzheimer's Disease: New Insights into Mechanisms of Neurodegeneration
D. Allan Butterfield and Debra Boyd-Kimball

Proteomics and Alcoholism
Frank A. Witzmann and Wendy N. Strother

Proteomics Studies of Traumatic Brain Injury
Kevin K. W. Wang, Andrew Ottens, William Haskins, Ming Cheng Liu, Firas Kobeissy, Nancy Denslow, SuShing Chen, and Ronald L. Hayes

Influence of Huntington's Disease on the Human and Mouse Proteome
Claus Zabel and Joachim Klose

Section V: Overview of the Neuroproteome

Proteomics—Application to the Brain
Katrin Marcus, Oliver Schmidt, Heike Schaefer, Michael Hamacher, AndrÅ van Hall, and Helmut E. Meyer

INDEX

Volume 62

$GABA_A$ Receptor Structure–Function Studies: A Reexamination in Light of New Acetylcholine Receptor Structures
Myles H. Akabas

Dopamine Mechanisms and Cocaine Reward
Aiko Ikegami and Christine L. Duvauchelle

Proteolytic Dysfunction in Neurodegenerative Disorders
Kevin St. P. McNaught

Neuroimaging Studies in Bipolar Children and Adolescents
Rene L. Olvera, David C. Glahn, Sheila C. Caetano, Steven R. Pliszka, and Jair C. Soares

Chemosensory G-Protein-Coupled Receptor Signaling in the Brain
Geoffrey E. Woodard

Disturbances of Emotion Regulation after Focal Brain Lesions
Antoine Bechara

The Use of *Caenorhabditis elegans* in Molecular Neuropharmacology
Jill C. Bettinger, Lucinda Carnell, Andrew G. Davies, and Steven L. McIntire

INDEX

Volume 63

Mapping Neuroreceptors at work: On the Definition and Interpretation of Binding Potentials after 20 years of Progress
Albert Gjedde, Dean F. Wong, Pedro Rosa-Neto, and Paul Cumming

Mitochondrial Dysfunction in Bipolar Disorder: From ^{31}P-Magnetic Resonance Spectroscopic Findings to Their Molecular Mechanisms
Tadafumi Kato

Large-Scale Microarray Studies of Gene Expression in Multiple Regions of the Brain in Schizophrenia and Alzeimer's Disease
Pavel L. Katsel, Kenneth L. Davis, and Vahram Haroutunian

Regulation of Serotonin 2C Receptor PRE-mRNA Editing By Serotonin
Claudia Schmauss

The Dopamine Hypothesis of Drug Addiction: Hypodopaminergic State
Miriam Melis, Saturnino Spiga, and Marco Diana

Human and Animal Spongiform Encephalopathies are Autoimmune Diseases: A Novel Theory and Its supporting Evidence
Bao Ting Zhu

Adenosine and Brain Function
Bertil B. Fredholm, Jiang-Fan Chen, Rodrigo A. Cunha, Per Svenningsson, and Jean-Marie Vaugeois

INDEX

Volume 64

Section I. The Cholinergic System
John Smythies

Section II. The Dopamine System
John Symythies

Section III. The Norepinephrine System
John Smythies

Section IV. The Adrenaline System
John Smythies

Section V. Serotonin System
John Smythies

INDEX

Volume 65

Insulin Resistance: Causes and Consequences
Zachary T. Bloomgarden

Antidepressant-Induced Manic Conversion: A Developmentally Informed Synthesis of the Literature
Christine J. Lim, James F. Leckman, Christopher Young, and AndrÉs Martin

Sites of Alcohol and Volatile Anesthetic Action on Glycine Receptors
Ingrid A. Lobo and R. Adron Harris

Role of the Orbitofrontal Cortex in Reinforcement Processing and Inhibitory Control: Evidence from Functional Magnetic Resonance Imaging Studies in Healthy Human Subjects
Rebecca Elliott and Bill Deakin

Common Substrates of Dysphoria in Stimulant Drug Abuse and Primary Depression: Therapeutic Targets
Kate Baicy, Carrie E. Bearden, John Monterosso, Arthur L. Brody, Andrew J. Isaacson, and Edythe D. London

The Role of cAMP Response Element–Binding Proteins in Mediating Stress-Induced Vulnerability to Drug Abuse
Arati Sadalge Kreibich and Julie A. Blendy

G-Protein–Coupled Receptor Deorphanizations
Yumiko Saito and Olivier Civelli

Mechanistic Connections Between Glucose/Lipid Disturbances and Weight Gain Induced by Antipsychotic Drugs
Donard S. Dwyer, Dallas Donohoe, Xiao-Hong Lu, and Eric J. Aamodt

Serotonin Firing Activity as a Marker for Mood Disorders: Lessons from Knockout Mice
Gabriella Gobbi

INDEX

Volume 66

Brain Atlases of Normal and Diseased Populations
Arthur W. Toga and Paul M. Thompson

Neuroimaging Databases as a Resource for Scientific Discovery
John Darrell Van Horn, John Wolfe, Autumn Agnoli, Jeffrey Woodward, Michael Schmitt, James Dobson, Sarene Schumacher, and Bennet Vance

Modeling Brain Responses
Karl J. Friston, William Penny, and Olivier David

Voxel-Based Morphometric Analysis Using Shape Transformations
Christos Davatzikos

The Cutting Edge of fMRI and High-Field fMRI
Dae-Shik Kim

Quantification of White Matter Using Diffusion-Tensor Imaging
Hae-Jeong Park

Perfusion fMRI for Functional Neuroimaging
Geoffrey K. Aguirre, John A. Detre, and Jiongjiong Wang

Functional Near-Infrared Spectroscopy: Potential and Limitations in Neuroimaging Studies
Yoko Hoshi

Neural Modeling and Functional Brain Imaging: The Interplay Between the Data-Fitting and Simulation Approaches
Barry Horwitz and Michael F. Glabus

Combined EEG and fMRI Studies of Human Brain Function
V. Menon and S. Crottaz-Herbette

INDEX

Volume 67

Distinguishing Neural Substrates of Heterogeneity Among Anxiety Disorders
Jack B. Nitschke and Wendy Heller

Neuroimaging in Dementia
K. P. Ebmeier, C. Donaghey, and N. J. Dougall

Prefrontal and Anterior Cingulate Contributions to Volition in Depression
Jack B. Nitschke and Kristen L. Mackiewicz

Functional Imaging Research in Schizophrenia
H. Tost, G. Ende, M. Ruf, F. A. Henn, and A. Meyer-Lindenberg

Neuroimaging in Functional Somatic Syndromes
Patrick B. Wood

Neuroimaging in Multiple Sclerosis
Alireza Minagar, Eduardo Gonzalez-Toledo, James Pinkston, and Stephen L. Jaffe

Stroke
Roger E. Kelley and Eduardo Gonzalez-Toledo

Functional MRI in Pediatric Neurobehavioral Disorders
Michael Seyffert and F. Xavier Castellanos

Structural MRI and Brain Development
Paul M. Thompson, Elizabeth R. Sowell, Nitin Gogtay, Jay N. Giedd, Christine N. Vidal, Kiralee M. Hayashi, Alex Leow, Rob Nicolson, Judith L. Rapoport, and Arthur W. Toga

Neuroimaging and Human Genetics
Georg Winterer, Ahmad R. Hariri, David Goldman, and Daniel R. Weinberger

Neuroreceptor Imaging in Psychiatry: Theory and Applications
W. Gordon Frankle, Mark Slifstein, Peter S. Talbot, and Marc Laruelle

INDEX

Volume 68

Fetal Magnetoencephalography: Viewing the Developing Brain *In Utero*
Hubert Preissl, Curtis L. Lowery, and Hari Eswaran

Magnetoencephalography in Studies of Infants and Children
Minna Huotilainen

Let's Talk Together: Memory Traces Revealed by Cooperative Activation in the Cerebral Cortex
Jochen Kaiser, Susanne Leiberg, and Werner Lutzenberger

Human Communication Investigated With Magnetoencephalography: Speech, Music, and Gestures
Thomas R. Knösche, Burkhard Maess, Akinori Nakamura, and Angela D. Friederici

Combining Magnetoencephalography and Functional Magnetic Resonance Imaging
Klaus Mathiak and Andreas J. Fallgatter

Beamformer Analysis of MEG Data
Arjan Hillebrand and Gareth R. Barnes

Functional Connectivity Analysis in Magnetoencephalography
Alfons Schnitzler and Joachim Gross

Human Visual Processing as Revealed by Magnetoencephalographys
Yoshiki Kaneoke, Shoko Watanabe, and Ryusuke Kakigi

A Review of Clinical Applications of Magnetoencephalography
Andrew C. Papanicolaou, Eduardo M. Castillo, Rebecca Billingsley-Marshall, Ekaterina Pataraia, and Panagiotis G. Simos

INDEX

Volume 69

Nematode Neurons: Anatomy and Anatomical Methods in Caenorhabditis elegans
David H. Hall, Robyn Lints, and Zeynep Altun

Investigations of Learning and Memory in Caenorhabditis elegans
Andrew C. Giles, Jacqueline K. Rose, and Catharine H. Rankin

Neural Specification and Differentiation
Eric Aamodt and Stephanie Aamodt

Sexual Behavior of the Caenorhabditis elegans Male
Scott W. Emmons

The Motor Circuit
Stephen E. Von Stetina, Millet Treinin, and David M. Miller III

Mechanosensation in Caenorhabditis elegans
Robert O'Hagan and Martin Chalfie

Volume 70

Spectral Processing by the Peripheral Auditory System Facts and Models
Enrique A. Lopez-Poveda

Basic Psychophysics of Human Spectral Processing
Brian C. J. Moore

Across-Channel Spectral Processing
John H. Grose, Joseph W. Hall III, and Emily Buss

Speech and Music Have Different Requirements for Spectral Resolution
Robert V. Shannon

Non-Linearities and the Representation of Auditory Spectra
Eric D. Young, Jane J. Yu, and Lina A. J. Reiss

Spectral Processing in the Inferior Colliculus
Kevin A. Davis

Neural Mechanisms for Spectral Analysis in the Auditory Midbrain, Thalamus, and Cortex
Monty A. Escabı' and Heather L. Read

Spectral Processing in the Auditory Cortex
Mitchell L. Sutter

Processing of Dynamic Spectral Properties of Sounds
Adrian Rees and Manuel S. Malmierca

Representations of Spectral Coding in the Human Brain
Deborah A. Hall, PhD

Spectral Processing and Sound Source Determination
Donal G. Sinex

Spectral Information in Sound Localization
Simon Carlile, Russell Martin, and Ken McAnally

Plasticity of Spectral Processing
Dexter R. F. Irvine and Beverly A. Wright

Spectral Processing In Cochlear Implants
Colette M. McKay

INDEX

Volume 71

Autism: Neuropathology, Alterations of the GABAergic System, and Animal Models
Christoph Schmitz, Imke A. J. van Kooten, Patrick R. Hof, Herman van Engeland, Paul H. Patterson, and Harry W. M. Steinbusch

The Role of GABA in the Early Neuronal Development
Marta Jelitai and Emı'lia Madarasz

GABAergic Signaling in the Developing Cerebellum
Chitoshi Takayama

Insights into GABA Functions in the Developing Cerebellum
Mo'nica L. Fiszman

Role of GABA in the Mechanism of the Onset of Puberty in Non-Human Primates
Ei Terasawa

Rett Syndrome: A Rosetta Stone for Understanding the Molecular Pathogenesis of Autism
Janine M. LaSalle, Amber Hogart, and Karen N. Thatcher

GABAergic Cerebellar System in Autism: A Neuropathological and Developmental Perspective
Gene J. Blatt

Reelin Glycoprotein in Autism and Schizophrenia
S. Hossein Fatemi

Is There A Connection Between Autism, Prader-Willi Syndrome, Catatonia, and GABA?
Dirk M. Dhossche, Yaru Song, and Yiming Liu

Alcohol, GABA Receptors, and Neurodevelopmental Disorders
Ujjwal K. Rout

Effects of Secretin on Extracellular GABA and Other Amino Acid Concentrations in the Rat Hippocampus
Hans-Willi Clement, Alexander Pschibul, and Eberhard Schulz

Predicted Role of Secretin and Oxytocin in the Treatment of Behavioral and Developmental Disorders: Implications for Autism
Martha G. Welch and David A. Ruggiero

Immunological Findings in Autism
Hari Har Parshad Cohly and Asit Panja

Correlates of Psychomotor Symptoms in Autism
Laura Stoppelbein, Sara Sytsma-Jordan, and Leilani Greening

GABRB3 Gene Deficient Mice: A Potential Model of Autism Spectrum Disorder
Timothy M. DeLorey

The Reeler Mouse: Anatomy of a Mutant
Gabriella D'Arcangelo

Shared Chromosomal Susceptibility Regions Between Autism and Other Mental Disorders
Yvon C. Chagnon index

INDEX

Volume 72

Classification Matters for Catatonia and Autism in Children
Klaus-Jürgen Neumärker

A Systematic Examination of Catatonia-Like Clinical Pictures in Autism Spectrum Disorders
Lorna Wing and Amitta Shah

Catatonia in Individuals with Autism Spectrum Disorders in Adolescence and Early Adulthood: A Long-Term Prospective Study
Masataka Ohta, Yukiko Kano, and Yoko Nagai

Are Autistic and Catatonic Regression Related? A Few Working Hypotheses Involving GABA, Purkinje Cell Survival, Neurogenesis, and ECT
Dirk Marcel Dhossche and Ujjwal Rout

Psychomotor Development and Psychopathology in Childhood
Dirk M. J. De Raeymaecker

The Importance of Catatonia and Stereotypies in Autistic Spectrum Disorders
Laura Stoppelbein, Leilani Greening, and Angelina Kakooza

Prader–Willi Syndrome: Atypical Psychoses and Motor Dysfunctions
Willem M. A. Verhoeven and Siegfried Tuinier

Towards a Valid Nosography and Psychopathology of Catatonia in Children and Adolescents
David Cohen

Is There a Common Neuronal Basis for Autism and Catatonia?
Dirk Marcel Dhossche, Brendan T. Carroll, and Tressa D. Carroll

Shared Susceptibility Region on Chromosome 15 Between Autism and Catatonia
Yvon C. Chagnon

Current Trends in Behavioral Interventions for Children with Autism
Dorothy Scattone and Kimberly R. Knight

Case Reports with a Child Psychiatric Exploration of Catatonia, Autism, and Delirium
Jan N. M. Schieveld

ECT and the Youth: Catatonia in Context
Frank K. M. Zaw

Catatonia in Autistic Spectrum Disorders: A Medical Treatment Algorithm
Max Fink, Michael A. Taylor, and Neera Ghaziuddin

Psychological Approaches to Chronic Catatonia-Like Deterioration in Autism Spectrum Disorders
Amitta Shah and Lorna Wing

Section V: Blueprints

Blueprints for the Assessment, Treatment, and Future Study of Catatonia in Autism Spectrum Disorders
Dirk Marcel, Dhossche, Amitta Shah, and Lorna Wing

INDEX

Volume 73

Chromosome 22 Deletion Syndrome and Schizophrenia
Nigel M. Williams, Michael C. O'Donovan, and Michael J. Owen

Characterization of Proteome of Human Cerebrospinal Fluid
Jing Xu, Jinzhi Chen, Elaine R. Peskind, Jinghua Jin, Jimmy Eng, Catherine Pan, Thomas J. Montine, David R. Goodlett, and Jing Zhang

Hormonal Pathways Regulating Intermale and Interfemale Aggression
Neal G. Simon, Qianxing Mo, Shan Hu, Carrie Garippa, and Shi-Fang Lu

Neuronal GAP Junctions: Expression, Function, and Implications for Behavior
Clinton B. McCracken and David C. S. Roberts

Effects of Genes and Stress on the Neurobiology of Depression
J. John Mann and Dianne Currier

Quantitative Imaging with the Micropet Small-Animal Pet Tomograph
Paul Vaska, Daniel J. Rubins, David L. Alexoff, and Wynne K. Schiffer

Understanding Myelination through Studying its Evolution
Rüdiger Schweigreiter, Betty I. Roots, Christine Bandtlow, and Robert M. Gould

INDEX

Volume 74

Evolutionary Neurobiology and Art
C. U. M. Smith

Section I: Visual Aspects

Perceptual Portraits
Nicholas Wade

The Neuropsychology of Visual Art: Conferring Capacity
Anjan Chatterjee

Vision, Illusions, and Reality
Christopher Kennard

Localization in the Visual Brain
George K. York

Section II: Episodic Disorders

Neurology, Synaesthesia, and Painting
Amy Ione

Fainting in Classical Art
Philip Smith

Migraine Art in the Internet: A Study of 450 Contemporary Artists
Klaus Podoll

Sarah Raphael's Migraine with Aura as Inspiration for the Foray of Her Work into Abstraction
Klaus Podoll and Debbie Ayles

The Visual Art of Contemporary Artists with Epilepsy
Steven C. Schachter

Section III: Brain Damage

Creativity in Painting and Style in Brain-Damaged Artists
Julien Bogousslavsky

Artistic Changes in Alzheimer's Disease
Sebastian J. Crutch and Martin N. Rossor

Section IV: Cerebrovascular Disease

Stroke in Painters
H. Bäzner and M. Hennerici

Visuospatial Neglect in Lovis Corinth's Self-Portraits
Olaf Blanke

Art, Constructional Apraxia, and the Brain
Louis Caplan

Section V: Genetic Diseases

Neurogenetics in Art
Alan E. H. Emery

A Naïve Artist of St Ives
F. Clifford Rose

Van Gogh's Madness
F. Clifford Rose

Absinthe, The Nervous System and Painting
Tiina Rekand

Section VI: Neurologists as Artists

Sir Charles Bell, KGH, FRS, FRSE (1774–1842)
Christopher Gardner-Thorpe

Section VII: Miscellaneous

Peg Leg Frieda
Espen Dietrichs

The Deafness of Goya (1746–1828)
F. Clifford Rose

INDEX

Volume 75

Introduction on the Use of the *Drosophila* Embryonic/Larval Neuromuscular Junction as a Model System to Study Synapse Development and Function, and a Brief Summary of Pathfinding and Target Recognition
Catalina Ruiz-Cañada and Vivian Budnik

Development and Structure of Motoneurons
Matthias Landgraf and Stefan Thor

The Development of the *Drosophila* Larval Body Wall Muscles
Karen Beckett and Mary K. Baylies

Organization of the Efferent System and Structure of Neuromuscular Junctions in *Drosophila*
Andreas Prokop

Development of Motoneuron Electrical Properties and Motor Output
Richard A. Baines

Transmitter Release at the Neuromuscular Junction
Thomas L. Schwarz

Vesicle Trafficking and Recycling at the Neuromuscular Junction: Two Pathways for Endocytosis
Yoshiaki Kidokoro

Glutamate Receptors at the *Drosophila* Neuromuscular Junction
Aaron DiAntonio

Scaffolding Proteins at the *Drosophila* Neuromuscular Junction
Bulent Ataman, Vivian Budnik, and Ulrich Thomas

Synaptic Cytoskeleton at the Neuromuscular Junction
Catalina Ruiz-Cañada and Vivian Budnik

Plasticity and Second Messengers During Synapse Development
Leslie C. Griffith and Vivian Budnik

Retrograde Signaling that Regulates Synaptic Development and Function at the *Drosophila* Neuromuscular Junction
Guillermo Marqués and Bing Zhang

Activity-Dependent Regulation of Transcription During Development of Synapses
Subhabrata Sanyal and Mani Ramaswami

Experience-Dependent Potentiation of Larval Neuromuscular Synapses
Christoph M. Schuster

Selected Methods for the Anatomical Study of *Drosophila* Embryonic and Larval Neuromuscular Junctions
Vivian Budnik, Michael Gorczyca, and Andreas Prokop

INDEX

Volume 76

Section I: Physiological Correlates of Freud's Theories

The ID, the Ego, and the Temporal Lobe
Shirley M. Ferguson and Mark Rayport

ID, Ego, and Temporal Lobe Revisited
Shirley M. Ferguson and Mark Rayport

Section II: Stereotaxic Studies

Olfactory Gustatory Responses Evoked by Electrical Stimulation of Amygdalar Region in Man Are Qualitatively Modifiable by Interview Content: Case Report and Review
Mark Rayport, Sepehr Sani, and Shirley M. Ferguson

Section III: Controversy in Definition of Behavioral Disturbance

Pathogenesis of Psychosis in Epilepsy. The "Seesaw" Theory: Myth or Reality?
Shirley M. Ferguson and Mark Rayport

Section IV: Outcome of Temporal Lobectomy

Memory Function After Temporal Lobectomy for Seizure Control: A Comparative Neuropsy chiatric and Neuropsychological Study
Shirley M. Ferguson, A. John McSweeny, and Mark Rayport

Life After Surgery for Temporolimbic Seizures
Shirley M. Ferguson, Mark Rayport, and Carolyn A. Schell

Appendix I
Mark Rayport

Appendix II: Conceptual Foundations of Studies of Patients Undergoing Temporal Lobe Surgery for Seizure Control
Mark Rayport

INDEX

Volume 77

Regenerating the Brain
David A. Greenberg and Kunlin Jin

Serotonin and Brain: Evolution, Neuroplasticity, and Homeostasis
Efrain C. Azmitia

Therapeutic Approaches to Promoting Axonal Regeneration in the Adult Mammalian Spinal Cord
Sari S. Hannila, Mustafa M. Siddiq, and Marie T. Filbin

Evidence for Neuroprotective Effects of Antipsychotic Drugs: Implications for the Pathophysiology and Treatment of Schizophrenia
Xin-Min Li and Haiyun Xu

Neurogenesis and Neuroenhancement in the Pathophysiology and Treatment of Bipolar Disorder
Robert J. Schloesser, Guang Chen, and Husseini K. Manji

Neuroreplacement, Growth Factor, and Small Molecule Neurotrophic Approaches for Treating Parkinson's Disease
Michael J. O'Neill, Marcus J. Messenger, Viktor Lakics, Tracey K. Murray, Eric H. Karran, Philip G. Szekeres, Eric S. Nisenbaum, and Kalpana M. Merchant

Using *Caenorhabditis elegans* Models of Neurodegenerative Disease to Identify Neuroprotective Strategies
Brian Kraemer and Gerard D. Schellenberg

Neuroprotection and Enhancement of Neurite Outgrowth With Small Molecular Weight Compounds From Screens of Chemical Libraries
Donard S. Dwyer and Addie Dickson

INDEX

Volume 78

Neurobiology of Dopamine in Schizophrenia
Olivier Guillin, Anissa Abi-Dargham, and Marc Laruelle

The Dopamine System and the Pathophysiology of Schizophrenia: A Basic Science Perspective
Yukiori Goto and Anthony A. Grace

Glutamate and Schizophrenia: Phencyclidine, N-methyl-D-aspartate Receptors, and Dopamine–Glutamate Interactions
Daniel C. Javitt

Deciphering the Disease Process of Schizophrenia: The Contribution of Cortical GABA Neurons
David A. Lewis and Takanori Hashimoto

Alterations of Serotonin Transmission in Schizophrenia
Anissa Abi-Dargham

Serotonin and Dopamine Interactions in Rodents and Primates: Implications for Psychosis and Antipsychotic Drug Development
Gerard J. Marek

Cholinergic Circuits and Signaling in the Pathophysiology of Schizophrenia
Joshua A. Berman, David A. Talmage, and Lorna W. Role

Schizophrenia and the α7 Nicotinic Acetylcholine Receptor
Laura F. Martin and Robert Freedman

Histamine and Schizophrenia
Jean-Michel Arrang

Cannabinoids and Psychosis
Deepak Cyril D'Souza

Involvement of Neuropeptide Systems in Schizophrenia: Human Studies
Ricardo Cáceda, Becky Kinkead, and Charles B. Nemeroff

Brain-Derived Neurotrophic Factor in Schizophrenia and Its Relation with Dopamine
Olivier Guillin, Caroline Demily, and Florence Thibaut

Schizophrenia Susceptibility Genes: In Search of a Molecular Logic and Novel Drug Targets for a Devastating Disorder
Joseph A. Gogos

INDEX

Volume 79

The Destructive Alliance: Interactions of Leukocytes, Cerebral Endothelial Cells, and the Immune Cascade in Pathogenesis of Multiple Sclerosis
Alireza Minagar, April Carpenter, and J. Steven Alexander

Role of B Cells in Pathogenesis of Multiple Sclerosis
Behrouz Nikbin, Mandana Mohyeddin Bonab, Farideh Khosravi, and Fatemeh Talebian

The Role of CD4 T Cells in the Pathogenesis of Multiple Sclerosis
Tanuja Chitnis

The CD8 T Cell in Multiple Sclerosis: Suppressor Cell or Mediator of Neuropathology?
Aaron J. Johnson, Georgette L. Suidan, Jeremiah McDole, and Istvan Pirko

Immunopathogenesis of Multiple Sclerosis
Smriti M. Agrawal and V. Wee Yong

Molecular Mimicry in Multiple Sclerosis
Jane E. Libbey, Lori L. McCoy, and Robert S. Fujinami

Molecular "Negativity" May Underlie Multiple Sclerosis: Role of the Myelin Basic Protein Family in the Pathogenesis of MS
Abdiwahab A. Musse and George Harauz

Microchimerism and Stem Cell Transplantation in Multiple Sclerosis
Behrouz Nikbin, Mandana Mohyeddin Bonab, and Fatemeh Talebian

The Insulin-Like Growth Factor System in Multiple Sclerosis
Daniel Chesik, Nadine Wilczak, and Jacques De Keyser

Cell-Derived Microparticles and Exosomes in Neuroinflammatory Disorders
Lawrence L. Horstman, Wenche Jy, Alireza Minagar, Carlos J. Bidot, Joaquin J. Jimenez, J. Steven Alexander, and Yeon S. Ahn

Multiple Sclerosis in Children: Clinical, Diagnostic, and Therapeutic Aspects
Kevin Rostásy

Migraine in Multiple Sclerosis
Debra G. Elliott

Multiple Sclerosis as a Painful Disease
Meghan Kenner, Uma Menon, and Debra Elliott

Multiple Sclerosis and Behavior
James B. Pinkston, Anita Kablinger, and Nadejda Alekseeva

Cerebrospinal Fluid Analysis in Multiple Sclerosis
Francisco A. Luque and Stephen L. Jaffe

Multiple Sclerosis in Isfahan, Iran
Mohammad Saadatnia, Masoud Etemadifar, and Amir Hadi Maghzi

Gender Issues in Multiple Sclerosis
Robert N. Schwendimann and Nadejda Alekseeva

Differential Diagnosis of Multiple Sclerosis
Halim Fadil, Roger E. Kelley, and Eduardo Gonzalez-Toledo

Prognostic Factors in Multiple Sclerosis
Roberto Bergamaschi

Neuroimaging in Multiple Sclerosis
Robert Zivadinov and Jennifer L. Cox

Detection of Cortical Lesions Is Dependent on Choice of Slice Thickness in Patients with Multiple Sclerosis
Ondrej Dolezal, Michael G. Dwyer, Dana Horakova, Eva Havrdova, Alireza Minagar, Srivats Balachandran, Niels Bergsland, Zdenek Seidl, Manuela Vaneckova, David Fritz, Jan Krasensky, and Robert Zivadinov

The Role of Quantitative Neuroimaging Indices in the Differentiation of Ischemia from Demyelination: An Analytical Study with Case Presentation
Romy Hoque, Christina Ledbetter, Eduardo Gonzalez-Toledo, Vivek Misra, Uma Menon, Meghan Kenner, Alejandro A. Rabinstein, Roger E. Kelley, Robert Zivadinov, and Alireza Minagar

HLA-DRB1*1501, -DQB1*0301, -DQB1*0302, -DQB1*0602, and -DQB1*0603 Alleles Are Associated with More Severe Disease Outcome on MRI in Patients with Multiple Sclerosis
Robert Zivadinov, Laura Uxa, Alessio Bratina, Antonio Bosco, Bhooma Srinivasaraghavan, Alireza Minagar, Maja Ukmar, Su yen Benedetto, and Marino Zorzon

Glatiramer Acetate: Mechanisms of Action in Multiple Sclerosis
Tjalf Ziemssen and Wiebke Schrempf

Evolving Therapies for Multiple Sclerosis
Elena Korniychuk, John M. Dempster, Eileen O'Connor, J. Steven Alexander, Roger E. Kelley, Meghan Kenner, Uma Menon, Vivek Misra, Romy Hoque, Eduardo C. Gonzalez-Toledo, Robert N. Schwendimann, Stacy Smith, and Alireza Minagar

Remyelination in Multiple Sclerosis
Divya M. Chari

Trigeminal Neuralgia: A Modern-Day Review
Kelly Hunt and Ravish Patwardhan

Optic Neuritis and the Neuro-Ophthalmology of Multiple Sclerosis
Paramjit Kaur and Jeffrey L. Bennett

Neuromyelitis Optica: New Findings on Pathogenesis
Dean M. Wingerchuk

INDEX

Volume 80

Epilepsy in the Elderly: Scope of the Problem
Ilo E. Leppik

Animal Models in Gerontology Research
Nancy L. Nadon

Animal Models of Geriatric Epilepsy
Lauren J. Murphree, Lynn M. Rundhaugen, and Kevin M. Kelly

Life and Death of Neurons in the Aging Cerebral Cortex
John H. Morrison and Patrick R. Hof

An In Vitro Model of Stroke-Induced Epilepsy: Elucidation of the Roles of Glutamate and Calcium in the Induction and Maintenance of Stroke-Induced Epileptogenesis
Robert J. DeLorenzo, David A. Sun, Robert E. Blair, and Sompong Sambati

Mechanisms of Action of Antiepileptic Drugs
H. Steve White, Misty D. Smith, and Karen S. Wilcox

Epidemiology and Outcomes of Status Epilepticus in the Elderly
Alan R. Towne

Diagnosing Epilepsy in the Elderly
R. Eugene Ramsay, Flavia M. Macias, and A. James Rowan

Pharmacoepidemiology in Community-Dwelling Elderly Taking Antiepileptic Drugs
Dan R. Berlowitz and Mary Jo V. Pugh

Use of Antiepileptic Medications in Nursing Homes
Judith Garrard, Susan L. Harms, Lynn E. Eberly, and Ilo E. Leppik

Differential Diagnosis of Multiple Sclerosis
Halim Fadil, Roger E. Kelley, and Eduardo Gonzalez-Toledo

Prognostic Factors in Multiple Sclerosis
Roberto Bergamaschi

Neuroimaging in Multiple Sclerosis
Robert Zivadinov and Jennifer L. Cox

Detection of Cortical Lesions Is Dependent on Choice of Slice Thickness in Patients with Multiple Sclerosis
Ondrej Dolezal, Michael G. Dwyer, Dana Horakova, Eva Havrdova, Alireza Minagar, Srivats Balachandran, Niels Bergsland, Zdenek Seidl, Manuela Vaneckova, David Fritz, Jan Krasensky, and Robert Zivadinov

The Role of Quantitative Neuroimaging Indices in the Differentiation of Ischemia from Demyelination: An Analytical Study with Case Presentation
Romy Hoque, Christina Ledbetter, Eduardo Gonzalez-Toledo, Vivek Misra, Uma Menon, Meghan Kenner, Alejandro A. Rabinstein, Roger E. Kelley, Robert Zivadinov, and Alireza Minagar

HLA-DRB1*1501, -DQB1*0301,-DQB1 *0302,-DQB1*0602, and -DQB1*0603 Alleles Are Associated with More Severe Disease Outcome on MRI in Patients with Multiple Sclerosis
Robert Zivadinov, Laura Uxa, Alessio Bratina, Antonio Bosco, Bhooma Srinivasaraghavan, Alireza Minagar, Maja Ukmar, Su yen Benedetto, and Marino Zorzon

Glatiramer Acetate: Mechanisms of Action in Multiple Sclerosis
Tjalf Ziemssen and Wiebke Schrempf

Evolving Therapies for Multiple Sclerosis
Elena Korniychuk, John M. Dempster, Eileen O'Connor, J. Steven Alexander, Roger E. Kelley, Meghan Kenner, Uma Menon, Vivek Misra, Romy Hoque, Eduardo C. Gonzalez-Toledo, Robert N. Schwendimann, Stacy Smith, and Alireza Minagar

Remyelination in Multiple Sclerosis
Divya M. Chari

Trigeminal Neuralgia: A Modern-Day Review
Kelly Hunt and Ravish Patwardhan

Optic Neuritis and the Neuro-Ophthalmology of Multiple Sclerosis
Paramjit Kaur and Jeffrey L. Bennett

Neuromyelitis Optica: New Findings on Pathogenesis
Dean M. Wingerchuk

INDEX

Volume 81

Epilepsy in the Elderly: Scope of the Problem
Ilo E. Leppik

Animal Models in Gerontology Research
Nancy L. Nadon

Animal Models of Geriatric Epilepsy
Lauren J. Murphree, Lynn M. Rundhaugen, and Kevin M. Kelly

Life and Death of Neurons in the Aging Cerebral Cortex
John H. Morrison and Patrick R. Hof

An In Vitro Model of Stroke-Induced Epilepsy: Elucidation of the Roles of Glutamate and Calcium in the Induction and Maintenance of Stroke-Induced Epileptogenesis
Robert J. DeLorenzo, David A. Sun, Robert E. Blair, and Sompong Sambati

Mechanisms of Action of Antiepileptic Drugs
H. Steve White, Misty D. Smith, and Karen S. Wilcox

Epidemiology and Outcomes of Status Epilepticus in the Elderly
Alan R. Towne

Diagnosing Epilepsy in the Elderly
R. Eugene Ramsay, Flavia M. Macias, and A. James Rowan

Pharmacoepidemiology in Community-Dwelling Elderly Taking Antiepileptic Drugs
Dan R. Berlowitz and Mary Jo V. Pugh

Use of Antiepileptic Medications in Nursing Homes
Judith Garrard, Susan L. Harms, Lynn E. Eberly, and Ilo E. Leppik

Age-Related Changes in Pharmacokinetics: Predictability and Assessment Methods
Emilio Perucca

Factors Affecting Antiepileptic Drug Pharmacokinetics in Community-Dwelling Elderly
James C. Cloyd, Susan Marino, and Angela K. Birnbaum

Pharmacokinetics of Antiepileptic Drugs in Elderly Nursing Home Residents
Angela K. Birnbaum

The Impact of Epilepsy on Older Veterans
Mary Jo V. Pugh, Dan R. Berlowitz, and Lewis Kazis

Risk and Predictability of Drug Interactions in the Elderly
René H. Levy and Carol Collins

Outcomes in Elderly Patients With Newly Diagnosed and Treated Epilepsy
Martin J. Brodie and Linda J. Stephen

Recruitment and Retention in Clinical Trials of the Elderly
Flavia M. Macias, R. Eugene Ramsay, and A. James Rowan

Treatment of Convulsive Status Epilepticus
David M. Treiman

Treatment of Nonconvulsive Status Epilepticus
Matthew C. Walker

Antiepileptic Drug Formulation and Treatment in the Elderly: Biopharmaceutical Considerations
Barry E. Gidal

INDEX

Volume 82

Inflammatory Mediators Leading to Protein Misfolding and Uncompetitive/Fast Off-Rate Drug Therapy for Neurodegenerative Disorders
Stuart A. Lipton, Zezong Gu, and Tomohiro Nakamura

Innate Immunity and Protective Neuroinflammation: New Emphasis on the Role of Neuroimmune Regulatory Proteins
M. Griffiths, J. W. Neal, and P. Gasque

Glutamate Release from Astrocytes in Physiological Conditions and in Neurodegenerative Disorders Characterized by Neuroinflammation
Sabino Vesce, Daniela Rossi, Liliana Brambilla, and Andrea Volterra

The High-Mobility Group Box 1 Cytokine Induces Transporter-Mediated Release of Glutamate from Glial Subcellular Particles (Gliosomes) Prepared from *In Situ*-Matured Astrocytes
Giambattista Bonanno, Luca Raiteri, Marco Milanese, Simona Zappettini, Edon Melloni, Marco Pedrazzi, Mario Passalacqua, Carlo Tacchetti, Cesare Usai, and Bianca Sparatore

The Role of Astrocytes and Complement System in Neural Plasticity
Milos Pekny, Ulrika Wilhelmsson, Yalda Rahpeymai Bogestål, and Marcela Pekna

New Insights into the Roles of Metalloproteinases in Neurodegeneration and Neuroprotection
A. J. Turner and N. N. Nalivaeva

Relevance of High-Mobility Group Protein Box 1 to Neurodegeneration
Silvia Fossati and Alberto Chiarugi

Early Upregulation of Matrix Metalloproteinases Following Reperfusion Triggers Neuroinflammatory Mediators in Brain Ischemia in Rat
Diana Amantea, Rossella Russo, Micaela Gliozzi, Vincenza Fratto, Laura Berliocchi, G. Bagetta, G. Bernardi, and M. Tiziana Corasaniti

The (Endo)Cannabinoid System in Multiple Sclerosis and Amyotrophic Lateral Sclerosis
Diego Centonze, Silvia Rossi, Alessandro Finazzi-Agrò, Giorgio Bernardi, and Mauro Maccarrone

Chemokines and Chemokine Receptors: Multipurpose Players in Neuroinflammation
Richard M. Ransohoff, LiPing Liu, and Astrid E. Cardona

Systemic and Acquired Immune Responses in Alzheimer's Disease
Markus Britschgi and Tony Wyss-Coray

Neuroinflammation in Alzheimer's Disease and Parkinson's Disease: Are Microglia Pathogenic in Either Disorder?
Joseph Rogers, Diego Mastroeni, Brian Leonard, Jeffrey Joyce, and Andrew Grover

Cytokines and Neuronal Ion Channels in Health and Disease
Barbara Viviani, Fabrizio Gardoni, and Marina Marinovich

Cyclooxygenase-2, Prostaglandin E_2, and Microglial Activation in Prion Diseases
Luisa Minghetti and Maurizio Pocchiari

Glia Proinflammatory Cytokine Upregulation as a Therapeutic Target for Neurodegenerative Diseases: Function-Based and Target-Based Discovery Approaches
Linda J. Van Eldik, Wendy L. Thompson, Hantamalala Ralay Ranaivo, Heather A. Behanna, and D. Martin Watterson

Oxidative Stress and the Pathogenesis of Neurodegenerative Disorders
Ashley Reynolds, Chad Laurie, R. Lee Mosley, and Howard E. Gendelman

Differential Modulation of Type 1 and Type 2 Cannabinoid Receptors Along the Neuroimmune Axis
Sergio Oddi, Paola Spagnuolo, Monica Bari, Antonella D'Agostino, and Mauro Maccarrone

Effects of the HIV-1 Viral Protein Tat on Central Neurotransmission: Role of Group I Metabotropic Glutamate Receptors
Elisa Neri, Veronica Musante, and Anna Pittaluga

Evidence to Implicate Early Modulation of Interleukin-1β Expression in the Neuroprotection Afforded by 17β-Estradiol in Male Rats Undergone Transient Middle Cerebral Artery Occlusion
Olga Chiappetta, Micaela Gliozzi, Elisa Siviglia, Diana Amantea, Luigi A. Morrone, Laura Berliocchi, G. Bagetta, and M. Tiziana Corasaniti

A Role for Brain Cyclooxygenase-2 and Prostaglandin-E_2 in Migraine: Effects of Nitroglycerin
Cristina Tassorelli, Rosaria Greco, Marie Therèse Armentero, Fabio Blandini, Giorgio Sandrini, and Giuseppe Nappi

The Blockade of K+-ATP Channels has Neuroprotective Effects in an *In Vitro* Model of Brain Ischemia
Robert Nisticò, Silvia Piccirilli, L. Sebastianelli, Giuseppe Nisticò, G. Bernardi, and N. B. Mercuri

Retinal Damage Caused by High Intraocular Pressure-Induced Transient Ischemia is Prevented by Coenzyme Q10 in Rat
Carlo Nucci, Rosanna Tartaglione, Angelica Cerulli, R. Mancino, A. Spanò, Federica Cavaliere, Laura Rombolà, G. Bagetta, M. Tiziana Corasaniti, and Luigi A. Morrone

Evidence Implicating Matrix Metalloproteinases in the Mechanism Underlying Accumulation of IL-1β and Neuronal Apoptosis in the Neocortex of HIV/gp120-Exposed Rats
Rossella Russo, Elisa Siviglia, Micaela Gliozzi, Diana Amantea, Annamaria Paoletti, Laura Berliocchi, G. Bagetta, and M. Tiziana Corasaniti

Neuroprotective Effect of Nitroglycerin in a Rodent Model of Ischemic Stroke: Evaluation of Bcl-2 Expression
Rosaria Greco, Diana Amantea, Fabio Blandini, Giuseppe Nappi, Giacinto Bagetta, M. Tiziana Corasaniti, and Cristina Tassorelli

INDEX

Volume 83

Gender Differences in Pharmacological Response
Gail D. Anderson

Epidemiology and Classification of Epilepsy: Gender Comparisons
John C. McHugh and Norman Delanty

Hormonal Influences on Seizures: Basic Neurobiology
Cheryl A. Frye

Catamenial Epilepsy
Patricia E. Penovich and Sandra Helmers

Epilepsy in Women: Special Considerations for Adolescents
Mary L. Zupanc and Sheryl Haut

Contraception in Women with Epilepsy: Pharmacokinetic Interactions, Contraceptive Options, and Management
Caryn Dutton and Nancy Foldvary-Schaefer

Reproductive Dysfunction in Women with Epilepsy: Menstrual Cycle Abnormalities, Fertility, and Polycystic Ovary Syndrome
Jürgen Bauer and Déirdre Cooper-Mahkorn

Sexual Dysfunction in Women with Epilepsy: Role of Antiepileptic Drugs and Psychotropic Medications
Mary A. Gutierrez, Romila Mushtaq, and Glen Stimmel

Pregnancy in Epilepsy: Issues of Concern
John DeToledo

Teratogenicity and Antiepileptic Drugs: Potential Mechanisms
Mark S. Yerby

Antiepileptic Drug Teratogenesis: What are the Risks for Congenital Malformations and Adverse Cognitive Outcomes?
Cynthia L. Harden

Teratogenicity of Antiepileptic Drugs: Role of Pharmacogenomics
Raman Sankar and Jason T. Lerner

Antiepileptic Drug Therapy in Pregnancy I: Gestation-Induced Effects on AED Pharmacokinetics
Page B. Pennell and Collin A. Hovinga

Antiepileptic Drug Therapy in Pregnancy II: Fetal and Neonatal Exposure
Collin A. Hovinga and Page B. Pennell

Seizures in Pregnancy: Diagnosis and Management
Robert L. Beach and Peter W. Kaplan

Management of Epilepsy and Pregnancy: An Obstetrical Perspective
Julian N. Robinson and Jane Cleary-Goldman

Pregnancy Registries: Strengths, Weaknesses, and Bias Interpretation of Pregnancy Registry Data
Marianne Cunnington and John Messenheimer

Bone Health in Women With Epilepsy: Clinical Features and Potential Mechanisms
Alison M. Pack and Thaddeus S. Walczak

Metabolic Effects of AEDs: Impact on Body Weight, Lipids and Glucose Metabolism
Raj D. Sheth and Georgia Montouris

Psychiatric Comorbidities in Epilepsy
W. Curt Lafrance, Jr., Andres M. Kanner, and Bruce Hermann

Issues for Mature Women with Epilepsy
Cynthia L. Harden

Pharmacodynamic and Pharmacokinetic Interactions of Psychotropic Drugs with Antiepileptic Drugs
Andres M. Kanner and Barry E. Gidal

Health Disparities in Epilepsy: How Patient-Oriented Outcomes in Women Differ from Men
Frank Gilliam

INDEX

Volume 84

Normal Brain Aging: Clinical, Immunological, Neuropsychological, and Neuroimaging Features
Maria T. Caserta, Yvonne Bannon, Francisco Fernandez, Brian Giunta, Mike R. Schoenberg, and Jun Tan

Subcortical Ischemic Cerebrovascular Dementia
Uma Menon and Roger E. Kelley

Cerebrovascular and Cardiovascular Pathology in Alzheimer's Disease
Jack C. de la Torre

Neuroimaging of Cognitive Impairments in Vascular Disease
Carol Di Perri, Turi O. Dalaker, Mona K. Beyer, and Robert Zivadinov

Contributions of Neuropsychology and Neuroimaging to Understanding Clinical Subtypes of Mild Cognitive Impairment
Amy J. Jak, Katherine J. Bangen, Christina E. Wierenga, Lisa Delano-Wood, Jody Corey-Bloom, and Mark W. Bondi

Proton Magnetic Resonance Spectroscopy in Dementias and Mild Cognitive Impairment
H. Randall Griffith, Christopher C. Stewart, and Jan A. den Hollander

Application of PET Imaging to Diagnosis of Alzheimer's Disease and Mild Cognitive Impairment
James M. Noble and Nikolaos Scarmeas

The Molecular and Cellular Pathogenesis of Dementia of the Alzheimer's Type: An Overview
Francisco A. Luque and Stephen L. Jaffe

Alzheimer's Disease Genetics: Current Status and Future Perspectives
Lars Bertram

Frontotemporal Lobar Degeneration: Insights from Neuropsychology and Neuroimaging
Andrea C. Bozoki and Muhammad U. Farooq

Lewy Body Dementia
Jennifer C. Hanson and Carol F. Lippa

Dementia in Parkinson's Disease
Bradley J. Robottom and William J. Weiner

Early Onset Dementia
Halim Fadil, Aimee Borazanci, Elhachmia Ait Ben Haddou, Mohamed Yahyaoui, Elena Korniychuk, Stephen L. Jaffe, and Alireza Minagar

Normal Pressure Hydrocephalus
Glen R. Finney

Reversible Dementias
Anahid Kabasakalian and Glen R. Finney

INDEX

Volume 85

Involvement of the Prefrontal Cortex in Problem Solving
Hajime Mushiake, Kazuhiro Sakamoto, Naohiro Saito, Toshiro Inui, Kazuyuki Aihara, and Jun Tanji

GluK1 Receptor Antagonists and Hippocampal Mossy Fiber Function
Robert Nisticò, Sheila Dargan, Stephen M. Fitzjohn, David Lodge, David E. Jane, Graham L. Collingridge, and Zuner A. Bortolotto

Monoamine Transporter as a Target Molecule for Psychostimulants
Ichiro Sora, BingJin Li, Setsu Fumushima, Asami Fukui, Yosefu Arime, Yoshiyuki Kasahara, Hiroaki Tomita, and Kazutaka Ikeda

Targeted Lipidomics as a Tool to Investigate Endocannabinoid Function
Giuseppe Astarita, Jennifer Geaga, Faizy Ahmed, and Daniele Piomelli

The Endocannabinoid System as a Target for Novel Anxiolytic and Antidepressant Drugs
Silvana Gaetani, Pasqua Dipasquale, Adele Romano, Laura Righetti, Tommaso Cassano, Daniele Piomelli, and Vincenzo Cuomo

$GABA_A$ Receptor Function and Gene Expression During Pregnancy and Postpartum
Giovanni Biggio, Maria Cristina Mostallino, Paolo Follesa, Alessandra Concas, and Enrico Sanna

Early Postnatal Stress and Neural Circuit Underlying Emotional Regulation
Machiko Matsumoto, Mitsuhiro Yoshioka, and Hiroko Togashi

Roles of the Histaminergic Neurotransmission on Methamphetamine-Induced Locomotor Sensitization and Reward: A Study of Receptors Gene Knockout Mice
Naoko Takino, Eiko Sakurai, Atsuo Kuramasu, Nobuyuki Okamura, and Kazuhiko Yanai

Developmental Exposure to Cannabinoids Causes Subtle and Enduring Neurofunctional Alterations
Patrizia Campolongo, Viviana Trezza, Maura Palmery, Luigia Trabace, and Vincenzo Cuomo

Neuronal Mechanisms for Pain-Induced Aversion: Behavioral Studies Using a Conditioned Place Aversion Test
Masabumi Minami

Bv8/Prokineticins and their Receptors: A New Pronociceptive System
Lucia Negri, Roberta Lattanzi, Elisa Giannini, Michela Canestrelli, Annalisa Nicotra, and Pietro Melchiorri

$P2Y_6$-Evoked Microglial Phagocytosis
Kazuhide Inoue, Schuichi Koizumi, Ayako Kataoka, Hidetoshi Tozaki-Saitoh, and Makoto Tsuda

PPAR and Pain
Takehiko Maeda and Shiroh Kishioka

Involvement of Inflammatory Mediators in Neuropathic Pain Caused by Vincristine
Norikazu Kiguchi, Takehiko Maeda, Yuka Kobayashi, Fumihiro Saika, and Shiroh Kishioka

Nociceptive Behavior Induced by the Endogenous Opioid Peptides Dynorphins in Uninjured Mice: Evidence with Intrathecal *N*-ethylmaleimide Inhibiting Dynorphin Degradation
Koichi Tan-No, Hiroaki Takahashi, Osamu Nakagawasai, Fukie Niijima, Shinobu Sakurada, Georgy Bakalkin, Lars Terenius, and Takeshi Tadano

Mechanism of Allodynia Evoked by Intrathecal Morphine-3-Glucuronide in Mice
Takaaki Komatsu, Shinobu Sakurada, Sou Katsuyama, Kengo Sanai, and Tsukasa Sakurada

(–)-Linalool Attenuates Allodynia in Neuropathic Pain Induced by Spinal Nerve Ligation in C57/Bl6 Mice
Laura Berliocchi, Rossella Russo, Alessandra Levato, Vincenza Fratto, Giacinto Bagetta, Shinobu Sakurada, Tsukasa Sakurada, Nicola Biagio Mercuri, and Maria Tiziana Corasaniti

Intraplantar Injection of Bergamot Essential Oil into the Mouse Hindpaw: Effects on Capsaicin-Induced Nociceptive Behaviors
Tsukasa Sakurada, Hikari Kuwahata, Soh Katsuyama, Takaaki Komatsu, Luigi A. Morrone, M. Tiziana Corasaniti, Giacinto Bagetta, and Shinobu Sakurada

New Therapy for Neuropathic Pain
Hirokazu Mizoguchi, Chizuko Watanabe, Akihiko Yonezawa, and Shinobu Sakurada

Regulated Exocytosis from Astrocytes: Physiological and Pathological Related Aspects
Corrado Calì, Julie Marchaland, Paola Spagnuolo, Julien Gremion, and Paola Bezzi

Glutamate Release from Astrocytic Gliosomes Under Physiological and Pathological Conditions
Marco Milanese, Tiziana Bonifacino, Simona Zappettini, Cesare Usai, Carlo Tacchetti, Mario Nobile, and Giambattista Bonanno

Neurotrophic and Neuroprotective Actions of an Enhancer of Ganglioside Biosynthesis
Jin-ichi Inokuchi

Involvement of Endocannabinoid Signaling in the Neuroprotective Effects of Subtype 1 Metabotropic Glutamate Receptor Antagonists in Models of Cerebral Ischemia
Elisa Landucci, Francesca Boscia, Elisabetta Gerace, Tania Scartabelli, Andrea Cozzi, Flavio Moroni, Guido Mannaioni, and Domenico E. Pellegrini-Giampietro

NF-kappaB Dimers in the Regulation of Neuronal Survival
Ilenia Sarnico, Annamaria Lanzillotta, Marina Benarese, Manuela Alghisi, Cristina Baiguera, Leontino Battistin, PierFranco Spano, and Marina Pizzi

Oxidative Stress in Stroke Pathophysiology: Validation of Hydrogen Peroxide Metabolism as a Pharmacological Target to Afford Neuroprotection
Diana Amantea, Maria Cristina Marrone, Robert Nisticò, Mauro Federici, Giacinto Bagetta, Giorgio Bernardi, and Nicola Biagio Mercuri

Role of Akt and ERK Signaling in the Neurogenesis following Brain Ischemia
Norifumi Shioda, Feng Han, and Kohji Fukunaga

Prevention of Glutamate Accumulation and Upregulation of Phospho-Akt may Account for Neuroprotection Afforded by Bergamot Essential Oil against Brain Injury Induced by Focal Cerebral Ischemia in Rat
Diana Amantea, Vincenza Fratto, Simona Maida, Domenicantonio Rotiroti, Salvatore Ragusa, Giuseppe Nappi, Giacinto Bagetta, and Maria Tiziana Corasaniti

Identification of Novel Pharmacological Targets to Minimize Excitotoxic Retinal Damage
Rossella Russo, Domenicantonio Rotiroti, Cristina Tassorelli, Carlo Nucci, Giacinto Bagetta, Massimo Gilberto Bucci, Maria Tiziana Corasaniti, and Luigi Antonio Morrone

Index

Volume 86

Section One: Hybrid Bionic Systems
EMG-Based and Gaze-Tracking-Based Man–Machine Interfaces
Federico Carpi and Danilo De Rossi

Bidirectional Interfaces with the Peripheral Nervous System
Silvestro Micera and Xavier Navarro

Interfacing Insect Brain for Space Applications
Giovanni Di Pino, Tobias Seidl, Antonella Benvenuto, Fabrizio Sergi, Domenico Campolo, Dino Accoto, Paolo Maria Rossini, and Eugenio Guglielmelli

Section Two: Meet the Brain
Meet the Brain: Neurophysiology
John Rothwell

Fundamentals of Electroencefalography, Magnetoencefalography, and Functional Magnetic Resonance Imaging
Claudio Babiloni, Vittorio Pizzella, Cosimo Del Gratta, Antonio Ferretti, and Gian Luca Romani

Implications of Brain Plasticity to Brain–Machine Interfaces Operation: A Potential Paradox?
Paolo Maria Rossini

Section Three: Brain Machine Interfaces, A New Brain-to-Environment Communication Channel
An Overview of BMIs
Francisco Sepulveda

Neurofeedback and Brain–Computer Interface: Clinical Applications
Niels Birbaumer, Ander Ramos Murguialday, Cornelia Weber, and Pedro Montoya

Flexibility and Practicality: Graz Brain–Computer Interface Approach
Reinhold Scherer, Gernot R. Müller-Putz, and Gert Pfurtscheller

On the Use of Brain–Computer Interfaces Outside Scientific Laboratories: Toward an Application in Domotic Environments
F. Babiloni, F. Cincotti, M. Marciani, S. Salinari, L. Astolfi, F. Aloise, F. De Vico Fallani, and D. Mattia

Brain–Computer Interface Research at the Wadsworth Center: Developments in Noninvasive Communication and Control
Dean J. Krusienski and Jonathan R. Wolpaw

Watching Brain TV and Playing Brain Ball: Exploring Novel BCL Strategies Using Real–Time Analysis of Human Intercranial Data
Karim Jerbi, Samson Freyermuth, Lorella Minotti, Philippe Kahane, Alain Berthoz, and Jean-Philippe Lachaux

Section Four: Brain-Machine Interfaces and Space
Adaptive Changes of Rhythmic EEG Oscillations in Space: Implications for Brain–Machine Interface Applications
G. Cheron, A. M. Cebolla, M. Petieau, A. Bengoetxea, E. Palmero-Soler, A. Leroy, and B. Dan

Validation of Brain–Machine Interfaces During Parabolic Flight
José del R. Millán, Pierre W. Ferrez, and Tobias Seidl

Matching Brain–Machine Interface Performance to Space Applications
Luca Citi, Oliver Tonet, and Martina Marinelli

Brain–Machine Interfaces for Space Applications—Research, Technological Development, and Opportunities
Leopold Summerer, Dario Izzo, and Luca Rossini

INDEX

Volume 87

Peripheral Nerve Repair and Regeneration Research: A Historical Note
Bruno Battiston, Igor Papalia, Pierluigi Tos, and Stefano Geuna

Development of the Peripheral Nerve
Suleyman Kaplan, Ersan Odaci, Bunyami Unal, Bunyamin Sahin, and Michele Fornaro

Histology of the Peripheral Nerve and Changes Occurring During Nerve Regeneration
Stefano Geuna, Stefania Raimondo, Giulia Ronchi, Federica Di Scipio, Pierluigi Tos, Krzysztof Czaja, and Michele Fornaro

Methods and Protocols in Peripheral Nerve Regeneration Experimental Research: Part I—Experimental Models
Pierluigi Tos, Giulia Ronchi, Igor Papalia, Vera Sallen, Josette Legagneux, Stefano Geuna, and Maria G. Giacobini-Robecchi

Methods and Protocols in Peripheral Nerve Regeneration Experimental Research: Part II—Morphological Techniques
Stefania Raimondo, Michele Fornaro, Federica Di Scipio, Giulia Ronchi, Maria G. Giacobini-Robecchi, and Stefano Geuna

Methods and Protocols in Peripheral Nerve Regeneration Experimental Research: Part III—Electrophysiological Evaluation
Xavier Navarro and Esther Udina

Methods and Protocols in Peripheral Nerve Regeneration Experimental Research: Part IV—Kinematic Gait Analysis to Quantify Peripheral Nerve Regeneration in the Rat
Luís M. Costa, Maria J. Simões, Ana C. Maurício and Artur S.P. Varejão

Current Techniques and Concepts in Peripheral Nerve Repair
Maria Siemionow and Grzegorz Brzezicki

Artificial Scaffolds for Peripheral Nerve Reconstruction
Valeria Chiono, Chiara Tonda-Turo, and Gianluca Ciardelli

Conduit Luminal Additives for Peripheral Nerve Repair
Hede Yan, Feng Zhang, Michael B. Chen, and William C. Lineaweaver

Tissue Engineering of Peripheral Nerves
Bruno Battiston, Stefania Raimondo, Pierluigi Tos, Valentina Gaidano, Chiara Audisio, Anna Scevola, Isabelle Perroteau, and Stefano Geuna

Mechanisms Underlying The End-to-Side Nerve Regeneration
Eleana Bontioti and Lars B. Dahlin

Experimental Results in End-To-Side Neurorrhaphy
Alexandros E. Beris and Marios G. Lykissas

End-to-Side Nerve Regeneration: From the Laboratory Bench to Clinical Applications
Pierluigi Tos, Stefano Artiaco, Igor Papalia, Ignazio Marcoccio, Stefano Geuna, and Bruno Battiston

Novel Pharmacological Approaches to Schwann Cells as Neuroprotective Agents for Peripheral Nerve Regeneration
Valerio Magnaghi, Patrizia Procacci, and Ada Maria Tata

Melatonin and Nerve Regeneration
Ersan Odaci and Suleyman Kaplan

Transthyretin: An Enhancer of Nerve Regeneration
Carolina E. Fleming, Fernando Milhazes Mar, Filipa Franquinho, and Mónica M. Sousa

Enhancement of Nerve Regeneration and Recovery by Immunosuppressive Agents
Damien P. Kuffler

The Role of Collagen in Peripheral Nerve Repair
Guido Koopmans, Birgit Hasse, and Nektarios Sinis

Gene Therapy Perspectives for Nerve Repair
Serena Zacchigna and Mauro Giacca

Use of Stem Cells for Improving Nerve Regeneration
Giorgio Terenghi, Mikael Wiberg, and Paul J. Kingham

Transplantation of Olfactory Ensheathing Cells for Peripheral Nerve Regeneration
Christine Radtke, Jeffery D. Kocsis, and Peter M. Vogt

Manual Stimulation of Target Muscles has Different Impact on Functional Recovery after Injury of Pure Motor or Mixed Nerves
Nektarios Sinis, Thodora Manoli, Frank Werdin, Armin Kraus, Hans E. Schaller, Orlando Guntinas-Lichius, Maria Grosheva, Andrey Irintchev, Emanouil Skouras, Sarah Dunlop, and Doychin N. Angelov

Electrical Stimulation for Improving Nerve Regeneration: Where do we Stand?
Tessa Gordon, Olewale A. R. Sulaiman, and Adil Ladak

Phototherapy in Peripheral Nerve Injury: Effects on Muscle Preservation and Nerve Regeneration
Shimon Rochkind, Stefano Geuna, and Asher Shainberg

Age-Related Differences in the Reinnervation after Peripheral Nerve Injury
Uroš Kovačič, Janez Sketelj, and Fajko F. Bajrović

Neural Plasticity After Nerve Injury and Regeneration
Xavier Navarro

Future Perspective in Peripheral Nerve Reconstruction
Lars Dahlin, Fredrik Johansson, Charlotta Lindwall, and Martin Kanje

INDEX

Volume 88

Effects Of Psychostimulants On Neurotrophins: Implications For Psychostimulant-Induced Neurotoxicity
Francesco Angelucci, Valerio Ricci, Gianfranco Spalletta, Carlo Caltagirone, Aleksander A. Mathé, and Pietro Bria

Dosing Time-Dependent Actions of Psychostimulants
H. Manev and T. Uz

Dopamine-Induced Behavioral Changes and Oxidative Stress in Methamphetamine-Induced Neurotoxicity
Taizo Kita, Ikuko Miyazaki, Masato Asanuma, Mika Takeshima, and George C. Wagner

Acute Methamphetamine Intoxication: Brain Hyperthermia, Blood–Brain Barrier, Brain Edema, and morphological cell abnormalities
Eugene A. Kiyatkin and Hari S. Sharma

Molecular Bases of Methamphetamine-Induced Neurodegeneration
Jean Lud Cadet and Irina N. Krasnova

Involvement of Nicotinic Receptors in Methamphetamine- and MDMA-Induced Neurotoxicity: Pharmacological Implications
E. Escubedo, J. Camarasa, C. Chipana, S. García-Ratés, and D.Pubill

Ethanol Alters the Physiology of Neuron–Glia Communication
Antonio González and Ginés M. Salido

Therapeutic Targeting of "DARPP-32": A Key Signaling Molecule in the Dopiminergic Pathway for the Treatment of Opiate Addiction
Supriya D. Mahajan, Ravikumar Aalinkeel, Jessica L. Reynolds, Bindukumar B. Nair, Donald E. Sykes, Zihua Hu, Adela Bonoiu, Hong Ding, Paras N. Prasad, and Stanley A. Schwartz

Pharmacological and Neurotoxicological Actions Mediated By Bupropion and Diethylpropion
Hugo R. Arias, Abel Santamaría, and Syed F. Ali

Neural and Cardiac Toxicities Associated With 3,4-Methylenedioxymethamphetamine (MDMA)
Michael H. Baumann and Richard B. Rothman

Cocaine-Induced Breakdown of the Blood–Brain Barrier and Neurotoxicity
Hari S. Sharma, Dafin Muresanu, Aruna Sharma, and Ranjana Patnaik

Cannabinoid Receptors in Brain: Pharmacogenetics, Neuropharmacology, Neurotoxicology, and Potential Therapeutic Applications
Emmanuel S. Onaivi

Intermittent Dopaminergic Stimulation causes Behavioral Sensitization in the Addicted Brain and Parkinsonism
Francesco Fornai, Francesca Biagioni, Federica Fulceri, Luigi Murri, Stefano Ruggieri, Antonio Paparelli

The Role of the Somatotrophic Axis in Neuroprotection and Neuroregeneration of the Addictive Brain
Fred Nyberg

INDEX

Volume 89

Molecular Profiling of Striatonigral and Striatopallidal Medium Spiny Neurons: Past, Present, and Future
Mary Kay Lobo

BAC to Degeneration: Bacterial Artificial Chromosome (Bac)-Mediated Transgenesis for Modeling Basal Ganglia Neurodegenerative Disorders
Xiao-Hong Lu

Behavioral Outcome Measures for the Assessment of Sensorimotor Function in Animal Models of Movement Disorders
Sheila M. Fleming

The Role of DNA Methylation in the Central Nervous System and Neuropsychiatric Disorders
Jian Feng and Guoping Fan

Heritability of Structural Brain Traits: An Endo-phenotype Approach to Deconstruct Schizophrenia
Nil Kaymaz and J. Van Os

The Role of Striatal NMDA Receptors in Drug Addiction
Yao-Ying Ma, Carlos Cepeda, and Cai-Lian Cui

Deciphering Rett Syndrome With Mouse Genetics, Epigenomics, and Human Neurons
Jifang Tao, Hao Wu, and Yi Eve Sun

INDEX

Volume 90

Part I: Introduction

Introductory Remarks on the History and Current Applications of TCS
Matthew B. Stern

Method and Validity of Transcranial Sonography in Movement Disorders
David Školoudík and Uwe Walter

Transcranial Sonography—Anatomy
Heiko Huber

Part II: Transcranial Sonography in Parkinsons Disease

Transcranial Sonography in Relation to SPECT and MIBG
Yoshinori Kajimoto, Hideto Miwa and Tomoyoshi Kondo

Diagnosis of Parkinson's Disease—Transcranial Sonography in Relation to MRI
Ludwig Niehaus and Kai Boelmans

Early Diagnosis of Parkinson's Disease
Alexandra Gaenslen and Daniela Berg

Transcranial Sonography in the Premotor Diagnosis of Parkinson's Disease
Stefanie Behnke, Ute Schroder and Daniela Berg

Pathophysiology of Transcranial Sonography Signal Changes in the Human Substantia Nigra
K. L. Double, G. Todd and S. R. Duma

Transcranial Sonography for the Discrimination of Idiopathic Parkinson's Disease from the Atypical Parkinsonian Syndromes
A. E. P. Bouwmans, A. M. M. Vlaar, K. Srulijes, W. H. Mess AND W. E. J. Weber

Transcranial Sonography in the Discrimination of Parkinson's Disease Versus Vascular Parkinsonism
Pablo Venegas-Francke

TCS in Monogenic Forms of Parkinson's Disease
Kathrin Brockmann and Johann Hagenah

Part III—Transcranial Sonography in other Movement Disorders and Depression

Transcranial Sonography in Brain Disorders with Trace Metal Accumulation
Uwe Walter

Transcranial Sonography in Dystonia
Alexandra Gaenslen

Transcranial Sonography in Essential Tremor
Heike Stockner and Isabel Wurster

VII—Transcranial Sonography in Restless Legs Syndrome
Jana Godau and Martin Sojer

Transcranial Sonography in Ataxia
Christos Krogias, Thomas Postert and Jens Eyding

Transcranial Sonography in Huntington's Disease
Christos Krogias, Jens Eyding and Thomas Postert

Transcranial Sonography in Depression
Milija D. Mijajlovic

Part IV: Future Applications and Conclusion

Transcranial Sonography-Assisted Stereotaxy and Follow-Up of Deep Brain Implants in Patients with Movement Disorders
Uwe Walter

Conclusions
Daniela Berg

INDEX

Volume 91

The Role of microRNAs in Drug Addiction: A Big Lesson from Tiny Molecules
Andrzej Zbigniew Pietrzykowski

The Genetics of Behavioral Alcohol Responses in *Drosophila*
Aylin R. Rodan and Adrian Rothenfluh

Neural Plasticity, Human Genetics, and Risk for Alcohol Dependence
Shirley Y. Hill

Using Expression Genetics to Study the Neurobiology of Ethanol and Alcoholism
Sean P. Farris, Aaron R. Wolen and Michael F. Miles

Genetic Variation and Brain Gene Expression in Rodent Models of Alcoholism: Implications for Medication Development
Karl Björk, Anita C. Hansson and W. olfgang H. Sommer

Identifying Quantitative Trait Loci (QTLs) and Genes (QTGs) for Alcohol-Related Phenotypes in Mice
Lauren C. Milner and Kari J. Buck

Glutamate Plasticity in the Drunken Amygdala: The Making of an Anxious Synapse
Brian A. Mccool, Daniel T. Christian, Marvin R. Diaz and Anna K. Läck

Ethanol Action on Dopaminergic Neurons in the Ventral Tegmental Area: Interaction with Intrinsic Ion Channels and Neurotransmitter Inputs
Hitoshi Morikawa and Richard A. Morrisett

Alcohol and the Prefrontal Cortex
Kenneth Abernathy, L. Judson Chandler and John J. Woodward

BK Channel and Alcohol, A Complicated Affair
Gilles Erwan Martin

A Review of Synaptic Plasticity at Purkinje Neurons with a Focus on Ethanol-Induced Cerebellar Dysfunction
C. Fernando Valenzuela, Britta Lindquist and Paula A. Zflmudio-Bulcock

INDEX

Volume 92

The Development of the Science of Dreaming
Claude Gottesmann

Dreaming as Inspiration: Evidence from Religion, Philosophy, Literature, and Film
Kelly Bulkeley

Developmental Perspective: Dreaming Across the Lifespan and What This Tells Us
Melissa M. Burnham and Christian Conte

REM and NREM Sleep Mentation
Patrick Mcnamara, Patricia Johnson, Deirdre McLaren, Erica Harris,Catherine Beauharnais and Sanford Auerbach

Neuroimaging of Dreaming: State of the Art and Limitations
Caroline Kussé, Vincenzo Muto, Laura Mascetti, Luca Matarazzo, Ariane Foret, Anahita Shaffii-Le Bourdiec and Pierre Maquet

Memory Consolidation, The Diurnal Rhythm of Cortisol, and The Nature of Dreams: A New Hypothesis
Jessica D. Payne

Characteristics and Contents of Dreams
Michael Schredl

Trait and Neurobiological Correlates of Individual Differences in Dream Recall and Dream Content
Mark Blagrove and Edward F. Pace-Schott

Consciousness in Dreams
David Kahn and Tzivia Gover

The Underlying Emotion and the Dream: Relating Dream Imagery to the Dreamer's Underlying Emotion can Help Elucidate the Nature of Dreaming
Ernest Hartmann

Dreaming, Handedness, and Sleep Architecture: Interhemispheric Mechanisms
Stephen D. Christman and Ruth E. Propper

To What Extent Do Neurobiological Sleep-Waking Processes Support Psychoanalysis?
Claude Gottesmann

The Use of Dreams in Modern Psychotherapy
Clara E. Hill and Sarah Knox

INDEX

Volume 93

Underlying Brain Mechanisms that Regulate Sleep-Wakefulness Cycles
Irma Gvilia

What Keeps Us Awake?—the Role of Clocks and Hourglasses, Light, and Melatonin
Christian Cajochen, Sarah Chellappa and Christina Schmidt

Suprachiasmatic Nucleus and Autonomic Nervous System Influences on Awakening From Sleep
Andries Kalsbeek, Chun-xia Yi, Susanne E. la Fleur, Ruud m. Buijs, and Eric Fliers

Preparation for Awakening: Self-Awakening Vs. Forced Awakening: Preparatory Changes in the Pre-Awakening Period
Mitsuo Hayashi, Noriko Matsuura and Hiroki Ikeda

Circadian and Sleep Episode Duration Influences on Cognitive Performance Following the Process of Awakening
Robert L. Matchock

The Cortisol Awakening Response in Context
Angela Clow, Frank Hucklebridge and Lisa Thorn

Causes and Correlates of Frequent Night Awakenings in Early Childhood
Amy Jo Schwichtenberg and Beth Goodlin-Jones

Pathologies of Awakenings: The Clinical Problem of Insomnia Considered From Multiple Theory Levels
Douglas E. Moul

The Neurochemistry of Awakening: Findings from Sleep Disorder Narcolepsy
Seiji Nishino and Yohei Sagawa

INDEX

Volume 94

$5\text{-}HT_6$ Medicinal Chemistry
Kevin G. Liu and Albert J. Robichaud

Patents
Nicolas Vincent Ruiz and Gloria Oranias

$5\text{-}HT_6$ Receptor Charactertization
Teresa Riccioni

$5\text{-}HT_6$ Receptor Signal Transduction: Second Messenger Systems
Xavier Codony, Javier Burgueño, Maria Javier Ramírez and José Miguel Vela

Electrophysiology of $5\text{-}HT_6$ Receptors
Annalisa Tassone, Graziella Madeo, Giuseppe Sciamanna, Antonio Pisani and Paola Bonsi

Genetic Variations and Association
Massimo Gennarelli and Annamaria Cattaneo

Pharmacokinetics of $5\text{-}HT_6$ Receptor Ligands
Angelo Mancinelli

INDEX

Volume 95

Introductory Remarks: Catechol-O-Methyltransferase Inhibition–An Innovative Approach to Enhance L-dopa Therapy in Parkinson's Disease with Dual Enzyme Inhibition
Erkki Nissinen

The Catechol-O-Methyltransferase Gene: its Regulation and Polymorphisms
Elizabeth M. Tunbridge

Distribution and Functions of Catechol-O-Methyltransferase Proteins: Do Recent Findings Change the Picture?
Timo T. Myöhänen and Pekka T. Männistö

Catechol-O-Methyltransferase Enzyme: Cofactor S-Adenosyl-L-Methionine and Related Mechanisms
Thomas Müller

Biochemistry and Pharmacology of Catechol-O-Methyltransferase Inhibitors
Erkki nissinen and Pekka T. Männisto

The Chemistry of Catechol-O-Methyltransferase Inhibitors
David A. Learmonth, László E. Kiss, and Patrício Soares-da-Silva

Toxicology and Safety of COMT Inhibitors
Kristiina Haasio

Catechol-O-Methyltransferase Inhibitors in Preclinical Models as Adjuncts of L-dopa Treatment
Concepció Marin and J. A. Obeso

Problems with the Present Inhibitors and a Relevance of New and Improved COMT Inhibitors in Parkinson's Disease
Seppo Kaakkola

Catechol-O-Methyltransferase and Pain
Oleg Kambur and Pekka T. Männistö

INDEX

Volume 96

The Central Role of $5\text{-}HT_6$ Receptors in Modulating Brain Neurochemistry
Lee A. Dawson

$5\text{-}HT_6$ Receptor Memory and Amnesia: Behavioral Pharmacology – Learning and Memory Processes
Alfredo Meneses, G. Pérez-García, R. Tellez, T. Ponce-Lopez and C. Castillo

Behavioral Pharmacology: Potential Antidepressant and Anxiolytic Properties
Anna Wesołowska and Magdalena Jastrzbska-Wisek

The 5-HT_6 Receptor as a Target for Developing Novel Antiobesity Drugs
David Heal, Jane Gosden and Sharon Smith

Behavioral and Neurochemical Pharmacology of 5-HT_6 Receptors Related to Reward and Reinforcement
Gaetano Di Chiara, Valentina Valentini and Sandro Fenu

5-HT_6 Receptor Ligands and their Antipsychotic Potential
Jørn Arnt and Christina Kurre Olsen

5-HT_6 Receptor Ligands as Antidementia Drugs
Ellen Siobhan Mitchell

Other 5-HT_6 Receptor-Mediated Effects
Franco Borsini

INDEX

Volume 97

Behavioral Pharmacology of Orofacial Movement Disorders
Noriaki Koshikawa, Satoshi Fujita and Kazunori Adachi

Regulation of Orofacial Movement: Dopamine Receptor Mechanisms and Mutant Models
John L. Waddington, Gerard J. O'Sullivan and Katsunori Tomiyama

Regulation of Orofacial Movement: Amino Acid Mechanisms and Mutant Models
Katsunori Tomiyama, Colm M.P. O'Tuathaigh, and John L. Waddington

The Trigeminal Circuits Responsible for Chewing
Karl-Gunnar Westberg and Arlette Kolta

Ultrastructural Basis for Craniofacial Sensory Processing in the Brainstem
Yong Chul Bae and Atsushi Yoshida

Mechanisms of Nociceptive Transduction and Transmission: A Machinery for Pain Sensation and Tools for Selective Analgesia
Alexander M. Binshtok

Peripheral and Central Mechanisms of Orofacial Inflammatory Pain
Barry J. Sessle

The Role of Trigeminal Interpolaris-Caudalis Transition Zone in Persistent Orofacial Pain
Ke Ren and Ronald Dubner

Physiological Mechanisms of Neuropathic Pain: The Orofacial Region
Koichi Iwata, Yoshiki Imamura, Kuniya Honda and Masamichi Shinoda

Neurobiology of Estrogen Status in Deep Craniofacial Pain
David A Bereiter and Keiichiro Okamoto

Macroscopic Connection of Rat Insular Cortex: Anatomical Bases Underlying its Physiological Functions
Masayuki Kobayashi

The Balance Between Excitation And Inhibition And Functional Sensory Processing in the Somatosensory Cortex
Zhi Zhang and Qian-Quan Sun

INDEX

Volume 98

An Introduction to Dyskinesia—the Clinical Spectrum
Ainhi Ha and Joseph Jankovic

L-dopa-induced Dyskinesia—Clinical Presentation, Genetics, And Treatment
L.K. Prashanth, Susan Fox and Wassilios G. Meissner

Experimental Models of L-DOPA-induced Dyskinesia
Tom H. Johnston and Emma L. Lane

Molecular Mechanisms of L-DOPA-induced Dyskinesia
Gilberto Fisone and Erwan Bezard

New Approaches to Therapy
Jonathan Brotchie and Peter Jenner

Surgical Approach to L-DOPA-induced Dyskinesias
Tejas Sankar and Andres M. Lozano

Clinical and Experimental Experiences of Graft-induced Dyskinesia
Emma L. Lane

Tardive Dyskinesia: Clinical Presentation and Treatment
P.N. van Harten and D.E. Tenback

Epidemiology and Risk Factors for (Tardive) Dyskinesia
D.E. Tenback and P.N. van Harten

Genetics of Tardive Dyskinesia
Heon-Jeong Lee and Seung-Gul Kang

Animal Models of Tardive Dyskinesia
S.K. Kulkarni and Ashish Dhir

Surgery for Tardive Dyskinesia
Stephane Thobois, Alice Poisson and Philippe Damier

Huntington's Disease: Clinical Presentation and Treatment
M.J.U. Novak and S.J. Tabrizi

Genetics and Neuropathology of Huntington's Disease: Huntington's Disease
Anton Reiner, Ioannis Dragatsis and Paula Dietrich

Pathogenic Mechanisms in Huntington's Disease
Lesley Jones and Alis Hughes

Experimental Models of HD And Reflection on Therapeutic Strategies
Olivia L. Bordiuk, Jinho Kim and Robert J. Ferrante

Cell-based Treatments for Huntington's Disease
Stephen B. Dunnett and Anne E. Rosser

Clinical Phenomenology of Dystonia
Carlo Colosimo and Alfredo Berardelli

Genetics and Pharmacological Treatment of Dystonia
Susan Bressman and Matthew James

Experimental Models of Dystonia
A. Tassone, G. Sciamanna, P. Bonsi, G. Martella and A. Pisani

Surgical Treatment of Dystonia
John Yianni, Alexander L. Green and Tipu Z. Aziz

INDEX

Volume 99

Seizure and Epilepsy: Studies of Seizure-disorders in *Drosophila*
Louise Parker, Iris C. Howlett, Zeid M. Rusan and Mark A. Tanouye

Homeostatic Control of Neural Activity: A *Drosophila* Model for Drug Tolerance and Dependence
Alfredo Ghezzi and Nigel S. Atkinson

Attention in *Drosophila*
Bruno van Swinderen

The roles of *Fruitless* and *Doublesex* in the Control of Male Courtship
Brigitte Dauwalder

Circadian Plasticity: from Structure to Behavior
Lia Frenkel and María Fernanda Ceriani

Learning and Memory in *Drosophila*: Behavior, Genetics, and Neural Systems
Lily Kahsai and Troy Zars

Studying Sensorimotor Processing with Physiology in Behaving *Drosophila*
Johannes D. Seelig and Vivek Jayaraman

Modeling Human Trinucleotide Repeat Diseases in *Drosophila*
Zhenming Yu and Nancy M. Bonini

From Genetics to Structure to Function: Exploring Sleep in *Drosophila*
Daniel Bushey and Chiara Cirelli

INDEX

Volume 100

Structural Properties of Human Monoamine Oxidases A and B
Claudia Binda, Andrea Mattevi and Dale E. Edmondson

Behavioral Outcomes of Monoamine Oxidase Deficiency: Preclinical and Clinical Evidence
Marco Bortolato and Jean C. Shih

Kinetic Behavior and Reversible Inhibition of Monoamine Oxidases—Enzymes that Many Want Dead
Keith F. Tipton, Gavin P. Davey and Andrew G. McDonald

The Pharmacology of Selegiline
Kálmán Magyar

Type A Monoamine Oxidase Regulates Life and Death of Neurons in Neurodegeneration and Neuroprotection
Makoto Naoi, Wakako Maruyama, Keiko Inaba-Hasegawa and Yukihiro Akao

Multimodal Drugs and their Future for Alzheimer's and Parkinson's Disease
Cornelis J. Van der Schyf and Werner J. Geldenhuys

Neuroprotective Profile of the Multitarget Drug Rasagiline in Parkinson's Disease
Orly Weinreb, Tamar Amit, Peter Riederer, Moussa B.H. Youdim and Silvia A. Mandel

Rasagiline in Parkinson's Disease
L.M. Chahine and M.B. Stern

Selective Inhibitors of Monoamine Oxidase Type B and the "Cheese Effect"
John P.M. Finberg and Ken Gillman

A Novel Anti-Alzheimer's Disease Drug, Ladostigil: Neuroprotective, Multimodal Brain-Selective Monoamine Oxidase and Cholinesterase Inhibitor
Orly Weinreb, Tamar Amit, Orit Bar-Am and Moussa B.H. Youdim

Novel MAO-B Inhibitors: Potential Therapeutic Use of the Selective MAO-B Inhibitor PF9601N in Parkinson's Disease
Mercedes Unzeta and Elisenda Sanz

INDEX

Volume 101

General Overview: Biomarkers in Neuroscience Research
Michaela D. Filiou and Christoph W. Turck

Imaging Brain Microglial Activation Using Positron Emission Tomography and Translocator Protein-Specific Radioligands
David R.J. Owen and Paul M. Matthews

The Utility of Gene Expression in Blood Cells for Diagnosing Neuropsychiatric Disorders
Christopher H. Woelk, Akul Singhania, Josué Pérez-Santiago, Stephen J. Glatt and Ming T. Tsuang

Proteomic Technologies for Biomarker Studies in Psychiatry: Advances and Needs
Daniel Martins-de-Souza, Paul C. Guest, Natacha Vanattou-Saifoudine, Laura W. Harris and Sabine Bahn

Converging Evidence of Blood-Based Biomarkers for Schizophrenia: An update
Man K. Chan, Paul C. Guest, Yishai Levin, Yagnesh Umrania, Emanuel Schwarz, Sabine Bahn and Hassan Rahmoune

Abnormalities in Metabolism and Hypothalamic–Pituitary–Adrenal Axis Function in Schizophrenia
Paul C. Guest, Daniel Martins-de-Souza, Natacha Vanattou-Saifoudine, Laura W. Harris and Sabine Bahn

Immune and Neuroimmune Alterations in Mood Disorders and Schizophrenia
Roosmarijn C. Drexhage, Karin Weigelt, Nico van Beveren, Dan Cohen, Marjan A. Versnel, Willem A. Nolen and Hemmo A. Drexhage

Behavioral and Molecular Biomarkers in Translational Animal Models for Neuropsychiatric Disorders
Zoltán Sarnyai, Murtada Alsaif, Sabine Bahn, Agnes Ernst, Paul C. Guest, Eva Hradetzky, Wolfgang Kluge, Viktoria Stelzhammer and Hendrik Wesseling

Stem Cell Models for Biomarker Discovery in Brain Disease
Alan Mackay-Sim, George Mellick and Stephen Wood

The Application of Multiplexed Assay Systems for Molecular Diagnostics
Emanuel Schwarz, Nico J.M. VanBeveren, Paul C. Guest, Rauf Izmailov and Sabine Bahn

Algorithm Development for Diagnostic Biomarker Assays
Rauf Izmailov, Paul C. Guest, Sabine Bahn and Emanuel Schwarz

Challenges of Introducing New Biomarker Products for Neuropsychiatric Disorders into the Market
Sabine Bahn, Richard Noll, Anthony Barnes, Emanuel Schwarz and Paul C. Guest

Toward Personalized Medicine in the Neuropsychiatric Field
Erik H.F. Wong, Jayne C. Fox, Mandy Y.M. Ng and Chi-Ming Lee

Clinical Utility of Serum Biomarkers for Major Psychiatric Disorders
Nico J.M. van Beveren and Witte J.G. Hoogendijk

The Future: Biomarkers, Biosensors, Neuroinformatics, and E-Neuropsychiatry
Christopher R. Lowe

SUBJECT INDEX

Volume 102

The Function and Mechanisms of Nurr1 Action in Midbrain Dopaminergic Neurons, from Development and Maintenance to Survival
Yu Luo

Monoclonal Antibodies as Novel Neurotherapeutic Agents in CNS Injury and Repair
Aruna Sharma and Hari Shanker Sharma

The Blood–Brain Barrier in Alzheimer's Disease: Novel Therapeutic Targets and Nanodrug delivery
Hari Shanker Sharma, Rudy J. Castellani, Mark A. Smith and Aruna Sharma

Neurovascular Aspects of Amyotrophic Lateral Sclerosis
Maria Carolina O. Rodrigues, Diana G. Hernandez-Ontiveros, Michael K. Louis, Alison E. Willing, Cesario V. Borlongan, Paul R. Sanberg, Júlio C. Voltarelli and Svitlana Garbuzova-Davis

Quercetin in Hypoxia-Induced Oxidative Stress: Novel Target for Neuroprotection
Anand Kumar Pandey, Ranjana Patnaik, Dafin F. Muresanu, Aruna Sharma and Hari Shanker Sharma

Environmental Conditions Modulate Neurotoxic Effects of Psychomotor Stimulant Drugs of Abuse
Eugene A. Kiyatkin and Hari Shanker Sharma

Central Nervous Tissue Damage after Hypoxia and Reperfusion in Conjunction with Cardiac Arrest and Cardiopulmonary Resuscitation: Mechanisms of Action and Possibilities for Mitigation
Lars Wiklund, Cecile Martijn, Adriana Miclescu, Egidijus Semenas, Sten Rubertsson and Hari Shanker Sharma

Interactions Between Opioids and Anabolic Androgenic Steroids: Implications for the Development of Addictive Behavior
Fred Nyberg and Mathias Hallberg

Neurotrophic Factors and Neurodegenerative Diseases: A Delivery Issue
Barbara Ruozi, Daniela Belletti, Lucia Bondioli, Alessandro De Vita, Flavio Forni, Maria Angela Vandelli and Giovanni Tosi

Neuroprotective Effects of Cerebrolysin, a Combination of Different Active Fragments of Neurotrophic Factors and Peptides on the Whole Body Hyperthermia-Induced Neurotoxicity: Modulatory Roles of Co-morbidity Factors and Nanoparticle Intoxication
Hari Shanker Sharma, Aruna Sharma, Herbert Mössler and Dafin Fior Muresanu

Alzheimer's Disease and Amyloid: Culprit or Coincidence?
Stephen D. Skaper

Vascular Endothelial Growth Factor and Other Angioglioneurins: Key Molecules in Brain Development and Restoration
José Vicente Lafuente, Naiara Ortuzar, Harkaitz Bengoetxea, Susana Bulnes and Enrike G. Argandoña

INDEX

Volume 103

Lost and Found in Behavioral Informatics
Melissa A. Haendel and Elissa J. Chesler

Biological Databases for Behavioral Neurobiology
Erich J. Baker

A Survey of the Neuroscience Resource Landscape: Perspectives from the Neuroscience Information Framework
Jonathan Cachat, Anita Bandrowski, Jeffery S. Grethe, Amarnath Gupta, Vadim Astakhov, Fahim Imam, Stephen D. Larson, and Maryann E. Martone

The Neurobehavior Ontology: An Ontology for Annotation and Integration of Behavior and Behavioral Phenotypes
Georgios V. Gkoutos, Paul N. Schofield, and Robert Hoehndorf

Ontologies for Human Behavior Analysis and Their Application to Clinical Data
Janna Hastings and Stefan Schulz

Text-Mining and Neuroscience
Kyle H. Ambert and Aaron M. Cohen

Applying *In Silico* Integrative Genomics to Genetic Studies of Human Disease: A Review
Scott F. Saccone

SUBJECT INDEX

Volume 104

Cross Species Integration of Functional Genomics Experiments
Jeremy J. Jay

Model Organism Databases in Behavioral Neuroscience
Mary Shimoyama, Jennifer R. Smith, G. Thomas Hayman, Victoria Petri, and Rajni Nigam

Accessing and Mining Data from Large-Scale Mouse Phenotyping Projects
Hugh Morgan, Michelle Simon, and Ann-Marie Mallon

Bioinformatics Resources for Behavior Studies in the Laboratory Mouse
Carol J. Bult

Using Genome-Wide Expression Profiling to Define Gene Networks Relevant to the Study of Complex Traits: From RNA Integrity to Network Topology
M.A. O'Brien, B.N. Costin, and M.F. Miles

Genetic and Molecular Network Analysis of Behavior
Robert W. Williams and Megan K. Mulligan

Large-Scale Neuroinformatics for *In Situ* Hybridization Data in the Mouse Brain
Lydia L. Ng, Susan M. Sunkin, David Feng, Chris Lau, Chinh Dang, and Michael J. Hawrylycz

Opportunities for Bioinformatics in the Classification of Behavior and Psychiatric Disorders
Elissa J. Chesler and Ryan W. Logan

SUBJECT INDEX

Volume 105

Optic Nerve Disease and Axon Pathophysiology
Alireza Ghaffarieh and Leonard A. Levin

Role of Electrical Activity of Neurons for Neuroprotection
Takeshi Morimoto

Molecular Control of Axon Growth: Insights from Comparative Gene Profiling and High-Throughput Screening
Murray G. Blackmore

Gatekeeper Between Quiescence and Differentiation: p53 in Axonal Outgrowth and Neurogenesis
Giorgia Quadrato and Simone Di Giovanni

Cyclin-Dependent Kinase 5 in Axon Growth and Regeneration
Tao Ye, Amy K. Y. Fu, and Nancy Y. Ip

Rho Signaling and Axon Regeneration
L. McKerracher, Gino B. Ferraro, and Alyson E. Fournier

Neuron-Intrinsic Inhibitors of Axon Regeneration: PTEN and SOCS3
Xueting Luo and Kevin K. Park

INDEX

Volume 106

Neurotrophic Factors and the Regeneration of Adult Retinal Ganglion Cell Axons
Alan R. Harvey, Jacob Wei Wei Ooi, and Jennifer Rodger

MBS: Signaling Endosomes and Growth Cone Motility in Axon Regeneration
Michael B. Steketee and Jeffrey L. Goldberg

Intrinsic Mechanisms Regulating Axon Regeneration: An Integrin Perspective
Richard Eva, Melissa R. Andrews, Elske H.P. Franssen, and James W. Fawcett

The Role of Serotonin in Axon and Dendrite Growth
Ephraim F. Trakhtenberg and Jeffrey L. Goldberg

Inflammatory Pathways in Spinal Cord Injury
Samuel David, Juan Guillermo Zarruk, and Nader Ghasemlou

Combinatorial Therapy Stimulates Long-Distance Regeneration, Target Reinnervation, and Partial Recovery of Vision After Optic Nerve Injury in Mice
Silmara de Lima, Ghaith Habboub, and Larry I. Benowitz

From Bench to Beside to Cure Spinal Cord Injury: Lost in Translation?
Andreas Hug and Norbert Weidner

SUBJECT INDEX

Volume 107

Neuromodulation: A More Comprehensive Concept Beyond Deep Brain Stimulation
Clement Hamani and Elena Moro

Computational Models of Neuromodulation
Christopher R. Butson

Neurophysiology of Deep Brain Stimulation
Manuela Rosa, Gaia Giannicola, Sara Marceglia, Manuela Fumagalli, Sergio Barbieri, and Alberto Priori

Neurophysiology of Cortical Stimulation
Jean-Pascal Lefaucheur

Neural Mechanisms of Spinal Cord Stimulation
Robert D. Foreman and Bengt Linderoth

Magnetoencephalography and Neuromodulation
Alfons Schnitzler and Jan Hirschmann

Current Challenges to the Clinical Translation of Brain Machine Interface Technology
Charles W. Lu, Parag G. Patil, and Cynthia A. Chestek

Nanotechnology in Neuromodulation
Russell J. Andrews

Optogenetic Neuromodulation
Paul S. A. Kalanithi and Jaimie M. Henderson

Diffusion Tensor Imaging and Neuromodulation: DTI as Key Technology for Deep Brain Stimulation
Volker Arnd Coenen, Thomas E. Schlaepfer, Niels Allert, and Burkhard Mädler

DBS and Electrical Neuro-Network Modulation to Treat Neurological Disorders
Amanda Thompson, Takashi Morishita, and Michael S. Okun

Neuromodulation in Psychiatric Disorders
Yasin Temel, Sarah A. Hescham, Ali Jahanshahi, Marcus L. F. Janssen, Sonny K. H. Tan, Jacobus J. van Overbeeke, Linda Ackermans, Mayke Oosterloo, Annelien Duits, Albert F. G. Leentjens, and LeeWei Lim

Ethical Aspects of Neuromodulation
Christiane Woopen

Subject Index

Volume 108

Tissue Engineering and Regenerative Medicine: Past, Present, and Future
António J. Salgado, Joaquim M. Oliveira, Albino Martins, Fábio G. Teixeira, Nuno A. Silva, Nuno M. Neves, Nuno Sousa, and Rui L. Reis

Tissue Engineering and Peripheral Nerve Reconstruction: An Overview
Stefano Geuna, S. Gnavi, I. Perroteau, Pierluigi Tos, and B. Battiston

Bone Marrow Mesenchymal Stem Cell Transplantation for Improving Nerve Regeneration
Júlia Teixeira Oliveira, Klauss Mostacada, Silmara de Lima, and Ana Maria Blanco Martinez

Perspectives of Employing Mesenchymal Stem Cells from the Wharton's Jelly of the Umbilical Cord for Peripheral Nerve Repair
Jorge Ribeiro, Andrea Gartner, Tiago Pereira, Raquel Gomes, Maria Ascensão Lopes, Carolina Gonçalves, Artur Varejão, Ana Lúcia Luís, and Ana Colette Maurício

Adipose-Derived Stem Cells and Nerve Regeneration: Promises and Pitfalls
Alessandro Faroni, Giorgio Terenghi, and Adam J. Reid

The Pros and Cons of Growth Factors and Cytokines in Peripheral Axon Regeneration
Lars Klimaschewski, Barbara Hausott, and Doychin N. Angelov

Role of Inflammation and Cytokines in Peripheral Nerve Regeneration
P. Dubový, R. Jančálek, and T. Kubek

Ghrelin: A Novel Neuromuscular Recovery Promoting Factor?
Raimondo Stefania, Ronchi Giulia, Geuna Stefano, Pascal Davide, Reano Simone, Filigheddu Nicoletta, and Graziani Andrea

Neuregulin 1 Role in Schwann Cell Regulation and Potential Applications to Promote Peripheral Nerve Regeneration
Giovanna Gambarotta, Federica Fregnan, Sara Gnavi, and Isabelle Perroteau

Extracellular Matrix Components in Peripheral Nerve Regeneration
Francisco Gonzalez-Perez, Esther Udina, and Xavier Navarro

Subject Index

Volume 109

The Use of Chitosan-Based Scaffold to Enhance Regeneration in the Nervous System
Sara Gnavi, Christina Barwig, Thomas Freier, Kirsten Haarstert-Talini, Claudia Grothe, and Stefano Geuna

Interfaces with the Peripheral Nerve for the Control of Neuroprostheses
Jaume del Valle and Xavier Navarro

The Use of Shock Waves in Peripheral Nerve Regeneration: New Perspectives?
Thomas Hausner and Antal Nógrádi

Phototherapy and Nerve Injury: Focus on Muscle Response
Shimon Rochkind, Stefano Geuna, and Asher Shainberg

Electrical Stimulation for Promoting Peripheral Nerve Regeneration
Kirsten Haastert-Talini and Claudia Grothe

Role of Physical Exercise for Improving Post-traumatic Nerve Regeneration
Paulo A.S. Armada-da-Silva, Cátia Pereira, SandraAmado, and António P. Veloso

The Role of Timing in Nerve Reconstruction
Lars B. Dahlin

Future Perspectives in Nerve Repair and Regeneration
Pierluigi Tos, Giulia Ronchi, Stefano Geuna, and Bruno Battiston

INDEX

Volume 110

The Relevance ofMetals in the Pathophysiology of Neurodegeneration, Pathological Considerations
Kurt A. Jellinger

Pantothenate Kinase-Associated Neurodegeneration (PKAN) and *PLA2G6*-Associated Neurodegeneration (PLAN): Review of Two Major Neurodegeneration with Brain Iron Accumulation (NBIA) Phenotypes
Manju A. Kurian and Susan J. Hayflick

Mitochondrial Membrane Protein-Associated Neurodegeneration (MPAN)
Monika Hartig, Holger Prokisch, Thomas Meitinger, and Thomas Klopstock

BPAN: The Only X-Linked Dominant NBIA Disorder
T.B. Haack, P. Hogarth, A. Gregory, P. Prokisch, and S.J. Hayflick

Neuroferritinopathy
M.J. Keogh, C.M. Morris, and P.F. Chinnery

Aceruloplasminemia: An Update
Satoshi Kono

Therapeutic Advances in Neurodegeneration with Brain Iron Accumulation
Giovanna Zorzi and Nardo Nardocci

The Neuropathology of Neurodegeneration with Brain Iron Accumulation
Michael C. Kruer

Imaging of Iron
Petr Dusek, Monika Dezortova, and Jens Wuerfel

The Role of Iron Imaging in Huntington's Disease
S.J.A. van den Bogaard, E.M. Dumas, and R.A.C. Roos

Lysosomal Storage Disorders and Iron
Jose Miguel Bras

Manganese and the Brain
Karin Tuschl, Philippa B. Mills, and Peter T. Clayton

Update on Wilson Disease
Aggarwal Annu and Bhatt Mohit

An Update on Primary Familial Brain Calcification
R.R. Lemos, J.B.M.M. Ferreira, M.P. Keasey, and J.R.M. Oliveira

INDEX

Volume 111

History of Acupuncture Research
Yi Zhuang, Jing-jing Xing, Juan Li, Bai-Yun Zeng, and Fan-rong Liang

Effects of Acupuncture Needling with Specific Sensation on Cerebral Hemodynamics and Autonomic Nervous Activity in Humans
Kouich Takamoto, Susumu Urakawa, Kazushige Sakai, Taketoshi Ono, and Hisao Nishijo

Acupuncture Point Specificity
Jing-jing Xing, Bai-Yun Zeng, Juan Li, Yi Zhuang, and Fan-rong Liang

Acupuncture Stimulation Induces Neurogenesis in Adult Brain
Min-Ho Nam, Kwang Seok Ahn, and Seung-Hoon Choi

Acupuncture and Neurotrophin Modulation
Marzia Soligo, Stefania Lucia Nori, Virginia Protto, Fulvio Florenzano, and Luigi Manni

Acupuncture Stimulation and Neuroendocrine Regulation
Jung-Sheng Yu, Bai-Yun Zeng, and Ching-Liang Hsieh

Current Development of Acupuncture Research in Parkinson's Disease
Bai-Yun Zeng, Sarah Salvage, and Peter Jenner

Acupuncture Therapy for Stroke Patients
Xin Li and Qiang Wang

Effects of Acupuncture Therapy on Alzheimer's Disease
Bai-Yun Zeng, Sarah Salvage, and Peter Jenner

Acupuncture Therapy for Psychiatric Illness
Karen Pilkington

Acupuncture for the Treatment of Insomnia
Kaicun Zhao

Acupuncture for the Treatment of Drug Addiction
Cai-Lian Cui, Liu-Zhen Wu, and Yi-jing Li

Acupuncture Regulation of Blood Pressure: Two Decades of Research
John C. Longhurst and Stephanie Tjen-A-Looi

Effect and Mechanism of Acupuncture on Gastrointestinal Diseases
Toku Takahashi

INDEX

Volume 112

An Introduction to the Clinical Phenomenology of Tourette Syndrome
Davide Martino, Namrata Madhusudan, Panagiotis Zis, and Andrea E. Cavanna

Functional Neuroanatomy of Tics
Irene Neuner, Frank Schneider, and N. Jon Shah

Functional Imaging of Dopaminergic Neurotransmission in Tourette Syndrome
Bàrbara Segura and Antonio P. Strafella

Nondopaminergic Neurotransmission in the Pathophysiology of Tourette Syndrome
Patrick T. Udvardi, Ester Nespoli, Francesca Rizzo, Bastian Hengerer, and Andrea G. Ludolph

Reinforcement Learning and Tourette Syndrome
Stefano Palminteri and Mathias Pessiglione

Genetic Susceptibility and Neurotransmitters in Tourette Syndrome
Peristera Paschou, Thomas V. Fernandez, Frank Sharp, Gary A. Heiman, and Pieter J. Hoekstra

Pharmacological Animal Models of Tic Disorders
Kevin W. McCairn and Masaki Isoda

Animal Models Recapitulating the Multifactorial Origin of Tourette Syndrome
Simone Macrì, Martina Proietti Onori, Veit Roessner, and Giovanni Laviola

Neuroendocrine Aspects of Tourette Syndrome
Davide Martino, Antonella Macerollo, and James F. Leckman

Clinical Pharmacology of Dopamine-Modulating Agents in Tourette's Syndrome
Sabine Mogwitz, Judith Buse, Stefan Ehrlich, and Veit Roessner

Clinical Pharmacology of Nondopaminergic Drugs in Tourette Syndrome
Andreas Hartmann

Antiepileptic Drugs and Tourette Syndrome
Andrea E. Cavanna and Andrea Nani

Clinical Pharmacology of Comorbid Obsessive–Compulsive Disorder in Tourette Syndrome
Valeria Neri and Francesco Cardona

Clinical Pharmacology of Comorbid Attention Deficit Hyperactivity Disorder in Tourette Syndrome
Renata Rizzo and Mariangela Gulisano

Emerging Treatment Strategies in Tourette Syndrome: What's in the Pipeline?
C. Termine, C. Selvini, G. Rossi, and U. Balottin

Tics and Other Stereotyped Movements as Side Effects of Pharmacological Treatment
Marcos Madruga-Garrido and Pablo Mir

INDEX

Volume 113

Autism Spectrum Disorder and the Cerebellum
Esther B.E. Becker and Catherine J. Stoodley

Contribution of Long Noncoding RNAs to Autism Spectrum Disorder Risk
Brent Wilkinson and Daniel B. Campbell

Identifying Essential Cell Types and Circuits in Autism Spectrum Disorders
Susan E. Maloney, Michael A. Rieger, and Joseph D. Dougherty

Connecting Signaling Pathways Underlying Communication to ASD Vulnerability
Stephanie Lepp, Ashley Anderson, and Genevieve Konopka

MET Receptor Tyrosine Kinase as an Autism Genetic Risk Factor
Yun Peng, Matthew Huentelman, Christopher Smith, and Shenfeng Qiu

Transcriptional Dysregulation of Neocortical Circuit Assembly in ASD
Kenneth Y. Kwan

Motor Skill in Autism Spectrum Disorders: A Subcortical View
Leanne Chukoskie, Jeanne Townsend, and Marissa Westerfield

Orchestration of Neurodevelopmental Programs by RBFOX1: Implications for Autism Spectrum Disorder
Brent R. Bill, Jennifer K. Lowe, Christina T. DyBuncio, and Brent L. Fogel

Immune Dysregulation in Autism Spectrum Disorder
Elaine Y. Hsiao

Autism Susceptibility Genes and the Transcriptional Landscape of the Human Brain
Shingo Miyauchi and Irina Voineagu

INDEX

Volume 114

Modern Concepts of Focal Epileptic Networks
Premysl Jiruska, Marco de Curtis, and John G.R. Jefferys

Neocortical Focus: Experimental View
Igor Timofeev, Sylvain Chauvette, and Sara Soltani

Malformations of Cortical Development and Neocortical Focus
Heiko J. Luhmann, Werner Kilb, and Hans Clusmann

Limbic Networks and Epileptiform Synchronization: The View from the Experimental Side
Charles Behr, Margherita D'Antuono, Shabnam Hamidi, Rochelle Herrington, Maxime Lévesque, Pariya Salami, Zahra Shiri, Rüdiger Köhling, and Massimo Avoli

Limbic Networks: Clinical Perspective
Aylin Y. Reid and Richard J. Staba

Modern Concepts of Seizure Modeling
Christophe Bernard, Sebastien Naze, Timothée Proix, and Viktor K. Jirsa

Mechanisms of Ictogenesis
Thomas Blauwblomme, Premysl Jiruska, and Gilles Huberfeld

Seizure Termination
Frédéric Zubler, Andreas Steimer, Heidemarie Gast, and Kaspar A. Schindler

Epileptic Focus and Alteration of Metabolism
Jakub Otáhal, Jaroslava Folbergrová, Richard Kovacs, Wolfram S. Kunz, and Nicola Maggio

Modern Techniques of Epileptic Focus Localization
Lukas Martinkovic, Hrvoje Hecimovic, Vlastimil Sulc, Radek Marecek, and Petr Marusic

From Treatment to Cure: Stopping Seizures, Preventing Seizures, and Reducing Brain Propensity to Seize
Ivan Pavlov and Stephanie Schorge

INDEX

Volume 115

Environmental Alterations of Epigenetics Prior to the Birth
Chiao-Ling Lo and Feng C. Zhou

Transgenerational Epigenetics and Brain Disorders
Nadia Rachdaoui and Dipak K. Sarkar

The Epigenetic Landscape of Alcoholism
Harish R. Krishnan, Amul J. Sakharkar, Tara L. Teppen, Tiffani D.M. Berkel, and Subhash C. Pandey

Epigenetic Regulatory Mechanisms in Stress-Induced Behavior
Sumana Chakravarty, Salil Saurav Pathak, Swati Maitra, Nitin Khandelwal, Karisetty Bhanu Chandra, and Arvind Kumar

Epigenetics of Schizophrenia: An Open and Shut Case
David P. Gavin and Christina Floreani

Epigenetic Mechanisms in Autism Spectrum Disorder
Adrian Zhubi, Edwin H. Cook, Alessandro Guidotti, and Dennis R. Grayson

MicroRNAs and Ethanol Toxicity
Rajesh C. Miranda

INDEX

Volume 116

Introduction to Sequencing the Brain Transcriptome
Robert Hitzemann, Priscila Darakjian, Nikki Walter, Ovidu Iancu, Robert Searles, and Shannon McWeeney

Analysis Considerations for Utilizing RNA-Seq to Characterize the Brain Transcriptome
Christina Zheng, Sunita Kawane, Daniel Bottomly, and Beth Wilmot

Data Integration and Reproducibility for High-Throughput Transcriptomics
Michael Mooney and Shannon McWeeney

Coexpression and Cosplicing Network Approaches for the Study of Mammalian Brain Transcriptomes
Ovidiu Dan Iancu, Alexander Colville, Priscila Darakjian, and Robert Hitzemann

Splicing in the Human Brain
Ammar Zaghlool, Adam Ameur, Lucia Cavalier, and Lars Feuk

Understanding Complex Transcriptome Dynamics in Schizophrenia and Other Neurological Diseases Using RNA Sequencing
Xi Wang and Murray J. Cairns

The Central Role of Noncoding RNA in the Brain
Boris Guennewig and Antony A. Cooper

Genetics of Gene Expression in CNS
Robert W. Williams and Ashutosh K Pandey

Transcriptomic Changes in Brain Development
Allissa A. Dillman and Mark R. Cookson

Gene Expression in the Addicted Brain
Zhifeng Zhou, Mary-Anne Enoch, and David Goldman

RNA-Seq Reveals Novel Transcriptional Reorganization in Human Alcoholic Brain
Sean P. Farris and R. Dayne Mayfield

Index

Volume 117

Learning-Induced Structural Plasticity in the Cerebellum
Hiroshi Nishiyama

Cerebellar Mechanisms of Learning and Plasticity Revealed by Delay Eyelid Conditioning
Michael D. Mauk, Wenke Li, Andrei Khilkevich, and Hunter Halverson

Cerebellar Long-Term Potentiation: Cellular Mechanisms and Role in Learning
Giorgio Grasselli and Christian Hansel

The Ontogeny of Associative Cerebellar Learning
John H. Freeman

Index

Volume 118

Neuroimmune Mechanisms of Alcohol and Drug Addiction
Changhai Cui, David Shurtleff, and R. Adron Harris

Neuroimmune Pathways in Alcohol Consumption: Evidence from Behavioral and Genetic Studies in Rodents and Humans
Gizelle Robinson, Dana Most, Laura B. Ferguson, Jody Mayfield, R. Adron Harris, and Yuri A. Blednov

Fetal Alcohol Spectrum Disorders and Neuroimmune Changes
Paul D. Drew and Cynthia J.M. Kane

Role of Microglia in Regulation of Ethanol Neurotoxic Action
Lucy Chastain and Dipak K. Sarkar

Functions of the Chemokine Receptor CXCR4 in the Central Nervous System and Its Regulation by μ-Opioid Receptors
Bradley Nash and Olimpia Meucci

Discovery of a Novel Site of Opioid Action at the Innate Immune Pattern-Recognition Receptor TLR4
Jonathan Henry W. Jacobsen, Linda R. Watkins, and Mark R. Hutchinson

Neuroimmune Basis of Methamphetamine Toxicity
Jennifer M. Loftis and Aaron Janowsky

Marijuana Use Brain Immune Mechanisms
Guy A. Cabral and Melissa Jamerson

Interactions of HIV and Drugs of Abuse: The Importance of Glia and Host Genetic Factors
Kurt F. Hauser and Pamela E. Knapp

Neuroimmune Basis of Alcoholic Brain Damage
Fulton T. Crews and Ryan P. Vetreno

Converging Actions of Alcohol on Liver and Brain Immune Signaling
Gyongyi Szabo and Dora Lippai

Opportunities for the Development of Neuroimmune Therapies in Addiction
Lara A. Ray, Daniel Roche, Keith Heinzerling, and Steve Shoptaw

Use of Addictive Substances and NeuroHIV
Sulie L. Chang, Kaitlyn P. Connaghan, Yufeng Wei, and Ming D. Li

INDEX

Volume 119

Adenosine Receptor Neurobiology: Overview
Jiang-Fan Chen, Chien-fei Lee, and Yijuang Chern

Adenosine Receptor PET Imaging in Human Brain
Masahiro Mishina and Kiich Ishiwata

An Overview of Adenosine A_{2A} Receptor Antagonists in Parkinson's Disease
Peter Jenner

Mode of Action of Adenosine A_{2A} Receptor Antagonists as Symptomatic Treatment for Parkinson's Disease
Akihisa Mori

Adenosine Receptors and Dyskinesia in Pathophysiology
Masahiko Tomiyama

Clinical/Pharmacological Aspect of Adenosine A_{2A} Receptor Antagonist for Dyskinesia
Tomoyuki Kanda and Shin-ichi Uchida

Interaction of Adenosine Receptors with Other Receptors from Therapeutic Perspective in Parkinson's Disease
Nicolas Morin and Thérèse Di Paolo

Effects of the Adenosine A_{2A} Receptor Antagonist on Cognitive Dysfunction in Parkinson's Disease
Shin-ichi Uchida, Takako Kadowaki-Horita, and Tomoyuki Kanda

Clinical Nonmotor Aspect of A_{2A} Antagonist in PD Treatment
Masahiro Nomoto, Masahiro Nagai, and Noriko Nishikawa

Adenosine Receptors and Huntington's Disease
Chien-fei Lee and Yijuang Chern

Adenosine Receptors and Epilepsy: Current Evidence and Future Potential
Susan A. Masino, Masahito Kawamura, Jr., and David N. Ruskin

Adenosine Receptor Control of Cognition in Normal and Disease
Jiang-Fan Chen

Adenosine Receptors in Cerebral Ischemia
Alessia Melani, Anna Maria Pugliese, and Felicita Pedata

Roles of Adenosine and its Receptors in Sleep–Wake Regulation
Zhi-Li Huang, Ze Zhang, and Wei-Min Qu

Involvement of Adenosine A_{2A} Receptors in Depression and Anxiety
Koji Yamada, Minoru Kobayashi, and Tomoyuki Kanda

The Adenosine Neuromodulation System in Schizophrenia
Daniel Rial, Diogo R. Lara, and Rodrigo A. Cunha

INDEX

Volume 120

The Story of "Speed" from "Cloud Nine" to Brain Gain
Andrew Lees, Katrin Sikk, and Pille Taba

Amphetamine-Type Stimulants: The Early History of Their Medical and Non-Medical Uses
Nicolas Rasmussen

Miracle or Menace?
Mike Jay

Psychostimulants: Basic and Clinical Pharmacology
Andrew C. McCreary, Christian P. Müller, and Małgorzata Filip

Epigenetic Mechanisms of Psychostimulant-Induced Addiction
Anti Kalda and Alexander Zharkovsky

Experimental Models on Effects of Psychostimulants
Sulev Kõks

Neurologic Complications of Psychomotor Stimulant Abuse
Juan Sanchez-Ramos

Neurobehavioral Sequelae of Psychostimulant Abuse
Atbin Djamshidian

Neuropsychiatric Adverse Effects of Amphetamine and Methamphetamine
Jaanus Harro

"Addicted to Euphoria": The History, Clinical Presentation, and Management of Party Drug Misuse
Jenny Bearn and Matthew O'Brien

"Natural Amphetamine" Khat: A Cultural Tradition or a Drug of Abuse?
Nilesh B. Patel

Methcathinone "Kitchen Chemistry" and Permanent Neurological Damage
Katrin Sikk and Pille Taba

"Legal Highs" – An Emerging Epidemic of Novel Psychoactive Substances
Jolanta B. Zawilska

Psychostimulants and Artistic, Musical, and Literary Creativity
Iain Smith

Opium as a Literary Stimulant: The Case of Samuel Taylor Coleridge
Neil Vickers

INDEX

Volume 121

Alzheimer's Disease: Genomics and Beyond
Fuhai Song, Guangchun Han, Zhouxian Bai, Xing Peng, Jiajia Wang, and Hongxing Lei

The Potential of Proteomics in Understanding Neurodegeneration
Ramavati Pal, Jan Petter Larsen, and Simon Geir Moller

Proteomics Approach to Identify Biomarkers in Neurodegenerative Diseases
Annapurna Nayak, Gregory Salt, Sunil K. Verma, and Uday Kishore

Uncovering Neurodegenerative Protein Modifications via Proteomic Profiling
Xavier Gallart-Palau, Aida Serra, and Siu Kwan Sze

Comparative Proteomics for the Evaluation of Protein Expression and Modifications in Neurodegenerative Diseases
Antonio Conti and Massimo Alessio

INDEX

Volume 122

Utility of Autoantibodies as Biomarkers for Diagnosis and Staging of Neurodegenerative Diseases
Cassandra DeMarshall, Abhirup Sarkar, Eric P. Nagele, Eric Goldwaser, George Godsey, Nimish K. Acharya, and Robert G. Nagele

Metabolomics of Neurodegenerative Diseases
Alejandro Botas, Hannah Moore Campbell, Xu Han, and Mirjana Maletic-Savatic

Parkinson's Disease: *In Vivo* Brain Metabolomics by MRS
Mario Rango

Recent Advances and Applications of Metabolomics to Investigate Neurodegenerative Diseases
Clara Ibáñez, Alejandro Cifuentes, and Carolina Simó

Lipidomics of Human Brain Aging and Alzheimer's Disease Pathology
Alba Naudí, Rosanna Cabré, Mariona Jové, Victoria Ayala, Hugo Gonzalo, Manuel Portero-Otín, Isidre Ferrer, and Reinald Pamplona

INDEX

FSC
www.fsc.org
MIX
Papier | Fördert gute Waldnutzung
FSC® C083411

Kolibri 360 GmbH
Ferdinand-Jühlke-Straße 7
99095 Erfurt, Deutschland
produktsicherheit@kolibri360.de